Dutch Fortifications

ALSO BY JEAN-DENIS G.G. LEPAGE AND FROM MCFARLAND

Military Trains and Railways: An Illustrated History (2017)

Hitler's Armed Forces Auxiliaries: An Illustrated History of the Wehrmachtsgefolge, 1933–1945 (2015)

Medieval Armies and Weapons in Western Europe: An Illustrated History (2005; paperback 2014)

An Illustrated Dictionary of the Third Reich (2014)

British Fortifications Through the Reign of Richard III: An Illustrated History (2012)

Castles and Fortified Cities of Medieval Europe: An Illustrated History (2002; paperback 2011)

The Fortifications of Paris: An Illustrated History (2006; paperback 2010)

*Vauban and the French Military Under Louis XIV:
An Illustrated History of Fortifications and Strategies* (2010)

French Fortifications, 1715–1815: An Illustrated History (2010)

Hitler Youth, 1922–1945: An Illustrated History (2009)

Aircraft of the Luftwaffe, 1935–1945: An Illustrated Guide (2009)

The French Foreign Legion: An Illustrated History (2008)

*German Military Vehicles of World War II: An Illustrated Guide to Cars, Trucks,
Half-Tracks, Motorcycles, Amphibious Vehicles and Others* (2007)

Dutch Fortifications

An Illustrated History from the Roman Era to the Cold War

Jean-Denis G.G. Lepage

McFarland & Company, Inc., Publishers
Jefferson, North Carolina

Library of Congress Cataloguing-in-Publication Data

Names: Lepage, Jean-Denis, author.
Title: Dutch fortifications : an illustrated history from the Roman era to the Cold War / Jean-Denis G.G. Lepage.
Description: Jefferson, North Carolina : McFarland & Company, Inc., Publishers, 2021 | Includes bibliographical references and index.
Identifiers: LCCN 2020048440 | ISBN 9781476680422 (paperback : acid free paper) ♾
ISBN 9781476640815 (ebook)
Subjects: LCSH: Fortification—Netherlands—History.
Classification: LCC UG429.N4 L46 2021 | DDC 725/.1809492—dc23
LC record available at https://lccn.loc.gov/2020048440

British Library cataloguing data are available

ISBN (print) 978-1-4766-8042-2
ISBN (ebook) 978-1-4766-4081-5

© 2021 Jean-Denis G.G. Lepage. All rights reserved

No part of this book may be reproduced or transmitted in any form or by any means, electronic or mechanical, including photocopying or recording, or by any information storage and retrieval system, without permission in writing from the publisher.

Front cover: The Muiderslot with moat in Muiden, The Netherlands (Shutterstock/ Dennis van de Water)

Printed in the United States of America

*McFarland & Company, Inc., Publishers
Box 611, Jefferson, North Carolina 28640
www.mcfarlandpub.com*

With thanks to Jeannette à Stuling, Herman Treu, Eltjo de Lang
and Ben Marcato, Nicole Lapaux, Siem Poppema,
Hans Sakkers, Association Menno van Coehoorn, Jan à Stuling,
Luit van der Tuuk (Dorestad Museum), Fritz Leroux, and Rudi Rolf.

Table of Contents

Introduction — 1
    Low Countries, Netherlands, Belgium, and Holland — 1
    Geography of the Netherlands — 1
    Water — 1

Chapter 1. The Romans in the Netherlands — 3
    Historical Background — 3
    Limes — 3
    Roman Fortifications — 4
    Urban Fortifications — 8
    End of Roman Rule — 15

Chapter 2. The Middle Ages ca. 500–1500 — 16
    The Netherlands in the Early Middle Ages — 16
    Fortifications in the Early Middle Ages — 19
    Medieval Castles — 22
    Medieval Cities — 42
    Urban Medieval Fortifications — 46
    Medieval Siege Warfare — 57
    Conclusion — 58

Chapter 3. Renaissance — 59
    The Burgundian Period — 59
    Habsburg and Spanish Rules — 59
    Firearms — 60
    Transitional Fortification — 60
    Italian Bastioned Fortification — 65
    Early Italian Bastioned Fortifications in the Netherlands — 70
    Coastal Fortification — 75
    Siege Warfare with Firearms — 76

Chapter 4. The Eighty Years' War — 78
    War with Spain — 78
    The Old Dutch System — 80
    Dutch Engineers — 94
    Defensive Waterlines — 106
    The Influence of Dutch ONS Fortifications in Northern Europe — 108
    Dutch Fortifications in the Colonies — 111

Chapter 5. The Golden 17th Century — 118
    A Blossoming Era — 118
    Dutch Fortifications in the Latter 17th Century — 118
    Siege Warfare — 124
    Menno van Coehoorn — 126
    New Dutch System — 130
    Menno Van Coehoorn's Works — 133
    Menno van Coehoorn's Influence — 136

Chapter 6. The 18th Century and the French Era — 139
    The 18th Century — 139
    The French Era — 145
    French Napoleonic Fortifications — 146
    French Realizations in the Netherlands — 146

Chapter 7. The 19th Century — 160
    The Industrial Revolution — 160
    The United Kingdom of the Netherlands — 160
    Dutch Fortifications in the Period 1830–1860 — 162
    The New Holland Waterline — 164
    The Rifled Artillery Crisis, ca. 1860 — 172
    Vestingwet 1874 — 177
    The Polygonal System — 180
    Crisis in the 1880s — 184
    The Stelling of Amsterdam, 1880–1920 — 187
    The Forts in the Stelling — 189
    Fortifications in the Colonies — 199

Chapter 8. The Period 1914–1940   202
    The First World War, 1914–1918   202
    Dutch Fortifications in the Interbellum, 1918–1939   204
    Fortifications of the Afsluitdijk (Ijsselmeer Dam)   207
    Lines of Defense, 1937–1940   212
    Interwar Standardized Pillboxes   214
    Assessment   223

Chapter 9. The Second World War, 1939–1945   225
    Defeat and Occupation   225
    The German Atlantic Wall, 1940–1945   226
    The Atlantic Wall in the Netherlands   245
    Liberation   256

Chapter 10. Dutch Fortifications in the Cold War   257
    Cold War   257
    Dutch Postwar Reconstruction   257
    Dutch Fortifications During the Cold War   257
    Remnants of Dutch Fortifications Today   264
    Conclusion   274

Bibliography   275

Index   277

Introduction

Low Countries, Netherlands, Belgium, and Holland

This book is about the history of fortifications in the Netherlands. First it must be clearly defined what the term *The Netherlands* means, and what territory it really covers.

Today when we say *The Netherlands* it refers to a well-determined region in northwestern Europe we also commonly (and wrongly) call *Holland*. However in the past these terms had a slightly different meaning. To begin with, *Holland* is not a country but a Dutch county and later a province today divided into two parts: *Zuid Holland* (south Holland) and *Noord Holland* (north Holland). At the beginning of modern times, the *Nederlanden* (Low Countries) included the present day Dutch and Belgian territories, a part of northern France, and the Grand Duchy of Luxembourg.

By the end of the 16th century the northern (Dutch) provinces proclaimed their independence from Spain during the so-called Eighty Years' War, and became known as the *Republic of the Seven United Provinces*. In the meantime the southern (Belgian) provinces remained under Spanish rule, and were called *Spanish Low Countries* (*Austrian Low Countries* in the 18th century). Beginning in 1815 both Dutch and Belgian countries were united for 15 years until the Belgians rebelled and obtained their independence from the Dutch in 1830. Since then there is no ambiguity among the terms the Netherlands, Holland, Belgium, and Luxembourg.

As for the term *Dutch,* which now designates the people of the modern nation "the Netherlands," it originates from the language that they speak—a language related to a branch of West Germanic Frankish language called *Old Dutsch* in the Middle Ages. A variation of Dutch spoken in northern Belgium (Flanders) is called Flemish. Another variant of Dutch is spoken in South Africa by the Afrikaners who descend from the Dutch and Huguenot settlers of the 17th century.

Geography of the Netherlands

Located around the delta formed by the mouths of the Meuse (Maas), the Scheldt, and the Rhine rivers, and a number of Rhine tributaries and bifurcations, the Netherlands is a very flat country composed of sediment deposits. More than a quarter of the Netherlands is below sea level. In addition to that, 50 percent of its land lies less than one meter above sea level. The international airport of Schiphol situated southwest of Amsterdam is actually 4.5 meters below sea level. The rivers, which flow across the country from the higher continent beyond, are at their mouths frequently below the level of the sea into which they have to be lifted by canals, locks, dams and dikes. Fortunately, the Netherlands is not in a tsunami-prone part of the world, and the Dutch people know how to build strong dikes.

Low rolling hills cover some of the central area, and in the far south, the land rises into the foothills of the Ardennes Mountains. Vaalserberg in the southern province of Limburg is the country's highest point, culminating at only 322 meters (1,053 feet).

Water

Water has always been a terrible and dangerous foe for the Dutch people. Over centuries, river floodings and sea invasions have regularly devastated the Netherlands, submerging large parts of the country and killing tens of thousands of people. Determined to save their homeland, reclaim it from the sea, and protect it from rivers, the Dutch have always waged a steady war against water. They built dams, and used countless windmills to pump water out of the so-called *polders*—man-made low-lying reclaimed lands. In the early 1930s the damming of the sea efforts continued when the *Afsluitdijk* (IJssel Dam) was built. The *Zuiderzee* (the bay of South Sea) was finally closed off from the North Sea and transformed into a large lake of fresh water now called the *IJsselmeer* (Lake of IJssel), and 550,000 acres of fertile land were reclaimed.

There are very few places in the world where the natural conditions have been so profoundly modified and transformed by human activities. The famous saying *"Deus mare, Batavus litora fecit"* (freely translated as *"God created the Sea, the Dutch built the shore"*) is actually true. The Dutch history is an everlasting combat against water. Geographically a difficult area to live, the ancient Netherlands had for the early inhabitants, Celtic and German tribes, one very important feature—safety. Indeed, wide rivers with strong current, lakes, wetlands, swamps, and thick forests were difficult obstacles for invaders. At the same time waterways and sea provided connections, and efficient means for developing local exchanges, regional trade, and international business making the Netherlands a commercially prosperous country.

Next to commerce, water was also used for defense. When the Dutch managed to dominate water, this element became an important ally. Together with man-made fortifications, inundations (the voluntary flooding of particular zones) formed the basis of the country's defense for centuries.

Note: The Dutch alphabet features an additional character composed of an "i" attached to a "j" (*ij* pronounced as English diphthong *ay* as in *say*, *bay* or *play*). This digraph can be used at the beginning of a noun for example *ijs* meaning ice, or *ijzer* (iron), or *ijzel* (glazed frost), or *ijverig* (diligent or industrious). In majuscule form both *i* and *j* must be capitalized as *IJ* for example in *IJsland* (Iceland) and *IJzer* (Yser River in Belgium).

This letter is regularly encountered in this book in names like *Nijmegen* (Nimegue), *Rijn* (Rhine River), the *IJ River* (flowing in Amsterdam), *IJmuiden* (mouth or River IJ), and *IJsselmeer* (Lake of River *IJssel*). There is the possibility to substitute *ij/IJ* for the Greek letter epsilon (y, Y) but the author has chosen to keep the original Dutch spelling *ij* and *IJ*.

Chapter 1

The Romans in the Netherlands

Historical Background

After the conquest of Gaul (France) in the 1st century BCE the Romans appeared in the region today known as the Netherlands. Then, that region was quite a difficult place to live, with impenetrable forests, wide rivers and large lakes, as well as low-lying lands regularly drowned in sea water. The Romans eventually conquered territories reaching to Utrecht and Nijmegen (Nimegue), but did not bother to advance further north, to conquer the place that would later become Amsterdam.

After several military campaigns that lasted from 57 to 12 BCE, the Romans annexed that part of the land south of the River Rhine and incorporated it into a large province designated Germania Inferior (Lower Germany). This administrative unit included present day's Luxembourg, southern Netherlands, parts of Belgium and parts of northwest Germany. The provincial capital was Colonia Claudia Ara Aggrippinensium (today Köln or Cologne in Germany). The Romans ruled the southern parts of the Netherlands from 12 BCE to ca. 406 CE. During that period of Pax Romana there were however a number of revolts. The most important was that of 69–70 CE. By then the Batavian tribes rose against Roman rule, and a number of Roman settlements were attacked and burnt. A year later, the Roman authorities deployed four legions and re-established their dominance.

The wide Old Rhine River and its tributaries constituted the northern frontier of the Roman Empire. North of the river the uncivilized world began: Germanic and Celtic tribes, notably the Frisians who were never subjected, but were influenced by the Romans through trade and lifestyle. To watch over those unruly and rough people called the Chamavii and Frisii (Frisians) the Romans fortified their border and established the so-called *limes* (frontier). The Roman Empire was at its greatest extent when the emperor Trajan died in CE 117.

Limes

First of all it should be noted that at first the frontiers of the Roman Empire were not really well defined. They were not strict and clear lines but temporary, more or less extending into hazy interval zones where the Romans indirectly or directly ruled. However, as the 1st century matured, the frontiers of the Roman Empire became increasingly fixed, and what had once been temporary limits became definitive borders. As a result the main task of the Roman legions was no longer conquest but keeping what had already been conquered, by policing the border tribes, mounting punitive raids, and preventing tax evasion and livestock rustling. Obviously one of the Romans' biggest problems was security at their borders. The emperor August had hoped to extent the limits of the empire to the Elbe River in the heart of Germany but several campaigns had turned to disaster so the northern frontier ran along the Rhine River.

The construction of the elaborate fortified defensive *limes* (frontier) along the rivers Rhine and Danube was started under the reign of the emperor Dominien (81–96). The fortification program was continued by Trajan (97–117), Hadrien (117–138) and many of their successors. The limes—in what later would become the Netherlands, was called the Limes Germanicus (Latin for Germanic frontier). It was a line of fortifications that bounded the Roman provinces of Germania Inferior, Germania Superior, Raetia, and Pannonia. From the years 83 to about 260 CE the limes were intended to protect the Roman Empire from unsubdued Germanic tribes.

The part in Germania Inferior in the Low Countries stretched from the North Sea outlet of the Rhine at Katwijk-aan-Zee, and ran along the then main Lower Rhine branches (modern Oude Rijn, Leidse Rijn, Kromme Rijn, Nederrijn) all the way down to near Castra Regina (today Regensburg) on the Danube. Both major north European rivers afforded natural protection from mass incursions into imperial territory. The natural obstacles were reinforced with man-made fortifications including signal posts, watchtowers, strongholds, palisades, earth embankments with ditches, and garrison camps—some becoming the sites of trading places that later became cities still flourishing today. Highways linked all settlements, towns and cities, and military roads ran along the borders so that troops could march quickly to any threatened point.

The Roman-fortified frontiers in what later became the Netherlands were intended primarily to keep people out, but also to control the passage of traders. There was indeed a great deal of trade moving through the frontier zones. Many local people settled just on the other side of the frontier in places where they could enjoy extensive and secure relations with the Romans beyond the border. So some of the natives living more or less peacefully along the frontier developed a familiarity with Roman ways and some attempted to emulate them. If needed, the Roman armies could fight the enemies of Rome but, in many cases, it proved simpler and more effective to win natives over, and enlist them as allies of the Roman state.

As a matter of fact the Roman Empire was so large and had such lengthy borderlines that the frontiers could only be patrolled and guarded, and not properly defended. There were way too few troops available. The Roman Limes' garrison could not repel a mass invasion by numerous aggressors. It was principally a deterrent, an early warning system and a delaying tactic. It could discourage marauders and small raiding gangs along, with keeping an eye on the neighboring people. There were always many traders and travellers moving from one side to another. The border fortification works were tools to control, regulate, encourage, and protect the flow of trade, and, naturally, to collect taxes.

They played a symbolic role too. The Roman Limes was the materialization of a mark that separated two different worlds, two distinct cultures, and two dissimilar societies. On one side were Roman order, law, organized markets and structured towns—civilization, in short. One the other side was a tribal and primitive society with technical backwardness, and illiteracy—barbarism, in short. At least that was how the Romans envisioned the whole thing. The Limes expressed Rome's power, and formed a physical frontier clearly delineating the borders of the Empire. The provision of a visible frontier secured the stability of disputed territory. The Limes marked a border, and constituted an explicit warning for bad-intentioned intruders, that beyond that line they were risking retribution from the Romans.

Roman Fortifications

Camps and forts

The large standing army was concentrated on the frontier and defended the interior of the empire against foreign invasions.

As far as was possible, the Roman imperial administration attempted to make the Roman armies as productive as possible. Some units operated brick and tile manufactories, lead and iron smelters, and many other enterprises. Troops were often allowed to remain quartered in the same garrison town almost permanently and often drew their recruits from the local population. Someone who had joined the Roman

Map of the Roman Empire, ca. 117 CE.

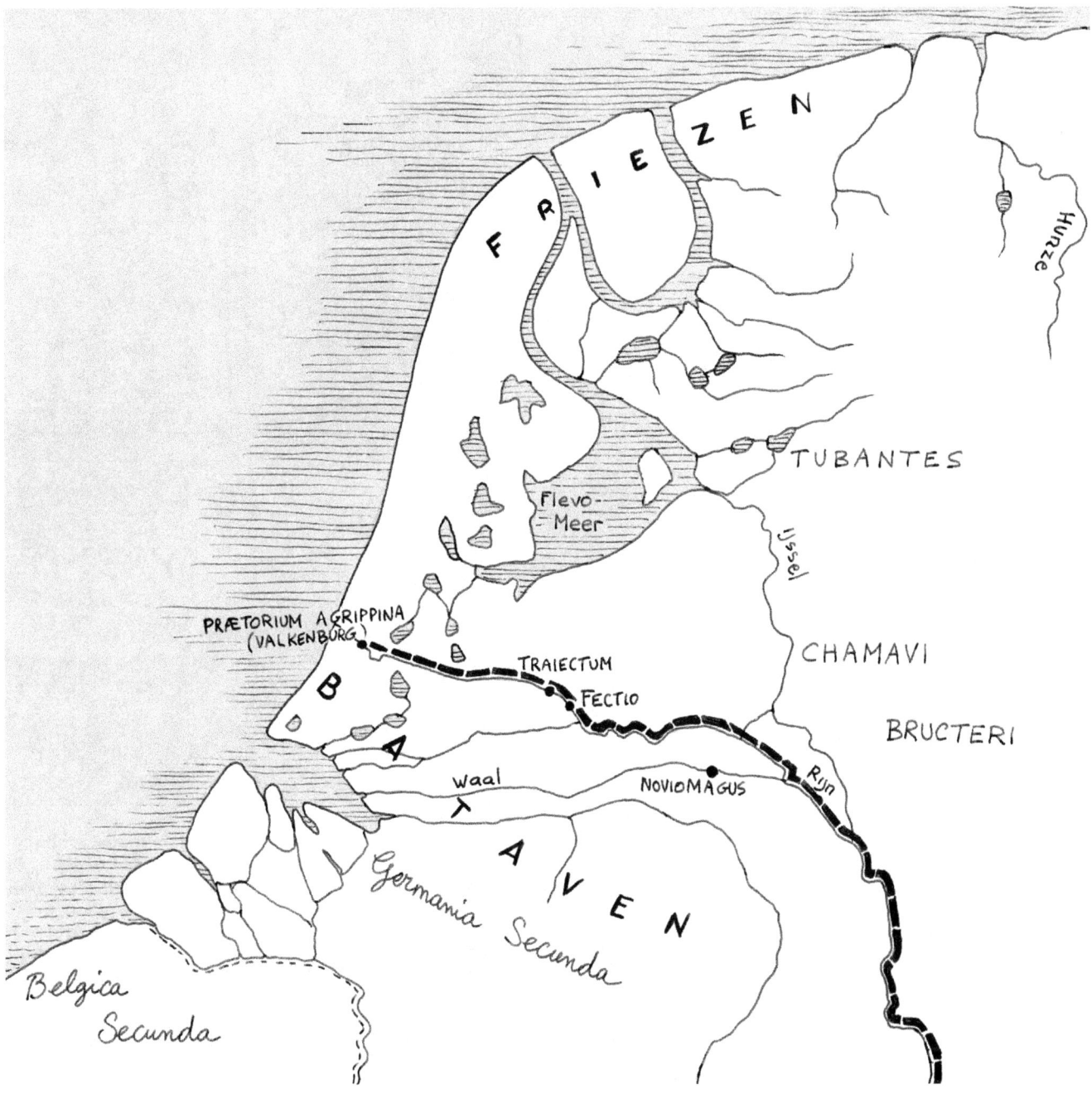

Map of the Roman Limes in the Low Countries. The Limes' main obstacle was the formidable Rhine River and its numerous branches, swamps and marshy lowlands and forests. In addition the Romans built military fortifications in the form of earth walls topped with palisades, ditches, watchtowers, and observation posts. As can be seen on the roadmap of Peutinger (a 13th century copy of an original Roman map), the fortifications ran parallel to the Rhine River from Colonia Ulpia Trajana (Xanten, Germany) in the east, via Traiectum (Utrecht) and Albaniana (Alphen aan de Rijn) to Lugdunum Batavorum (Katwijk) and Praetorium Agrippina (Valkenburg) on the coast of today's North Sea. To accommodate their soldiers, administrative personnel and traders, the Romans created posts, permanent camps, fortlets and forts. Some of those garrison places later became villages, towns and cities.

army had decided upon his life's work since the standard enlistment was for 25 years. Since many recruits came from poor and isolated regions far from the centers of Roman life, the army literally taught them from the bottom up. They learned to dress properly, to speak a bit of Latin, to practice personal hygiene, and to learn at least one and perhaps more than one trade. Along with this, however, it was also important that soldiers learn of the greatness of Rome, of the majesty of its institutions, and of numerous divinities who were honored by great rituals.

Consequently, Roman auxiliary soldiers invested their spare time and effort in turning the towns that sprang up along their fortresses into little Romes, or at least close to what the soldiers believed the essence of Rome to be.

Stationed on the frontier, they were set to the task of creating the transportation and communication networks—roads, bridges, beacons, canals, ports, aqueducts as well as numerous other public works throughout the Empire. Wherever it was sent or wherever it was settled, the Roman army provided local inhabitants an outstanding example of *Romanitas*, the sense of belonging to a great civilization. When Rome fell into disorder and civil war or when the imperial administration had descended into the depths of corruption or ineffectiveness, the army's reverence for the ideal of Rome remained undiminished even though they might acclaim their general as emperor and march upon Rome to clear up the mess there.

The Roman conquest marked the introduction to what is today known as the Netherlands of a whole range of fortifications, constructed according to standard patterns. No matter where in the Roman Empire, fortifications (but also urban layout, organization, bathhouses, amphitheaters, arenas, official buildings, religious temples, bridges, and roads) tended to follow a regular uniformity. Although there were indeed many variations from site to site, and from land to land, general patterns and regular designs were endlessly repeated throughout the Empire. For a fortified camp, the basic shape was a rectangle with rounded corners, often described by modern archaeologists as *playing-card* form. This basic Roman military enclosure was represented in several main types including: the temporary or semi-permanent (tented) camp; the legionary fortress; and the fort.

Other more or less standardized works included fortlets and watchtowers; frontier works and lines; coastal forts; and urban fortification. In the countryside, the Roman upper class and wealthy landowners constructed a typical (often luxurious and large) farming center known as a *villa rustica*.

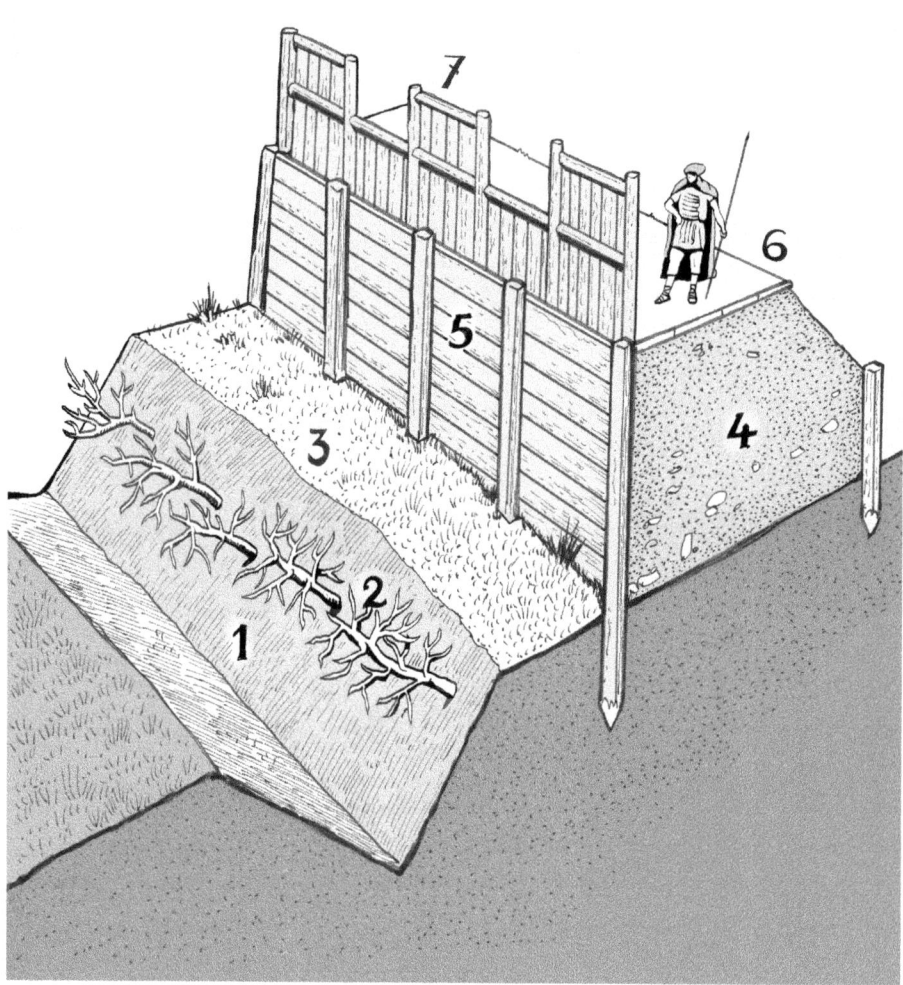

Roman wall cross-section: 1: V-shaped ditch. 2: Cippe (obstacle made of sharpened tree branches). 3: Berm (patrol path). 4: Rampart. 5: Timber wall. 6: Wallwalk or walkway. 7: Lorica aka parapet or breastwork—a man high defensive screen fitted with open parts called crenels or pinnea, and standing parts called merlons. The parapet could also be composed of an earth bank or a palisade (stakes) or made of wattle (interwoven twigs). An earth embankment with the exterior made steep by revetments of sods or hurdle-work had a double advantage over a thorny hedge: besides being a better obstacle against assault, it gave the defenders an advantage of position in a hand-to-hand fight.

Palisade. Walls were often built of alternate layers of stones, and earth. Timber being plentiful, parapets were frequently made of aligned tree trunks.

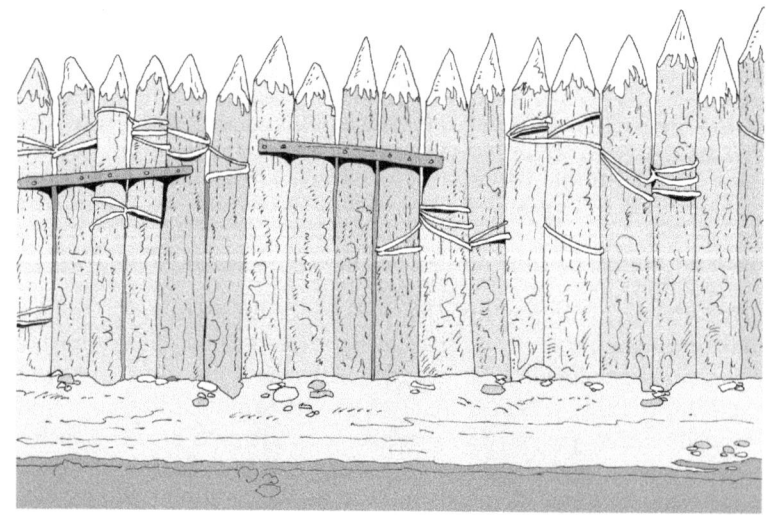

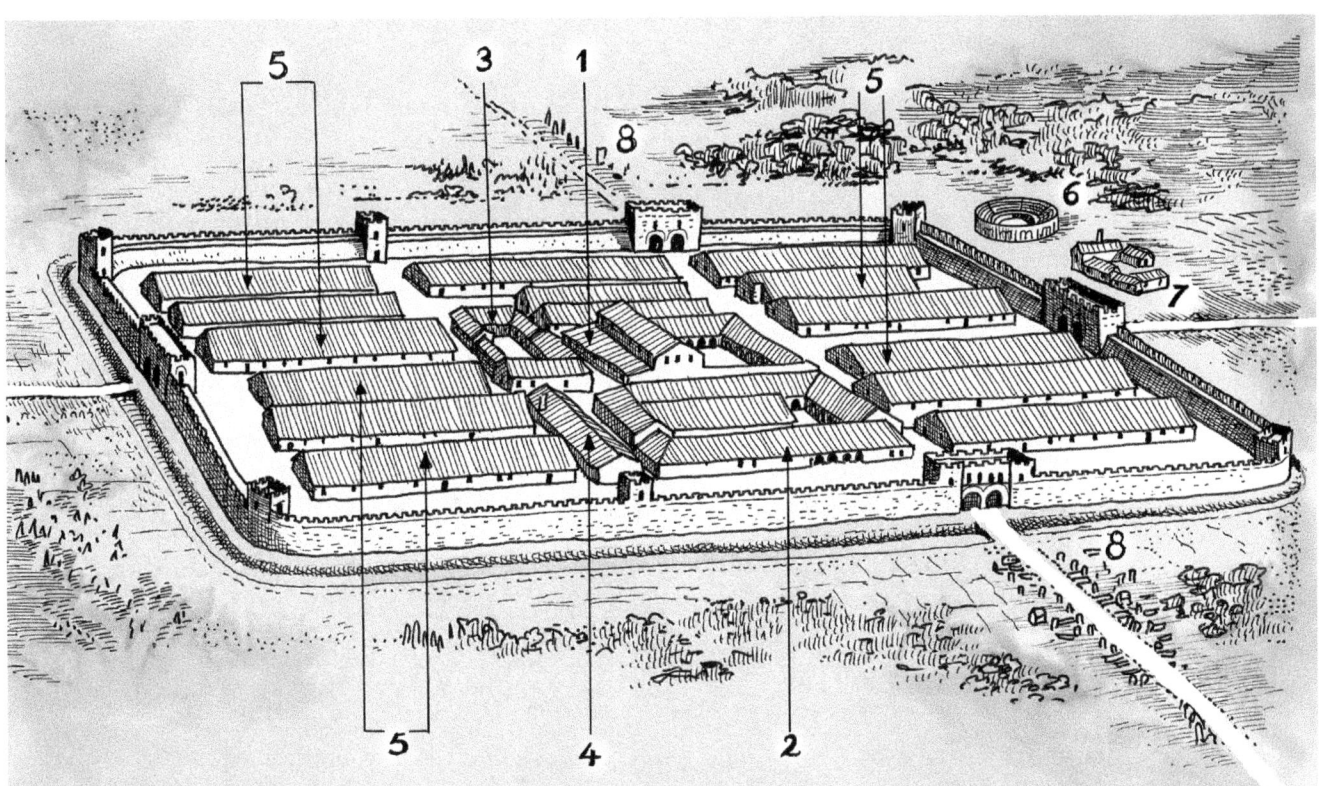

Castrum: A place for auxiliary troops, it comprised the headquarters building called principia (1) placed in the center. The commanding officers' house called praetorium (2) was next to the headquarters building. The rest of the fort was divided into strigae (subdivisions) filled with various buildings. The valetudinarium (3) was the military hospital. The horrea (4) included various storeplaces, a granary, and workshops. There were rows of barracks (5) for the troops. Outside the fort there was often an amphitheater (6)—an open, circular or oval building with a central space for the presentation of dramatic or sporting events surrounded by tiers of seats for spectators. There was often a public bathhouse (7). Necropoles aka cemeteries (8) were located outside the fortress along the access roads, because death was regarded as impure.

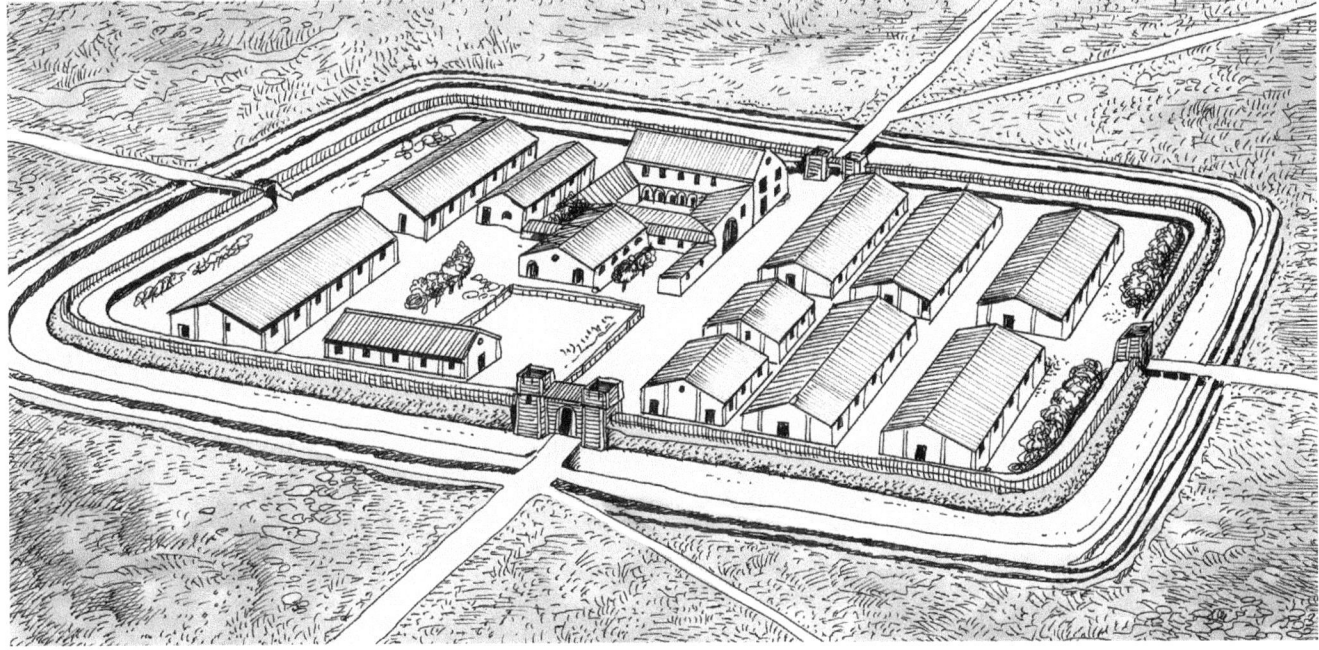

Castellum. A castellum was generally smaller than a castrum, a castellum could house 500 to 800 men, infantry, and cavalry.

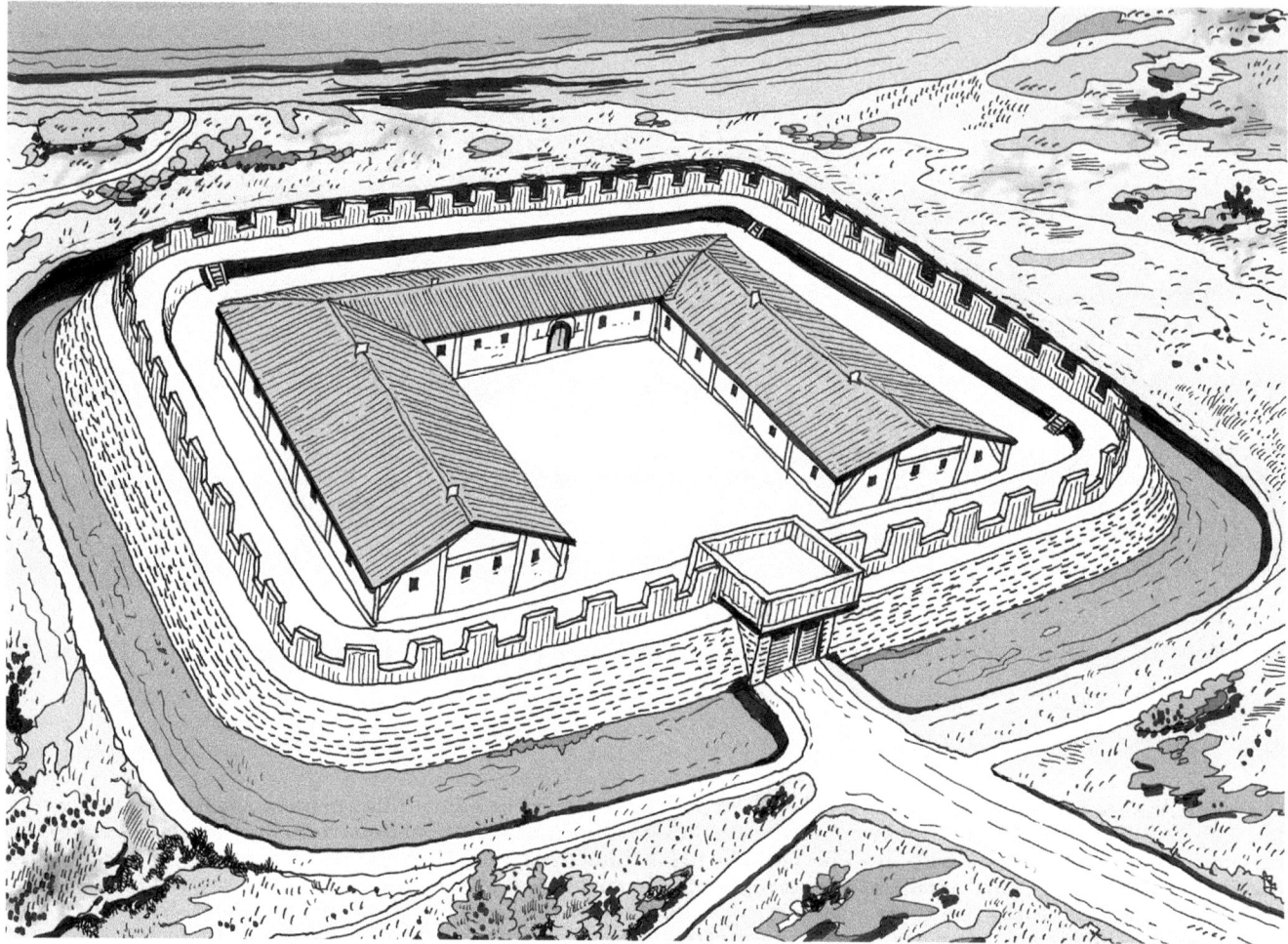

Ockenburg fortlet—a small fortified building placed on a strategic position. Ockenburg located in the dunes west of The Hague (province of South Holland) was built about 150 CE. Intended to watch the coast, it probably housed a cavalry squadron of about 16 mounted soldiers. The fortlet included three U-shaped conturbernia (barracks), a part of which was arranged as stalls for the horses. This main building was fortified with an earth wall reveted with stones, a ditch, and a gatehouse.

Worthy of mention in a land where water plays such an important role is the fact that the Romans brought the art of constructing artificial waterways, drainage ditches, dams, and dikes. For example the military commander Drusus ordered the digging of a dyked canal connecting the Old Rhine to the River Vecht. Later the general Corbulo had a waterway excavated to link the mouths of the rivers Maas and Rhine.

Urban Fortifications

Roman towns

The most obvious characteristic of Roman Netherlands was the development of towns, which formed the basic of Roman administration and civilization. Many cities grew out of previous native settlements, but also from Roman military camps or market centers. Basically there were three different sorts of towns founded by the Romans: a few *coloniae*—towns populated by a majority of Roman settlers and discharged veterans; the *municipiae*—a number of cities in which the whole native population was given Roman citizenship; and a majority of *civitates* (singular *civitas*), which included the old local tribal communities, their towns, villages and settlements, through which the Romans administrated the native population in the countryside.

Genuine Italian Roman soldiers, administrators and officials were few in that remote part of the Empire. They did their best to reconstruct the backgrounds of sunny Italy in this misty, swampy and wet clime. The newly created towns were built with timber, stone and brick, and their streets intersected at right angles forming regular blocks occupied by dwellings, markets and shops. In the middle there was generally a *forum* (main square) with public baths, temples and basilica. Towns were connected by a road network that favored trade and circulation of troops.

Roman urban fortifications

At first, not all Roman settlements had defensive walls, but that changed after several local tribes revolted and attacks occurred, notably the Batavian revolt in CE 69–70. Some veterans' coloniae—the most reliable towns—became semi-militarized fortified centers. Then, from the end of the 2nd century to the end of the 3rd century CE, almost every town in the Netherlands, as well as many villas in the countryside were fitted with fortifications, suggesting a decision taken at imperial level. At first many of these were no more than palisade and earthworks, but gradually many towns had stone walls with towers and gatehouses.

As long as *Pax Romana* (Roman Peace) reigned, urban fortifications apparently had more of a symbolic function than a real military significance. The gate generally had more the appearance of an arch de triumph than that of a fortified point. As for the towers, they were often constructed against the rear of the wall, which means that they did not protrude in the moat, and therefore could not provide flanking fire along the outer curtain. Besides some important and large buildings such as arenas, many theaters and amphitheaters were built outside the defended perimeter.

In the first two centuries CE confidence and aggression were the hallmarks of Roman policy along the huge borders. In the 3rd and 4th centuries, however, because of the ever increasing pressure of the "barbarian" tribes at the frontiers—and to the enormous burden of maintaining large military forces—forts, fortresses, and fortified towns were no longer springboards for attack and conquest, but purely defensive strongholds.

During their occupation of the Low Countries the Romans founded a number of important settlements, some of which have survived. The main cities and towns with Roman origins, or which were extensively developed by them included: Lugdunum Batavorum (today Katwijk),

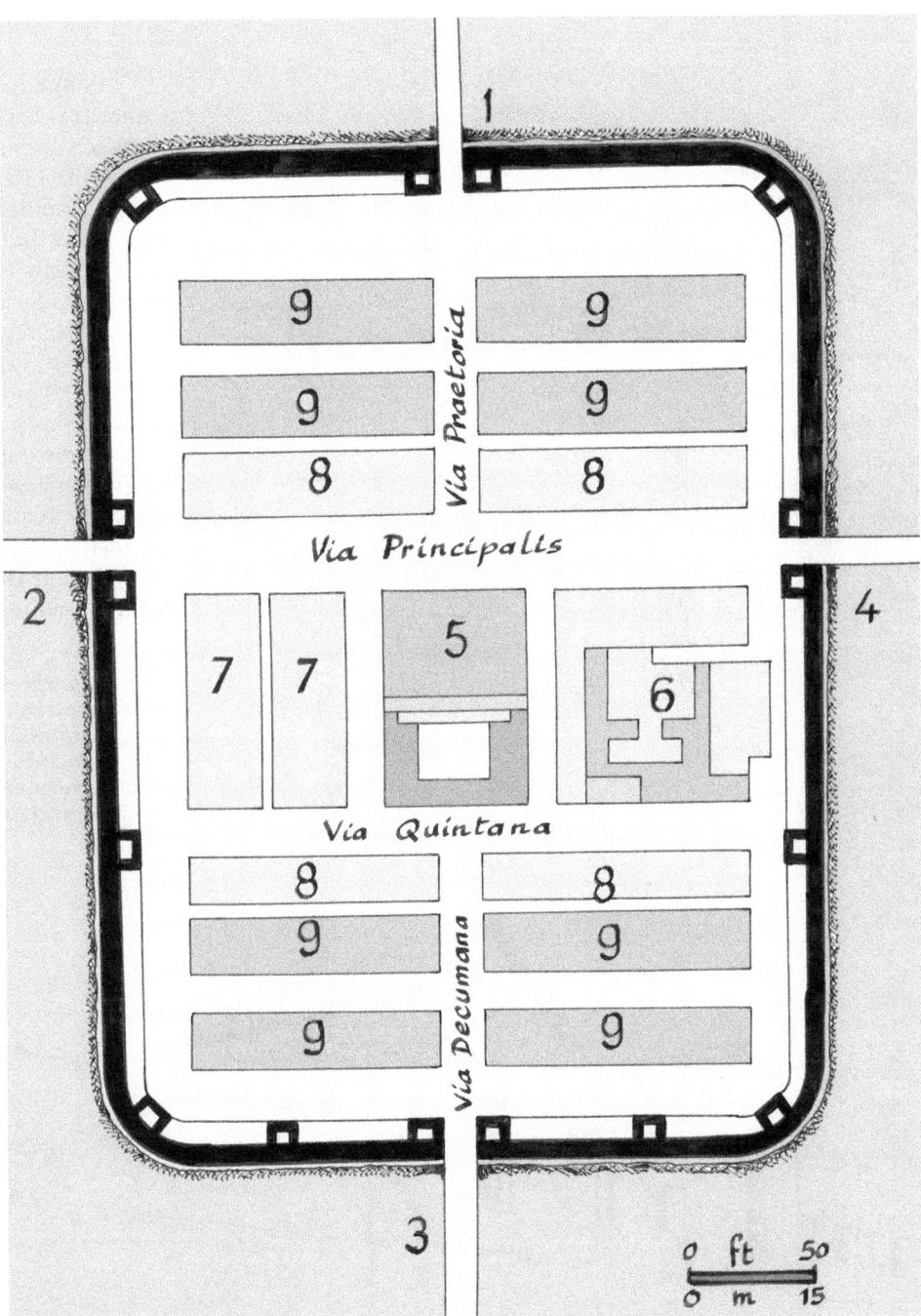

Plan fort. In ground plan Roman temporary camps, castella, permanent castra and auxiliary fort had a similar shape, looking like a playing card whenever possible, with a porta (entrance) on each side. Most military place usually included the following standard elements. 1: Porta Praetoria (North Gate); 2: Porta Principalis Sinistra (West Gate); 3: Porta Decumana (South Gate); 4: Porta Principalis Dextra (East Gate); 5: Principia (Headquarters, religious and administrative buildings); 6: Praetorium (Commanding officers' residence); 7: Horrea (supply stores); 8: Fabricae and Stabuli (workshops and stables); 9: Barracks for the troops.

Noviomagus Batavorum (Nijmegen), Ceuclum (Cuijk), Praetorium Agrippinae (Valkenburg), Albaniana (Alphen aan de Rijn), Traiectum (Utrecht), Forum Hadriani (Voorburg), Levefanum (Wijk bij Duurstede), Coriovallum (Heerlen), Nigrum Pullum (Zwammerdam), Traiectum ad Mosam (Maastricht), and several others.

Dutch Fortifications

Voorburg

Municipium Aelicum Cananefatum (later known as Forum Hadriani) is today the modern city of Voorburg and constitutes a suburb of Den Haag (The Hague) in the province of South Holland. Founded right after the uprising of 69–70 CE, it was the civitas (capital) of the Cananefates, a Germanic tribe who lived in the Rhine delta. Given its location it was an important town, a port, a significant economic market place and a military strategic stronghold in the Roman limes.

In 121 CE emperor Hadrian, on an inspection tour in that part of the Roman Empire, renamed it for himself: Forum Hadriani. In the 270s CE the inhabitants were victims of various plagues and raids by Saxon pirates. It became pointless to try to defend the town, and the weakened Romans left Forum Hadriani. After the collapse of the Roman Empire during the so-called Dark Age

Left: Turris—(watchtower), a small post often defended by a V-shaped ditch and a palisade.
Below: Cross-section of a Roman wall: 1: Foundation. 2: External revetment. 3: Internal revetment. 4: Core of the wall made of rubble and mortar.

Chapter 1. The Romans in the Netherlands

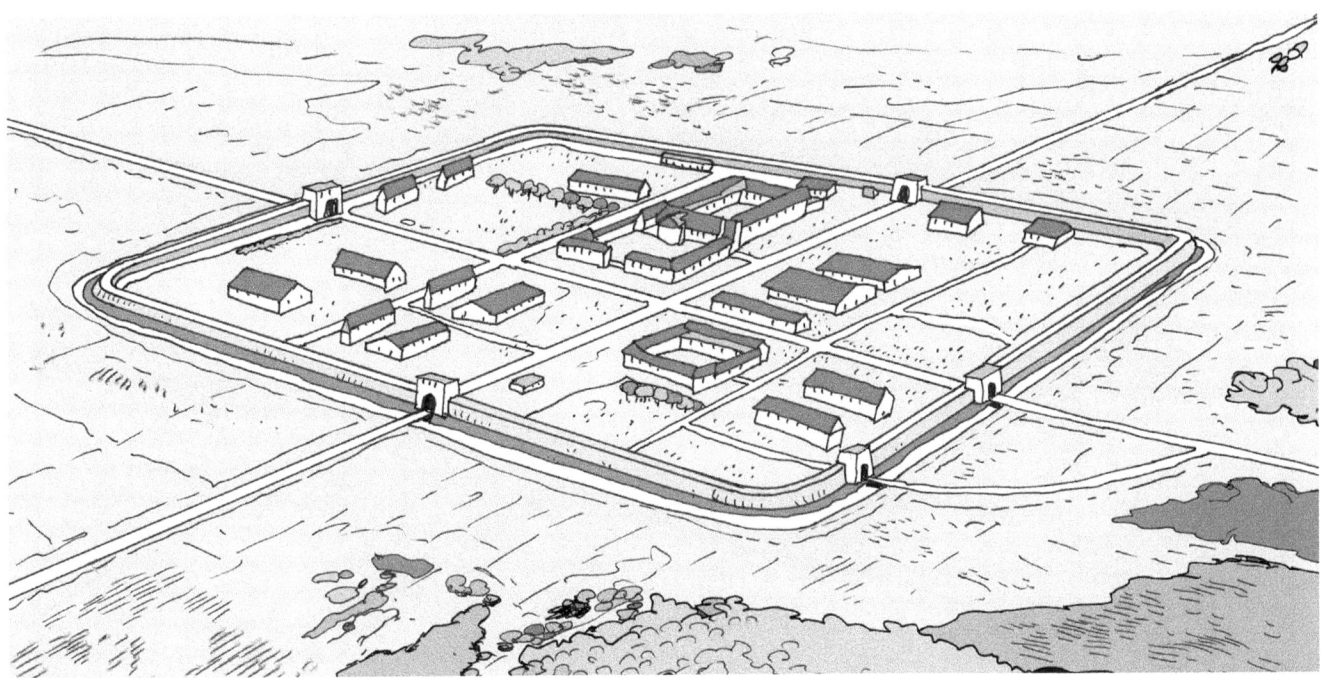

Voorburg (Forum Hadriani)

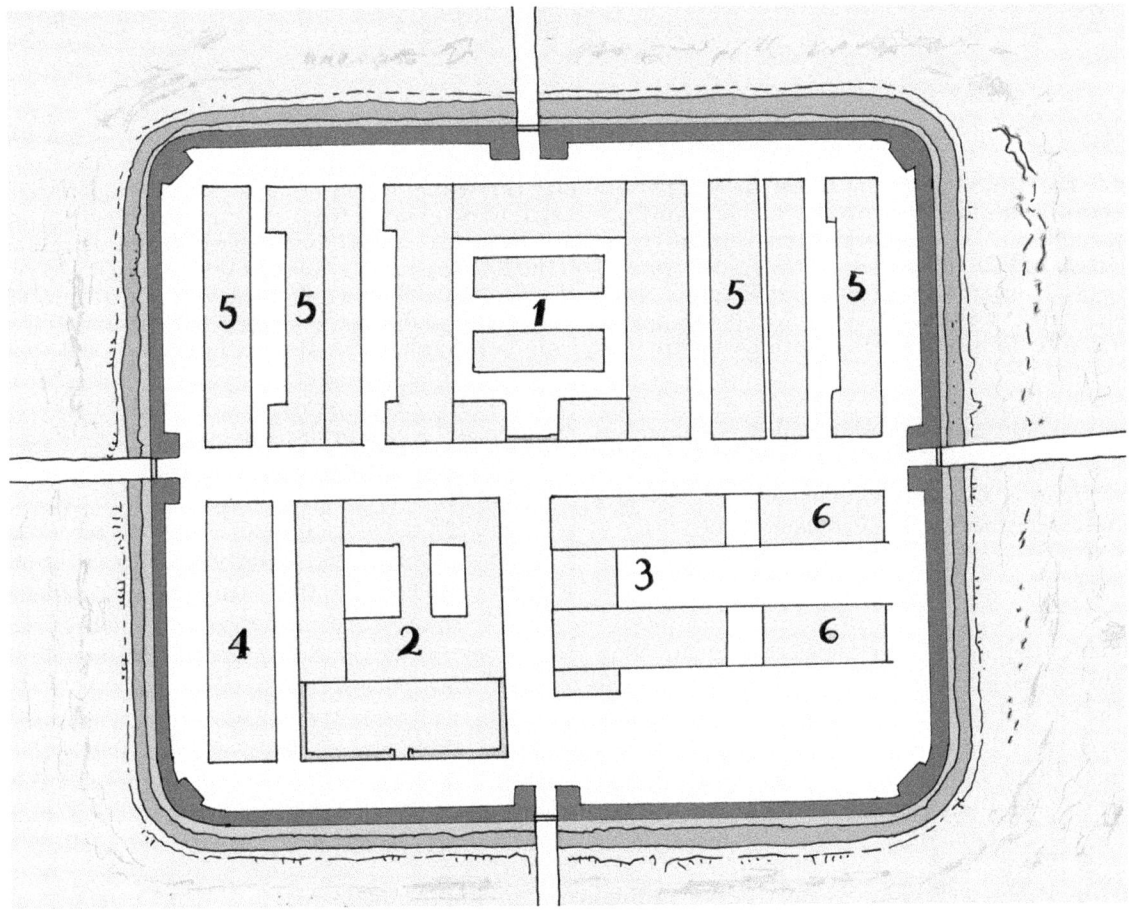

Valkenburg (Praetorium Aggrippinae). Founded in 39–40 CE by order of Emperor Caligula Praetorium, Aggrippinae was destroyed during the Batavian uprising in 69–70 CE. It was later rebuilt and finally abandoned when the Romans left the country. 1: Headquarters. 2: Prefect Residence. 3:Stables. 4: Hospital. 5: Barracks. 6: Supply Stores.

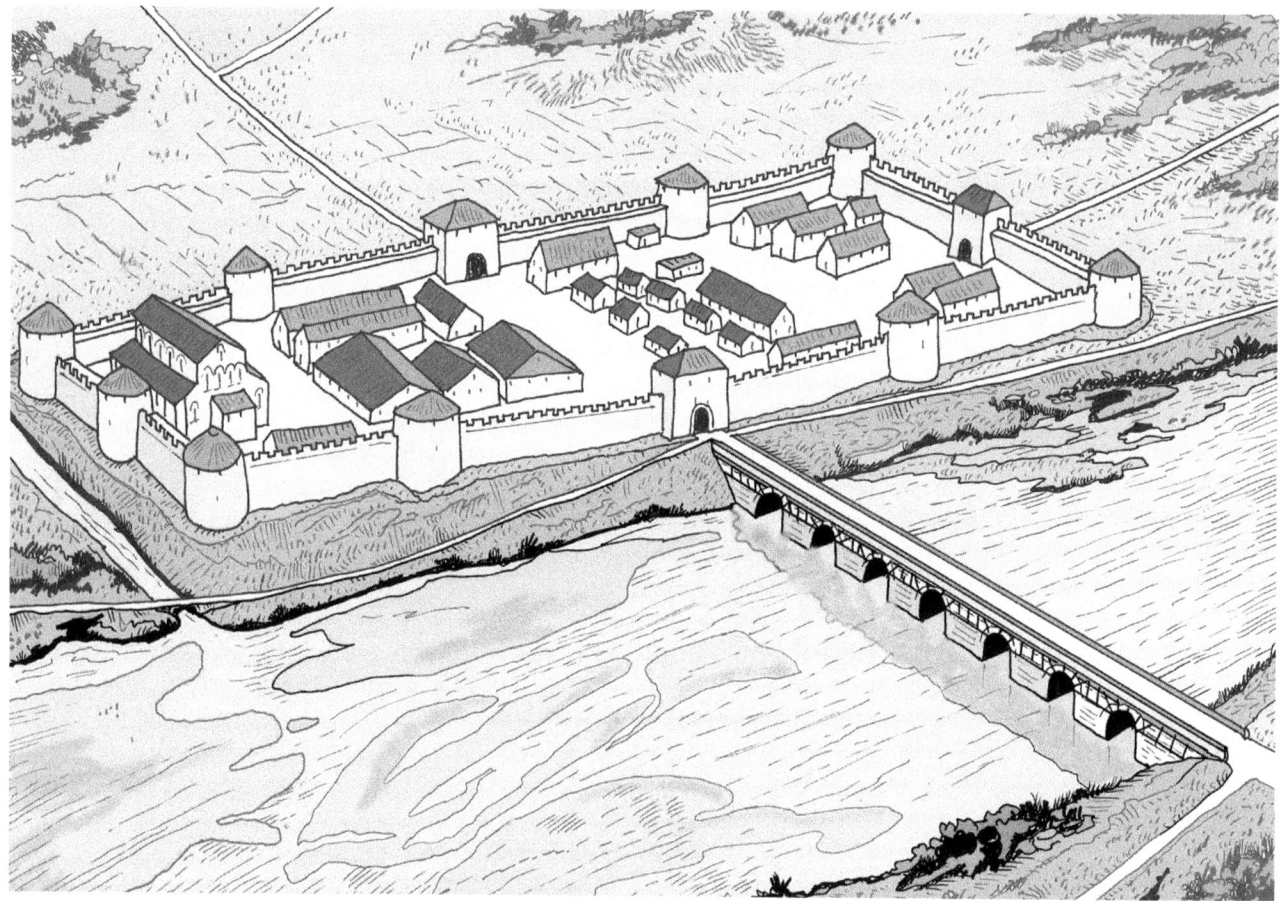

Maastricht. The capital of the southern province of Limburg (known in Roman time as Mosa Trajectum or Trajectum ad Mosam meaning "ford on the Meuse River"), Maastricht originated from a castrum protecting an important bridge (pontem Mosea fluminis) crossing the river. The bridge was made of masonry pillars and timber arches. Founded in ca. 10 BCE, the fort was strategically placed on the Roman highway (called Via Belgica) linking Cologne in Germany to Tongeren, and Tournai in Belgium, and further to Valencienne and Boulogne in northern France. In ca. 270, Maastricht was destroyed by German raiders. It was rebuilt and enlarged in 333, this time with stonewalls, gates, circular towers, and a bridgehead castellum on the opposite bank of the river, later known as the borough of Wijk.

(early medieval era) the place was gradually depopulated and finally abandoned. Houses and walls made of stone were used as quarries, and after decades of plunder and centuries of neglect what remained disappeared underground.

Alphen

Albaniana was the name of a Roman castellum (fort) in modern-day Alphen aan den Rijn in the province of South Holland. The fort was located on the banks of the Old Rhine River between two other strongholds: the castella of Matilo and Nigrum Pullum. Castellum Albaniana (meaning Fort by the White Waters) was constructed in the years 41–54 CE as part of the Lower Germanic Limes. A rather isolated outpost, the fort included an outer earth wall (ca. 100 × 80 meters), wooden towers and a timber gatehouse hemmed by a wet ditch. During the Rebellion of the Batavi (between CE 69 and 70), the fort was destroyed by the rebels. In the 160s CE, it was rebuilt in brick, and included a garrison estimated to about 400 soldiers with their families. Gradually the fort grew in size and became the center of a vicus (village) with a mixed population composed of soldiers and local civilian merchants, craftsmen and peasants.

In the middle of the third century, presumably 270 CE, the castellum was evacuated and the vicus greatly depopulated. By then German intruders (notably the Franks) crossed the Rhine and started to disrupt the Roman Empire. The invaders could not be stopped, and Castellum Albaniana

Opposite, top: **Vechten (Fechio).** Located at the confluence of the Rhine and Vecht rivers near Utrecht, Vechten was originally a Roman castellum probably founded in 5 CE by order of Emperor Augustus. After the Batavian uprising of 69–70, castellum Fechio was reconstructed in stone. It seems that Fechio was abandoned ca. 275 CE.
Opposite, bottom: **Velsen.** Situated near IJmuiden in the province of North Holland, Velsen was originally a tented camp (probably named Flavum) and a harbor enclosed by a timber palisade located on the River IJ built in 39–47 CE by the Roman army.

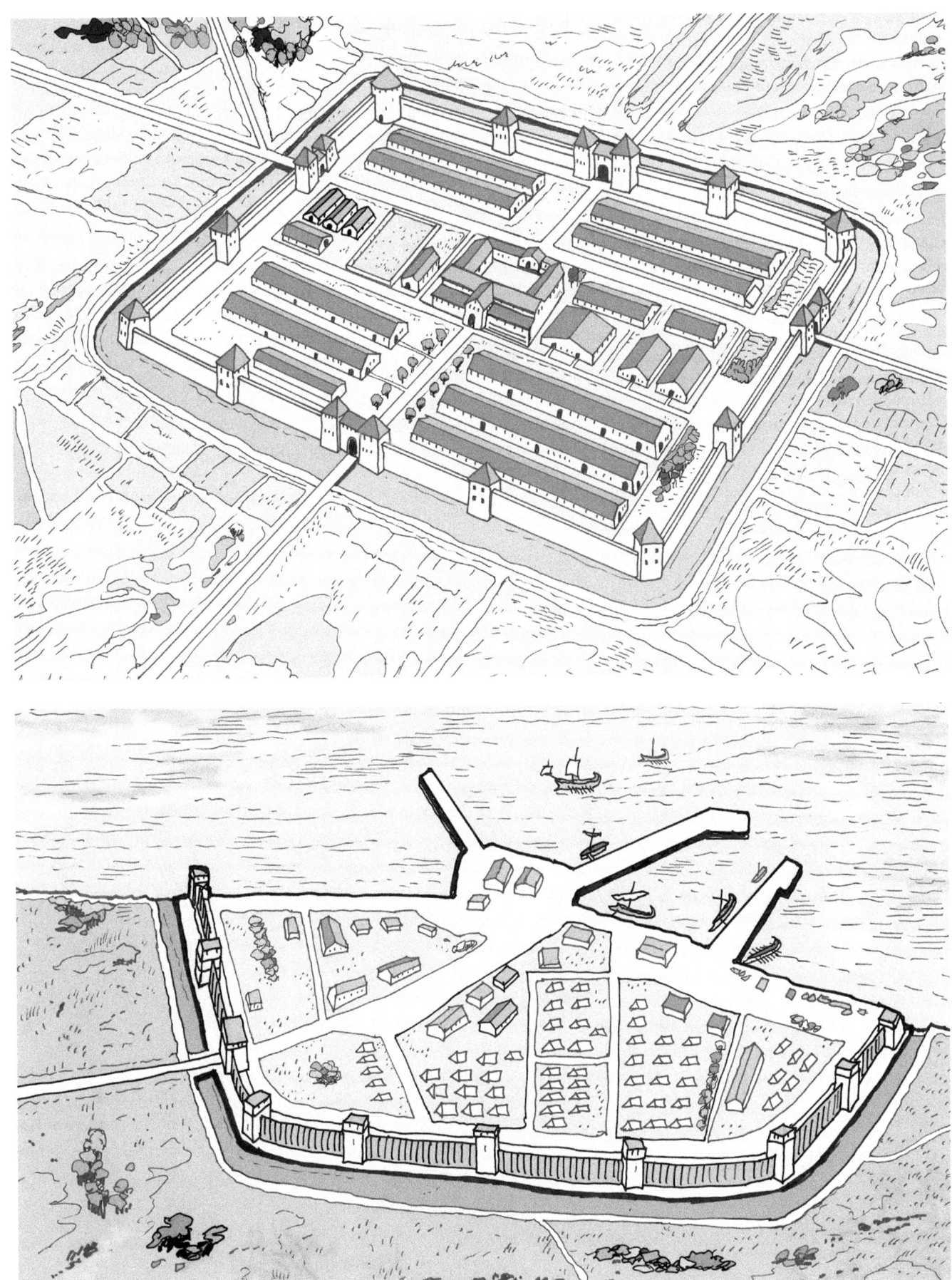

(and the rest of the limes) was attacked, taken, pillaged and devastated. In medieval times the fort was used as a stone quarry, and most vestiges disappeared. A few remnants and artifacts were rediscovered in the 20th century.

Utrecht

Utrecht (the capital city of the province of the same name) originates from a Roman timber castellum (fort) founded by the Roman army during the reign of Emperor Claudius (41–54 CE). Called Traiectum, the stronghold was intended to guard a fordable passage across the main branch of the Rhine River. It was a part of the defensive limes at the northern border of the Roman Empire in the remote province of Germania Inferior. In CE 69–70 during the revolt of the Batavi the fort was attacked and greatly damaged. Once the Romans had restored their rule in the region, the fortress was reconstructed.

Between CE 70 and CE 270 the fort was rebuilt in masonry, improved and enlarged four times. At its greatest extension (from the end of the 2nd century to the middle of the 3rd) the fort of Traiectum had a size of 125 by 150 meters (410 by 492 feet), roughly spreading around present day Domplein (Cathedral Square) in the Binnenstad (Old City) of Utrecht. It included a masonry enceinte, wall towers, four gates each flanked by a clavicula (semi-circular defensive outwork), and a moat. The fortress contained the principia (headquarters building) placed inside a rectangular courtyard bordered by a portico (colonnade), barracks for the troops, stores and several ancillary buildings. As time went by, civilian craftsmen and merchants established themselves in villages east and west of the castellum.

The settlements and castellum of Traiectum were finally abandoned in the CE 270s, when the Franks infiltrated the Roman Empire. During the Dark Age like most Roman urban settlements the site was poorly populated. However, in the Carolingian era—with a strong and determined line of bishops—Traiectum was re-born as Utrecht. It became the first and most important Christian center in the Netherlands, the see (seat of authority) of the prince-bishops, and a flourishing medieval trading town.

Nijmegen

Located in the Dutch province of Gelderland, on the Waal River, Nijmegen is the oldest city in the Netherlands,

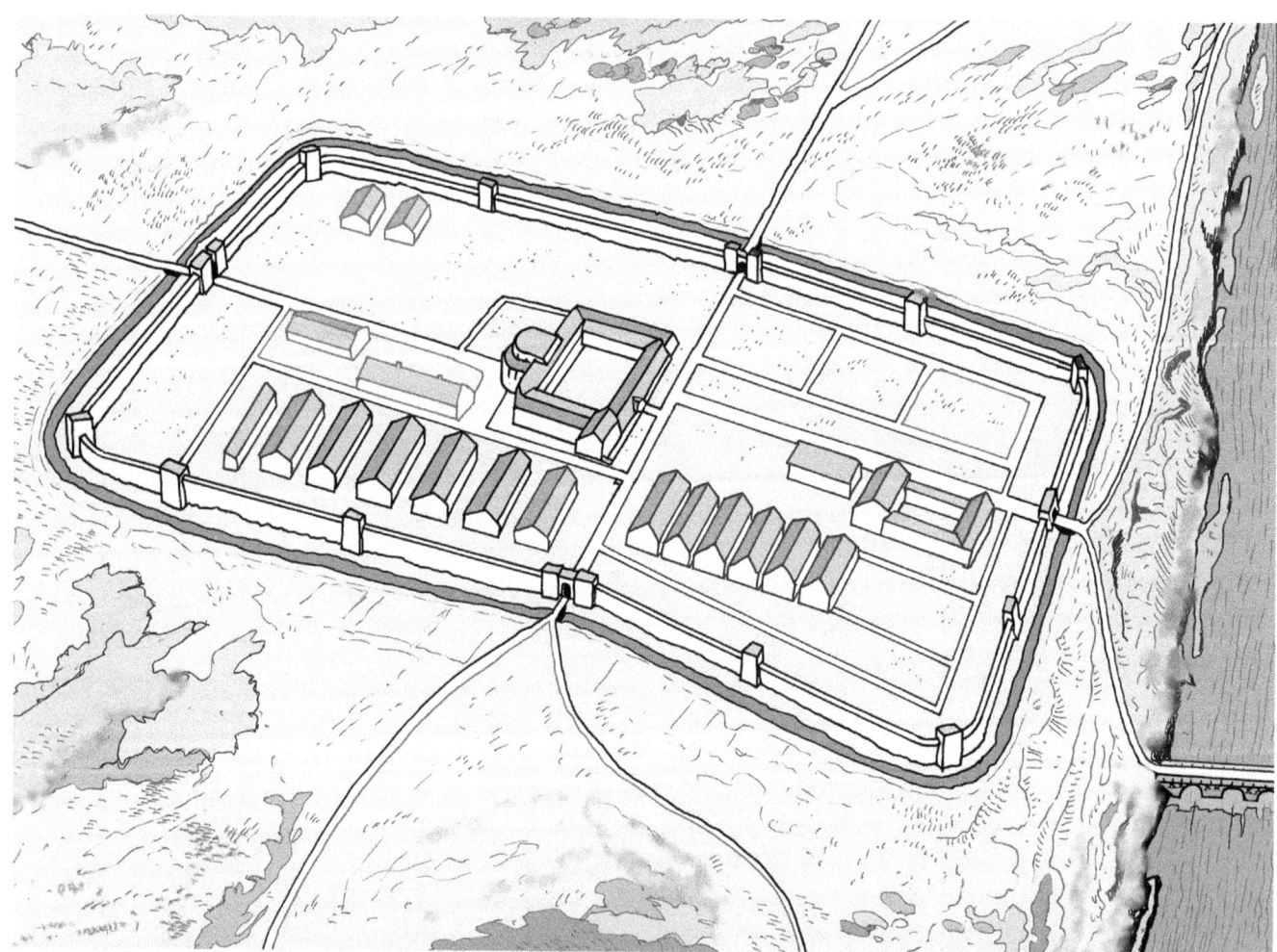

Nimegue (Noviomagus)

having celebrated in 2005 two thousand years of existence. The town originates from a Roman military camp located on a small ridge dominating the Waal and Rhine valley, thus a site with a great strategic value. Around the camp grew a small village called Oppidum Batavorum, which was destroyed during the Batavi uprising in 69 CE. After the Romans had restored their authority, the camp was re-built, called Novio Magus, and garrisoned by Julius Caesar's famous Legio X Gemina. In the late 90s CE another civilian vicus grew near the military camp. In 104, Emperor Trajan gave the settlement a new name: Ulpia Noviomagus Batavorum, Noviomagus for short (from which the current name Nijmegen originates).

According to the Dutch historian Paul van der Heijden (*Romeinse Wegen in Nederland*, Matrijs Pub. 2016) the garrison of Nijmegen was estimated to have 5,000 soldiers. After the collapse of the Roman Empire the town lost all significance for centuries. However, it regained importance during the 8th century as part of the Frankish Carolingian kingdom. The mighty Emperor Charlemagne had a *Pfalz* (a castle, an official residence or imperial palace) called Valkhof built at Nijmegen. With its location on the Waal River, Nijmegen's trading activities flourished in the Middle Age, particularly with neighboring Germany.

End of Roman Rule

At the beginning of the Christian era the region around the mouths of the Rhine and Meuse was densely populated. However in the following centuries water defeated the inhabitants, as the whole western part of the Netherlands became an almost uninhabitable peat bog regularly flooded by the sea. This was partly caused by natural sea invasion and flooding but also by the decline of Roman power. Indeed from the 3rd century, the Roman Empire entered into a serious crisis. In the Low Lands the organization of labor necessary to keep water at bay by erecting and maintaining dikes did not work any longer. With many more people leaving the flooded region, the Romans collected less tax, and the occupation of the Northern provinces only became an insufferable and useless burden.

As a result the northern frontiers became increasingly vulnerable to invasion, particularly from Germanic tribes and Frankish marauders who in successive waves of migration entered the Roman Empire. According to tradition the Roman rule in the Netherlands came to an end in about 406 CE. Germania Inferior no longer existed and later became a part of the Merovingian and Carolingian Frankish kingdoms—the largest and most powerful post–Roman realms in Western Europe between ca. 406 and 843.

Chapter 2

The Middle Ages ca. 500–1500

The term Middle Ages designates the era of European history from the collapse of the Roman Empire in the West (ca. 5th century) to the conquest of Constantinople by the Ottoman Turks in 1453.

The early part of the period (ca. 500–1000) is sometimes labeled the Dark Ages, while the later part (ca. 1000–1453) is often termed the High Middle Ages. The whole epoch is denoted by the gradual appearance of separate realms, the revival and growth of trade and urban life, the increase in power of monarchies and the domination of the Roman Catholic Church. The 15th century Renaissance is regarded as a transition period announcing what many call Modern Times.

The Netherlands in the Early Middle Ages

The Franks

As the Roman state got weaker, barbaric northern Germanic tribes started to infiltrate the Empire. The Roman Empire did not "fall" to murderous hordes of savage barbarians. The invaders who toppled the empire in the West were nomadic, often primitive and illiterate, it is true. But it should be noted that then the term *barbarian* meant *foreigner* or *outsider*. Relatively few in numbers, they did not intend to destroy the Roman Empire. What they actually wanted was to share Rome's wealth and achievements. Some were already Christians, many had had contact with the Romans and had become partially Romanized, and some were *federati* (military allies).

The Franks (actually a conglomerate of various Germanic people) formed the foundation of the medieval European world. The Germanic Franks inhabited the delta lands (today approximately the province of Zeeland) at the mouths of the Rhine and Scheldt rivers. In about 350, they were allowed to occupy lands south of the Rhine, in what is now the southern Netherlands and northern Belgium. The Frisians in the north maintained their position, and the east of the Lower Countries became the land of the Germanic Saxons.

After the collapse of the Roman Empire, large parts of the Latin civilization disappeared, and consequently the Celto-Germanic tribal and agricultural culture overwhelmed the Low Countries.

It would appear that sea level varies over time, and the higher or lower water level has a great effect upon low-lying lands such as those the Franks inhabited. At the height of the Roman Empire, the sea level was low and this particular region was rich in agricultural products and active in trade and commerce between the Romans and the Germanic tribes. As time passed, however, the sea began to encroach, and the area became a great marsh. Indeed a marine transgression (sea level rise), and a decline in population led to Roman activity stopping and Roman institutions withdrawing. The coastal lands remained largely unpopulated for the next two centuries.

From their base in present-day northern Belgium the Franks slowly pushed down into the southern lands (Roman Gaul that later became France). They renounced any allegiance to the collapsing Roman Empire, and pursued their own aims. They were pragmatic about things. Rather than chasing vague ambitions of imperial power, the Frankish kings were generally content to enjoy the fruits of their own estates and levy tribute upon submitted people. One of their kings named Chlodwig (Clovis in French) converted to Christianity and established the Merovingian dynasty—named after the semi-legendary King Merowig who supposedly was Clovis's grandfather.

Merovingians

With the conversion of Clovis, the Franks enjoyed the support of the Catholic Church. The Church provided them with the skilled and educated personnel they needed. The Franks could call upon the clergy for administrative services whenever they needed and, when they began to expand into non–Christian lands, church missionaries worked with the Frankish kings in pacifying, evangelizing and educating these new subjects.

After Clovis's death in 511 the kingdom of the Franks was divided among his four sons. The Low Countries then became a part of a large realm called Austrasia. In the 7th and 8th centuries these lands were evangelized by several missionaries (Wilfred, Willibrord and Bonifacius), and the town of Utrecht became a celebrated Christian center. The Merovingian dynasty ruled from 481 to 751 CE, but not without blood feuds, family strife, civil wars, and fratricide conflicts. The Merovingians were an example of barbarian kinship in the post–Roman era. In the absence of a strong central government their society was held together by kinship, private vengeance and religion.

As early as 639 the power gradually passed to senior officers each known as a Majordomus (Mayors of the Palace or royal chamberlains). Eventually in 751, the Mayor of the Palace Peppin the Short deposed the last Merovingian king, Childeric III, and seized the throne. Peppin had himself anointed by the Anglo-Saxon missionary Bonifacius because he wanted to demonstrate with ostentation that his kingship was placed in the tradition of the Roman imperial line with the benediction of the Christian God. The chaotic Merovingian dynasty was brought to an end, and a new dynasty (Carolingian) came to power.

Carolingians

The Frankish Carolingian dynasty was created by Peppin the Short, but it took its name from its most illustrious member: Carolus Magnus meaning Charles the Great, better known as *Charlemagne* (742–814). From 741 to 987, the Carolingian kings and emperors reigned not only over France but also a large part of Western Europe conquered by war, including regions of Germany, Austria, Switzerland, the Netherlands, Belgium, and parts of Northern Italy. The Carolingians sought to establish their power on three pillars: war leadership; Christian rule; and the legacy of Rome.

Although it lasted only for a short while the Carolingian Empire helped shape the face of Europe. Charlemagne is often regarded as the founder of the idea of Europe as a cultural and political expression. Charles I the Great was king of the Franks from CE 768, king of the Lombards from 774, and Holy Roman Emperor from 800, the first recognized emperor in Western Europe since the fall of the Western Roman Empire three centuries earlier. Charlemagne expanded the Frankish territories, and in order to protect the northern border of the realm, fought against the Frisians in the actual Netherlands, defeated them and supported the evangelization by Christian missionaries.

In Nijmegen Charlemagne had built one of his palaces. Tradition says that Nijmegen was his favorite residence, while Aix-la-Chapelle (today Aachen in North Rhine-Westphalia, Germany) was the empire's administrative capital. At the borders of the conquered territories, military defenses were organized into *Marken* (Marches), fortified buffer zones placed under the leadership of appointed loyal officers called *Margraves* from *Mark*, border, and *Graf*, count. Other territories were organized following the traditional Merovingian division into counties and duchies (and bishoprics when conquered territories were evangelized).

Each of these territories was headed and administrated by appointed loyal officers named *Graven* or *Comes* (counts, literally "companions" or earls), and *Herzogen* or *Dux* (dukes). The duties of these noblemen (and their local supporters, personal servants, advisors, military counselors, friends, clients, and allies) were to defend the Carolingian multinational Empire, to submit and administer the newly conquered population, render justice, and collect taxes. The Carolingian period can best be understood as an era of a slow transition from the civilization of the early Germanic tribes and the Roman heritage to the civilization of the Middle Ages.

After the death of Charlemagne in 814, his son Louis the Pious succeeded him. The Empire founded by Charlemagne was too large and much too heterogeneous to survive as a united Europe. It entered into a quick and dramatic decline right after the Emperor's death. In August 843 Louis the Pious's sons and successors agreed (Treaty of Verdun) to partition the Carolingian empire into three distinct realms.

Charles II the Bald received Francia Occidentalis (Western Francia, later known as France) including the territories to the west of the rivers Scheldt, Meuse, Saône, and Rhône.

Louis the German was given Francia Orientali (later known as Germany)—the vast area east of the Rhine and north of the Alps.

As for the third brother, Lothair, he kept the title of Emperor and acquired a middle realm stretching between Western and Eastern Francia, including provinces from Frisia (the Netherlands) in the north to the old Lombard duchies south of Rome, Italy. This central kingdom was called after him: Lotharingen or Lotharingia, from which the name Lorraine evolved. This incoherent, heterogeneous and indefensible middle kingdom rapidly collapsed after Lothair's death in 855. It broke into several independent entities known as Frisia, Low Countries, Lorraine, Champagne, Burgundy, Arles, Provence and northern Italy.

In 925, the king of Germany Heinrich I succeeded in subjecting the Low Countries to his rule. Officially this whole region remained part of the so-called German Holy Roman Empire until the Treaty of Münster in 1648.

Viking attacks

The 9th and 10th centuries saw the weakening of the European realms. Internal conflicts, dynastic quarrels, and local civil wars were worsened by attacks by new and dreadful external foes. Viking, Northmen or Norsemen were the names given to raiders from Scandinavia (Denmark, Sweden and Norway). Between ca. 850 and ca. 1000, they came on their excellent ships from the North as traders but also as plunderers. Why those men from the north suddenly appeared in western Europe remains unclear.

After the death of Charlemagne when central authority weakened and crumbled, the Vikings raided most of the European coasts. They were adventurous sailors, fierce warriors and barbaric pagans who did not hesitate to plunder churches and monasteries. Most of the Vikings who attacked the Low Countries came from Denmark. At first they made attacks, and returned home with stolen booty, but by the end of the 9th century they established fortified bases as winter quarters in Frisia and in the southern province of Zeeland from which they further launched their aggressive operations to France and England. Subsequently the Vikings made Europe unsafe, until the 1000s.

From that time on they gradually ceased to be a nuisance in the Low Countries, the more so as many of them had converted to Christianity. Some Vikings established themselves permanently, notably in Kennemerland (the region of Amsterdam) and created a dynasty soon becoming the family of the counts of Holland.

The central governments proved incapable of fighting off local bandits, marauders, and foreign raiders and invaders. Unable to maintain a centralized rule, the collapse of the authorities resulted in a vacuum of power, and correspondingly increased the power and influence of the provincial earls, dukes, and regional counts and barons, and local lords and noblemen. Gradually a new social and political order known as *feudalism* was established.

Feudalism

Feudalism is a term invented in the 16th century by royal lawyers—primarily in England—to describe the decentralized and complex social, political, and economic society that came into being during the decline of the Carolingian Empire. Feudalism was an improvised, specific and decentralized arrangement that arose when central authority could not perform its functions, and when it could not prevent the rise of local powers.

Charlemagne's Empire had been administrated and controlled through a system of personal relationship, based on lordship and loyalty between the ruler and powerful men upon whom the Emperor relied for support and for carrying out his commands. Originally counts, dukes, marquis as well as the local petty lords and rulers could be installed, removed, revoked or transferred at the Emperor's will. After the weakening of the central Carolingian power, noble magnates were more and more tending to become landholders ignoring instructions and orders coming from the emperor. Office holding and the possession of land became less a reward for loyal service than a hereditary family right. Gradually as the Carolingian monarchy was on the verge of dissolution, the regional governors, administrators and other public officers clung to their charges and bequeathed their power to their descendants.

The administrated region became an hereditary private property, authority became dispersed, and power was exercised personally and locally. The external threat, insecurity and general chaos caused by internal feuds, local quarrels, and the Viking incursions only facilitated and sped up this process. This all together gradually led to the development of a class of militarized magnates who created numerous independent and dynastic principalities. This change became a characteristic feature of the western European world in the Middle Ages. It was to affect every layer of society in a form of fragmentation and privatization of public authority.

The later Carolingians were forced to accede to these social upheavals, first in order to buy support, and later, because they no longer had the means to counter them. The feudal society, from top to bottom, was based on personal ties and hierarchy. The ruler (called lord or later suzerain) bound men to himself as *vavasours*. These men at their turn had allies known as vassals through a special oath of loyalty. The vavasours and the vassals were obliged to serve their lords especially in time of war. In return the lord granted them many privileges, and a beneficium, a piece of land called a *fief* (hence the terms fiefdom and feudal) with hereditary right.

The upper class became a non-working aristocracy totally devoted to ruling and military tasks, a privileged social group of landowners—as agriculture still was the main source of livelihood. This social "nobility elite" (whose power rested solely on coercion) formed a fighting force on horse living in castles and manors. The nobility's domination through dependent relationship and tenures was imposed by force to the less fortunate, poorer, and weaker members of society known as the Commoners. The linking of fief with vassalage gradually became a method of military, police, social, economical and political organization, but it was not a stable system.

Actually it was not a system at all. Lords, vavasours, and vassals, seeing the weakness of royal or imperial authority and beset by local conflicts and civil wars, tended to attach themselves more closely to their immediate overlords. This improvised practice, which became a tradition—was thus based on the primitive principle "Might is Right," and on the exchange of services. It was guaranteed only by sworn loyalty, verbal promises and allegiances, vows and oaths.

Of course, given human nature, this practice raised formidable problems of conflicting loyalties, betrayed agreements, and shifting alliances caused by personal greed, individual dislike, ambition, feuds and vengeance. Problems were most of the time resolved by pressure, coercion, threat, or forced or arranged marriages. Very often outright force was used resulting in kidnapping, assassination, murder, and war. Feudalism became a strict social relationship that, however, varied considerably in different parts of Europe and over periods of time.

Principalities in the Low Countries

In the Low Countries a number of great vassals, nominally subject to the Emperor of Germany but in fact ignoring him, established themselves as quasi-sovereign hereditary princes especially when they managed to force submission upon large territories by war or merging by marriage and inheritance. The lands that form the provinces of the present-day Netherlands originated around the year 1000. Basically the 10th and 11th centuries form a period of uncertainty, impetuousness and reorganization after the chaos of the early Middle Age. There was then no unified Netherlands, but gradually and slowly a number of large, and powerful (often antagonistic and rival) principalities emerged in Flanders, Brabant, Holland, Zeeland and Gelderland (aka Guelders).

The region of Utrecht (called Sticht) also included the Oversticht, the two northern provinces of Overijssel, and Drenthe. The Sticht counties were possessed by a Roman Catholic episcopate ruled by elected bishops. Those religious leaders evolved into a position of prime secular power. There were also many monasteries, abbeys, convents housing large communities of monks and nuns. As it moved west, monasticism shifted away from its eremitic roots (that focused upon solitude, withdrawal from the world, spirituality and contemplation) to regulated communal ways of life emphasizing *ora et labora* (prayer and work), and eventually intellectual activities like reading and copying of manuscripts. Although their members lived a secluded life, monastic communities became an important social force in shaping nature, education, culture and economy.

The northern province of Frisia was not ruled by one noble family but was fragmented into numerous small entities, demesnes, manors, and lordships. Noblemen, dukes, counts and even abbots and bishops (and soon independent wealthy towns) fought each other for greater power, influence, and independence. Feuds and conflicts were numerous, long lasting, and often sprang up over disputed successions and territorial conquests.

After the deep crisis of the 9th and 10th centuries, however, violence was still rife but collaboration, exchanges, and trade developed. A slow, gradual but strong economical urban development started to make the Low Countries one of the richest areas in Europe. Dikes (that had been neglected since the collapse of the Roman Empire) were elevated again and maintained along the seashores and riverbanks, reclamation of land was carried out, and there was a gradual resurgence of towns. New towns appeared due to agriculture development along with crafts and commerce. Important trading links reaching as far as Scandinavia and northern Italy, gradually transformed the Low Countries into an area where a relative safety of movement and economical activity was established, making sustained growth possible.

Fortifications in the Early Middle Ages

Burgen

Very little has come down to us from the Merovingian and Carolingian era, which has often been termed the *Dark Ages*. On the subject of fortifications in the Netherlands during the Dark Ages we are stepping into a period of great doubt and obscurity. Historical written sources are scanty and often unreliable. As a result no precise conclusions may be drawn with certainty.

In a background of brutality, fortifications were not unknown, and fortified settlements and defended farms of course existed. A few archeological findings strongly suggest that they consisted of primitive works with ditch, and earth wall topped with a timber palisade—a copy of earlier Roman practice, and inheritance of Celto-Germanic methods of fortification. Basically these limited military options included a defensive position, an inner perimeter encircled by a ditch, and an earth wall topped by a palisade. Access to the perimeter was through a fortified timber tower or gateway. This kind of early medieval defenses has been designated *walburgen*.

A variant of this primitive form was the so-called *ringwalburg*—a fortified circle-shaped settlement. Archeological vestiges of ringwalburgen are principally encountered in the Dutch Southern province of Zeeland and along the shores of the North Sea.

Walburgen and ringwalburgen were particularly constructed for protection against the Viking raiders. They offered a relatively secure refuge for the people of the vicinity when bandits, marauders or Viking raiders made the countryside unsafe. The fortified settlements also provided a base from which cavalry (and on the coast, ships) could operate against thieves, pirates, and the dreaded Norsemen. They also provided for the commercial exchange, control of the economy, and promotion of trade. For example the trading port and town of Dorestad (located near the present day Wijk bij Duurstede in the province of Utrecht), flourished between the 7th and 9th centuries. Dorestad, originally founded by the Romans (then called Vicus Portus), was strategically situated in a fork in the Rhine River, and dominated international trade between southern Scandinavia, central western Germany, southern Netherlands, northern France and England. Controlled by the Carolingian Franks, it declined considerably after repeated Viking attacks in the 9th century. Fortified burgen also constituted embryonal administrative centers as well as religious centers, places of pilgrimage, and spiritual sanctuaries under the protection of the Roman Catholic Church.

Burgen were not constructed according to a rigid pattern or a standardized model, but as local conditions permitted. Technically speaking a typical *walburg* might have been a medium sized village or a small, fortified settlement or a port. It was enclosed by a ditch whose excavated dirt was used to elevate an earth wall. This wall was flattened at the top and fitted with a kind of breastwork. This could be either a palisade made of aligned tree trunks or a fence made of rods and stakes interlaced with twigs and branches. The entrance was defended with a timber gatehouse or a kind of wooden tower fitted with a thick gate. For crossing the moat there was a destructible access or a movable drawbridge.

Unfortunately for posterity walburgen were made of earth, wood, timber, thatch, wattle and daub. The depredations of time have left nothing of these perishable fortified settlements. Of the walburgen that have survived as later medieval towns, little remains to be seen of them. Only a few surviving examples of foundations have been excavated at den Burg (on the island of Texel), at Heimenberg (near Rhenen in the province of Utrecht), at the Hunneschans (at Udel near Apeldoorn), and at Lichtenvoorde (near Groenlo in Gelderland). Ringwalburgen have been discovered and excavated principally in the provinces of Zeeland (o.a. at Oost-Souburg, and Middelburg), and in South Holland (at Maasland near Delft, and Naaldwijk near The Hague).

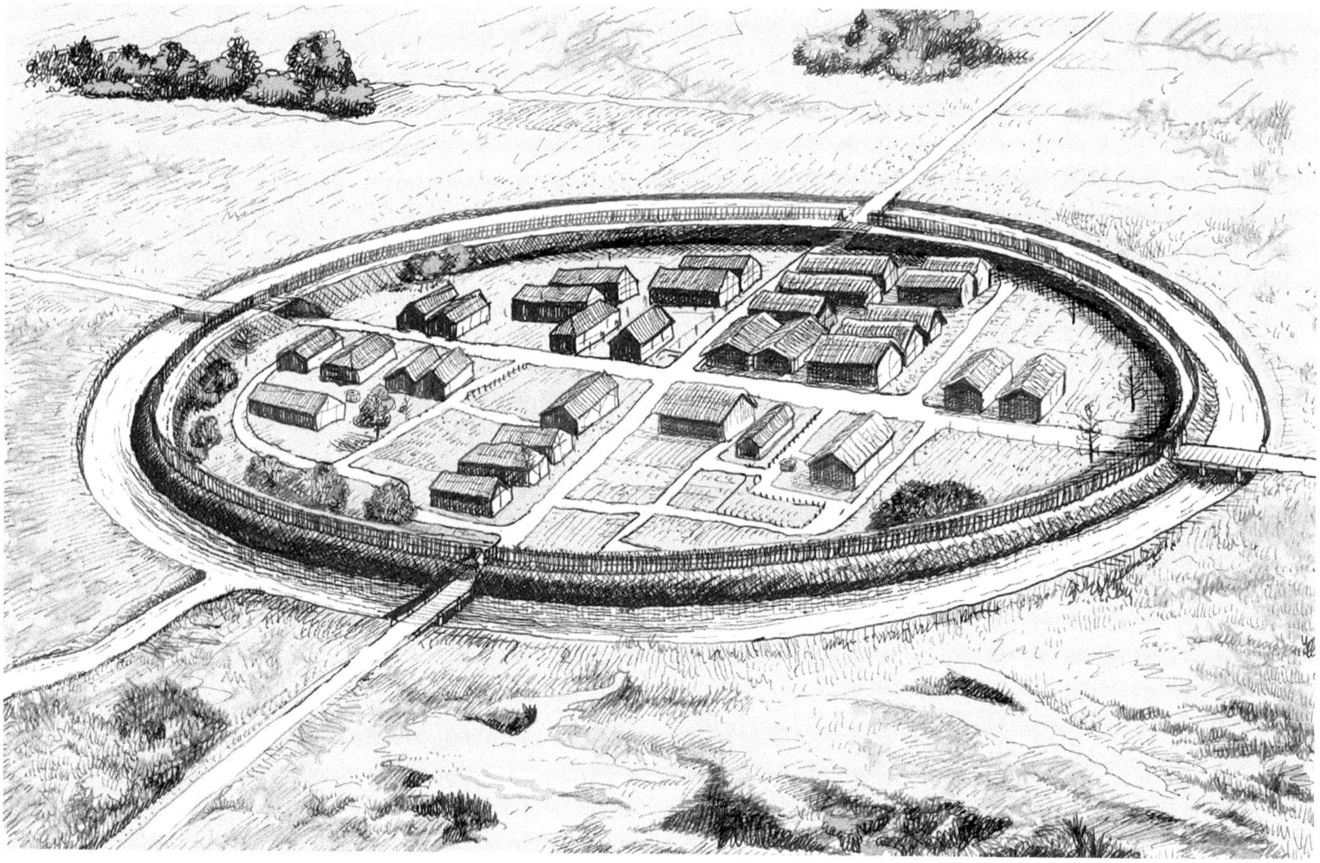

20   Dutch Fortifications

Chapter 2. The Middle Ages ca. 500–1500 21

Hunneschans. Situated at the edge of the Lake of Uddel west of Apeldoorn in the Veluwe (province of Gelderland), the fortified place was probably erected in the 8th century. The horseshoe-shaped Hunneschans was a settlement, a port, and a trading center. It was enclosed with a 4 m high earthwall about 100 m in diameter surrounded by a 3.5 m deep ditch filled with the water of the lake. The wall was topped with a palisade. The small fortress was used as a refuge for the local people and occupied until the 13th century.

Opposite, top: Oost Souburg. Located on Walcheren Island near Vlissingen in the province of Zeeland, the municipality of Oost Souburg originates from a ringwalburg established in the Carolingian era to protect the inhabitants from Viking raiders.

Opposite, bottom: Earth and timber wall. A palisade is a fence, an obstacle, or a wall or a breastwork composed of a row of closely aligned wooden stakes or sharpened tree-trunks. As a defensive structure, a palisade was often used in conjunction with a ditch whose deblai (excavated earth) was used to constitute an earthwall. The top of the earthwall was invariably flattened and arranged as an allure (aka wallwalk or chemin de ronde), a walkway allowing for patrol and making an adequate observatory, and a combat emplacement protected by a breastwork (parapet).

Medieval Castles

Definition and function of a castle

The fragmentation of public authority locally exercised by a minority of warriors, as well as the general insecurity caused by the Vikings, and other thieves, marauders and raiders, bore particularly heavily upon the defenseless peasantry. So people at large turned to any local powerful man who could defend them more effectively than the remote and powerless German Emperor. One result of these periods of permanent danger was a steady increase in the number of fortifications, as one answer to vulnerability was the construction of a fortified refuge that became known as a castle. The appearance and development of castles greatly added to the *de facto* power of the fighting upper class of warriors. Indeed they alone as wealthy landowners could afford the expense of their construction, the equipment and weapons needed for war, and the maintenance of a garrison.

Even when the Vikings' attacks ended, the knights and the ruling class's favorite occupation was combat. Feudal barons would fight to keep their vassals in hand. Landowners would try to gain additional territories from their neighbors, or would follow their lords to battle because it was their feudal duty or in the hopes of sharing in the booty. In this background of violence, a stronghold or a place of safety played a central role.

A castle (*kasteel* or *slot* in Dutch)—from Latin *castellum* (a Roman fortified camp) is a defensive structure seen as one of the main symbols of the Middle Ages. The term has a history of scholarly debate surrounding its exact meaning, but it is regarded as being quite distinct from the terms fort, fortress, citadel, and palace. A medieval castle had several functions. First of all it was a military fortified private residence owned, built and occupied by a local nobleman. It was a symbol and display of feudal authority, and an expression of the lord's power and prestige. It was also a place of social and cultural life; an administrative and economic center commanding and living off a specific territory (fief) within a social organization (feudalism). It was a fortified place designed for defense but also a base from which military operations into enemy territory were prepared and launched; and a place of refuge for the civilian peasants of the vicinity in times of crisis.

In its simplest terms, the accepted definition of a medieval castle among academics is: "A private fortified residence combining military, residential, administrative, fiscal, and economic functions." However some medieval castles can be termed *fort* as they had a primarily military purpose rather than a residential function. For example some castles/forts were built to control a passage or dominate a river ford, to collect tolls and taxes, or to prevent enemies and smugglers from sailing up a river.

The earliest recorded structures universally acknowledged by historians as "castles" were motte-and-bailey castles (described below). Wood and earth structures were gradually replaced with stone towers and donjons during the following centuries. Castles grew in size, capacity and complexity all through the Middle Ages until the introduction of firearms in the late 15th century limited their effectiveness, rendered them militarily useless in 16th century, and led to the rise of new forms of fortification. This said, medieval castles and fortified towns remained significant strategic points in the wars of early modern Europe. From the Renaissance onward, the loosening of military importance allowed for a more aesthetic and comfortable approach to design and construction. Then castles were opened up and expanded into prestigious pleasure residences, their "castle" designations, relics of the feudal age, often remained attached to the dwelling, resulting in many non-military castles, residential palaces and châteaux.

From the 18th century to the 20th century, as a manifestation of a romantic interest in the medieval period, and as part of the broader Gothic revival in architecture, several so-called castles were built. These impressive buildings had no military function and no defensive purpose at all, but they incorporated stylistic elements of earlier periods, such as a moat, a gatehouse with a drawbridge, crenellation, battlements, arrow slits, corbelled machicoulis, pinnacles, turrets and sturdy towers—all purely as ostentatious decorative ornaments.

Motte-and-bailey castle

Before the 13th century most castles were made of wood, and a frequent early form of castle was the so-called *motte-and-bailey castle* (in Dutch called a *mottekasteel*). The motte-and-bailey castle seems to originate from France, as a reaction against Vikings and other marauding raiders. Widespread by the year 1000 CE this early form of castle was naturally numerous in northwest Europe in the regions exposed to Scandinavian attacks. A typical motte-and-bailey castle was built around an artificial raised earthwork (called motte) or on a natural hill. On the flattened top there was a timber tower (often quadrangular), with both a military and a residential character. This structure was known as a *keep*, although other terms were used: tower, great-tower, donjon, odel, dunio, domus, domicilium or castellum. The wooden keep usually included one to three stories featuring a hall or living room, sleeping accommodations, supplies, food and water-stores. The entrance to the tower was above ground level and could be reached only by a removable ladder.

The wooden keep was surrounded by a protective palisade or later a masonry enclosure, and in that case it was called a *shell-keep*. The earth for the mound was taken from a ditch excavated around the motte and around the whole castle. At the foot of the mound there was the bailey, a kind of courtyard, typically enclosed by another wooden palisade. It was usually oval, round or horseshoe-shaped, and used as a living and working area. Inside the bailey a small rural community lived in autarchy including servants, peasants, and skilled artisans—e.g., a blacksmith, a miller, carpenters and other necessary craftsmen of this age.

Castle Middelburg at Alkmaar (province of North Holland). It is not known when this castle was created, perhaps in the 13th century. Built by order of the counts of Holland, it was actually a fort housing a garrison headed by a bailiff. It was intended to control a strategic trade road connecting Kennemerland (coastal region northwest of Amsterdam) and West Frisia. From the 11th century onwards, the counts of Holland followed an aggressive policy and came to dominate the archipelago of Zeeland and the western part of the Frisian country. In 1256, they conquered the region of river Amstel and its main town Amsterdam and established their capital in The Hague. The county of Holland became the most significant political power in the Low Countries in the 14th century onwards.

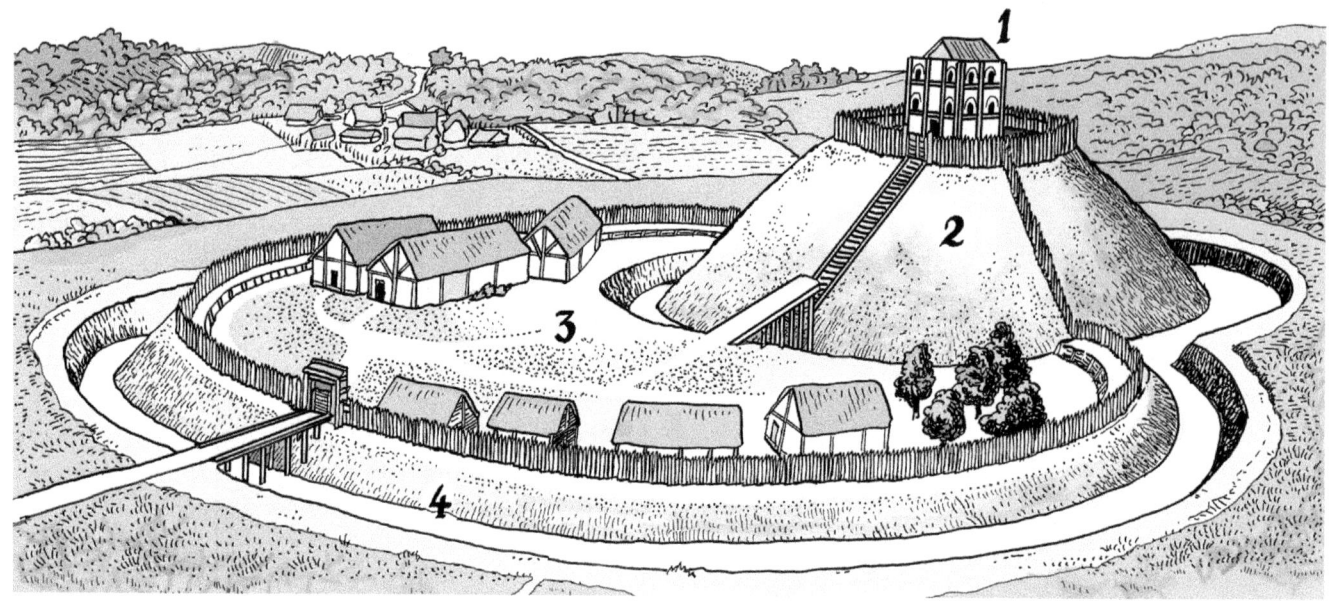

Motte-and-bailey castle: 1: Tower. 2: Motte. 3: Bailey. 4: External enceinte and ditch.

Borsele motte-and-bailey castle. Situated in the peninsula of Zuid-Beveland near Vlissingen (Flushing) in the province of Zeeland, the castle of Borsele in the 11th century included the following: 1: Motte (artificial mound). 2: Tower (residence for the lord and his family). 3: Access bridge. 4: Bailey (containing stables, storeplaces and huts for the servants). 5: Moat.

Doorwerth. Located on the Rhine River near the city of Arnhem (province of Gelderland), Doorwerth Castle was first mentioned as a timber castle on a motte in 1260 and 1280. In the 14th and 15th centuries the castle was continually enlarged, modified and reinforced and reached its present appearance in the 1560s. At the end of the 18th century the castle was no longer inhabited, and as a result remained in a neglected state until 1910, when it was purchased and restored by a retired artillery officer named Frederic Adolph Hoever. After restoration, it was used as a Dutch Artillery Museum. During the battle of Arnhem (Operation Market Garden in September 1944) and during the liberation of Arnhem (April 1945), Doorwerth castle was severely damaged. After World War II, it was restored to its function as a museum, and today constitutes a major tourist attraction in Gelderland.

Burcht at Leiden. The burcht at Leiden (province of South Holland) was built around the year 1000 on an artificial motte erected between two arms of the Rhine River. The motte is about 12 m high; its top was flattened and originally included a palisade. By the end of the 11th century the palisade was replaced with an oval-shaped stonewall. It is rather similar to the shell-keeps encountered in England (e.g., at Restormel in Cornwall or the Round Tower at Windsor Castle in Berkshire). The burcht still exists today, and its stonewall has a diameter of about 35 m, it includes a wallwalk resting on arches, the breastwork is fitted with crenels, there is a well in the hemmed perimeter, and the entrance is defended by a small tower. At the foot of the motte there was a bailey (enclosed courtyard) with various ancillary buildings. The whole place was originally enclosed with a palisade and a moat.

Opposite, top: Castle Horn. Located near Leudal in the province of Limburg on the Meuse River, Horn castle probably originated from a 10th century motte castle that was enlarged in the 12th, 13th and 15th centuries. Abandoned for a long time, the castle was restored between 1954 and 1957.

Opposite, bottom: Castle Oostvorne. This castle (also known as Burcht van Oostvoorne or Jacoba Burcht) stood in the village of Oostvoorne, in the province of South Holland. Originating from a motte castle built in the 12th century to control a strategic spot at the mouth of the Maas River, it was rebuilt in stone and bricks in the 13th century. In the 14th century it became an important residence for the counts of Holland. After 1503 the castle was abandoned and between 1534 and 1552 it was demolished. (Speculative reconstitution based on an ancient print.)

Doornenburg. The woontoren Doornenburg located in the province of Gelderland (originating from a 9th century fortified manor) was constructed in the 13th century. Through the centuries, the castle was expanded further into its current form with a chapel, several service buildings, a courtyard and a walled enclosure with towers. The castle included a large moat and its main access was defended by a voorburcht (independent advanced barbican). Devastated in 1945, the castle was rebuilt in the period 1947 to 1968. (After a 1875 painting by Carl Hilgers.)

The bailey would have looked like a small village with huts and quarters, stables for domestic animals, storehouses, workshops and other ancillary buildings, perhaps a small chapel and a hall, as well as a garden and small meadows for the horses and cattle when space was available. The palisaded bailey was most of the time surrounded by a moat making an extra external line of defense. The bailey was connected to the keep on the motte by a timber drawbridge, which could be destroyed or dismantled in time of crisis. There was in many cases another drawbridge at the entrance of the bailey that could similarly be destroyed or raised for protection.

Many motte-and-bailey castles were built in Europe and in the Netherlands in the 11th and 12th centuries. They were favored as a relatively cheap but effective defensive fortification that could repel most attackers that were not too numerous or equipped with sophisticated siege machines. A motte-and-bailey castle could be built relatively rapidly with readily available materials and without highly skilled designers and laborers. The simple structure of moat-and-bailey castles allowed many designs, shapes, dimensions, arrangement, ground plans and layout. Each of them was a unique system specifically intended to meet the needs and requirements of a particular geographical and political situation, and to be within the budget of the lord.

Opposite, top: **Castle Loenersloot.** Located in the village of Loenen in the province of Utrecht, the castle originally consisted of a simple woontoren built about 960. In ca. 1250 a five-stories high cylindrical keep was constructed. Later several buildings, and a broad moat were added around the donjon. The castle was transformed into a comfortable residence in the 18th century.

Opposite, bottom: **Kampen.** Located on a convenient place on the busy trade route between the Zuiderzee and the Rhine in the province of Overijssel was founded probably around 1150. Its inhabitants obtained a charter with city rights in 1236. From then on Kampen developed into a prosperous port, and a member of the Hanseatic League. At first, the town was defended by a palisade and a moat. In 1325 a new enceinte comprising stone walls, towers and gatehouses made of brick was erected. It was enlarged in 1465.

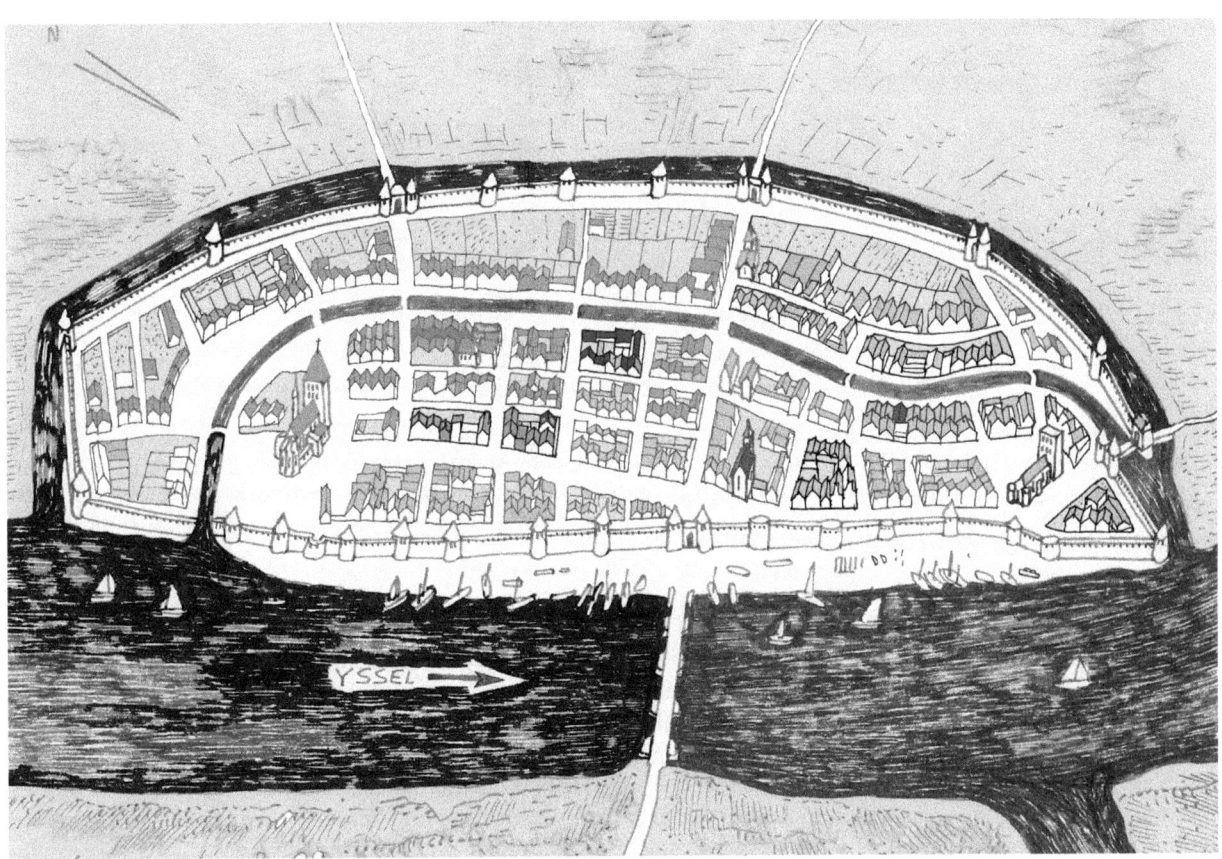

Zwolle. The name that comes from the word Suolle meaning "hill"; Zwolle is today the capital city of the Overijssel province. It was formally founded in ca. 800 CE by Frisian merchants. The city was granted city right by the bishop of Utrecht in 1230, and became a member of the Hanseatic League. Today moats enclosing the old city still clearly show the tracé of the ancient fortifications. A number of fragments of the medieval city wall, including two strong towers, still stand along the waterside of the Thorbeckegracht.

A number of Dutch medieval castles originated from an early *mottekasteel*, for example at De Ooij (near Berg en Dal, Gelderland), Barlham (near Doetinchem province of Gelderland), Gulpen (province of Limburg), Maurik (Brabant), Borsele (Zeeland), Oostvoorne (South Holland), and Bronkhorst (Gelderland). The best preserved example can still be seen today at Leiden in the province of South Holland.

There is an interesting reconstruction of an 11th century motte-and-bailey castle located between Domburg and Oostkapelle in the province of Zeeland. Completed in 2011, this duplicate of an early Middle Age castle comprises an enclosed perimeter inside which there is a palisaded 4 meter high mound topped with a 10 meter square roofed tower made of timber. The motte-and-bailey castle is part of the Nature and Landscape Museum Terra Maris. In the close vicinity tourists can also visit the 13th century castle of Westhove, which today is open to the public, and houses a café, a restaurant, a hotel, and gardens.

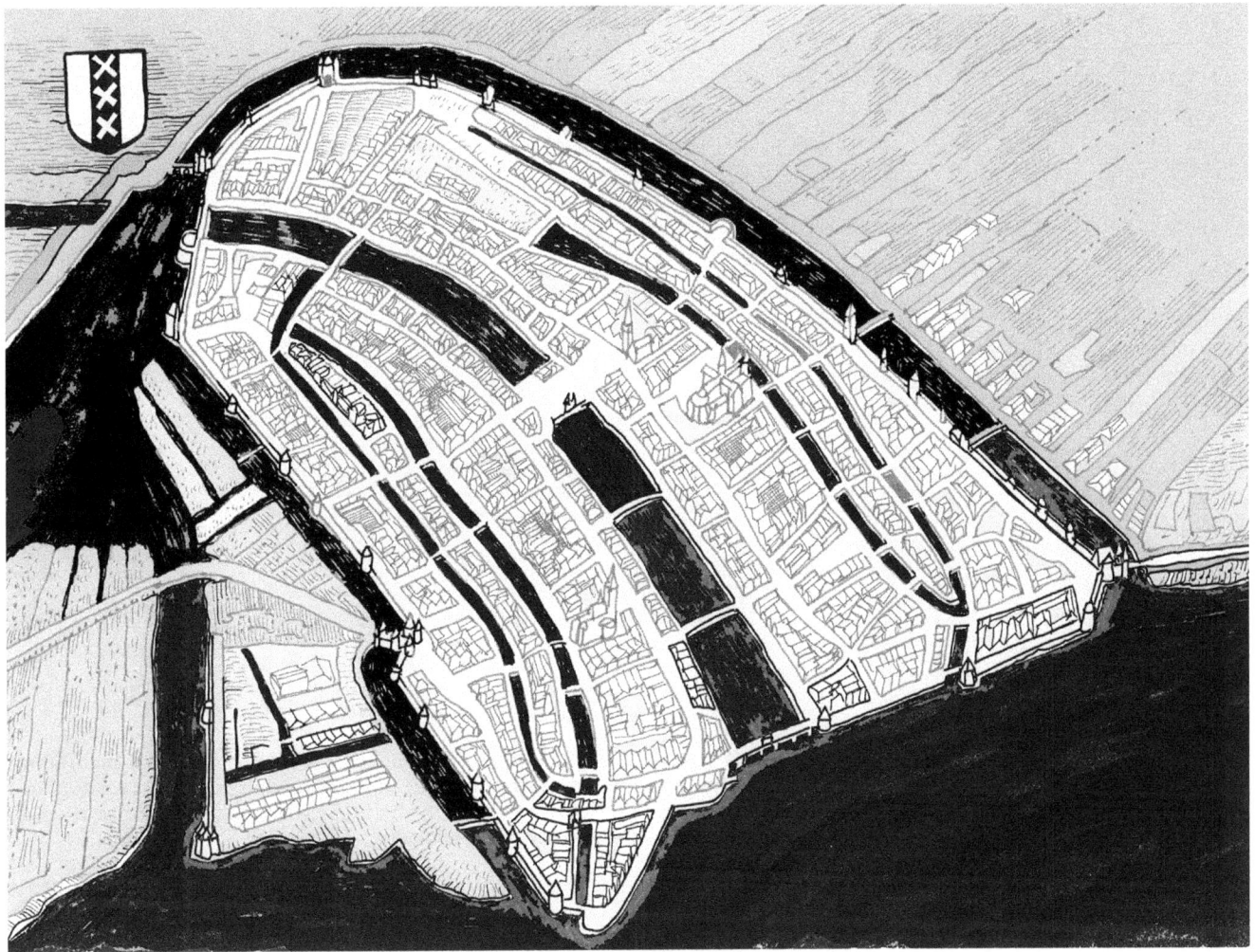

Amsterdam. Today the capital of the Netherlands, the city began as a small village on a dam and dikes built on the IJ and Amstel rivers. In 1275 the Count of Holland granted the citizens of Amsterdam a toll-free status. After 1380 Amsterdam flourished and grew rapidly as a center of trade, winning control over the sea trade with Scandinavia.

Stone Donjon and Woontoren

The rather simple motte-and-bailey castle made of timber was of course a perishable structure vulnerable to fire. So the most powerful and wealthiest barons, lords, castelans, bishops, and other ruling noblemen wanted a stronger residence. A logical development was to construct the defensive elements in stone or brick. It should be noted that natural stones are not always available everywhere in the Netherlands. This has long been leading to the confection and use of an alternative affordable building material: the brick, composed of clay dried and often fired in a kiln. Usually laid flat and bonded, bricks form a stable, strong, well isolating and durable building material, sometimes referred to as artificial stone.

In the 11th and 12th centuries, the boisterous feudal society expanded and matured. A new departure of great importance was made in the seigniorial castle, which restored for some centuries a definite superiority to the defense. Gradually a new building called a *donjon* or *woontoren* (literally living tower) appeared. Originating from the Latin term *dominium* (meaning house of the lord) the donjon was a fortified place—in a way an early form of castle. In the northern territories of what later would become the Netherlands, such a manor or stone house was called a *stiens* in the province of Frisia, and a *borg* in the province of Groningen. In the early Middle Ages the donjon was not only a place of refuge; local magnates and their families, servants and retinue, and a few men-of-arms permanently lived in those fortified homes.

The *woontoren* was not always built on a (natural or artificial) motte, and often the timber structure was replaced with masonry and bricks (or stones when available) reinforced with metal cramps. The tower was characterized by sturdiness, height, and verticality. It was several stories high, and was either square, or rectangular, or occasionally octagonal or round in plan. It had massive walls pierced with only a few tiny, narrow windows. It included several superposed rooms linked by ladders or a narrow staircase.

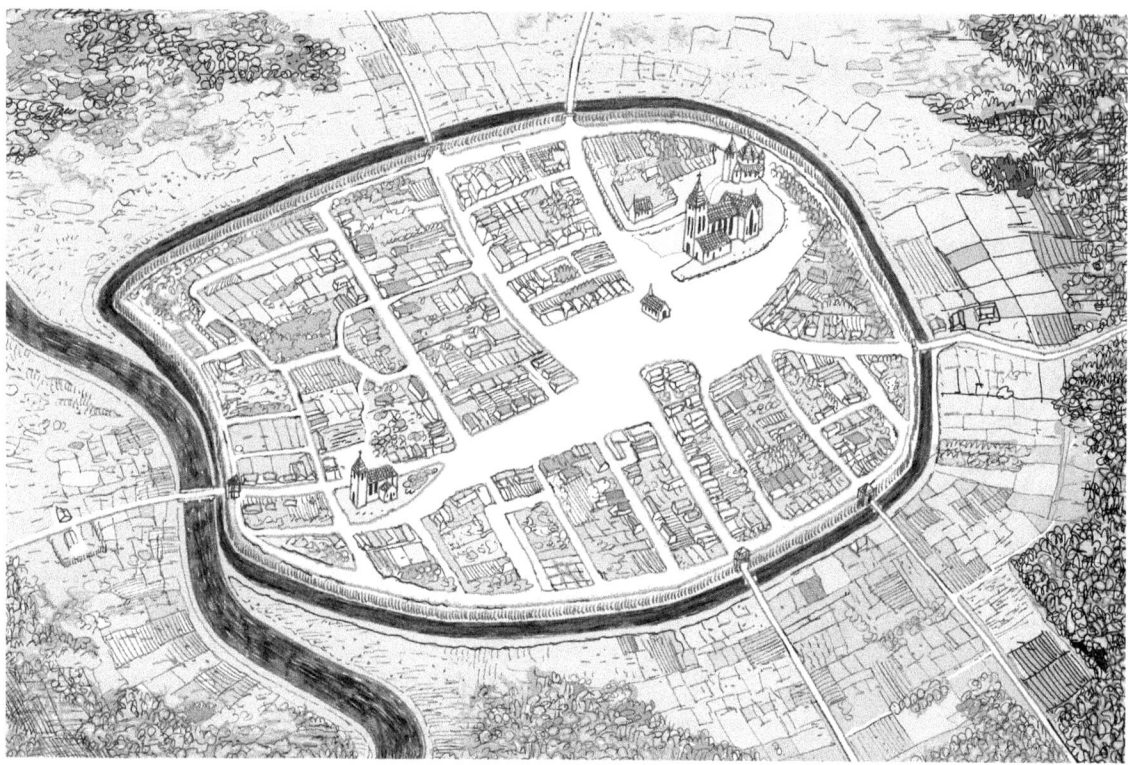

Groningen. Located in the northern province of Groningen, the city of Groningen was a possession of the bishop of Utrecht. Since the 11th century, the small settlement was populated by merchants, peasants and fishermen and defended by a moat, and an earthen wall topped by a palisade. It was only in the late 13th century that the citizens could afford a moated enceinte, towers and gatehouses made of strong masonry.

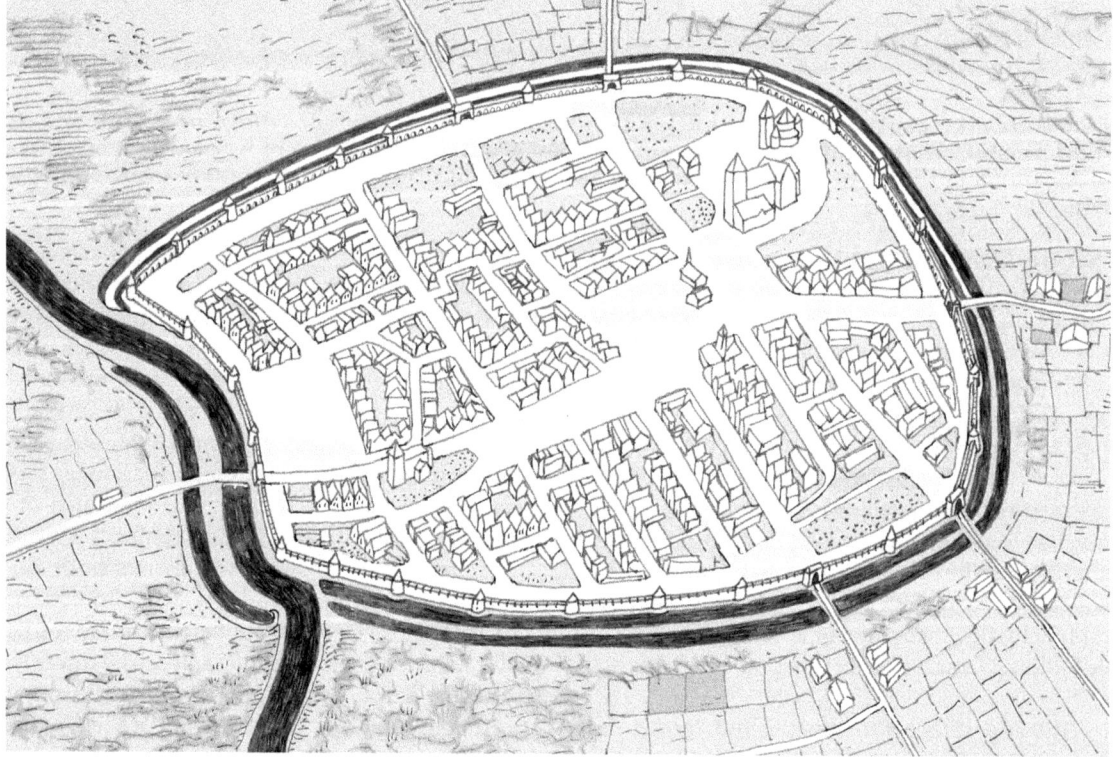

Groningen 1260. Until 1260 the town of Groningen was defended by a ditch, and an earth wall topped with a palisade. The construction of the first stonewall with moats, towers and gatehouses was started in 1260 and completed in ca. 1337.

Dussen castle. Located in the North Brabant town of Dussen, this castle was originally a fortified house. In 1330, the dungeon was built, and in 1387, permission was granted to extend the modest keep into a real waterburcht (moated fortress). Like many Dutch castles it exchanged hands several times and suffered many severe destructions, notably in 1300, 1421, and 1573. It was rebuilt, enlarged and restored on several occasions. In 1931, the castle was put up for sale with plans for the structure to be demolished. Fortunately the municipality stepped in, purchased the property and began restoration work. The castle was severely damaged in 1944 during World War II. In 1980, renovations were once again carried out. Today, Dussen Castle serves primarily as an event venue for weddings, conferences, and business meetings. (A speculative reconstruction of how the castle might have looked in the 1650s after a drawing by the artist Roelant Roghman.)

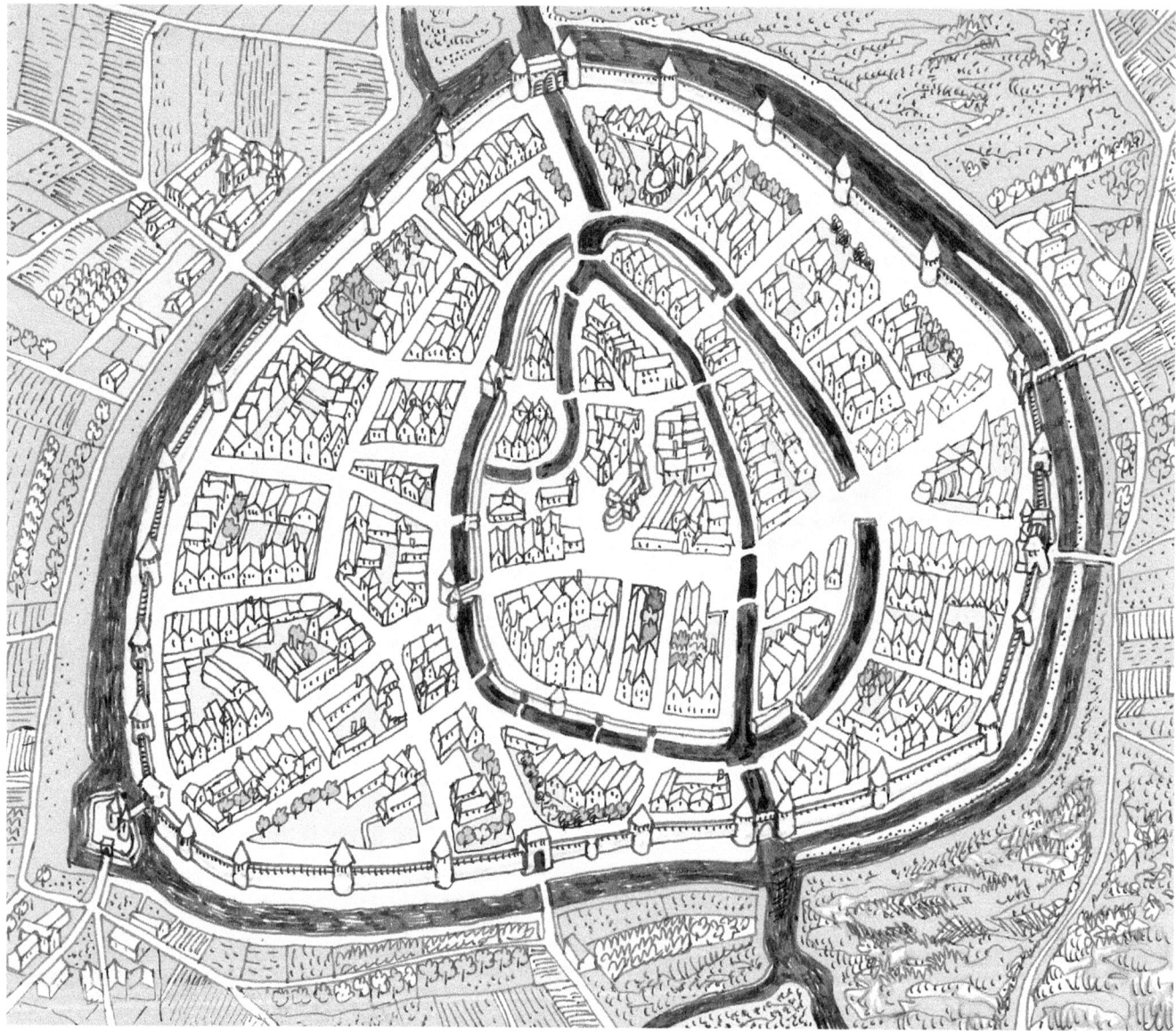

Amersfoort (named after a ford across the Amer River) appeared in the 11th century and grew to an important economic center. Granted city rights in 1259 by the bishop of Utrecht, the small town was defended by a first moat, and a wall made out of brick completed around 1300. Later, the need for enlargement of the city became apparent and in about 1380 the construction of new fortifications (walls, towers, gatehouses and moats) was begun, and completed around 1450.

Staircases inside towers usually turned clockwise to allow for the defenders to fight with their right hand and restricting the swing of any attacker. Typically (for obvious security reason) the entrance was a simple door placed high above the ground floor and only accessible via a ladder or a wooden staircase. The tower included a blind vaulted chamber at ground level for storing supplies; the aula was a multi-purpose living-hall on the first floor. The other stories featured rooms, accommodations and often an oratory (a small chapel). Survival was impossible without water, so each castle included a well or a cistern collecting rainwater.

The top of the edifice was arranged as a combat and observation platform fitted with a battlement—a protective breastwork formed of crenels (voids) and merlons (solid upright section offering cover) from which the defenders could shoot and drop projectiles down on attackers. The site of a woontoren could be chosen purely for its natural strength, without regard (except as a secondary consideration) to the protection of anything outside it; and as its area was small it was often easy to find a natural position entirely suited for the purpose.

Because of the low ground level in the Netherlands, most Dutch and Flemish fortresses were very often *waterburchten* (water-castles) surrounded by a wide ditch filled with water. Although primitive, woontoren and donjons were rather expensive to build and only the wealthiest rulers could afford them. Woontoren primarily provided safety, and in addition they were seen from miles around in the flat Dutch countryside, and thus also played an important psychological role: they clearly showed the local ruler's strength, prestige and authority. However as a residence, they were certainly cramped, cold, drafty, dark, gloomy, and uncomfortable. In many cases the stone tower or donjon was a portent of what later was to become the typical medieval castle.

Stone castles

Depending on many factors, the compact donjon and the cluster around it could grow into a more spacious and elaborate place. There was of course no standardization, and each castle evolved and presented an original layout and appearance, but Dutch castles had many common features.

A typical Dutch medieval castle was characterized by a high vertical crenelated enceinte (composed of stone walls called *curtains*), high flanking towers, often an imposing gatehouse, a residential hall, a chapel, and sometimes a high main tower or a pre-existing keep. Obviously where the enemy's approach was easiest, the walls were higher, flanking towers stronger and ditches wider and deeper. Over the centuries the Dutch used their eternal nemesis—water as an instrument of deterrence and defense. Invariably, as already said, Dutch castles were surrounded with a deep and broad moat (ditch filled with water), and often termed a *waterburcht* (water or moated fortress). The moat could only be crossed by a fixed bridge and a mobile part called a drawbridge. A castle built under such conditions was practically impregnable, and this was one of the causes of the independence of local barons, and petty lords in the 11th and 12th centuries.

The 13th and 14th centuries marked a period of balance and prosperity considered as the apogee of the Western medieval civilization.

This seigniorial castle built primarily as a private stronghold for a local magnates or for small bodies of warriors dominating a conquered country, restored for some centuries (until firearms became effective in the Renaissance), was definitely superior on the defense. It was extremely dangerous to attempt assaulting such strong vertical structures, so castles could only be reduced by attrition and blockade, but for many reasons a siege of long duration was very difficult to carry out in the feudal age.

In the 12th and 13th centuries many influences were at work in the development of castellar fortification. The Crusades in Palestine enabled the West to learn much of Byzantine and Muslim Middle East warfare, architecture, and military engineering. Intercourse with the continent of Europe was constant, and new ideas, practices and designs developed in the Middle East were brought back home. Siege warfare and European fortifications underwent a significant evolution in the 13th and 14th centuries.

The experience of sieges and fortress design gained during the Crusades by contact with the Muslim civilization

Patrician stone house (cross-section).

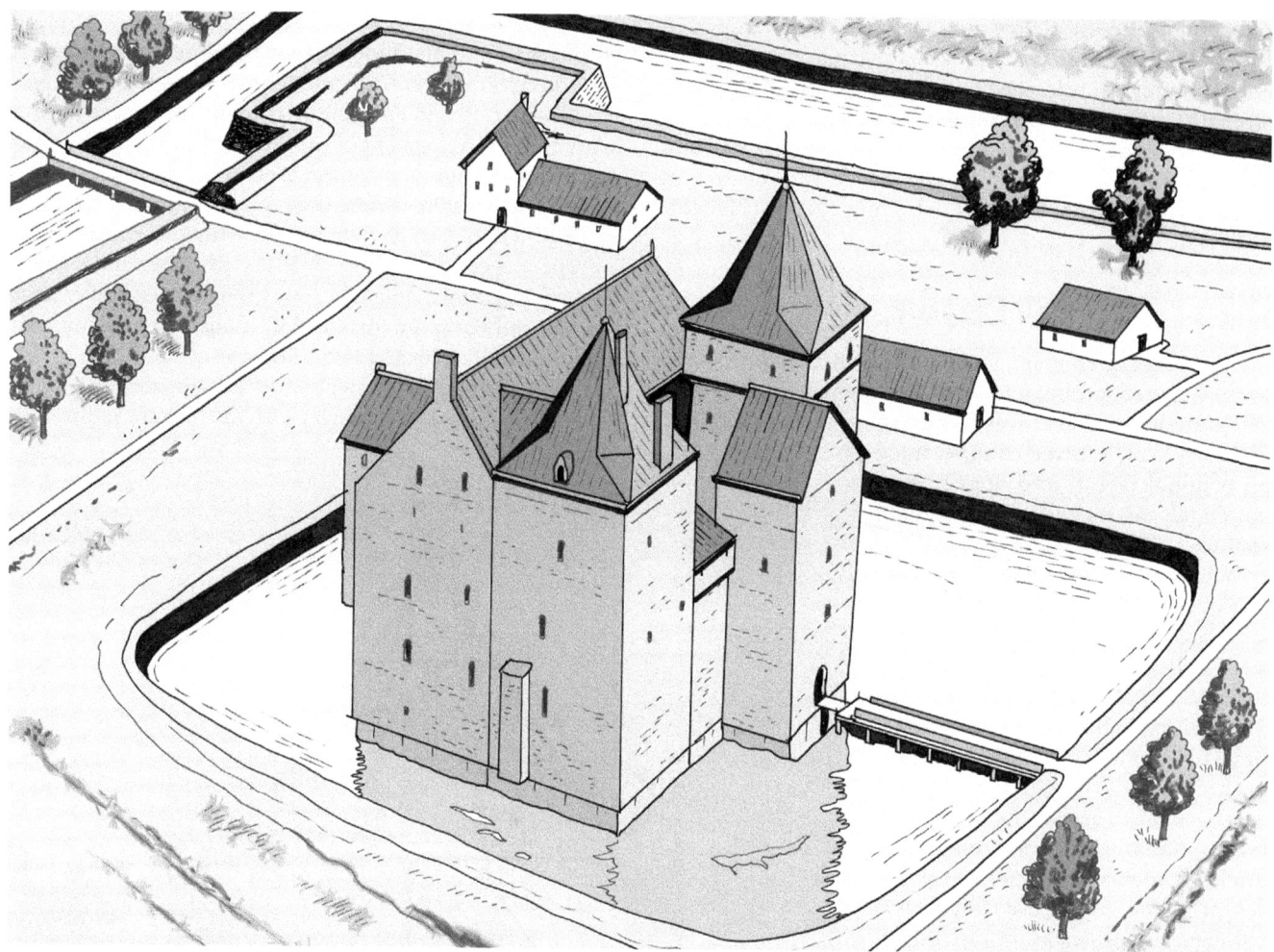

Loevestein. Located in the municipality of Zaltbommel in the province of Gelderland, Loevestein Castle was placed at a strategic location where the Maas and Waal rivers come together. It probably originated from a small fortified base built by the Vikings in the 9th century. Between 1357 and 1368 the castle was constructed by the lord Dirc Loef van Horne, hence "Loef's stein" (stonehouse). At first it was a fort—a simple square brick building, used to charge tolls from trading vessels using the rivers. By 1372, the castle was enlarged by order of the counts of Holland. During the Eighty Years' War in 1575 it was modernized as a fortress surrounded by earthen fortifications with stone bastions, moats, an arsenal, a powder tower, a house for a senior commanding officer, and barracks for the soldiers. From 1619 the castle became a prison for political prisoners. One famous inmate was the eminent lawyer, poet and politician Hugo de Groot (Hugo Grotius). Until World War II Loevestein Castle was part of the Holland Waterline, the main Dutch defense line that was based on flooding an area of land south and east of the western provinces. See Chapter 7. Today the castle is used as a medieval museum.

and the Byzantine world were important. Besides, more money and thus larger garrisons were at the disposal of the great feudal lords. The ruling class also had a growing interest and a better understanding of fortification, and this led to more complex and more sophisticated designs. Gradually, important progress was done in enlarging and refining the space for living, and improving the elements of defense—notably arrangements for *flanking* (guard, observe, and defend with enfilading fire from the side). The effect of flanking was improved by giving more projection to the towers, and by the creation of more numerous and more appropriate combat emplacements (e.g., brattice, turrets, bartizans) allowing for wider observation and better arcs of fire.

The size and height of towers and walls were often increased as was the thickness and quality of the masonry. In many cases the central core expanded, and was surrounded with various ancillary buildings placed inside a walled enclosure.

In the 13th and 14th centuries appeared the so-called rectangular castle. If irregular ground plans were in some cases imposed by natural sites, the tendency was to build castles following a regular, symmetrical, geometrical and rigorous layout. A regular rectangle or square castle was a homogeneous, compact, comprehensive, and sophisticated fortress composed of four straight strong and high walls joined by cylindrical towers at each corners. Towers offered refuges for the garrison in case of a successful escalade, and from them the walk along the top of the walls could be enfiladed. Access to the stories inside

Castle Ammersoyen. Located near the Meuse River next to the village of Ammerzoden in Gelderland, this castle was built in the 12th century, and greatly enlarged in the 1350s. It is a waterburcht connected to the separate bailey by a bridge as the entire site is moated. Severely damaged in 1944, Ammersoyen castle has been renovated after World War II, and today it is one of the best-preserved medieval fortresses in the Netherlands.

towers was by means of ladder or masonry spiral staircases. It is worth noting that most medieval staircases and passageways in castles were deliberately very narrow, so that one single defender could hold it against several attackers.

The castle had a powerfully defended gatehouse generally placed in the middle of one curtain. Often the keep became redundant and was discarded. There was a tendency towards complexity and multiplication of lines of defenses in order to guard against surprise and to localize successful assaults. Great attention was paid to the "defence in depth" with detached elements while easy communications allowed for rapid movement for the garrison. Flanking towers were sometimes cut off from their walls thereby ensuring separate resistance. Complicated entrances with traps and many doors were arranged.

Almost all defense was from the tops of the walls and towers, the loopholes on the lower stories being mainly for light, ventilation, and observation. Machicoulis galleries (for vertical defense) were protected either by stonewalls built out on corbels, or by strong timber non-permanent hoardings. Brattices, and bartizans (aka guérites—small observation turrets) as well as machicolations on stone brackets became standard elements. Loopholes and crenels were carefully placed, and protected by movable wooden shutters.

Great attention and much ingenuity were expended on details of all kinds, and this influence lasted long after medieval times. Some sophisticated castles were furnished with deceiving elements and cunning traps concealed in unexpected and dark places. These included hidden pits in floors, dead-end staircases, fake posterns, genuine secret passages from which the defenders might sally forth upon intruders, labyrinthian corridors, rear-yard or chicanes where confused attackers were ambushed and delayed.

Step by step defense was rationalized by passive obstacles (practically always a broad wet ditch), sometimes an advanced barbican was placed in front of the gatehouse. Inside the castle the whole garrison could move rapidly to reinforce any threatened point. The defenders could position war machines in the courtyard, and hurled their projectiles

Brattice. A timber or masonry gallery built on top of the rampart, a brattice projected forward from the parapet. Its floor was fitted with machicoulis allowing the defenders a better field of vertical fire. Brattices were very often placed above an access, a gate, a postern or a door.

Brattice

above the walls. The corner towers offered an efficient flanking from numerous active combat emplacements installed in well-protected works. With all this, the attack was inferior to the defense, and these great improvements, reviving the essential principles of Ancient and Roman fortification, enabled to build castles in sites totally deprived of natural defenses.

Ancillary buildings, residential quarters, hall, chapel, wells and cisterns, huts for servants, quarters for soldiers, stables for domestic animals, storehouses, workshops, and other facilities related to the castle community's life were placed against the walls, leaving space for a central courtyard, called bailey or base court and even a garden. This kind of classical regular rectangular (or square) castle was sometimes called courtyard-castle or quadrangular castle.

An excellent example of such a fortress is Muiderslot (Muiden Castle), which still exists today. It is situated southeast of Amsterdam where the river Vecht flows into the former Zuiderzee (Southern Sea). This river was the main north-south route for more than a thousand years. It had been used by the Romans and the Vikings before the medieval merchants. The castle of Muiden was actually a fort, as it was intended to control the river that was vital for the city of Utrecht, until the 15th century the most important town of the northern Netherlands, and main rival of Holland. For almost 400 years the bishops of Utrecht and the counts of Holland fought bitter wars over this strategic region and the river. In the end, Holland won.

Opposite, top: **Barbican.** A barbican or basteja was an independent advanced defensive outwork generally placed in front of a gate in a castle or a walled city. Its shape, size, and strength could vary a lot from an enormous D-shaped tower to a small palisade enclosing outpost.

Opposite, bottom: General view of a 14th century castle.

Chapter 2. The Middle Ages ca. 500–1500

Curtain: 1: Wallwalk resting on arcades. 2: Merlon. 3: Crenel. 4: Roof. 5: Stair. 6: Machicolation.

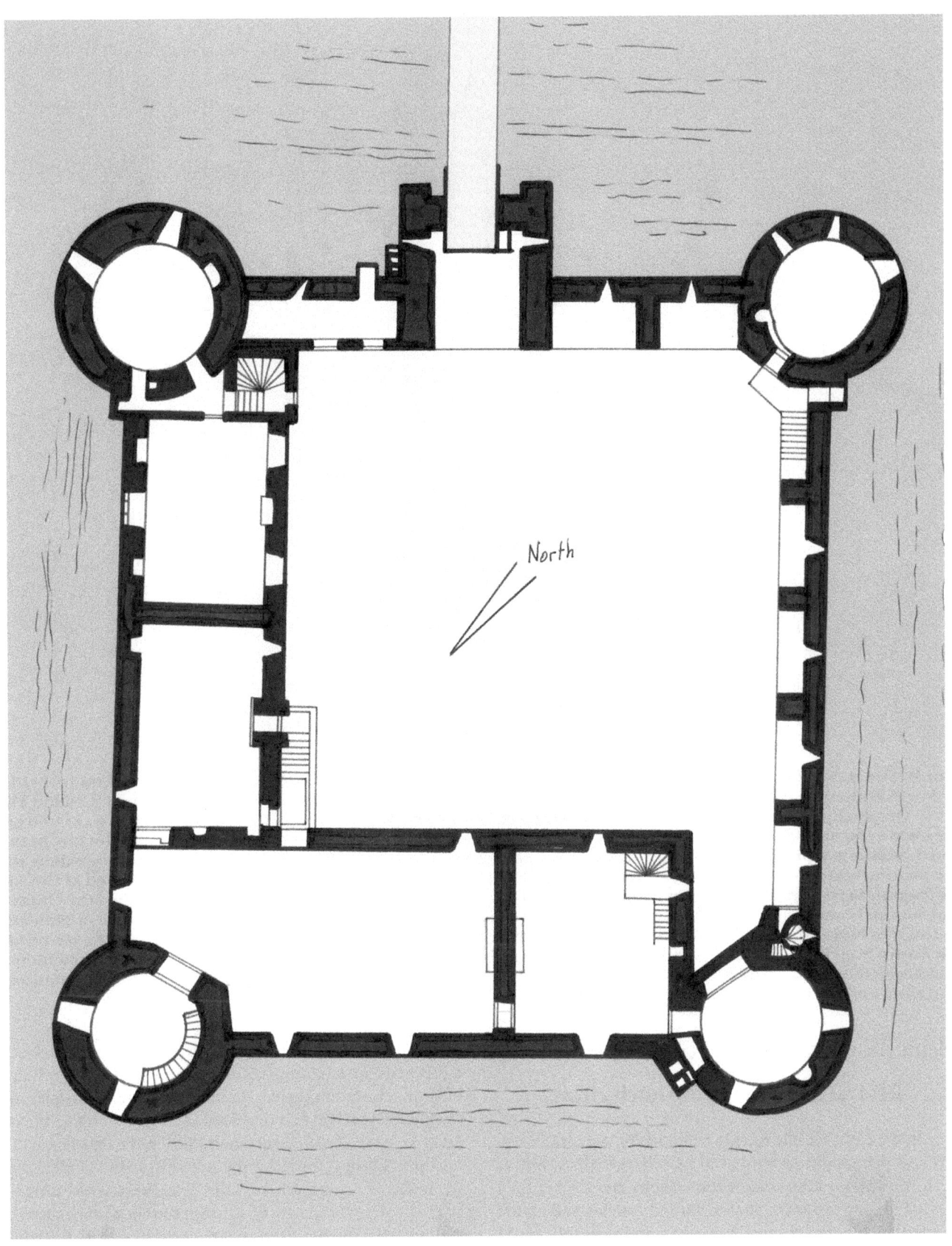

Muiderslot plan. In plan the castle of Muiden shows many similarities with other square castles like Helmond Castle, Ammersoyen Castle, and the partially remaining Radboud Castle. Muiderslot is a typical example of regular castle without a keep. It includes a moat, a rectangular perimeter with four walls enclosing a courtyard and buildings, a gatehouse, and strong round roofed towers at the four corners.

Muiderslot. Located at the mouth of the Vecht River near Muiden some 15 km southeast of Amsterdam (province of North Holland), the castle of Muiden was built in the 1280s by order of the Count of Holland and Zeeland Floris V. Originally a small stronghold for enforcing a toll on merchants sailing on the Vecht River, it was enlarged between 1370 and 1386 by order of Duke Albert I of Bavaria. All through the centuries, the Muiderslot has been a court building, a prison and the official residence of castle bailiffs. One of the most famous of these wardens was the prolific historian, playwright, author and poet Pieter Corneliszoon Hooft (1581–1647). Hooft lived at the castle from 1609 to 1647. During this period, he invited many friends and prominent intellectuals. By then the Muiderslot was an important meeting center for literature, science and art known as the Muiderkring (Circle of Muiden). In the 17th and 18th centuries the vicinity of the medieval castle was modernized with the addition of a bastioned enclosure built in Old Dutch System style. In the late 18th century Muiderslot was used as a prison, then abandoned and neglected, until restoration was carried out in the 19th century. Today the celebrated Muiderslot is a national museum, and a much visited tourist attraction.

Medieval Cities

Revival and growth of Dutch cities

In the early Middle Ages period, castles, manors, counties and principalities tended to be self-sufficient, but as the lords, earls, counts, dukes and bishops grew richer they needed luxury products, manufactured articles and exotic goods that their own local craftsmen and peasants could not produce. To supply these wants, interregional, and later international trade flourished.

In the 12th and 13th centuries, an important change took place in Western civilization. Gradually populations grew, the amount of money in circulation increased, and a primitive form of capitalist economy based on exchange of goods started to thrive. Accelerated by the Crusades, the commercial revival first occurred in Italy (notably in the cities of Venice, Amalfi, Genoa and Pisa) in the 11th century. In northern Europe, the rich province of Belgian Flanders (e.g., the towns of Bruges, Ghent, Lille, Ypres and Arras) progressively became the center of a prosperous economy based on trading with the Mediterranean world, the British Islands, and Scandinavia. In the Low Countries, the Meuse and Scheldt rivers, as well as the Rhine with their numerous navigable tributaries served as convenient inland water routes deep into Belgium, France and Germany.

Castle Brederode. Literally "wide wood," this castle is located near Santpoort-Zuid and Haarlem in the province of North Holland. It was founded in the second half of the 13th century by William I van Brederode (1215–1285). The castle formed part of the lordship Brederode, which had been given in loan in the 13th century to the lords of Brederode by the count of Holland. Originating from a simple stone tower, the castle was enlarged about 1300 by Dirk II van Brederode. The Brederode family played an important role in the history of the Netherlands. At its maximum extent Brederode castle included a large moat, a fortified advanced building, and a main rectangular body comprising walls, an inner bailey, a gatehouse, a large residential wing, and four corner towers. As a base of strategic importance controlling the road to the Kennemerland region, the castle was involved into numerous conflicts, and was several times besieged, damaged, plundered, destroyed, and rebuilt. The castle is in ruins since the 19th century, and today displays its impressive and romantic remnants in a peaceful and green landscape. (Conjectural reconstitution after an 1854 lithography by the artist PA Schipperus.)

Helmond Castle. Situated east of Eindhoven in the province of North Brabant, this castle was constructed in the 1320s on the site of an older 10th century fortified settlement. It is another example of a regular rectangular castle—a fairly strong waterburcht. From the 16th century onwards, the castle lost its military function and was turned into a comfortable residence with the addition of many windows. Since 1921 the castle is the property of the municipality of Helmond and open to the public.

The revival of international trade stimulated a new growth of urban life from the 12th century onward. A constant stream of merchants from every corner of Europe found their way to fairs for the exchange and purchase of goods of all kinds. By then the outlines of the European economic system, as it persisted until the discovery of the American continent in 1492, gradually came to existence.

Medieval Dutch cities were rather small in comparison with the great metropolises of Ancient civilizations (e.g., Babylon, Nineveh, Athens or Rome) or our modern conurbations today. They were not numerous, as the population of the Netherlands remained predominantly rural, but because of their increasing wealth, the cities came to exercise an influence on economic and political life well out of proportion to their number of inhabitants.

Charter of city rights

Wealthy traders and merchants resented having to bear the feudal yoke of lay or cleric overlords. Sometimes they revolted or they bargained with their masters who had to—or had profit from—granting them *stadsrecht* (municipal rights and privileges). The city right was confirmed by a "charter of freedom"—an official written agreement concluded between a lord and representatives of the urban inhabitants. The charter gave a town some independence in legal, administrative and trading matters in return for taxation. It often offered the following privileges:

The right to erect a defense wall around the inhabited area; the right to hold commercial markets; the right to store particular goods; the right to charge tolls and levy taxes; the right to mint city coinage; the right to organize official commercial weighing; personal freedom for all citizens including freedom of mobility; judiciary and law-making within the town's boundaries; self-governance through elected burgomasters, magistrates, city counselors, and guilds (trade associations that laid down the rules for hours worked, prices and quality of the products, for apprenticeship, and examinations of new members).

In certain cases (Groningen for example) wealthy towns had such a great degree of autonomy that they became quasi-independent oligarchic merchant republics.

In this context, trading towns could grow, and merchants and traders increased their wealth by developing commerce, crafts and local artisanal production. The growth of towns gradually changed the organization of society. The

Chapter 2. The Middle Ages ca. 500-1500

Den Ham. Situated near the village of Vleuten (province of Utrecht) the tower of Den Ham was first mentioned in 1325. The keep, built on a little islet, partly dates from the 13th century. It is now 27 m high with 7 stories, 10 meters long and 9 meters wide. The thickness of its walls varies from 1.40 m to 1.80 m. In 1870 the entire castle was demolished except the keep we still see today. The castle stood empty for many years but is now privately owned and not open to the public. (After a 1744 drawing by the artist J. de Beijer.)

availability of more money greatly altered the traditional relationship between lord and vassals, replacing feudal services with cash payments.

The growth of towns in the Low Countries saw the appearance of a new social, oligarchic, materialistic, profit-minded middle class known as *burgerij* (bourgeoisie), composed of rich traders and businessmen who became political players alongside the other power groups—the almighty Catholic clergy and the aristocratic landowning nobility. The terms *bourgeoisie* and *bourgeois* then had no political or social class significance in the modern Marxist sense. Today the term is often derogatory, and suggests small-minded persons whose beliefs, attitudes, and practices are conservative and conventionally lower middle-class. In the Middle Ages a bourgeois was simply a trader or a shopkeeper who lived and worked inside a town. Whoever lived one year and one day inside a town was granted the status of free citizen with specific rights, and advantages, as well as obligations and duties, which were quite distinct from the peasant's social position in society. After about 1200 the Dutch towns set the tone in politics, economy and culture.

Hanseatic League

The Hanseatic League was a unique phenomenon of North European history. The cooperation and mergers of merchants for the promotion of their trade abroad gave rise to a town covenant, which in its heyday comprised many ports at sea and inner land cities.

A number of trading cities in the Low Countries (including for example Harderwijk, Elburg, Deventer, Kampen and Groningen) were members of the Hanseatic League. Created in the 1240s, the group centered on Lübeck as a commercial and defensive association of northern German harbor cities growing into existence in the 13th century. The association gradually dominated maritime trade in the North Sea and Baltic Sea, and also functioned as an independent political and economical power. The wealthy League rivaled the naval power of the various Scandinavian, Dutch, German and English realms.

The Hanseatic cities raised their own armies for mutual defense, and their wealth allowed them, and required them, to invest in fortifications. From the 13th to the middle of the 15th century, the Hanseatic League largely dominated the exchange of goods between the north-east and the north-west of Europe by covering the raw material and food supply needs of the West from the East, which had been opened by the German colonization, and providing the East with western products.

In the late Middle Ages the League included about 190 cities in 16 countries. It maintained trading posts in all the commercial centers of the north, e.g., at London (England), Bruges (Belgium), Bergen (Sweden), Rostock (in German Mecklenburg), and as far as Novgorod (in Northwest Russia). The League continued to operate well into the 17th century, but its golden age was roughly 1200 to 1500. Thereafter it failed to take full advantage of the wave of maritime exploration to the west, south and east of Europe.

Urban Medieval Fortifications

Urban Enceinte

An important prerogative stipulated by the charter of freedom was the right for the citizens to construct defenses around their town. This was of great significance not only for security reasons, but also as a display of pride, wealth and sovereignty. The urban *enceinte* constituted a juridical border, it defined the urban space—quite distinct from the surrounding countryside—and indicated the space to be populated. Urban fortifications were a collective property, financed by all citizens. They were always very expensive but the urban wall was also a worthwhile investment: it facilitated the collection of import and export taxes, it allowed control of entry and exit, and in many cases could work as a dike or a dam against dangerous river floodings.

For centuries urban fortifications were made of earth (elevated walls and excavated ditches) and timber (palisades and towers). Obviously, lesser towns with limited budget were much slower to build stone walls than rich cities. Earth ramparts and wooden palisades were still in use in many places during the 12th century and even later. Fortifications were continuously enlarged following the growth of a town, and constantly improved to match current levels of military development.

Urban wall. Quite similar to the castle's curtain, an urban wall included a wallwalk, often (but not always) resting on arches, and a crenelated breastwork.

Battlements: *Left:* Arrow slit seen from inside. *Middle:* Wallwalk and entrance to a tower. *Right:* Battlement and machicolation.

Loophole shapes.

Entrance at Leiden.

Left: Arrowloop (niche in the wall with slit seen from inside).

Opposite, top: Castle Ter Does. This castle stood at Leiderdorp near Leiden (province of South Holland), and was constructed in the late 1290s. The castle was demolished in 1740. The conjectural reconstruction is based upon an etching by the artist A. Rademaker made in 1725.

Opposite, bottom: Montfoort. Located at Montfoort in the province of Utrecht, this castle was built in the 1160s by order of the bishop of Utrecht, Godefried van Rhenen. Like most Dutch castles, Montfoort has had a complicated and violent history. Heavily damaged by the French invaders in 1672, the castle was never rebuilt. Today only a part of the gate still exists and houses a restaurant. (This speculative reconstruction is made after an 1184 print.)

Siege warfare with still powerful hurling machines (catapults and trebuchets) and new attacking techniques (belfry and mining) was of course the main spur to develop and upgrade fortification. The evolution of urban military architecture was directly connected to castle fortification, and most elements (ditches, crenelated walls, towers and gatehouses) used to protect castles were applied in the defense of towns.

Until the introduction of reliable firearms in the Renaissance, the concept of defense focused on two major ideas in order to counteract the main means of attack. First, verticality and height were the keys to oppose assault by ladder and allowed the defenders to benefit from a high position. Secondly, sturdiness and thickness of masonry or brick walls offered resistance to the battering ram and underground mining.

Collective urban fortifications thus featured the same elements as a private castle including walls, ditches, towers, and gates, but the defense of a town presented quite different problems. A private castle was a compact structure housing a lord, his family, a group of combatants and their servants within a limited space. The location (on top of a hill with steep sides or in the middle of a swamp for example) and the development of a private castle would be chosen in the manner most suitable for defense. By contrast a town comprised a large built-up area for a great number of non-combatant inhabitants whose lives and activities necessitated a substantial perimeter. Therefore the general configuration of a fortified town was the result of difficult compromises.

On one side, the requirements of defense demanded inaccessibility, difficult approaches, and compactness. On the other side, living, working and trading conditions tended to push out or develop the living space, to provide easy accessibility with many convenient access points, and to connect the city to commercial routes, rivers or canals. A town had thus many more weak points in its defense than a castle. Of course, there were no standard rules, and all towns had different fortifications but all had the following common features: vertical walls, high towers and strong gatehouses. Masonry was often made of brick.

Tower at Amsterdam. The Schreierstoren, located on the Prins Hendrikkade, was originally a part of the medieval city wall of Amsterdam built in 1487.

Walls

The walls, also called *curtains* (from French *courtine* and Latin *cortina*), were frequently straight rather than curved, in order to reduce blind spots (areas below and beyond which the ground cannot be seen and fired at by the defenders). The top of the wall included a continuous *wall walk* (allure) protected by a *battlement* (crenelated breastwork or parapet). The allure was used for patrolling, observing, and fighting. It could be covered with a roof for extra protection against enemy projectiles and against rain and wind. Staircases gave access from the terre-plein to the wall walk.

Towers

At regular interval (equaling the range of bow and crossbow), walls were flanked by projecting towers of various strength, designs and shapes (cylindrical square or U-shaped). Towers rose higher than the wall itself, providing maximum visibility, and fighting emplacements pierced with narrow vertical openings known as arrowslits. The towers generally included several stories arranged for various purposes, e.g., supply-store, combat emplacements, shelter or quarters for the guards. The stories were linked up by staircases and ladders. The top of the tower was always arranged as a platform (open or covered with a roof, sometimes fitted with hoarding) whose high position offered a convenient observatory, and a dominating combat emplacement.

Walls and towers formed the scarp, and at their foot there was often a berm (a flat, continuous, bordering strip of ground). As already discussed, most Dutch castles and towns were surrounded by a dry ditch or a wet moat with variable depth and width. It could be crossed by using a drawbridge leading to a gatehouse.

Lunenburg Tower. Lunenburg Tower, located at Neerlangbroek near Wijk bij Duurstede in the province Utrecht, was built in the 1270s. It is characteristic of the many local petty lords' fortified stone residences. The woontoren was defended by a moat crossed by a wooden bridge. The total height is 15.5 m, width is 8.4 m, length 9.3 m and the external walls are 1.6 m thick. The Lunenburg tower was damaged during World War II, and restored afterwards. Today it is a private property, not open to the public.

Zierikzee. The small town of Zierikzee is located in the southwest Netherlands, 30 km southwest of Rotterdam in the municipality of Schouwen-Duiveland in the province of Zeeland. The city was granted city rights in 1248. Above shows the imposing Zuidhavenpoort (South Harbor Gate) built in the 14th century that still exists today.

Gates

At the junction of the existing main streets and the enceinte, accesses were pierced. The gates formed weak points in the city's wall, and naturally were heavily fortified by powerful gatehouses. The gatehouse included a wide arched portal—large enough to let a loaded horse-drawn cart through. The portal was arranged either into a rectangular building, or a mural tower through which the entryway passed, or deeply recessed between a pair of strong flanking towers. Invariably it featured a drawbridge, and strong and heavy wooden doors. The gates often included a *portcullis*—a strong wooden or iron barrier or grating, which could be vertically raised by a windlass placed on the first floor of the building. In case of emergency the gateway could be blocked instantly by releasing the very heavy grating that would slide down on grooves on each side of the gateway. Active combat emplacements comprised battlements, crenellated turrets, and firing-chambers equipped with loopholes allowing archers and bowmen to defend the entry against attack.

The gatehouse was, however, more than just a military defended access. It included a custom-office where taxes and tolls were levied on all persons and all goods coming in or going out the city. Taxes and tolls were major forms of income for medieval cities. The gatehouse was obviously a highly guarded point, and might also include a store-place or an arsenal for weapons and ammunitions, quarters and lodgings for civil servants, tax collectors and guards. It could serve as prison too. The city gatehouse also played a prestigious and symbolical role: its imposing defenses displayed the strength and power of the city, and ornaments showed with ostentation to foreign visitors and travellers the pride, wealth and importance of its citizens.

For security reason accesses to a city were as limited as possible resulting in many occasions in annoying traffic-jams particularly on market days; therefore secondary accesses—called *posterns*—were arranged and opened in peacetime. Posterns, also called sally ports were back or

Helpoort at Maastricht (Limburg). Also known as the Jekerpoort, the Helpoort (Hell Gate) was a part of the defenses built in the late 1220s. In the 1330s owing to the city growth, a second urban enceinte was constructed. It had a total length of 2.5 km, and included a new wall made of stone, towers, a moat, and thirteen gates.

Groningen gatehouse called Apoort built in the late 1510s. (Conjectural reconstitution after a print by the 17th century artist Cornelis Pronk.)

Chapter 2. The Middle Ages ca. 500–1500

side entrances, small exit points, or tiny gateways having two functions. In peacetime they were doorways allowing to entering and leaving the castle or the town thereby saving the opening of the main gate. In wartime, from posterns the defenders could undertake a sally; specially intended for this military purpose, some sally ports were hidden or at least well concealed.

Commercial and trading towns in the Low Countries were often placed along or across a river. River-banks were reinforced by dikes, while waters running in moats were retained by dams, and impounded by batardeaus (cofferdams or embankments or small dikes constructed across a moat). In many cases urban fortifications included one or more watergates. A watergate was a passage arranged in the curtain to allow barges and inland-boats navigation. It often took the form of a masonry gatehouse with one or more arches. In time of crisis the passage could be blocked by sliding down one or more strong and heavy grating called portcullises.

From dusk to dawn, gates, watergates, and posterns were closed.

Left: Kampen. The Koornmarktpoort (wheatmark gate) was completed in 1385.
Below: **Hoarding (also called propugnacla) is a temporary projecting wooden shed-like construction placed on the external top of the ramparts and towers of a medieval castle or a town threatened by a siege. It was fitted with openings in the floor for the purpose of allowing the defenders to improve their field of fire along the length of wall and towers, and most particularly, directly downwards to the wall base without the need to expose themselves to danger. In peacetime, hoardings were stored as prefabricated elements. The rapid installation of hoarding in time of war was facilitated by putlog holes that were left on purpose in the top masonry of castle walls.**

Jeker tower at Maastricht (Limburg). This tower still exists. It was part of the first wall dating from the middle of the 13th century. It was named after the Jeker River that flows through the neighborhood into the Meuse.

Sassenpoort at Zwolle. The Sassenpoort, constructed by the end of the 14th century is the sole remaining city gate, but is one of the largest and most impressive gates in the Netherlands.

Opposite, top: Zutphen (Gelderland) Watergate. Zutphen originated from a Germanic settlement in ca. 300 CE established on the strategic confluence of the IJssel and Berkel rivers. The settlement received town rights in the 1190s, making it one of the oldest towns in the Netherlands. As a member of the Hanseatic League, the city was fortified in 1312. The fortifications were dismantled in 1874, but fortunately many medieval remnants still exist. The depicted Berkelpoort (watergate upon the Berkel River) was first mentioned in 1424. It was restored in 1888, and in 1952.

Opposite, bottom: Koppelpoort Watergate at Amersfoort. The famous Koppelpoort ("coupled gate" as it combined land and water gate), was part of Amersfoort's second wall completed in the 1450s. The building was restored in the 1880s by the architect Pierre Kuipers who added non-original battlements in order to enhance the medieval character.

Chapter 2. The Middle Ages ca. 500–1500 55

Militia

Originally all able-bodied men had the obligation to take arms for the defenses of their settlement, village or town. Gradually this duty was reserved to rich burghers (citizens)—the only ones who could afford to have good weapons, who had time to drill, and who were strongly motivated to defend their property and wealth. So emerged the *schuttersgilde* (municipal militia) exclusively drawn from influent and wealthy traders and craftsmen. The urban militia was organized in companies, and each unit, headed by a captain, was allotted to man a part of the defensive walls. The militia was also used as a police force called *wacht* in Dutch. Members of the wacht were tasked with guarding the gates during daytime, and patrolled the (unsafe) streets at night during the curfew from dusk to dawn. The wacht was also used to maintain order and repress tumults, revolts, and riots as the chasm between rich and poor was often important and violent. The urban militia disposed of a hall for meeting, celebrations and feats, and a special ground for drilling known as *shuttersdoelen* where the men would train in the use of various weapons including sword, spear, bow, and crossbow. The militia could also be mobilized, and engaged for an offensive expedition outside the town. As time went by, however, many rich citizens found that their lives (and that of their sons) were too precious to be exposed to the hazards of military activity, and their time too valuable to be wasted in dull patrolling and boring guarding duties. Instead they gladly accepted to pay for replacement. The municipality then recruited mercenaries (professional hired soldiers) to defend their wealth, man their walls and gates, and launched attack on rival cities.

Late medieval fortifications

Late medieval realizations were sometimes astonishing by their gigantism, sophistication, and the introduction of elaborate defensive systems.

The sophisticated late-medieval castles and urban defenses were so expensive that they were only within the reach of wealthy kings, princes, dukes, counts, and rich earls, abbots and bishops. Dispositions that have been described above concerned merely a few fortresses. Besides, each castle and each town had its own development depending on natural site, strategical situation, and greatly depended upon its owner's wealth and social position. Medieval fortifications developed at the mercy of circumstances without rule or guiding line. Individualism, particularism and tradition too

Murder holes and portcullis: 1: Murder hole. 2: Portcullis. 3: Winch for raising the portcullis. 4: Drawbridge.

were very strong in certain regions and many late medieval works were erected following conventional custom designs. Moreover, many castles and urban enceintes remained modest because of financial limitation as many local lords, and citizens simply could not afford the burden of building and maintaining huge fortifications. At the same time, many noblemen and city authorities preferred to pay their suzerain in cash money rather than in time of military service. In the 13th century vassals were gradually beginning to change into tenants. Feudalism, the use of land in return for armed service was weakening, but it took a very long while before it disappeared completely.

Medieval Siege Warfare

As for siege warfare in the Dark Ages, there is a lack of reliable written sources so the ways in which early *burgen* were attacked and defended remain largely conjectural. Siege warfare was more than probably quite primitive and rudimentary, particularly compared to the sophisticated ancient Assyrian, Greek and Roman use of siege machines and siege technics. It probably differed very little from tactics developed in the Iron Age. Attackers would naturally use treachery, intrigue, and pressured negotiation. A small party could climb the palisaded wall at night or early in the morning when guards were tired after a night watch. In the case of a direct military attack, attackers would simply attempt to smash the gate with a battering ram (often no more than a tree trunk), and both sides would launch missiles (using bows, later crossbows).

Given the relatively small height of earthwalls and palisades, attackers would directly assault these defenses using scaling ladders, or by heaping up a ramp of faggots or other portable materials, or by trying to destroy the palisades or gates by fire. One of the main disadvantages of early fortifications was that there was no flanking at all, so there were never enough warriors to man the whole defensive perimeter.

When the increasing height of walls made simple escalade too difficult, other means of attack were invented, and new siegecraft were introduced. Defenders would shoot arrows, throw javelins and spears, drop rocks and stones upon the attackers, or would sally on horse or on foot for hand-to-hand combat outside their walls. Both sides would use catapult or trebuchet as a sort of primitive artillery rediscovered from Ancient times. Attackers would assault the top of the wall using large ladders or a *belfry*—a high, wheeled, wooden tower allowing attackers to set foot on the parapet for hand-to-hand combat with the defenders. Warfare in the medieval period was the preserve of the well armed and high-born and wealthy nobility. However, when needed, poorly armed local peasants, villagers, and urban militiamen (often reinforced with contingents of remunerated professional mercenaries) participated in offensive operations and in the defense of the ramparts.

All medieval fortifications were essentially local, private

Machicolation front view (*left*) and cross section (*right*). Machicolation worked actually just like hoarding, but had a permanent character. Made of masonry it was a strongly built overhanging projection supported by corbels. The floor was pierced with openings (machicolations) so that missiles and projectiles could be thrown or dropped on attackers.

or communal. All able-bodied men were involved. They were obliged by customary law to take part personally and physically (and later to contribute in money) to the building, repairing and maintenance of defense works.

Conclusion

Although denied, scorned, and condemned by many historians in the 19th century, the achievements of the Middle Ages were considerable. About CE 500 Western Europe was a complete chaos of barbaric rival tribes squabbling in the ruins of the Roman Empire. By CE 1500 they had created a relatively advanced civilization that prepared to embark on westernization of large parts of the world. In technology, for example, the medieval world displayed an extraordinary creativity. Vast regions (particularly in the Low Countries) were reclaimed from swamps, and bogs while forests, fallow and waste spaces were turned into fertile useful farmland.

Late medieval inventions that were to change culture and warfare included for example the development of ocean-going ships (allowing far travelling and discovery of new continents), printing (enormously spreading literacy, knowledge and education), and the introduction of black powder and firearms (substantially changing warfare). All this went together with cleverness, proficiency, ability, passion, and inventiveness. At the same time, however, the Western Europeans were moved by naive superstitions, credulous fallacies, unquenchable greed, haughty egoism, arrogant disdain, vain pride, and using the false excuse of being the chosen few executors of a divine scheme. The Middle Ages was the cradle of European self-awareness and announced Europe's later global domination.

The Renaissance and the following modern era saw the Western Europeans impose by sheer force their civilization, biased views, intolerant principles, narrow-minded way of life, unfounded religion and bigotry, far-fetched beliefs, rigid superstitions, criminal racism, partisan prejudices, excessive imperiousness, and hubristic conceptions to large parts of the rest of the world. The European feeling of superiority culminated in 19th century colonialism.

Chapter 3

Renaissance

The Burgundian Period

The term Renaissance (meaning *rebirth* in French) designates the revival of European art and literature under the influence of classical models in the 14th–16th centuries. It also indicates a culture, a style of art and architecture, an elitist way of life, new developments in reasoning and science, and more widely an enormous curiosity and a renewed interest in many matters, an original view on life, and a fresh conception of mankind.

The Renaissance started in Florence, Italy, with an artistical revival of interest in classical Greece and Rome. The period from the end of the 15th century has been termed the High Renaissance, when Venice and Rome started to be involved in cultural developments. Renaissance culture spread to the rest of Europe from the early 16th century, and was predominant for the next hundred years.

All through the Middle Ages the rich Low Countries were coveted by both France and Germany, but they were not able to subdue them. By the end of the Middle Age, however, a new power rose in France: Burgundy. Today the French province is particularly renowned for the production of top quality wine, and good food. It approximately corresponds to the present-day départements of Côte-d'Or, Saône-et-Loire, Yonne, and Nièvre. It holds its name from the 4th century Burgunds, a Germanic tribe, who may have come from a coastal region near the Baltic Sea. In the 6th century the kingdom of Burgundy was conquered by the Franks, became a part of the Carolingian Empire, and later a duchy of the Capetian kingdom of France. During the Hundred Years' War (1337–1454), King John II of France (b. 1319, reign 1350–1363) gave the duchy to his youngest son, Philip, as an *appanage* (a provision made for the maintenance of the younger children of kings and princes).

Burgundy soon became a major rival to the French crown, and the ducal court in Dijon outshone the royal Capetian court in Paris both economically and culturally. By heritage and marriage the Duke of Burgundy Philip II the Bold acquired important territories including Charolais, Artois, Franche Comté in France and the rich county of Flanders in the Low Countries. His successors John the Fearless, Philip V the Good, and Charles I the Bold continued that intelligent and fruitful policy based on marriage and heritance, and also on pressure, outright war, and annexations. No doubt the dukes of Burgundy wanted to become sovereigns independent from France. They attempted to revive the 9th century Lotharingia—the ancient Carolingian realm of Lothair from CE 843. In the 15th century they almost succeeded. By then they had become powerful, ambitious and dangerous rivals to the kings of France. They ruled over vast territories including Burgundy, a part of northern France and the Low Countries (Netherlands, Belgium, and Luxembourg).

Under Burgundian rule the Low Countries began a process of unification. There was a loss of independence, notably for the free cities, which at the same time started to suffer competition from England in the cloth business. However, under the pressure of the Burgundian ambition to create a unified and centralized state distinct from France and Germany, the various provinces, principalities, bishoprics, dominions and free cities started to consider themselves as members of a common nation. Besides the dukes were Frenchmen, and most important Renaissance men. They lived luxuriously surrounded by lawyers, writers, and artists, they indulged in pleasures and parties, they introduced a new spirit, and without doubt contributed to a general development of the native population. Under the last Burgundian Dukes' rule the medieval world was slowly but surely decreasing. For the first time in its history the Low Countries became a single, administratively autonomous unit under an individual ruler.

The Dukes of Burgundy introduced the function of *stadhouder* (meaning *lieu-tenant* = place-holder) an official regent or deputy appointed to represent the legal ruler.

Habsburg and Spanish Rules

After the death of Charles I the Bold in 1477, the Burgundian power collapsed. The Duchy of Burgundy itself was re-annexed by France by King Louis XI, and returned to the status of a French province. However, following the marriage of Charles the Bold's 19-year-old daughter and heiress Mary of Burgundy to Archduke Maximilian von Habsburg, the northern part of the Burgundian empire was taken over by Austria. It was an astonishing example of how a royal marriage could affect the lives of millions of ordinary people. Now the Low Countries were irretrievably involved with the power and politics of the Austrian Habsburgs dynasty. Through the marriage of Maximilian's son Philip the Handsome to Queen Joanna of Castile in 1498, Maximilian established the Habsburg dynasty in

Spain, which allowed his grandson Charles to hold the thrones of both Castile and Aragon. In 1516 Charles I of Spain was elected Emperor of Germany.

Charles V

Charles V (1519–1558) was sovereign of both the Spanish realm as Charles I from 1516, and the Holy Roman Empire as Charles V from 1519. He also ruled over a part of the lands of the former Duchy of Burgundy from 1506. He brought together under his rule extensive territories in western, central, and southern Europe (notably Naples and Sicily), as well as the Spanish viceroyalties in the recently discovered Central and South Americas and Asia. As a result, his territories spanned nearly 4 million square kilometers (1.5 million square miles). Charles V's possession was known as "the empire on which the sun never sets."

Charles V was born in February 1500 in the Flemish city of Ghent. He spoke fluent French and Dutch and later learned Spanish. He had a good understanding of, and a profound attachment to, the Low Countries. As heir of three of Europe's leading dynasties (the houses of French Valois-Burgundy, the Austrian Habsburg, and Spanish Trastamara), he was a busy man with numerous enemies. Charles V spent his whole life at war simultaneously fighting against three major nemeses: The Capetian Valois kings of France (who felt choked by Charles's territorial encirclement of France); the Ottoman Muslim Turks (dangerously advancing in the Mediterranean Sea); and the Protestant Reformation (that since 1517 was regarded as endangering the unity of European Roman Catholic Christianity).

Firearms

The end of the Middle Ages and the Renaissance were marked by the development and use of black powder and firearms. Cannons, it seems, appeared in Europe in the 1320s. At first, those new noisy and smoking weapons were not effective. They took too long to load, were inaccurate, and fired with too little force projectiles that were too light. Gradually, however, artillery pieces and the quality of gunpowder were improved. Slowly but ineluctably firearms played an increasingly important role, notably in siege warfare. In the 15th century such advances were made that a serious crisis in fortification arose. The heavily fortified capital of the Eastern Roman (Byzantine) Empire, Constantinople, was captured by the Ottoman Turks in 1453 mainly by the use of large artillery pieces.

In 1494 King Charles VIII of France launched a successful expedition in Italy. He had a modern siege train consisting of mobile artillery on wheel. The astonishing rapidity with which Italian medieval castles and fortified towns fell before him proved the uselessness of the old Middle Ages defenses. It was then clear that a new system of defenses had to come.

After 1500 firearms became basic attack weapons. Cannons of various shapes and calibers started to be widely used—at least for those kings, princes and rulers who were wealthy enough to afford them. At the same time small portable individual guns were developed for infantry use, later to become muskets, rifles, and pistols.

Transitional Fortification

In the 15th century the vertical medieval military architecture entered into a progressive crisis. High walls, multistory keeps and tall towers intended to be impassible obstacles gradually became vulnerable targets. Changes brought by the use of artillery did not strike like a sudden revolution but followed a slow and gradual evolution. The remarkable castles, forts, and urban enceintes of the 13th and 14th centuries did not become obsolete overnight. Thick masonry, high walls and towers were targets for only a few attackers—emperors, kings, dukes, and senior rulers who were rich enough to afford powerful artillery pieces, black powder and experienced artillerymen to operate them. The advent of effective artillery was inextricably bound up with the rise of the modern centralized state. The introduction of gunpowder and artillery in no way diminished the role of medieval castles, citadels and fortified cities as strongholds, bases for operations, quarters for troops, armory and supply store. Vertical medieval fortifications were still capable of resisting a gang of marauders, but it was obvious that something had to be done by castle builders against modern artillery.

To meet the needs of artillery warfare, it was impossible to find immediate and efficient solutions because of lack of experience and because of traditionalism and conservatism. Although ideas certainly circulated, transitional fortifications remained tied to local medieval traditions.

Master-builders' theoretical considerations and practical realizations in the second half of the 15th century essentially aimed at adapting and modernizing pre-existing fortresses. The number of existing fortresses that had to be modernized was enormous, and so was the number of attempts made to find solutions to the issues. The so-called transition fortification (between the medieval vertical system and the horizontal angled bastioned system) developed without basic principles or clearly defined theory, each stronghold being individually adapted to firearms. Transition fortification, covering approximately the period from 1450 to 1530 was an attempt to conciliate two essential demands: (1) to resist the destructive effects of heavy artillery and hand-held guns by passive means; and (2) actively to use as efficiently as possible defensive fire-arms.

Passive elements

To resist artillery, castle builders did their best to improve the quality of masonry. As a result their first reaction was to increase the thickness of existing works by adding external layer of masonry. This solution, apparently logical, was in practice a costly impasse. Another manner of increasing the wall's resistance was to avoid the exposure of flat surfaces

Chapter 3. Renaissance

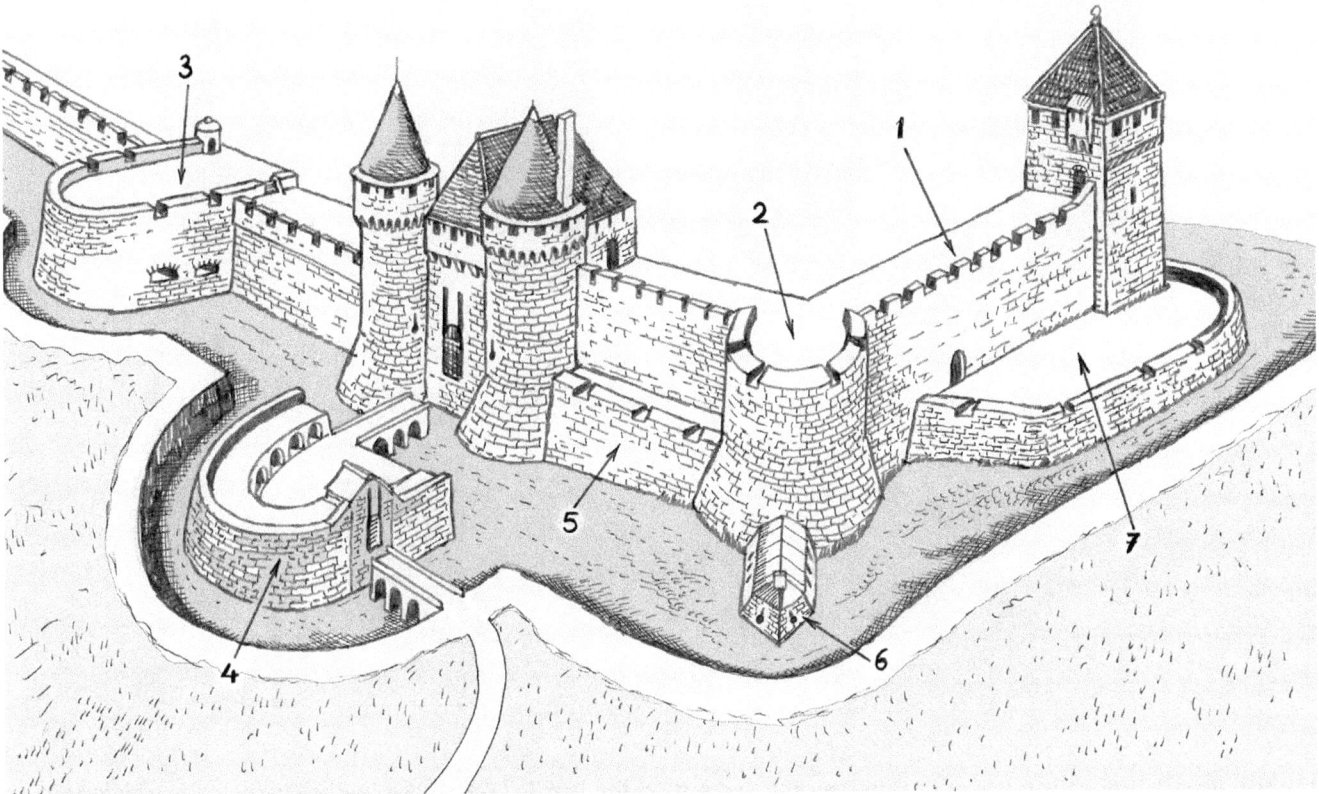

Transitional fortification: 1: Ramparted medieval wall. 2: Lowered medieval corner tower. 3: Roundel or casemated artillery tower. 4: Barbican. 5: Fausse-braie. 6: Caponier. 7: Artillery bulwark.

to enemy shot at right angle. This could be done by favoring curved outlines, or D-plans or horseshoe shapes. In order to deflect projectiles the upper surfaces of the walls (parapets) could also be inclined or given a thick roundish profile.

At the same time another method was used: the rampart, composed of a thick layer of earth heaped up between an ancient wall and a new built wall. This arrangement was relatively cheap and very efficient. Indeed earth was available and its smothering elasticity allowed for absorbing the impact of cannonballs just like a cushion. Besides, the rampart was rather wide thus enabling the placement of defensive artillery pieces.

Analyzing the principle and effect of grazing fire (sweeping close to the surface it defends), castle designers and master builders came to the conclusion that height was a mixed blessing: high curtains and elevated towers had become convenient targets for enemy gunners. To diminish the exposed surfaces and to maximize direct or grazing fire master-builders tended to reduce the height of the works. Therefore towers were often cut down to the same level of the curtains and ramparts. Their roofs were suppressed in favor of terraces where artillery could be emplaced. Simultaneously in order to keep inaccessibility to the place, ditches were made deeper (for adding protection) and wider (for keeping enemy's guns at a distance). Reducing the wall height and increasing the moat depth resulted in a half-sunk fortification, which became one of the basic features of the future bastioned fortification.

Loopholes and embrasure

To utilize firearms in a defensive purpose, master-builders brought adaptations and improvements to existing castles, citadels and urban enceintes. They made a distinction between small arms (for close range) and heavy cannons (for medium range).

Traditional arrow loopholes and crossbow slits were numerous in previously built castles but they were narrow and vertical, thus rather inappropriate for small arms, portable guns, pistols, harquebuses and other long-barreled musket-type weapons. To allow the discharging of those early firearms, loopholes and crosslets were adapted. This adjustment usually took the form of a round hole at the base or in the middle of the arrow-slit, which becomes known as a *cross-and-orb*. This rudimentary modification allowed the use of firearms and bows, both sorts of weapon remaining in the late-medieval arsenal. Through time, openings were developed for the exclusive utilization of firearms. As a result the vertical slits were reduced in size or simply omitted. In ground plan they often widened out or were given an X-shaped plan to increase traverse (sideways movement).

For heavy artillery, breastworks, casemates and firing chambers were fitted with full-scale embrasures (large openings in walls or parapets). Blocked with wooden shutters when not in use, the inward part was beveled and splayed wide so as to enable gunners to traverse their weapon.

Firing holes for portable individual weapons and

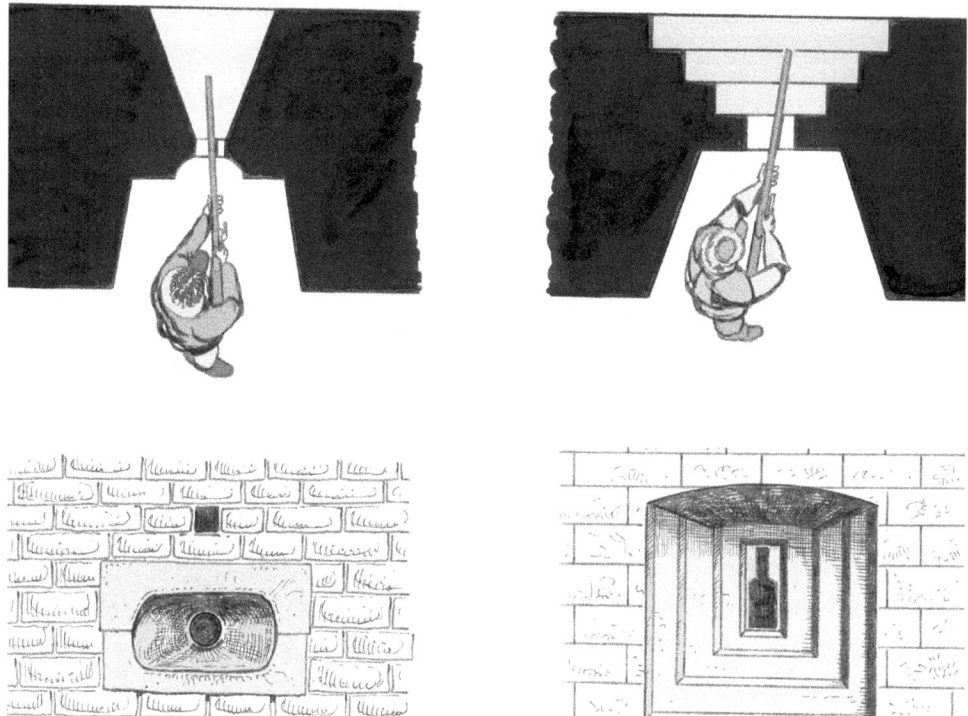

Embrasures. Numerous types of beveled, or splayed out on the inside, apertures (embrasures) were designed for the use of small firearms and medium guns. *Top:* Seen in plan. *Below:* Front view.

Embrasures (front view): *Top:* Wide gunport for artillery. *Bottom:* Medieval arrow splits adapted to the use of small firearms.

embrasures for crew served cannons were generally placed in the side of salient structures in order to flank the adjacent elements.

Active guns emplacements

Active elements for heavy cannons set complex problems and demand structural arrangements. A gun is heavy and cumbersome, and its service imposes a rather large emplacement not only for muzzle-loading, ammunitions and accessories, but also for the *recoil* force (the violent backward movement as a reaction to firing a shot). The release of toxic smoke was always a problem, and gun emplacements had to be easily accessible for supply purpose. Obviously not all citadel, town or castle combat emplacements were suitable for mounting a cannon. The wall walk on top of a medieval curtain was often too narrow to permit safe recoil, or tower floors were too weak to bear the considerable weight, or the stoutest roofed and easily strengthened part of the building may not command a good field of fire.

New suitable artillery emplacements were therefore needed, and the so-called bulwark appeared. The *bulwark* was a kind of platform built outside the place, reinforcing a vulnerable point, and specially designed to house artillery. Bulwarks came in various shapes and sizes, and presented many advantages. The dimensions were calculated in order to place, supply and operate artillery. Situated outside the enceinte, the bulwark created an additional line of defense and increased the range of the guns. Its profile was generally low in order to maximize grazing fire. It offered space where the besieged might regroup for a withdrawal or a sally. It worked as a shield protecting the scarp of the main enceinte. It placed the inside of the castle—or the suburb of a city—further away from enemy attacking artillery.

The bulwark was relatively cheap to build when constituted of rampart (somewhat thin masonry retaining a thick mass of earth). When urban fortifications were dismantled in later period, the bulwark got its modern meaning: a boulevard or an avenue, a wide lane generally with trees alongside.

The bulwark might also be an existing medieval tower, which had been lowered or a totally new work constructed to reinforce a weak part of the wall. In that case the bulwark was a low, strong and squat artillery tower. This work was possibly circular or U-shaped, and projecting in order to flank curtain and ditch. It was called *roundel, rondelle, bastei* or *basteja* in northern Europe, *bastillon* in France and *torrionne* in Italy. Its summit was arranged as a platform with gun embrasures and sometimes fitted with crenellation and machicolation as medieval traditions were still strong.

The artillery tower included one or more stories fitted with flanking casemates. A *casemate* was a vaulted, enclosed gun-chamber pierced with a firing embrasure that gave an excellent protection to gun, gunners and ammunitions. The thickness of its wall, however, allowed only a limited observation and a reduced field of fire. Also firing guns produced toxic fumes. In spite of ventilation draughts, chimneys, ducts, vents and shafts, after a few shots the chamber was full of choking smoke. In peacetime, the casemate was generally an obscure, humid, musty, and draughty place.

Gun Casemate Ventilation. The cross-section shows the following: 1: Firing chamber. 2: Embrasure. 3: Ventilation duct and chimney. 4: Passage.

Casemate (cross-section). A small firing chamber placed in the wall of a fortress, with openings from which firing weapons could be used.

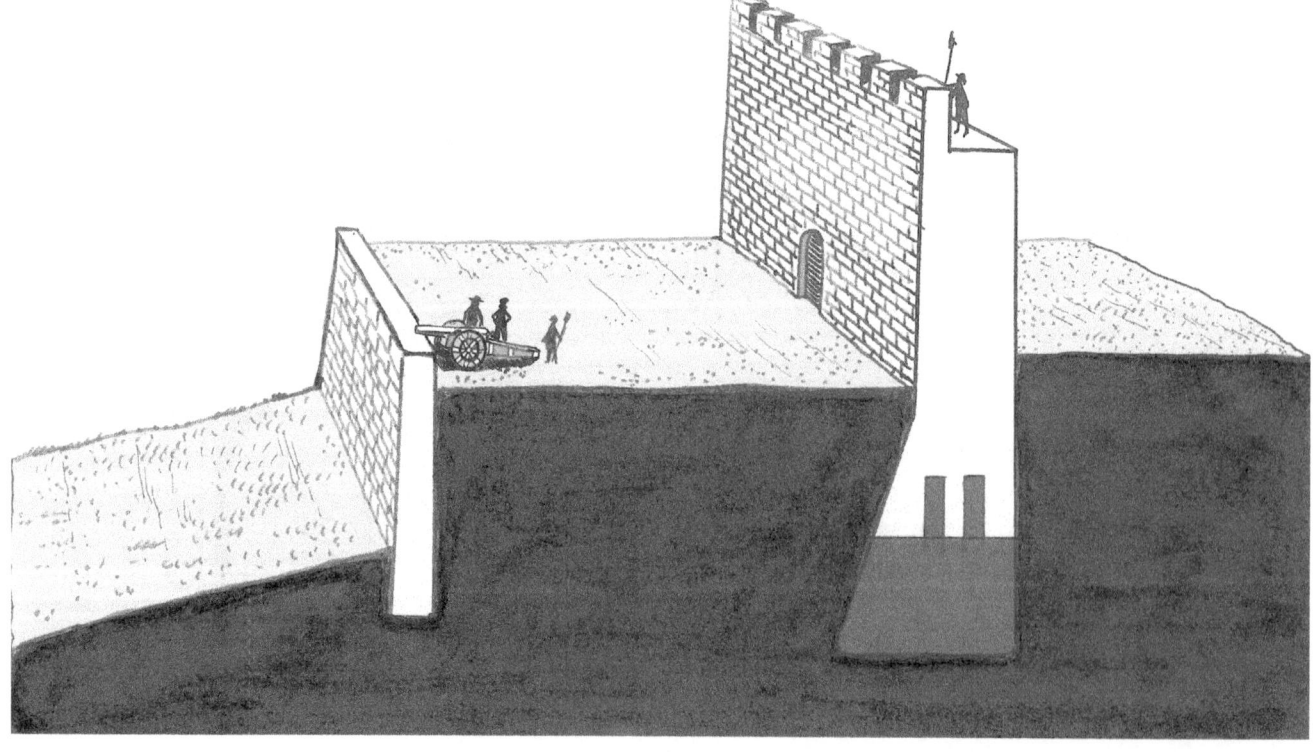

Cross-section of a bulwark. Projecting from an existing medieval wall, a bulwark was established, offering a gun platform.

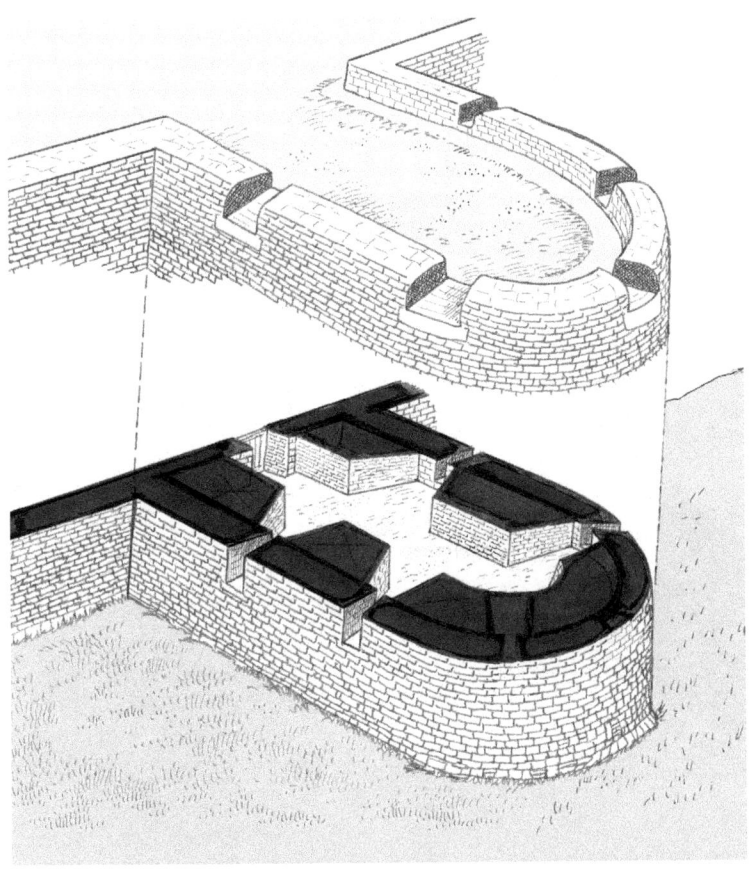

Schematic view of an artillery tower. Also called bulwark, boulevard, rondelle, bastillon, basteja, bastei or torrionne, the artillery tower was a squat and protruding structure. Particularly designed for flanking, a typical artillery tower was made up of thick walls, and included a top platform with embrasures in a roundish parapet, as well as one or more casemated stories.

The roundish shape of the 15th century artillery tower left blind spots at its foot. For a better flanking of the ditch, master-builders of the transition time designed two special works: the fausse-braie and the caponier.

The *fausse-braie* was a kind of low bulwark, a lower under-wall constructed outside and alongside the main enceinte generally between two towers. It was often open and fitted with embrasures. The *caponier*, also called *moineau*, was a small low profiled work running at right angle across a dry ditch; it was projecting at the foot of a wall or of a tower. The caponier included closed, vaulted, casemates fitted with small firing-holes through which musketry or light cannon fire could be directed against any enemy advancing in the dry ditch. Used in the second half of the 15th century the caponier was still used in the bastioned system and became the main flanking element in the 19th century polygonal fortification.

The entrance to a town or a castle was also influenced by the new weaponry. *Gatehouses* were fitted with embrasures, arrow-splits were modified, while barbicans were adapted to firearms. In the 15th century barbicans tended to be powerful bulwarks, formidable artillery towers with strong masonry, ramparted walls, terre-plein fitted with gun emplacements behind thick parapets and stories furnished with gun casemates. The barbican's shape was various but it was often a strong U-shaped work projecting in the moat ahead of the gatehouse; as medieval traditions still remained strong, the work could retain obsolete machicolation and crenellation. The barbican allowed the regrouping and withdrawing of a sallying party. It was always surrounded by its own outer ditch and fitted with its own drawbridge.

Some of the elements of the "transitional fortification" described above were rather useful and efficient; many others though were only temporary expedients, improvisations without a future. In Italy radical modifications in designs appeared and soon a new system was developed.

Italian Bastioned Fortification

Bastion

The crisis of fortification and all problems generated by firearms were finally solved by an Italian invention: the bastioned system.

In 1495 the success of the French King Charles VIII's powerful and modern artillery spurred many designers to search for new fortifications. By ca. 1500 the Renaissance had made scholars and military engineers well aware of mathematics and geometry necessary for their trade. Significant Italian military engineers and architects such as the family San Gallo or Michele San Micheli further developed the early theorists' works. They were devoted to military study and experimented with new fortification methods, and this resulted in the introduction of the bastion.

Bastioned fortifications appeared in the beginning of the 16th century, but who actually invented it is not clearly known. Some historians adhere to Renaissance specialist

John Hale's theory according to which the bastion is without doubt an Italian invention. Another camp argues that the bastion originated from indigenous experiments and the synthesis of cross-development, and mutual influence in other parts of Europe. In fact the question of who actually invented the bastion is of very little importance. What really matters is its introduction.

The earliest Italian bastions were modified bulwarks with straight faces and flanks, attached to the main wall, for which the old medieval towers often acted as keeps. At first the terms bulwark and bastion were more or less interchangeable. Towards the end of the 16th century the term bulwark began to be reserved for banks of earth thrown up a little distance in front of the main wall to protect it from breaching fire, and it thus reverted to its original advanced defensive intention. The term "bastion" henceforth denoted a specific artillery position connected by flanks to the main wall.

A bastion was a protruding, terraced platform generally as high as the main wall. It was distinguishable from any previous artillery tower because of two essential characteristics: a low ramparted profile and a pentagonal arrow-headed ground plan.

The bastion profile was ramparted, which meant that it was constituted by a rampart: two relatively thin masonry walls (called revetments) retaining a thick mass of earth absorbing the smashing impact of cannonballs. The

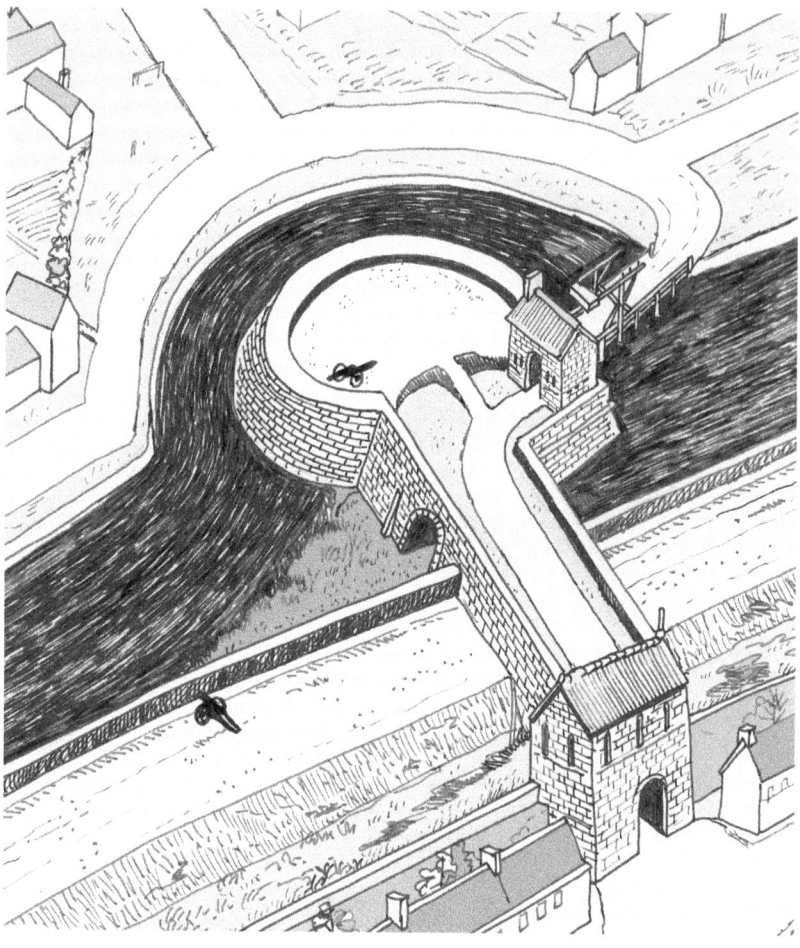

Roundel. The Boteringerondeel was a low artillery tower or roundel (rondeel in Dutch) constructed in 1547 to protect the Boteringe gatehouse, in the north of the town of Groningen.

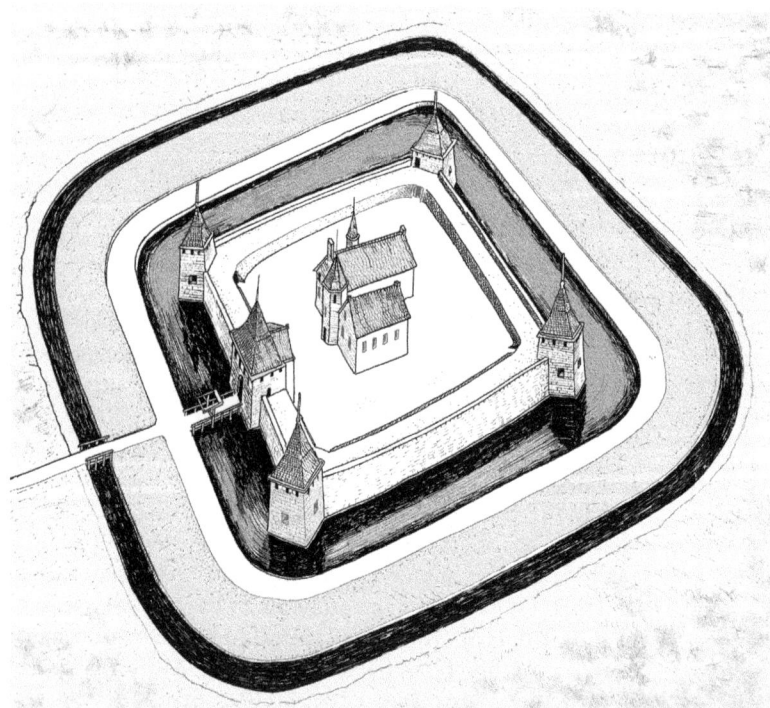

Castle Wedde. Situated west of Groningen near the border with Germany, the castle was originally a rural manor built in 1370 by the local lord Van Addinga. Its origin was a rectangular stone woontoren with a width of 8.50 and a length of 10.50 m, including three stories. In the 1460s, the castle was adapted to the use of firearms with the addition of a rectangular enclosure of four ramparted walls flanked with four square artillery towers at the corners, a wet moat, a glacis and a second external wet ditch. Located on a strategic passage at the border between the Netherlands and Germany, the Wedde Borg saw a lot of turmoil. It was heavily damaged in 1478, reconstructed in 1486, modernized in 1530, and captured and restored by order of king of Spain and German emperor Charles V in 1536. The stronghold was devastated again during a siege ordered by Count Willem-Lodewijk of Nassau in October 1593. Wedde was captured by the German bishop of Münster in 1665 and again in 1672. It was occupied by the French between 1795 and 1814. The castle was then abandoned and left in ruins until a restoration took place between 1955 and 1958.

Ewssum. The waterburcht Ewssum is situated near the village of Middelstum in the north of the province Groningen. The castle, built about 1278 by the local lord Ewe in den Oert, was the center of a small domain. Note the typical bulb roof and the special drawbridge, characteristic of northern Europe. In 1472, lord Onno van Ewssum added a low flanking artillery tower. Today the castle has disappeared and only the artillery tower remains in the middle of the wide wet ditch.

bastion was rather low above the ground in order to avoid being an easy target, while the depth of the moat prevented scaling. Bastions and curtains included a thick breastwork with embrasures protecting gun emplacements, a banket (firestep) for infantry soldiers fitted with small arms and a wall walk (chemin de ronde) broad enough and suitable for supplying, firing, and accommodating the recoil of artillery.

The bastion's pentagonal plan was formed of two *faces* outwards turned to the enemy; both faces joined at the jutting-out *salient*. They were connected to the curtain by two sections of wall called *flanks*; the meeting point of face and flank was called *shoulder*. The gorge was the inner space turned to the inside of the city or fort. The surface enclosed by those five lines was called *terre-plein*. To increase the defenders' safety, Italian bastions were often fitted with an *orillon* (aka *ear*), which was composed of a recess and

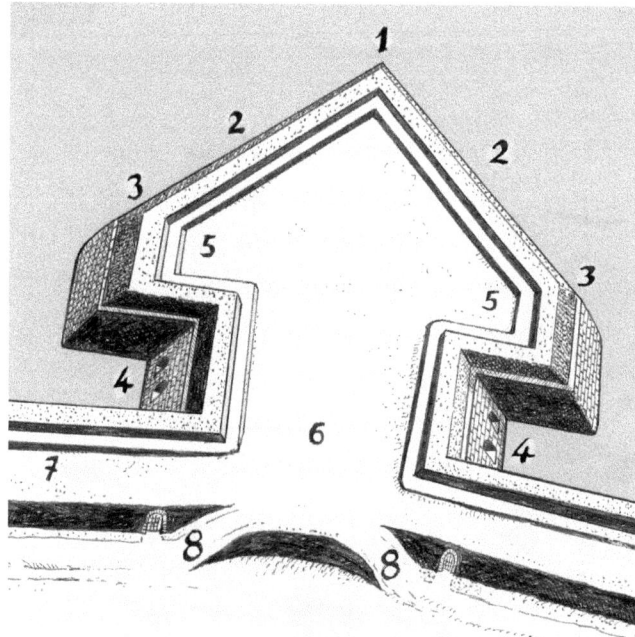

The main parts of a bastion: 1: Salient, the point of the bastion, an echauguette or sentrybox for observation could be installed on top of it. 2: Faces. 3: Shoulder, point of junction of the flank and the face. 4: Flank featuring casemates (or open platforms) enfilading the moat. 5: Ear or orillon protecting the recessed flank; the ear was either round or square. 6: Gorge of the bastion. 7: Curtain, the wall between two bastions; curtains and bastions formed the scarp (inner edge of the ditch). 8: Sloping ramps allowed cannons and supply-carts to be brought on the wide wallwalk on top of the curtain, which was both a communication and a place to install the artillery. This depicted bastion was said to be "full" as its terre-plein was filled with earth.

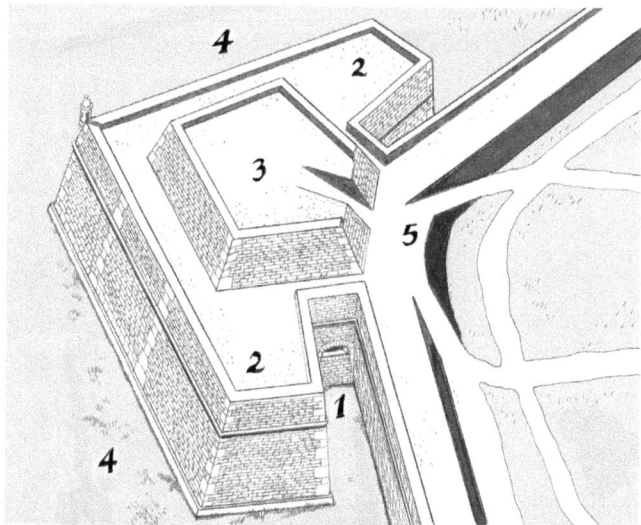

Italian bastion with orillon. Note the casemated recessed flank (1) screened by an ear or orillon (2). Access to the lower story was done via a postern and a gallery passing under the curtain. Some bastions could be fitted with a raised structure called a cavalier (3). Its tracé was similar to that of the bastion; the purpose of this inner work was to gain observation possibility, to give additional firepower and increased height to the bastion so as to command the surroundings. The cavalier also acted as a kind of huge shield preventing enfilade-fire and protecting buildings in the town or fort. The moat (4) could be either dry or filled with water. In the gorge (5) ramps (sloping accesses) were installed.

a protruding screen built on the shoulder protecting the defenders in the flank from oblique enemy bombardments, but allowing them to enfilade the ditch. In plan the ear was typically round or square—shapes that gave bastions their characteristic arrowhead or ace of spades form. The question of the arrangement of flanks and ears was one of the main preoccupations of engineers.

In the bastioned system, the ditch became essential and characteristic. Serving as it did for the double purpose of supplying earth for filling the rampart and allowing the wall to be sunk for concealment, it was found also to have a definite use as a formidable obstacle that actually increased the depth of the wall.

The bastioned *tracé* constituted a decisive revolution in military architecture. Indeed no assault party could approach the curtain without being fired at from all sides.

Bastioned front

The new Italian bastioned system, largely resulting from experimentations conducted by architects and military engineers like Jacomo Castriotto, Girolame Maggi and Francesco de Marchi, was a great improvement. Its practical base was the bastioned front composed of one curtain and two projecting half-bastions allowing to house more

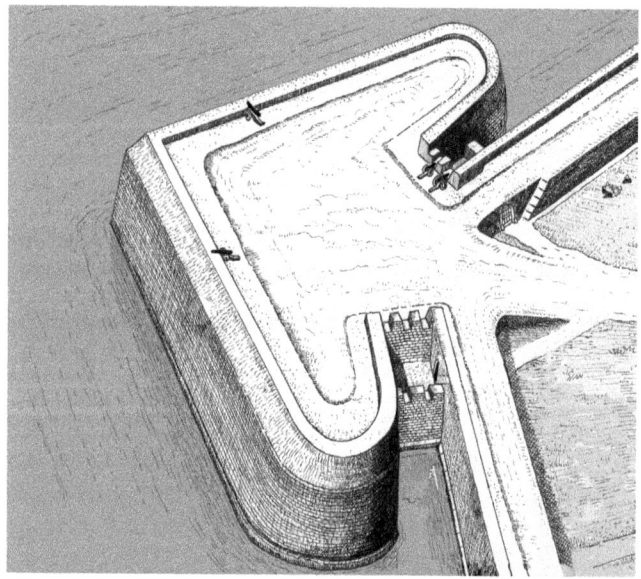

Bastion with roundish ears (orillons).

guns. But the main improvement was the essential notion of flanking: each bastion flank protected not only the curtain but also each face of the neighboring bastion. The new bastioned system thus suppressed all blind spots—zones below and beyond which the ground cannot be seen and defended. Every part of the fortress was always covered by fire coming from neighboring parts.

Within a few decades, the New Italian bastioned *tracé* restored the balance of arm in favor of the defense. It permitted a maximum efficiency with a relative few number of defenders. The new Italian system was further improved by the creation of an outer work called *ravelin* (also named *demi-lune* or half-moon) placed in front of the curtain as a triangular independent island. Another important feature was the creation of the *covered way*, a continuous broad lane placed on top of the counterscarp all around the fortress. It formed a first advanced line of combat because it was covered by an uninterrupted breastwork or a palisade.

The Italian bastioned front was an ensemble of elements related by rules and geometrical ratios. The basic unit might be repeated at will to form an urban enceinte or a fort. Five fronts constitute a pentagonal citadel; the same front with a varying salient angle could be used to form an urban enceinte adapted to natural conditions. As discussed, a fort was an independent, enclosed work defendable at every side, manned only by military personnel. It is thus quite different from a *vesting*, a fortified city whose defenses were constructed to protect a civilian population.

Endlessly, the five lines of the bastion, and the size given to the bastioned front could vary in length and be connected

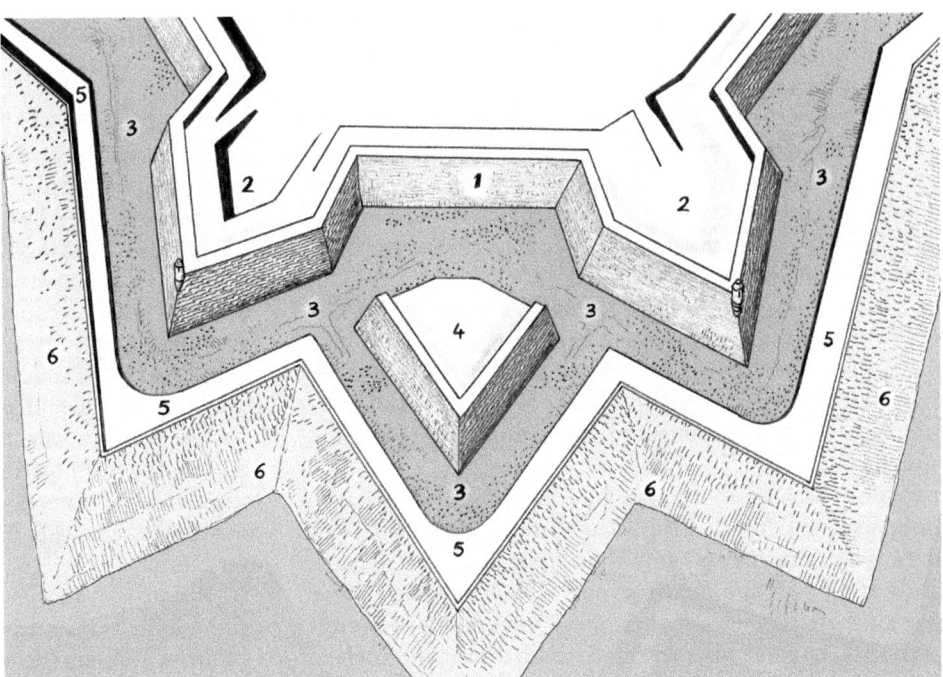

Bastioned front. The bastioned fortification dominated military architecture for nearly 300 years. The sketch shows the flanking fire offered by the new Italian bastioned system reducing all blind spots. The basic elements of the new system were: The curtain (1); the large bastion (2); the ditch (3); and the ravelin, aka demi-lune (4). On the counterscarp (outer edge of the ditch) there was the covered way (5), and the bare and flat glacis (6) that denied any cover to attackers.

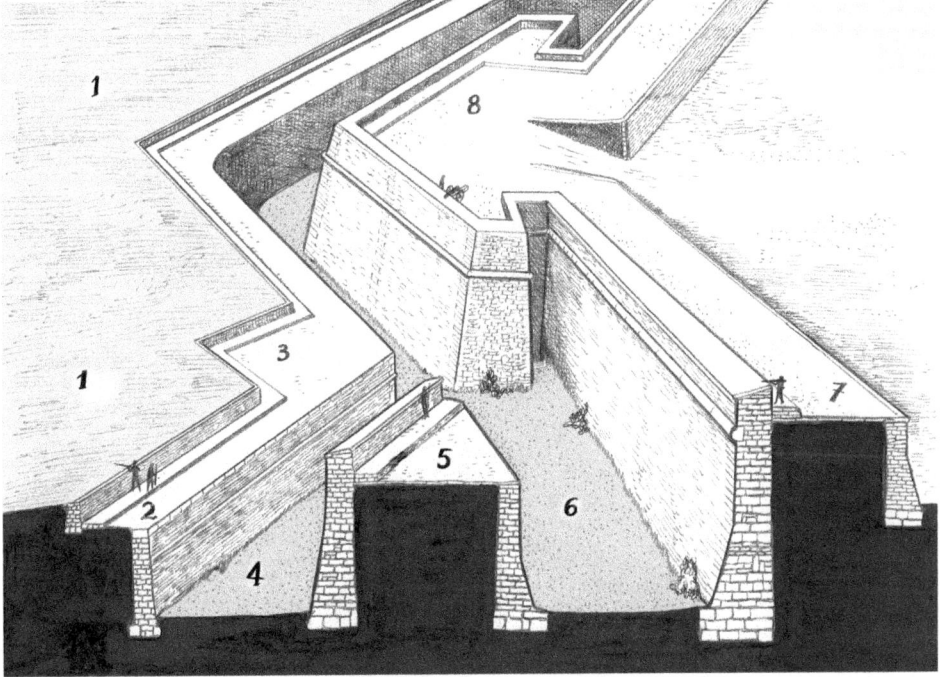

Cross-section bastioned system. The cross-section shows the flat and bare glacis (1); on the counterscarp the covered way (2) and the place of arms (3), fitted with a breastwork and a banket; the ravelin ditch (4); the ravelin (5); and the main moat (6). The scarp was composed of the thick ramparted curtain (7), which strengthened the wall, and offered enough space for the operation of artillery; and the flanking bastion (8).

with various angles. These very numerous variations were determined by engineers according to local conditions to adapt fortifications to the site, but also caused by a sort of fashion or style created by currents, schools or movements. This phenomenon gave birth to countless theoretical bastioned fronts each with its own style. Of course there were endless passionate disputes among engineers of opposing cliques, each engineer or school of engineers asserting that their method was the best.

The art of fortification became a purely military science with a specialized corps of engineers with knowledge in artillery, construction, geometry and mathematics. The adoption of the bastioned system announced the standardization of military architecture, which progressively became a state monopoly.

The bastioned system had only one drawback: immense cost. While Middle Age towns were owners of their walls and towers, with the appearance of expensive guns and bastioned fortifications, cities could no longer pay for these expenses; they received funds from the central authority. In return the rulers demanded control, tutelage and then exclusivity over defenses and only allowed fortification of places with a strategic value for the security of the state.

In the 16th, 17th and 18th centuries, the bastioned system had a large impact on urbanism. The wide space occupied by concentric lines of bastions, outworks, ditches, covered way and glacis were of such massive scale that the formation of suburbs was very difficult or even impossible—because they were forbidden. As a result few cities could grow and expand until the Industrial Revolution in the 19th century. Then new methods of fortification appeared as a replacement of the obsolete continuous bastioned enceinte.

Early Italian Bastioned Fortifications in the Netherlands

During most of the 16th century, Italian engineers dominated European fortifications. François I and Henri II in France, the German emperor Charles V in Spain, Germany, Italy and the Low Countries, the Knights Hospitaler in Malta, as well as the Tudor sovereigns in Britain all relied upon Italian architects to build strongholds, citadels, forts, coastal forts and urban enceintes. The involvement of Italian specialists in military architecture was certainly not intentional. There was simply a shortage of local military architects, which caused the need for foreign experts.

The development of the Italian monopoly was also facilitated by the introduction of mechanical template printing. Indeed many architects wrote treatises, and books about fortifications were easily produced and widely distributed all over Europe.

According to the Dutch historian Charles van den Heuvel the Italian tracé was used for the first time in the Low Countries at Breda (province of Brabant) when a new urban enceinte was built between 1531 and 1536. Designed by the Italian engineer Francesco de Marchi, it included Old Italian styled bastions. In 1540 other Italian engineers, Donatto de Boni Pellezuoli from Bergamo and Piedro da Trente, designed the citadel of Ghent (Flanders).

In 1542 the same Pellezuoli started the construction of a new bastioned enceinte around Antwerp. In 1544 Pellezuoli designed the citadel of Kamerijk (Cambrai today in Northern France), in 1546 the fortifications of Mariemburg near Liège, and took part to the construction of Fort Rammekens near Vlissingen (Zeeland) in 1547. Between 1537 and 1558 de Boni Pelluzuoli worked intermittently on the fortifications of Utrecht together with another Italian engineer

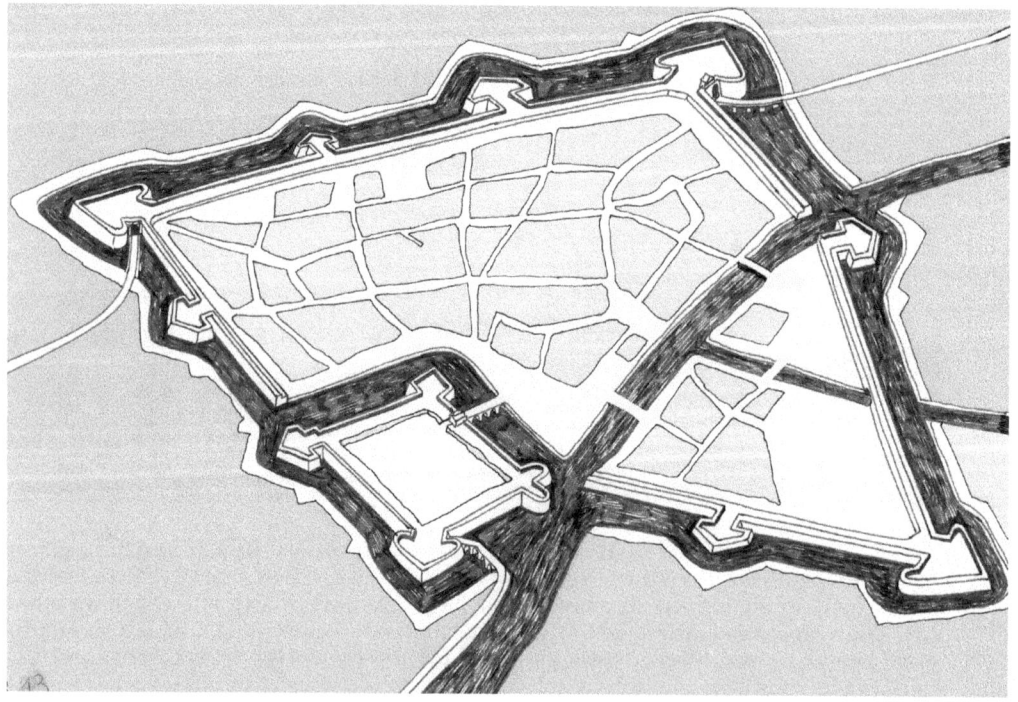

Breda, ca. 1550. Breda is located in southwestern Netherlands in the province of North Brabant at the confluence of the Mark (Merk) and Aa rivers. In the Middle Ages it was a fief of the duchy of Brabant. The town passed to the house of Nassau in 1404, and ultimately to William I of Orange (1533–84). The new bastioned fortifications were designed by the Italian engineer Francesco de Marchi, and built between 1531 and 1536 by order of Count Henry III of Nassau. Breda remained an important strategic Dutch fortress on the Mark River until the 19th century.

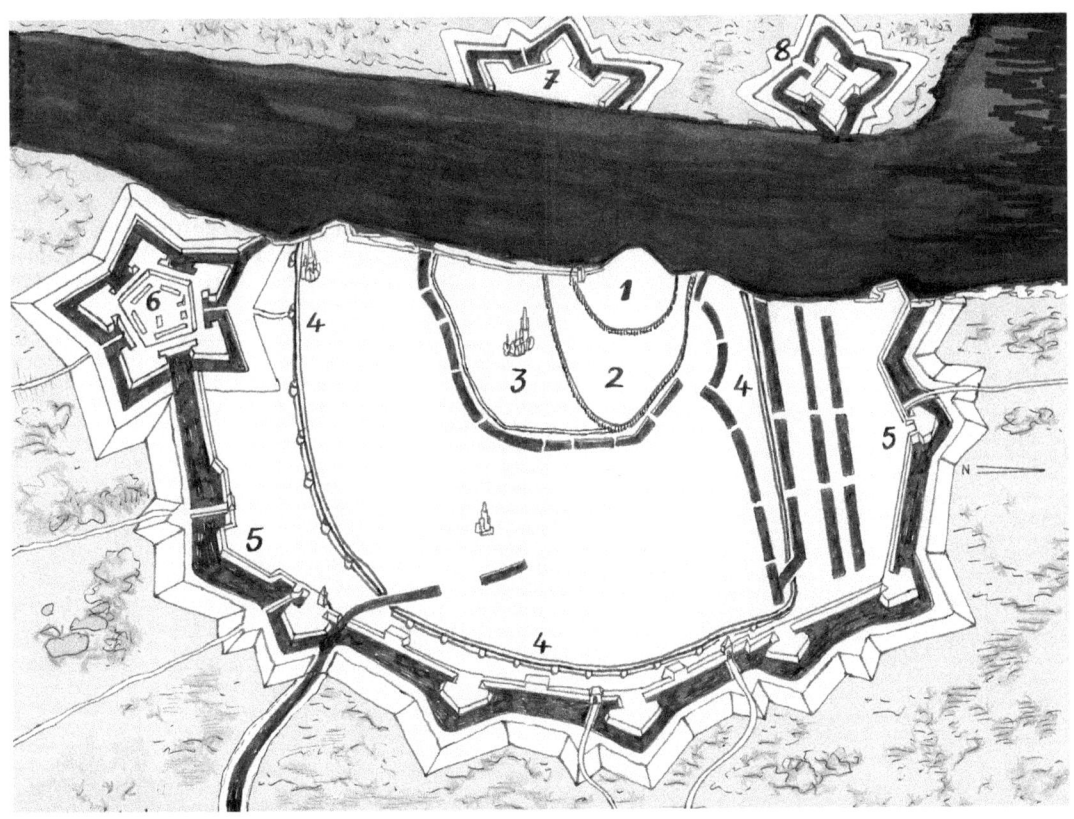

Antwerp, ca. 1610. Located on the Escaut (Scheldt) River, Antwerp is the second main Belgian city after Brussels. Originating from a Benedictine abbey founded in the 3rd century CE, it became a small fortified village (1) called Burcht dating from about 980 CE. In the 11th century it was enlarged with a suburb called Ruienstad (2). A third wall (3) was constructed between 1240 and 1291. A fourth enceinte (4) with a length of 5.50 km with 52 towers was built in the 1410s. By order of Emperor Charles V, new Italian styled bastions (5) designed by the engineer Donato Boni di Pellezuolo were added in 1540. A bastioned citadel (6) was constructed between 1567 and 1569 by the engineers Francesco Pacciotto and Bartholomeo Scampi. On the opposite bank of the Scheldt River two strongholds completed the defenses of the town and harbor: Fort Flander Headbridge (7) and Fort Isabella (8).

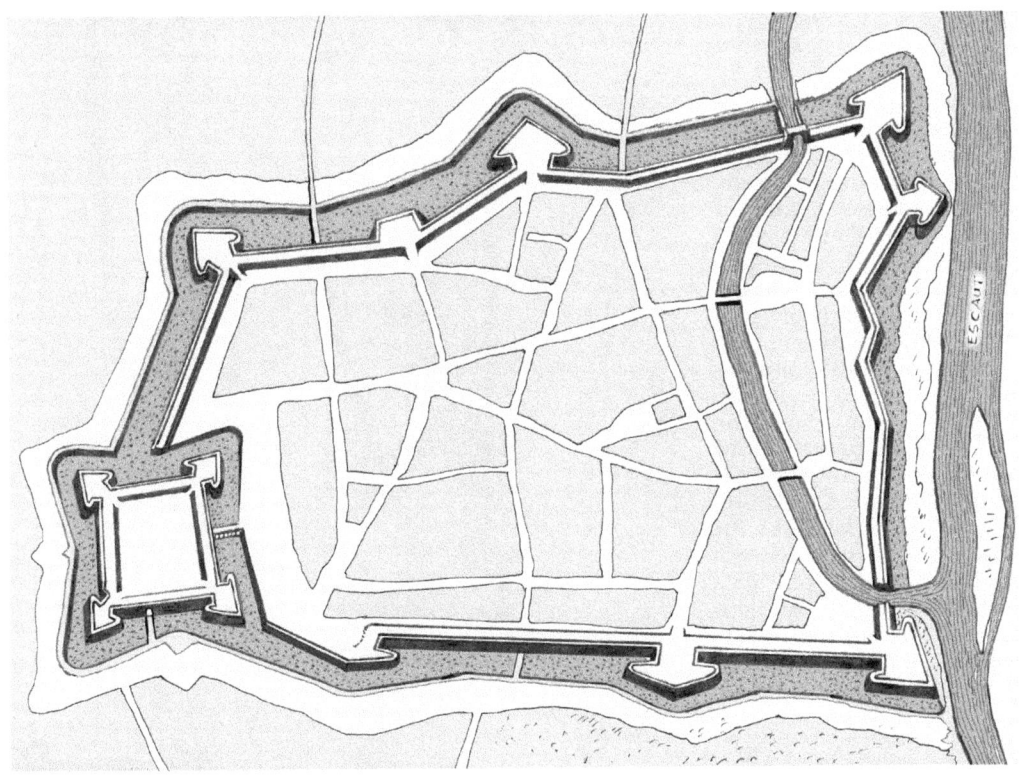

Cambrai, ca. 1554. Located in the Nord département, Hauts-de-France région, northern France, Cambrai (Kamerijk in Flemish) lies along the Escaut River, south of Roubaix. The town was called Camaracum under the Romans, and its bishops were made counts by the German king Henry I in the 10th century. Cambrai was long a bone of contention among its neighbors—the counties of Flanders and Hainaut, the kingdom of France, and the Holy Roman Empire—and it frequently changed hands. It is not known who designed the 16th century bastioned fortifications, but the citadel was designed in 1544 by the Italian engineer Pellezuoli. Cambrai eventually was assigned to France by the Treaty of Nijmegen (1678).

Dutch Fortifications

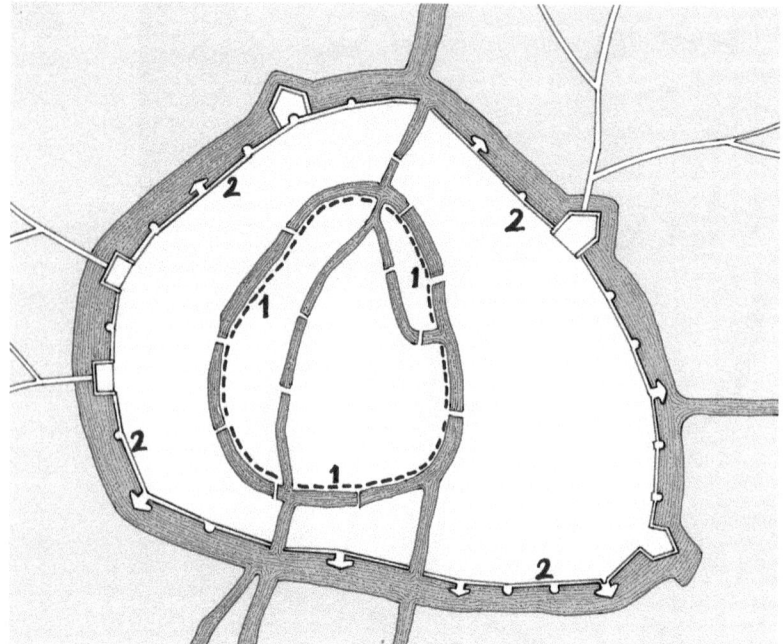

Amersfoort, ca. 1594. Located in the central part of the Netherlands, Amersfoort is the second largest city in the province of Utrecht. The city appeared in the 11th century and city rights were granted in 1259 by the bishops of Utrecht. In the 1300s the city was defended by a brick wall (1). When the need for enlargement of the town became apparent around 1380 the earlier wall was dismantled, and the construction of a new wall with tower (2) was begun and completed around 1450. This second wall had a length of 2,850 m. To these existing defenses, small Italian-styled bastions were added in the period 1560 to 1570.

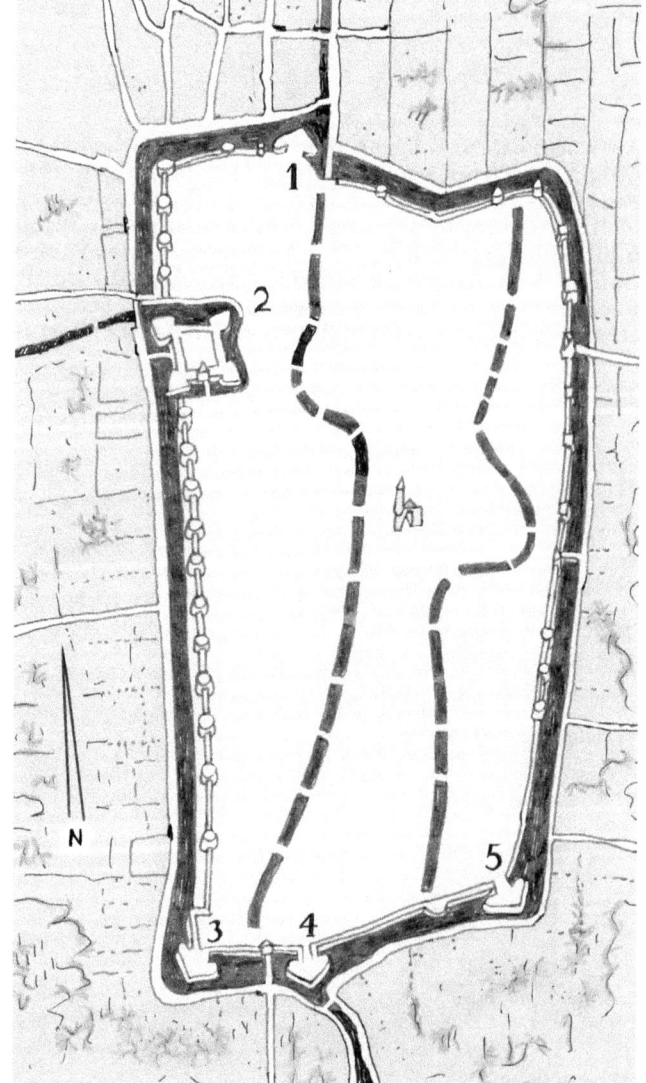

Utrecht, ca. 1550. During the Middle Ages Utrecht was an important city in the Netherlands as a major ecclesiastic, political, and commercial center. The bishops of the Catholic Church were secular lords, not only in the present province of Utrecht but also in parts of Guelders and Overijssel. The bishops often quarreled with the counts of Holland and Gelderland, but finally it was Burgundy that gained control over Utrecht in the 15th century. In 1528 the city lost its independence, and the bishop was no longer a secular prince. Between 1537 and 1558 four new bastions were added to the medieval walls. They were designed by the Italian engineers Pelluzuoli and Marco da Verona. The new bastions marked a curious fusion of Dutch local practice and Italian innovation. Built between 1543 and 1558 bastion Zonnenburg (3), for example, included an internal open courtyard in the gorge, and small houses with stepped gables placed in the salient and in the flanks for housing artillerymen, soldiers and ammunition. Note the citadel Vredenburg (2) built between 1528 and 1535 placed in overlapping position across the urban fortifications: 1: Bastion Morgenster. 2: Citadel Vredenburg. 3: Bastion Sterrenburg. 4: Bastion Mannenburg. 5: Bastion Zonnenburg.

called Marco da Verona. The same da Verona designed bastioned fortification in the south of Amsterdam in 1548.

In the meantime Alessandro Pasqualini from Bologna designed fortifications for the castle of Buren in 1543, and at Amsterdam in 1545. In the second half of the 16th century Italian engineers, e.g., Chippino Vitelli, Gabrio Serbeloni, Bartolomeo Campi and Francesco de Marchi were working in the Low Countries. The famous Francesco Paciotto designed the citadels of Antwerp in 1567, and Vlissingen in 1571.

Dwangburchten (citadels)

In the 16th century, a new form of fort appeared in the Low Countries: the urban citadel, called a *dwangburcht* in Dutch. The term *dwangburcht* comes from the verb *dwingen* meaning to compel or to coerce, and the substantive *burcht* (castle). The term citadel comes from French *citadelle*, and from Italian *cittadella*, based on Latin *civitas* (city).

The Burgundian, and later Habsburg and Spanish rulers were never fully accepted by the population of the Low Countries. The foreign rulers were perfectly aware of this hostility. In order to prevent and if need be repress any popular riot or rebellion, they constructed citadels in the most important cities.

A citadel was indeed a very particular kind of detached work. Like a fort, a citadel was purely military but it was a fortress built within a fortified city. The citadel was always placed on overlapping position across the urban fortifications, which allowed its access to be independent from public city-gates. The citadel was accessible by a main gate facing toward the city and a secondary access leading directly to the countryside. In certain cases, the citadel was an old medieval urban castle that was modernized, or a former work adapted to modern warfare with firing weapons.

If the work was entirely new, a geometrical bastioned form was chosen, often a regular pentagon, which fulfilled military demands, left no dead angles and offered a practical and efficient internal organization. Between the city and the citadel a wide and bare space was established; this space, called *esplanade* served as open field of fire and could be used as military training and drilling ground. Citadel, esplanade and new fortified enceinte cost a lot of money and, sometimes, required the destruction of existing houses or even larger urban neighborhoods.

The citadel fulfilled three distinctive roles. The first function was logistics. The citadel contained everything needed in order to resist a long siege, such as barracks, food, water and foraging stores, arsenal, powder-house,

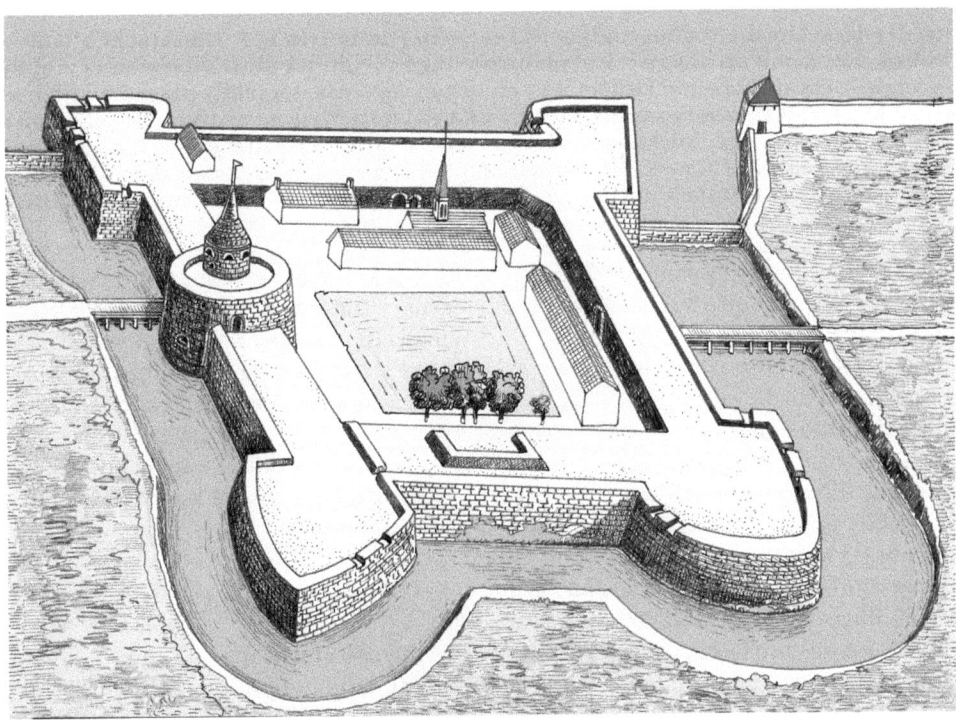

Vredenburg (Citadel of Utrecht). The dwangbucht (citadel) Vredenburg at Utrecht was constructed in 1528 by order of Emperor Charles V in order to control and eventually repress the boisterous newly conquered inhabitants of the city. Set astride the urban wall, the citadel was designed by the engineer Jean de Terremonde and the brothers Rombout and Marcellis Kelderman. Built between 1528 and 1535, it was a rectangular masonry work surrounded by a moat, and fitted with Italian-style casemated amandel-shaped roundels at each corner. The masonry walls had an average thickness of 3.70 m. The construction of the dwangbucht and the maintenance of the 150 garrison soldiers (about 300,000 florins per year) were paid by the 10,000 or so citizens. In 1577 the population of Utrecht joined the anti–Spanish rebellion, and the hated Vredenburg citadel was attacked, and captured after negotiation. That symbol of Spanish tyranny was soon demolished. Today the name of the citadel is used for a music venue established at the same place (Muziekcentrum Tivoli-Vredenburg).

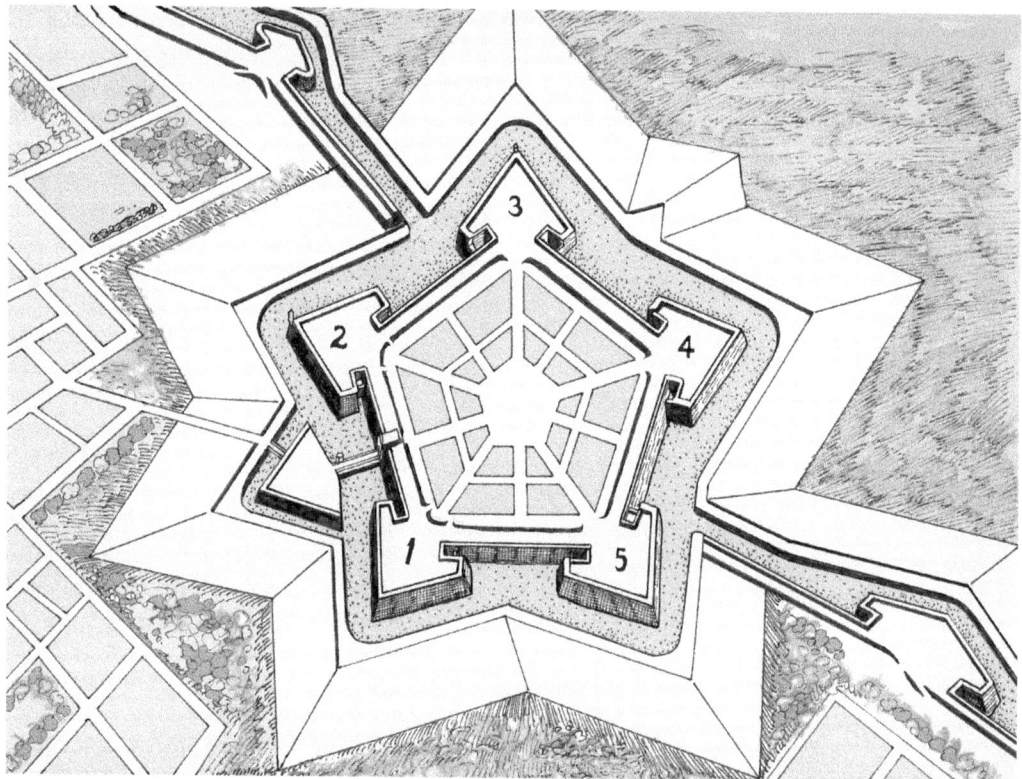

Citadel of Antwerp. Designed by the Italian engineer Francesco Paciotto in 1567 the citadel of Antwerp, aka Zuiderkasteel (South Castle), was a perfect geometrical plane pentagon with five straight curtains and five bastions with ears placed at the angles. The pentagonal citadel was an urban fortress, typically placed on high ground dominating a city. Completed in 1572, it was surrounded by a moat and contained everything the garrison needed: powder houses, food and ammunition stores, workshops, an arsenal, a chapel, a residence for the governor, houses for the officers, and barracks for the troops. The citadel was demolished between 1874 and 1881 when new modern urban fortifications were constructed: 1: Duke's bastion. 2: Hernando bastion. 3: Toledo bastion. 4: Alba bastion. 5: Paciotto bastion.

workshops and so on. It was also a supply point for armies in campaign, it could be used as winter quarters, and was often a military administrative center.

Secondly, the citadel was a powerful military stronghold. Just like the keep in the medieval castle, it acted as a refuge, a final fallback position, a *réduit* (redoubt) from which to continue the defense even when the rest of the town was conquered. For this purpose, and when the natural situation was suitable, the citadel was preferably built on a high position in order to both command and protect the city. It was bristling with weapons, and strongly fortified with powerful bastions, ditches with outworks, and covered way. This display of strength was also meant to deter enemies to lay siege, and to impress the local population.

The third and most important role was indeed political. A citadel was intended to subjugate, control and overawe short-dated conquered populations with questionable loyalty or rebellious propensity. A part of the weapons were directed towards the esplanade and the city to repress insurrections. Its garrison might venture forth to subdue or bombard dissidents at any time, and could also discourage by force the inhabitants from surrendering at a premature stage in the eventuality of a siege.

Very often the construction of the expensive citadel as well as the occupying garrison's pay was financed by citizens' money as a punishment. For all these reasons, the citadel represented a threat, it was often an unpopular and hated place, sometimes an object of terror and dictatorship, and always an enormous financial burden. As soon as relationships between the occupiers and the newly conquered population smoothed, urban authorities firmly asked for its dismantlement or at least that the military take over the expenditures.

Decline of the Italian monopoly

By the end of the 16th century the Italian fortification design monopoly was in decline. Too obsessed by theory, system, symmetry, and geometry (instead of adapting to local conditions), and selling their experience as mercenaries, Italian engineers were gradually replaced with national military architects.

According to the historian Charles van den Heuvel, the first Dutch military engineers were Marcellis Keldermans, Willem van Noort and Sebastiaan van Noijen in the 1550s. During the war of liberation in the Netherlands from 1568 to 1648, a new generation of Dutch engineers created a bastioned style they designated the Old Dutch System. In 1575 at the instigation of Prince William of Orange, a university was created at Leiden where students could study theology, law, medicine, philosophy and mathematics as well as the "free arts" (grammar, dialectics, rhetorics, geometry, music and astronomy). Curiously, engineering at Leiden was a part of the fencing curriculum, but originally military engineers were civilian architects or officers formed in pragmatic manner.

The bastioned system was universally adopted in a multitude of styles. It proved its worth and was in use until the first half of the 19th century. The general adoption of the bastioned system had a clear result. The days of the castle (private fortified residence) were over, and henceforth all fortifications were purely military, decided, funded and built by governments and sovereigns in order to protect their countries as a whole.

Coastal Fortification

With the introduction of cannons and portable firearms a new form of defense appeared: military coastal fortifications intended to repulse attacks at the shoreline and to fire at ships offshore. Of course, before the introduction of guns, coastal defenses had existed. As early as Roman times, coastal fortifications included early warning systems consisting of watchtowers, forts and strongholds placed along the coast or at the mouths of navigable rivers that could alert local naval or ground forces of an impending attack.

Later in Carolingian time, protection against Viking raiders took the form of burgen, and fortified villages were established along navigable rivers to prevent raiders from sailing inland. There were as well seacoast watchers whose

Fort Rammekens. Constructed in 1547, by order of Maria of Hungary (Charles V's s sister and governess of the Low Countries), this fort was strategically located on the Westerschelde (west Scheldt) near Vlissingen (Flushing) in the province of Zeeland, it was a strong coastal fort intended to control the waterway leading to the port of Antwerp. It was designed by the Italian engineer Donato de Boni Pellizuoli, and the construction was conducted by the master-builder Peter Fransz from Antwerp. The fort, situated on a bend of the dike was given a lozenge-shape configuration, and included a large casemated frontal bastion at the side of the river, and at the rear two casemated half-bastions facing the inland, and a wet ditch. Fort Rammekens still exists today, and represents a unique example of early Italian fortification in the Netherlands. The fort has had a particularly long military career. It was taken by the Dutch insurgents in 1573, and used as a rear base by Stadhouder Maurice of Nassau during the battle of Nieuwpoort in 1600. Temporarily occupied by the British in 1809, it was reshaped by Napoleon in 1811, and modernized as a Dutch coastal battery later in the 19th century. During the German occupation of the Netherlands during World War II (1940–1945), Fort Rammekens was turned into an Atlantic Wall Stützpunkt (coastal strong point) called StP. Rommel in 1943.

duty it was to warn the local militia and navy, which would attempt to intercept intruders and raiders.

When firearms technology was developed with cannons having adequate range in the 16th century, coastal fortification and coastal artillery became important subdivisions of the armed forces. Anti-ship artillery deployed in coastal batteries could then defend with lethal effectiveness strategically important places, deny the use of straits and sea-lanes, and hinder, damage or even sink enemy attacking ships. Because an aggressor usually targeted coastal cities, ports or harbors, coastal fortifications were naturally established around such facilities, and sites on the littoral where landings could take place.

However, coastal fortification has always been a difficult matter. There are shores that do not need artificial man-made defenses, where nature provides steep cliffs, shoals, dangerous currents and reefs. On the other hand there are also countless suitable sites where an aggressor can choose the time and place to land and attack, particularly on easily accessible long and gently sloping beaches. These favorable places for an attacker were (and are) still so numerous that defending them all is a difficult if not an impossible task.

Siege Warfare with Firearms

Siege fortifications

At the end of the Middle Age, early guns and portable firearms progressively played a more and more important role particularly in siege warfare. Although primitive and unreliable, these new weapons were far more effective and less cumbersome than ancient and medieval hurling machines. Gradually siege warfare was dominated by the deadly clash of artillery, unless a small and stealthy party could infiltrate and open the gate to the rest of their comrades. Indeed just like in the centuries before the introduction of firearms, a bastioned fortress could still be taken by using surprise, menace, pressure, treason, blockade and attrition.

A good example of a tricky operation reminiscent of the Greek Trojan horse was the capture of the town of Breda held by Spanish troops in March 1590 by the Dutch Prince Maurice of Nassau. The cunning Prince had a small assault combat group hidden inside a peat barge that was allowed to enter the city. Once inside the place, the Dutch soldiers captured a gate in a fast surprise attack and let their comrades in. It was in the actual storming of a place that firearms brought radical changes.

Underground mines (now filled with explosive) could be used by attackers to smash walls, and destroy gates, while heavy siege cannons could make a breach by delivering an uninterrupted series of hammer blows until a part of the stonework collapsed.

Gates and doors could still be smashed by a ramming party but gunpowder allowed for radical new and quick destructive possibilities in the form of explosive devices (known as petards) that could blow to pieces wooden doors and even section of stonewalls.

Siege works

Of course, the besieged also used firearms. In the actual storming of a place, the advantages of firearms lay with the defenders, as discharges of guns and arquebuses made uncovered approaching very dangerous. The assailing party then reacted by developing a siege fortification mostly made of earth—and accordingly resorted to digging—usually at night. Henceforth pickaxes, spades, and wheelbarrows became weapons of war too.

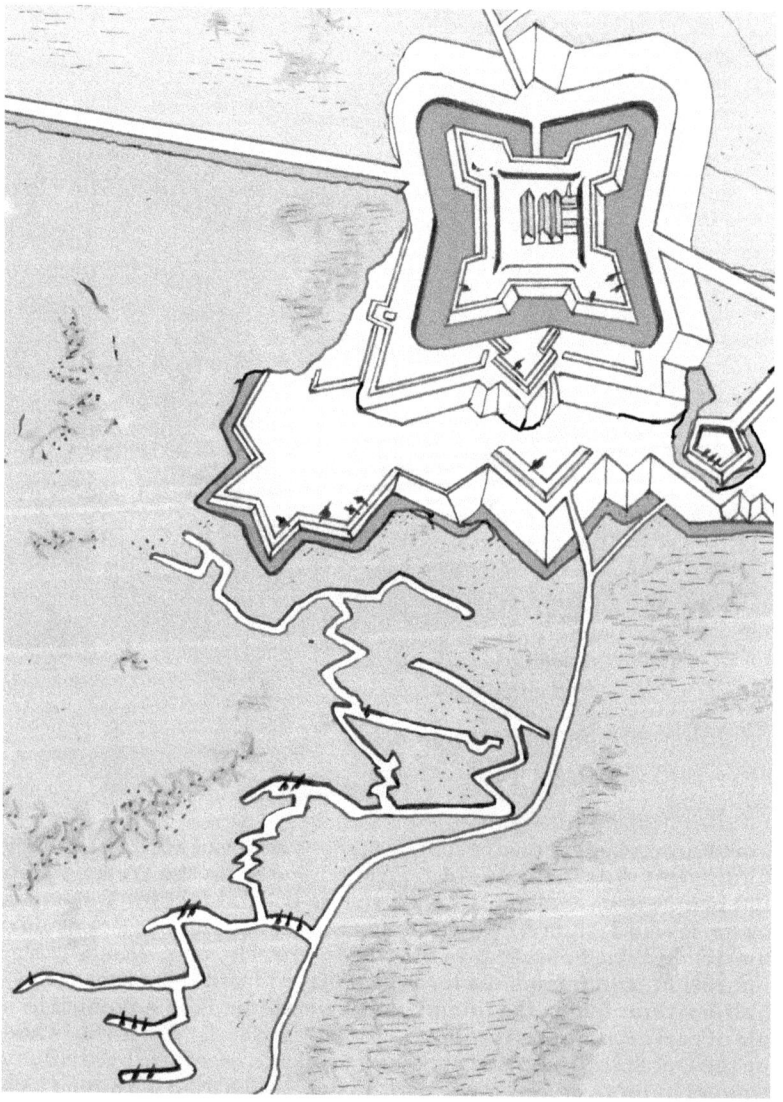

Siege works. The attacking party made use of trenches, zigzagging saps, and siege gun batteries.

The useful range of siege cannons being about 50 meters, attacking methods consisted of bringing artillery as close as possible to the defensive walls by digging a network of trenches in zigzag patterns (in order to avoid enfilading fire directed along a line from end to end). The attackers also used gabions and fascines for protection. Trenches and saps connected batteries, fortlets, sconces and redoubts. These field fortifications made of piled earth and palisades were built to serve as command post, supply-magazines and places where assaulting parties were regrouped, and where wounded were gathered.

The numerous approaching trenches and earthworks were made by civilians and peasants of the neighborhood who were arbitrarily rounded up. Indeed 16th and early 17th centuries men-of-arms would feel dishonored to handle spades, shovels, pickaxes and wheelbarrows. Before the great French engineer Vauban, siege-warfare was commonly led on an unskillful and clumsy way. Because of careless and foolhardy leaders, ill-disciplined soldiers and lack of tactical system, siege operations cost lots of lives, and military engineering personnel and forced-labor civilian diggers were particularly exposed. Success was generally achieved through the defenders' weakness rather than the besiegers' merit.

The first phase of the siege was an artillery duel. The attackers bombarded the defenders with cannons (installed in protected siege batteries) that tried to make a breach in the walls. Mortars (short and squat guns firing in high-curved ballistic trajectories) launched bombs (explosive devices), carcasses (incendiary projectiles) and anti-personnel shrapnel. Mortar-gunners often fired at random to cause blind destruction, panic and terror. The defenders riposted with counter-fire aiming to kill workers digging approaching trenches and to destroy breach batteries.

When the breach was made (either by bombarding the walls or by exploding subterranean mines under the enemy fortifications) assaulting foot soldiers stormed in a rush and fought in a bitter hand-to-hand combat in the smoking ruins. As ever the final assault was a crucial battle for both parties and possibly the turning point of the siege. A repulsed assault often cost a lot of casualties and could cause the collapse of the attacking party. On the other hand a successful assault could result in pillage, rape, destruction, arson, and massacre of the defenders. To avoid this terrible predicament, the defenders often chose to negotiate an honorable capitulation before the general situation went that bad.

View of a siege. In the thunder of explosions and thick smoke clouds of artillery, the approaching sector was a swarm of activities, a coming-and-going of workers, suppliers, dead and wounded who were evacuated while fresh troops moved on to the front line. Suddenly this zone could become a bloody battlefield when the defenders launched a counter-attack. A counter-attack (aka sortie or sally) was a surprise strike intended to disorganize and drive the attackers back. Sorties and counter-attacks were quite important for the defenders' morale, and of course, tactically a successful sortie might turn the tide of the siege.

Chapter 4

The Eighty Years' War

War with Spain

Philip II

In 1555 King and Emperor Charles V, worn out by immense responsibilities, abdicated and retired, and granted the Netherlands to his son, Philip II, king of Spain. While Charles V, born in Ghent, had achieved some respect as he regarded the Low Countries with some consideration, the very bigoted Catholic and 100 percent Spanish new King Philip II considered the Low Countries only as a valuable piece of property. These rich territories were merely a good source of money of which he was always in considerable need for his widespread military and power activities.

Since 1517 and during the reign of Charles V, Protestantism had made headway in the Low Countries, particularly Jean Calvin's doctrine. Protestantism was a political and religious reaction against the scandalous corruption, oppressive intolerance and doctrinal incoherence of Roman Catholicism. Among many things Protestantism rejected the tyrannical authority of the Roman papacy living in outrageous luxury, the cult of the Virgin Mary and the Saints, and some sacraments; perhaps a bit naively Protestants demanded a strict return to the original words of Christ and back to the genuine message of the Bible. Protestantism is popularly considered to have begun in Germany in 1517 with Martin Luther, but the reformed doctrine found a part of its roots in the Low Countries owing to such humanist intellectuals, theologians and philosophers as Geert Groote, Jan van Ruysbroek, Thomas à Kempis, Roelof Huysman, Wessel Gansfort, and Desiderius Erasmus. For the merchants in trading cities of the Low Countries whose ancient charters and privileges had been limited by the Burgundians and Habsburgs, Protestantism was an excellent excuse for fuelling the upcoming great revolt.

By the middle of the 16th century—with growing anti-Spanish feelings, increasing high taxation, fear and hatred against the criminal Spanish Inquisition, oppressive methods of administration, and dislike of the ministers around Philip II's natural daughter and regent Margret Duchess of Parma, the Low Countries had become a powder keg ready to explode.

Rebellion

In the 1560s the Netherlands rebelled against their Spanish masters in what became known as the Dutch Revolt, the Dutch War of Independence or the Eighty Years' War—as indeed the conflict began in 1568 and ended 80 years later in 1648. The exceptional length of the war shows the difficulty encountered by the rebels to obtain their independence. It also demonstrates that Spain although still the richest and most powerful realm in Europe, was already in decline as they could not quell the revolt of the tiny Netherlands. Indeed throughout the 16th century, the Spanish army had a reputation of invincibility. Spain was at the head of a huge overseas empire that stretched across the Americas and the Pacific, and controlled vast territories in Europe. The Dutch revolt was a daring and major challenge to mighty Spain's world dominance.

At first things went bad for the rebels who had enormous difficulties in uniting themselves against Spain. The early phase of the rebellion only showed up the divisions and inexperience of the insurgents. The war started as a series or minor local revolts against the Spanish authorities. Religious toleration and freedom from the attentions of the Inquisition were among the demands most commonly made. But the Protestant cause was not well served by the intemperate behavior of some of the extremist Calvinists.

Enraged and radical mobs went on the rampage in August 1566, smashing the treasures of many cathedrals and churches in the Low Countries. Hearing of such events, the infuriated Catholic Philip II resolved upon severe measures. He ordered the general Fernando Álvarez de Toledo, Duke of Alba (1507–1582), a veteran of many campaigns, to march north with an army from Italy. The dreadful Duke's mission was to restore order in the Netherlands regardless of what measures might be required. Rapidly the Duke's terror, retaliation, and atrocities temporarily regained control of the revolted provinces. At the same time the Duke's exactions intensified the hatred and determination of the insurgents.

Under the leadership of the Prince of Orange, William "the Silent" (1533–1584), the rebels gradually made headway. After many difficulties and setbacks in the 1560s and 1570s they managed to take control of the north of the Low Countries. From the 1580s the conflict attained a more regular character than a common revolt. In 1579 the northern provinces of the Netherlands signed the Union of Utrecht, in which they promised to support each other in their defense against the Spanish army. The Union of Utrecht is now regarded as the foundation of the Republic of the Seven United Provinces. This was followed in 1581 by the Act of

Abjuration, the declaration of independence of the provinces from Spain.

The Dutch Act of Abjuration is often deemed as a model for the 1776 American Declaration of Independence. The new nation was called *Republiek der Zeven Verenigde Nederlanden* (the Republic of the Seven United Provinces) including Holland, Zeeland, Utrecht, Gelderland, Overijssel, Frisia and Groningen. The Dutch Republic was actually a loose federal nation headed by an assembly called the General States dominated by a small but powerful and wealthy merchant oligarchy. The Dutch confederal republic lasted until French revolutionary forces invaded in 1795 and set up a new republic, called the Batavian Republic.

Each province was governed by its local provincial states and by a *stadthouder* (chief executive, or lieu-tenant literally "place holder"—at first representing the King of Spain). The southern (Catholic) provinces that were conquered by the Dutch in the final stages of the war were federally governed by the General States. They were not represented, and called *Generaliteitslanden* (Generality Lands) including Staats-Brabant (present-day North Brabant), Staats-Vlaanderen (present-day Zeeland-Flander) and Staats-Limburg (the southern region around Maastricht).

By the end of the 16th century, the general situation was confused and complicated, and the war bogged down into a deadlock. After the assassination of William of Orange (10 July 1584) his son Maurits (1567–1625) succeeded him as leader of the Dutch revolt. Maurits died in 1625, and was succeeded by his younger half-brother Frederick Hendrik (1584–1647). Both Maurits and Frederik Hendrik were capable leaders, competent generals, and skilled statesmen who brought the Dutch rebellion to a resounding success.

In 1609, a truce was agreed that lasted for 12 years. The truce gave the northern republic a lull in which to develop two outstanding Dutch skills: seafaring and commerce. Hostilities were resumed in 1621, but it was clear to all parties involved in the conflict that Spain would never succeed in restoring her rule over the territories north of the Meuse-Rhine Rivers. It was also evident that the newly created Dutch Republic did not have the strength to conquer the southern part of the Low Countries (roughly corresponding to present-day Belgium) tightly held by Spain. Indeed the South remained predominantly Catholic under Spanish control.

The war led to the separation of the north and south, and to the formation of two distinct entities: the independent Republic of the United Provinces of the Netherlands in the North, and the Spanish Low Countries (Belgium) in the South.

The military stalemate and the political deadlock were settled only in 1648 by the Treaty of Münster in Westphalia that at the same time put an end to the cataclysmic Thirty Years' War (1618–1648) that had ravaged large parts of Germany.

Reluctantly Spain finally had to recognize the sovereignty of the Republic in the North. The tiny Dutch Republic emerged from the war as a strong independent state, and the economic powerhouse of Europe. Feelings of national identity developed in the Netherlands during this war.

The atrocious Eighty Years' War (just like the Thirty Years' War in Germany and Central Europe) was a complicated conflict with many issues. It was a ruthless and sectarian war of religion opposing Christian Catholics to Christian Protestants in the name of the same God. It was also a conflict for independence, and a civil war, in which many of the armies were mercenary. Hostilities were waged with great brutality, as wars fought against rebels, or against followers of a different religious creed normally involved far more intolerance and savagery. Belligerents displayed ruthlessness, cruelty, and unusual harshness towards each other because both sides stubbornly believed they were punishing the enemy of their supposed God.

Religious beliefs remained a permanent aspect of the Dutch rebellion, however, it should not be overstated. When needed, both sides readily traded doctrinal religious convictions for military, political and economic advantages. There were always strong interrelations, intertwinings and compromises depending upon fluctuations in warfare, economy, society and religion.

The successful rebellion of the Northern Netherlands was without doubt a remarkable achievement of the 16th and 17th centuries. Other revolts upset powerful states of this era, but none gave birth to a new, independent, republican, and sovereign country. The Dutch War of Independence was one of the defining conflicts of its era. It marked the end of Spanish supremacy, and secured the triumph of the Protestant Reformation in northwestern Europe (notably in Britain, Scandinavia and northern Germany). The conflict also durably reconfigured the geopolitics of the Continent. Amazingly the independent Republic of the Seven United Provinces became in a few decades a global power.

Nature of the war

The Eighty Years' War began with a series of low-intensity guerrilla campaigns, including skirmishes, raids, naval privateering, a few classical battles in open field, but increasingly siege warfare. It was fought by regular soldiers, urban militias and mercenaries armed with long pikes and muskets. Technically, the development of artillery and the introduction of firearms like the musket had deeply transformed not only strategy and tactics, but also the composition of armies. Since the introduction of firearms at the end of the Middle Ages, gangs of peasants with homemade farming weapons and dashing noble knights clad in armor could no longer stand against infantry and artillery. Neither could musketeers and pikemen function without strict discipline and full-time drilling. Soldiering became a profession, armies were larger, weapons more powerful and lethal, their maintenance costs skyrocketed, and the damage they wreaked became devastating.

In the Low Countries, as the revolt and its suppression centered largely on issues about taxation, control of commerce, and religious freedom, the conflict necessarily involved not only combatants but also civilians at all levels of society. This entanglement of many sectors of the Dutch population into the conflict was one reason for the resolve and subsequent successes of the Dutch rebels in defending cities. Many of the characteristics of the Eighty Years' War were precursors of the modern concept of "total war," most notably the fact that Dutch cities and civilian populations were considered important targets.

Both the rebels and the occupying Spanish forces heavily relied upon mercenaries. Indeed even the most advanced states of the 16th and 17th centuries lived from hand to mouth, improvising armies and navies to suit particular and temporary situations. Recruiting, paying, equipping, feeding and maintaining a permanent national force were feats beyond the power of any government.

On the whole, mercenaries in service of the Dutch were paid on time owing to a good management of funds, as well as standardization, regulation, budgeting and bookkeeping introduced by Prins Maurice of Orange.

On the other hand, mighty Spain—then still a formidable military power whose empire was global, could not cope with these extremely high costs. The Spanish coffers were often empty, although Spain ruthlessly exploited the richness of her colonies, notably in South America. When their conditions became unbearable, soldiers and mercenaries in Spanish service revolted. Few fighting forces could boast of as many mutinies and tumults as the Spanish army of Flanders. Between 1572 and 1607 it was shaken by no fewer than 46 mutinies, some relatively insignificant, others involving thousands of men, giving rise to serious politic and domestic repercussions, and causing great desolation. For example unpaid, infuriated and neglected soldiers in Spanish service plundered Delfshaven in April 1572, Mechelen in October 1572, Naarden in December 1572, Antwerp in 1576, and Oostende in 1604 just to name a few.

The Republic of the Dutch United Provinces won its independence from the mighty Spanish empire also by fighting on water. During the long Eighty Years' War, the Dutch fleet played a determining role. In coastal and inland waters several decisive battles were fought including the seizure of the port of Den Briel in 1572 by the so-called Watergeuzzen ("Sea Beggars" privateers in service of the insurgents). Another important battle was the siege of Leiden in 1574 that was broken by a fleet sailing across flooded polders.

Over the oceans the Dutch navy started the conquest of a new colonial empire spreading from the Caribbean islands, to Brazil, Africa, Indonesia and Japan, thereby increasing national income to such levels that it could partly finance the expensive conflict at home against Spain. The Dutch also made money by using privateers that attacked Portuguese and Spanish ships. For example the capture of the Spanish silver fleet off Cuba in 1628 yielded vast profits.

Wealth by oversea trade and privateering in the Republic of the United Provinces (principally Holland and Zeeland) accumulated at an extraordinary rate during the 17th century. It created an entirely new social ruling Protestant class, made up of greedy, energetic and determined merchants who regarded themselves as realizing a divine scheme with eminently practical business concerns with great significance for later centuries.

The Old Dutch System

Development of the ONS

The medieval castles lost some of their importance when artillery was developed. In the leisurely wars of the 17th and 18th centuries, when roads were few and bad, fortified towns, large and small, played an important role great in resisting the march of enemy armies. The war between Spain and Dutch insurgents was largely dominated by the natural terrain intersected with flat marshy grounds, rivers, waterways and canals. Military operations were always doomed to

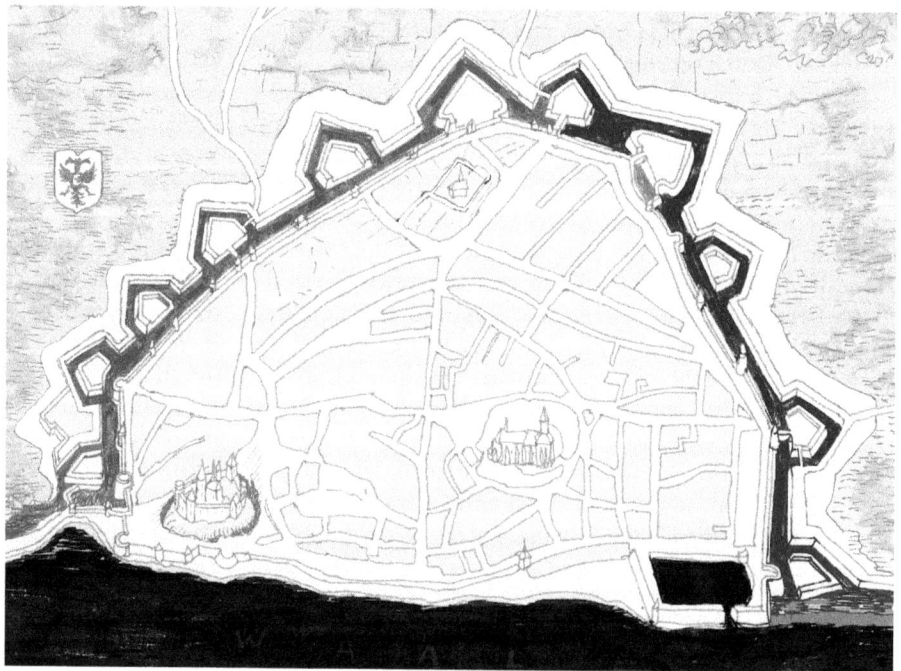

Nijmegen. In the 1560s, the fortifications of Nijmegen (Nimegue) were improved, by order of the Duke of Alva, by the Spanish engineer Horologio. Ahead of the medieval walls and towers advanced ravelins were constructed in the moat. In 1591 the Dutch rebels captured the city, and Prince Maurice commissioned the Dutch engineers Adrian Anthonisz, Andries LeRoy, and David van Orliens to design a new bastioned enceinte that was completed around 1605.

Chapter 4. The Eighty Years' War

Tiel. The medieval fortifications of Tiel in the province of Gelderland were hastily modernized in the 1580s by the Dutch engineers Johan van Rijswijk and Adrian Anthonisz with large bastions placed ahead of the gatehouses.

lose momentum, the more so as the country was (and still is) one of the most urbanized regions of Europe.

Because the country was dotted with numerous fortified towns, and cities, the Eighty Years' War was characterized by siege warfare. The main targets were cities and their inhabitants so both sides made a wide use of the newly created bastioned system of fortification to defend towns, cities, villages, ports, passages and key locations. The Dutch developed a unique and empiric method. At first they built in an improvised fashion what later became known as the *Oude Nederlandsche Stelsel* (ONS, Old Dutch System)—designated such when a new method appeared later in the second half of the 17th century called the New Dutch System.

According to the Dutch historian Bouko de Groot, the basic system of what became the ONS was designed in the 1590s by the engineer and mathematician Simon Stevin by request of Prince Maurice of Orange. The old Dutch system of fortification was directly derived from early 16th century Italian bastioned fortification, but it was adapted to the Low Countries' natural conditions, and to the modesty of the rebels' treasury.

Indeed three specifications determined the development of the early Dutch defenses, namely lack of time, shortage of money, and abundance of water. When the Dutch insurgents started their rebellion against Spain, they would certainly have been pleased to construct strong and expensive masonry fortresses directly based on Italian models. But there was neither time nor money for such refinements. The Dutch insurgents had to improvise, and quickly came to the idea that cheap, broad, and wet ditches could replace expensive masonry walls. Everywhere water was near the surface, and swamps, lakes, rivers or canals were available for creating impassable obstacles like moats and inundations. The wide defensive *gracht* (ditch filled with water) had already proved its excellent hindering worth in the medieval castle aka *waterburcht*.

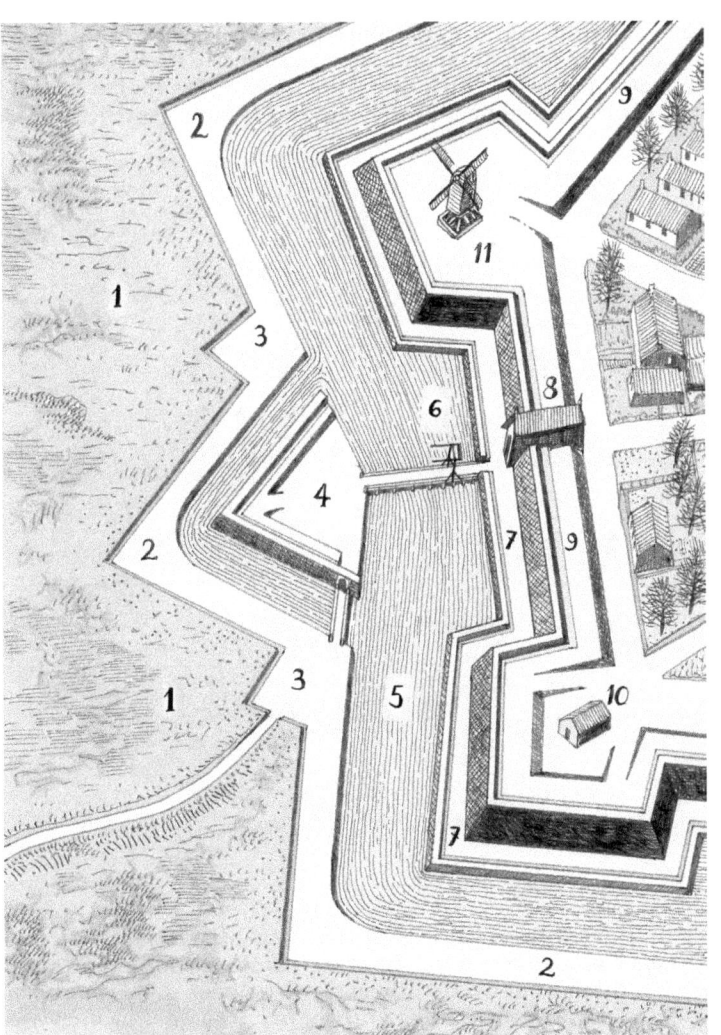

Old Dutch bastioned system: 1: Glacis. **2:** Covered way. **3:** Place of arms. **4:** Ravelin (aka demi-lune). **5:** Moat—ditch filled with water. **6:** Entrance with bridge and drawbridge. **7:** Continuous berm and fausse-braie (lower wall). **8:** Gatehouse. **9:** Remparted curtain. **10:** Empty bastion. **11:** Full bastion.

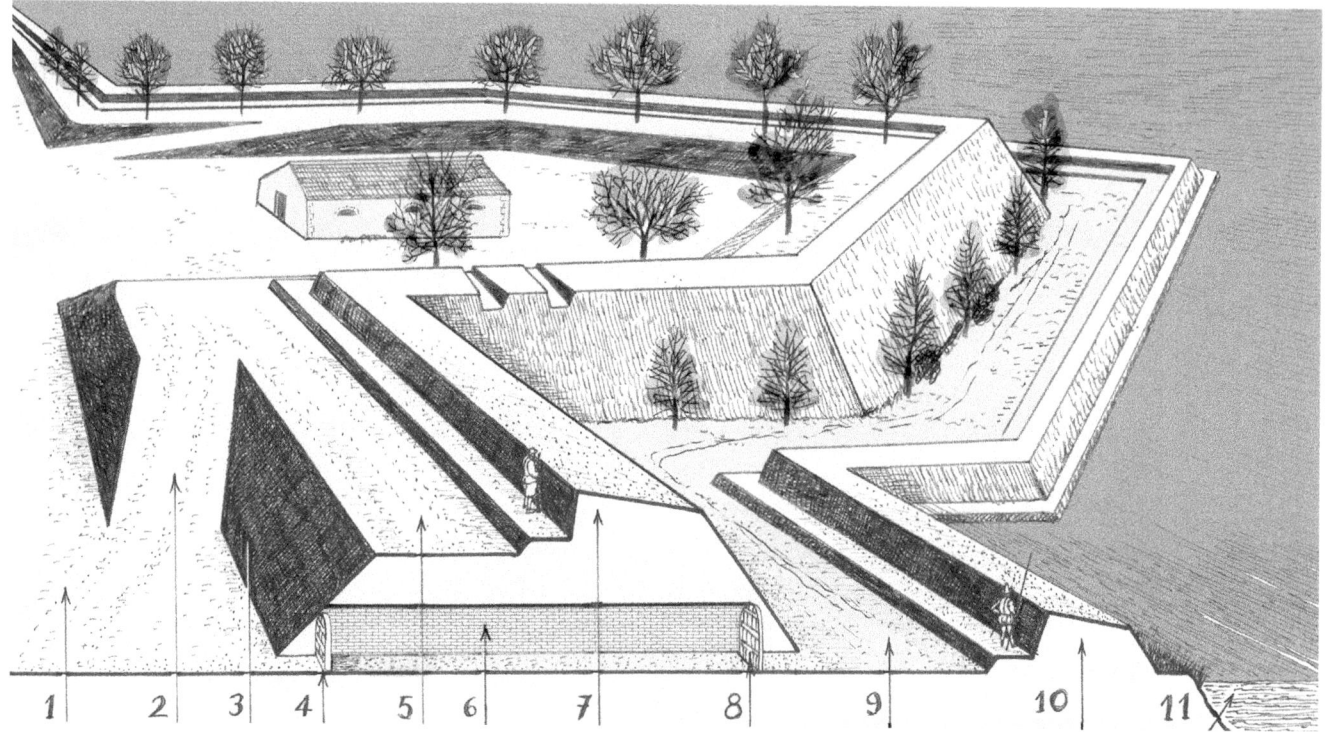

Cross-section of an Old Dutch fortification: 1: Terre-plein. 2: Ramp, access to rampart. 3: Interior slope of rampart. 4: Door to the sortie. 5: Wallwalk, large enough to place artillery. 6: Sortie (vaulted passage in the form of a tunnel made of masonry) allowing communication between the terre-plein and the lower outer lane. 7: Parapet or breastwork. 8: Door to the berm. 9: Berm or lower outer lane. 10: Onderwal or fausse-braie (outer lower wall). 11: Wet ditch.

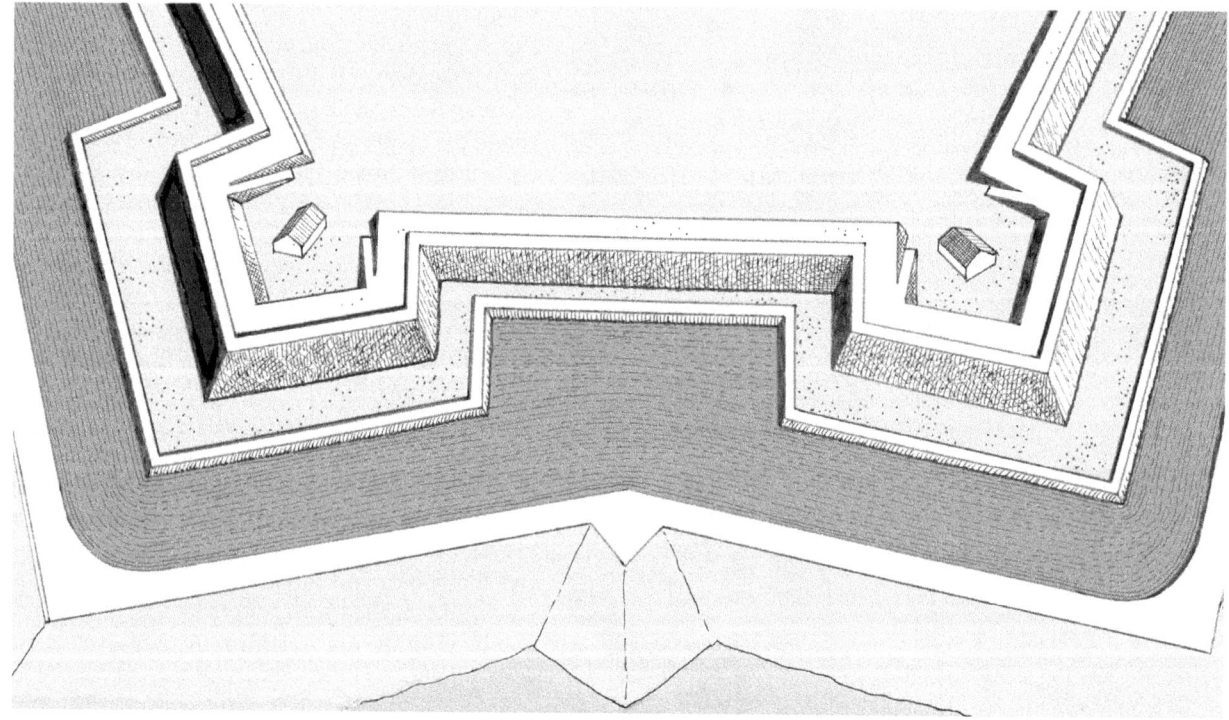

Fausse-braie. The fausse-braie (aka onderwal = "under wall" or lower wall) was a thick man-high breastwork featuring a firing step, and serving several purposes. It was desirable that the weight of the rampart should be drawn back a little from the edge of the ditch, and the fausse-braie filled what would otherwise have been dead ground at the foot of the rampart. The onderwal also offered a low combat position for grazing fire over the ditch, which was very important, and which defenders posted on the main rampart would support by flanking and plunging fire.

ONS Bastion. A typical ONS bastion was made of earth. It included right 90 degrees flanks most often without orillon. A bastion was said void (or empty) when it was terraced only along its revetments; this hollow and protected terre-plein was suitable to place a powder-house for example. A bastion was called solid (or full) when its terre-plein was completely filled with earth. On the upper surface of a solid-bastion, a windmill could be placed or a cavalier. The depicted bastion was empty (not filled with earth) and could house a powder magazine. Note the arkel (small observation turret) at the salient.

Arkel (aka sentinel or schildwachthuisje)—a sentry-box or a small turret, often made of wood and placed at the salient or shoulder of a bastion. The purpose was to have an observation post to watch over the ditch.

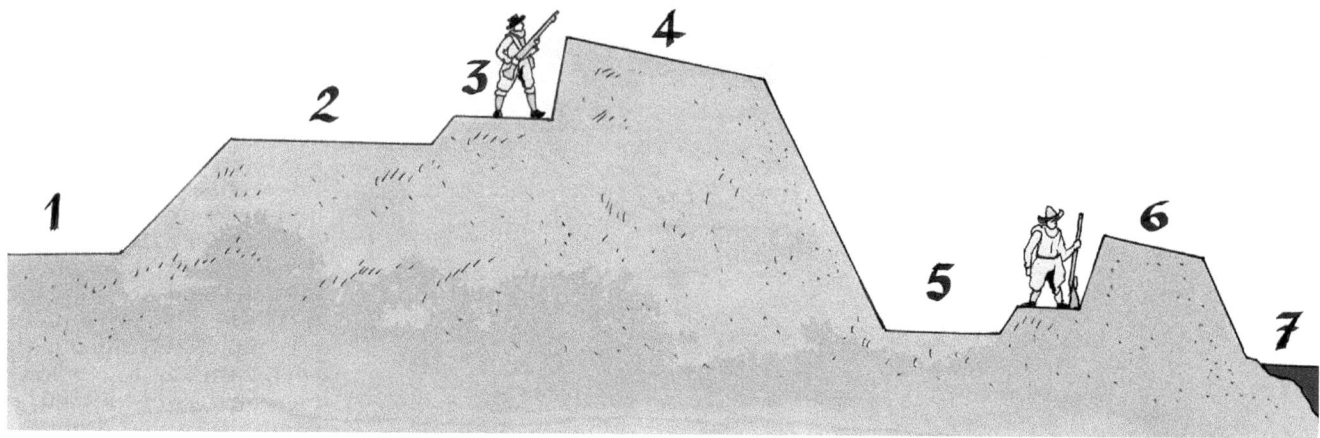

Cross-section of a rampart: 1: Terre-plein. 2: Chemin de ronde (wallwalk). 3: Firestep. 4: Plongée (parapet or breastwork). 5: Berm. 6: Onderwal. 7: Moat.

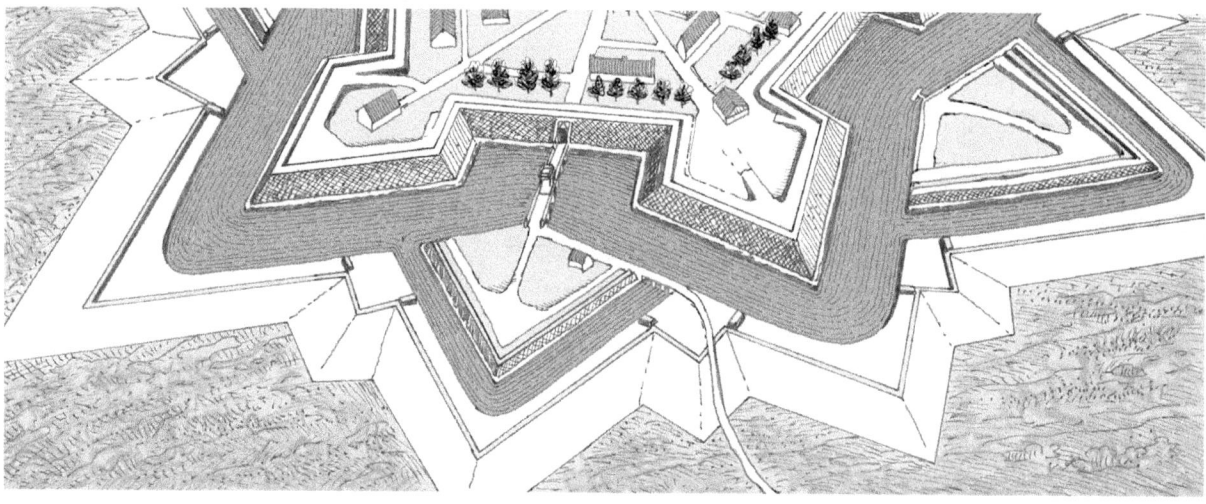

Ravelin (also called demi-lune) was the most important outwork. It was a kind of island placed in the ditch, in front of the curtain, between two bastions. The ravelin was systematically placed in front of a gate and also built in the gorge or ahead of a crownwork and of a hornwork as we shall further see. The ravelin thus shielded a curtain or the entrance to a fort, a citadel or a city. It also covered the flank of the bastion and formed an additional obstacle before the main work. It was very often triangular, composed of two faces protruding towards the enemy.

What the Dutch actually did at first in fortifying their towns and constructing forts displayed no evidence of an elaborate system. At the start of the rebellion in 1568, very few towns in the Netherlands were defended by modern bastioned fortifications, and improvements were piecemeal. Within a few decades all that had changed, first in the provinces of Holland and Zeeland and soon in all provinces that rallied the revolt.

Starting as a rule from an existing enceinte, most of the times an obsolete medieval wall, they would first lower towers and arrange their tops as bulwarks and artillery platforms. High brick wall that had become so vulnerable to artillery fire were lowered, eventually replaced or reinforced with thick ramparts made of earth that lessened the destructive impact and vibration of the shot.

At the same time flanking elements made of earth were added at vulnerable points including bulwarks, bastions, redans and lunettes. The designers would very often add to or widen a broad wet ditch with extra protection provided by small triangular fortified islands called ravelins. On the counterscarp (the outer side of the moat) they set up a covered way protected by a man-high breastwork intended to connect the outworks, and permit the launching of sorties (outward raids). Beyond these advanced positions there was a bare gently sloping glacis denying any cover to possible enemies. Wherever possible impassable flooded zones were established within or ahead of the glacis, preventing enemy advance. Inundations prepared on purpose were used wherever natural conditions permitted.

The Dutch realized that flooding low-lying areas formed an excellent defense against enemy troops, as was demonstrated on several occasions—for example, during the sieges of Alkmaar in 1573 and Leiden in 1574. In the 1590s Prince Maurits of Nassau, and later his half-brother Frederick

Hendrik, gave the order to prepare a large military defensive flooding zone stretching from Muiden to Vreeswijk in order to defend Amsterdam. This primitive waterline (termed the Old Holland Waterline) was successfully used in 1672 against the invading armies of the French King Louis XIV.

At the same time and with good insight the Dutch recognized the inherent weaknesses of wet ditches and inundations. For instance when it froze in the winter, ditches and flooded zones turned into ice, on which enemies could advance—and thus then no longer provided any obstacle. So they arranged for running waters through sophisticated hydrologic systems of sluices, batardeaux, basins, cunets, drainage ducts and channels. Dams, dikes and embanking were established to regulate and control the flow of waterways, canals, streams, and rivers, swamps and lakes.

Characteristics of the ONS

At first the elements of fortifications (bastions, walls, ditches, ravelins etc.) were quite irregular in their design and relation to each other. As the Eighty Years' War dragged on and on, Old Dutch fortifications matured and became a system in earnest with clearly defined standard elements related to each other by fixed rules and geometrical ratios.

The ONS was characterized by low bastioned profile and by earth works with no or very little expensive masonry. Bastions were neither fitted with casemates nor with orillons; the flank angle was mostly 90°; the ditch was wide, always filled with water and often defended by a continuous *onderwal* (aka fausse-braie or lower wall) at the foot of the scarp (the inner side of the moat). Further the ONS included the standard elements of bastioned fortifications like ravelins—triangular islands protecting the curtains and gatehouses; a covered way, and a glacis established on the counterscarp. There could also be another advanced ditch at the foot of the glacis.

Gatehouse

Just as in a medieval castle the gate to a city or a fort was the most vulnerable part. Thus the number of access points to forts, citadels, and fortified cities were limited as much as possible. The gate was practically always placed in the middle of a curtain in order to be defended from both adjacent bastions' flanks. In many instances it was a stone building including a tunnel passing under the rampart. For security

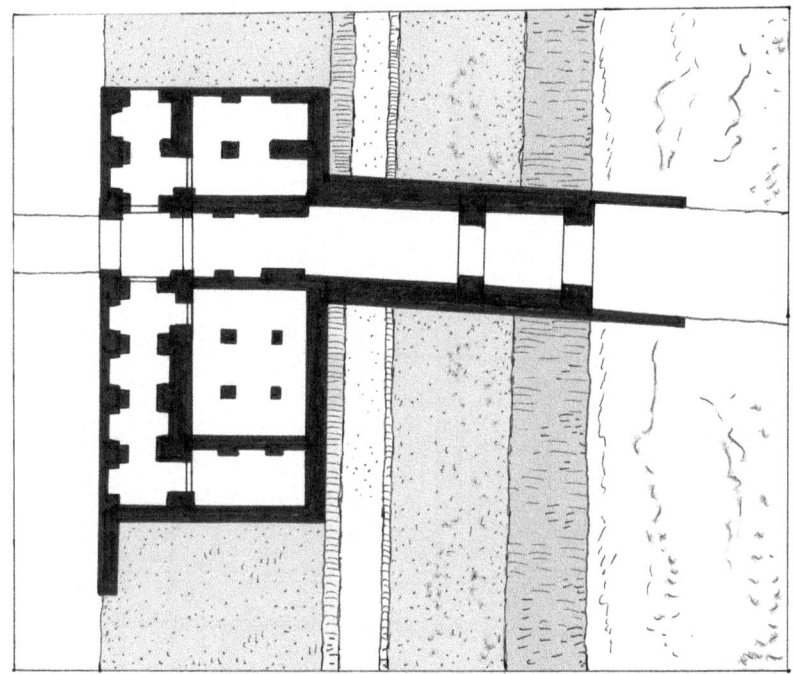

Gatehouse. Cross-section (*top*) and plan (*bottom*). Just as in the Middle Ages, the gatehouse included a drawbridge, and strong and heavy wooden doors. It also featured a custom-office where taxes and tolls were levied on all persons and all goods coming in or going out of the city. Taxes and tolls remained major forms of income for cities. The gatehouse was obviously a strongly guarded passage. It might also include an arsenal for weapons and ammunitions, quarters and lodgings for civil servants, tax collectors and guards. It could serve as a prison. The city gatehouse also played a prestigious and symbolical role. In the ONS it was one of the few masonry elements, and displayed a few ornaments expressing to foreign visitors and travellers the wealth and importance of the town.

reason its width allowed passage for only one wagon or cart at a time. The building, called *poort* (gatehouse), included various premises available for guards (and urban toll-officers in the case of a fortified city).

At both ends, the tunnel was closed by a heavy double-leaved door reinforced by metal parts, huge nails and locked by a strong transversal beam. Above the doorway hung a *hamei* (also called herse or organ). This adaptation of the medieval *portcullis* was a heavy barrier made of strong wooden balks, which could be raised by winching machinery placed in a chamber in the first floor of the gatehouse. Should the occasion arise, the *hamei* could be very rapidly slid down by means of its own weight and side-grooves.

In the ONS the gatehouse was often the only stone

Buiten Apoort Gate, ca. 1623 (Groningen)

building in the fortifications as bastions, curtains and other works were made of earth. It was therefore the only place where some kind of decoration was possible in order to express greatness and power. Marking the entrance of a town or of a fort, gatehouses were sometimes adorned with decorative finery although the Calvinist religion forbade any worldly ostentation. Dutch gatehouses of the period were on the whole rather soberly decorated, but some could display front sides nicely proportioned, columns and pilasters framing the entrance with an ornamented pediment on top.

Gates were always given a name, sometimes one honoring a noble leader or a famous patriot. Most often it was named after a neighborhood, a cardinal direction, or the road leading to another province or town such as *Westerpoort* (West Gate), *Haarlemmerpoort* (Harlem Gate), *Friese Poort* (Frisian Gate), *Munsterse Poort* (Münster Gate).

Advanced works

Advanced outworks were added generally in the form of horn- and crownworks. They were intended to multiply the defensive positions and obstacles opposing the progression of besiegers. Indeed the series of outworks pushed forward took the first brunt of the attack. If the enemy managed to conquer one of them, he should find that there was another one covering it from the rear, so the time of final victory would be frustratingly postponed. The outline, alignment, and positioning of outworks were

Opposite, top: Crownwork and hornworks. These were projecting outworks occupying a portion of terrain in front of the main defense. They were advanced combat positions designed to force the besiegers to begin a siege from a greater distance, and to occupy and cover parts of the ground not easily seen from the main wall. The crownwork (1) was formed of two bastioned fronts thus comprising two curtains, one bastion, two half bastions and two small wings. A hornwork (2) was an outwork composed of a bastioned front including a curtain and two half bastions and two long parallel wings. Both the crownwork and the hornwork always had their own ditch, and were often reinforced by one or more ravelin at their front and/or gorge.

Opposite, bottom: Outworks. The real strength of the bastioned fortification was in its geometric design, which allowed defenders to place weapons at key points to provide enfilading fire against every possible angle of attack. The various open areas around a bastioned fortress were in fact killing grounds where besiegers would come under crossfire. The fortification in the 17th century was based on curtains, ditches and bastions but it also included several outworks, all including weapons emplacements and protecting each other: 1: Bastion with ears. 2: Curtain. 3: Bastion with cavalier. 4: Tenaille. 5: Caponier. 6: Ditch (or moat filled with water). 7: Counterguard. 8: Ravelin (aka Demi-lune). 9: Crownwork. 10: Hornwork. 11: Covered way and places d'armes. 12: Glacis. 13: Flèche (Advanced arrow aka redan).

Chapter 4. The Eighty Years' War

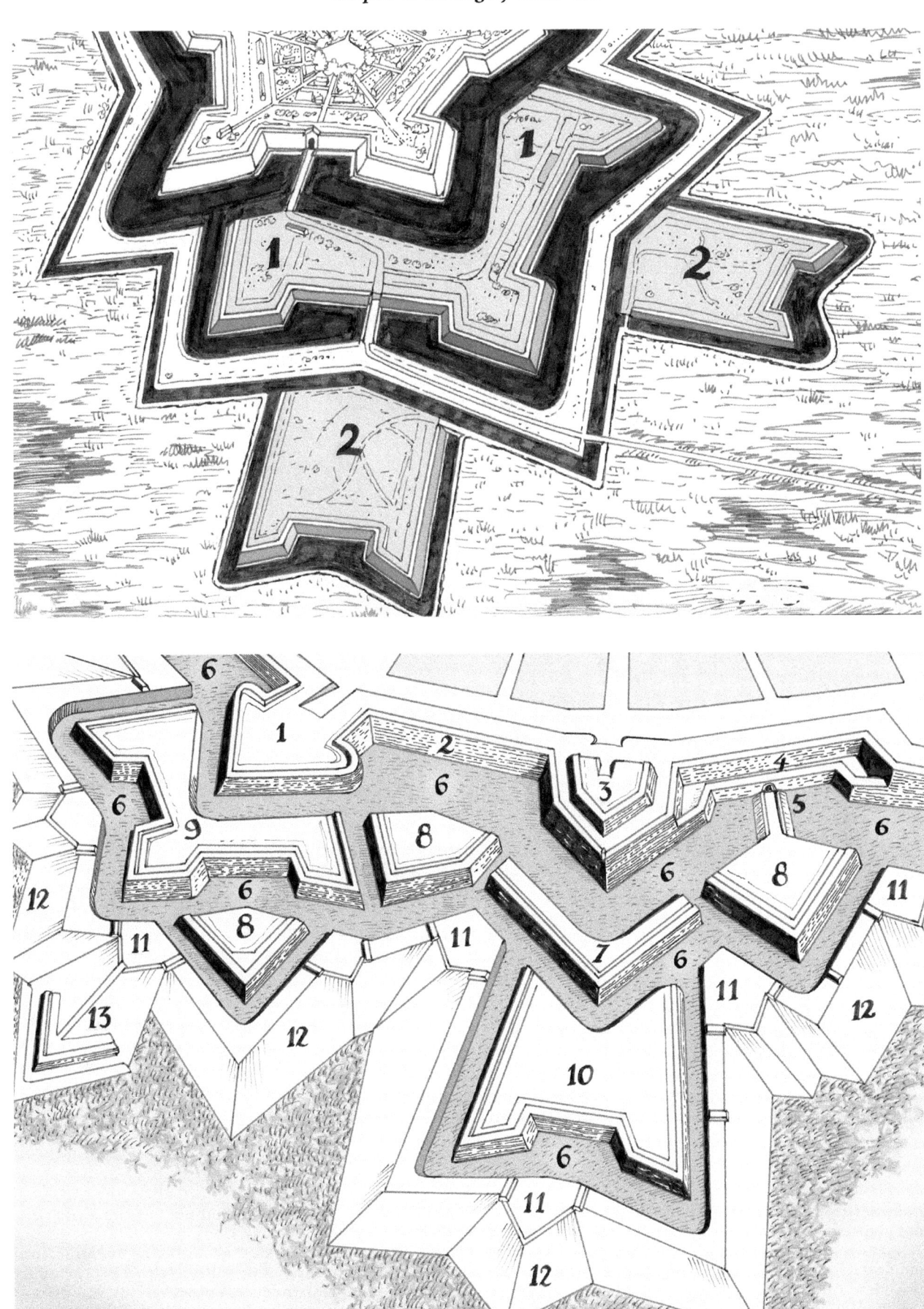

Entrance. The entrance to a fortified city or a fort was by definition a weak and vulnerable spot. It was thus guarded and defended by an outwork (very often a ravelin), a drawbridge, a gatehouse, and a corps de garde (guard house).

Opposite, top: Drawbridge. By pulling down the chain (1), the mobile counterweighted part of the gaffs (lifting arms) pivoted down (2), lifting up the hinged mobile bridge roadway (3).

Opposite, bottom: Covered way. The so-called covered way was a continuous broad walkway or lane running all around a fortress along the top of the counterscarp and following the outline of the ditch. It was protected from enemy view and fire by a parapet (breastwork). Together with the placement of arms the covered way was an essential element of the bastioned fortification. They were actually the defender's ears and eyes, and allowed troops to be posted as a first advanced line of defense at the periphery of the fortress.

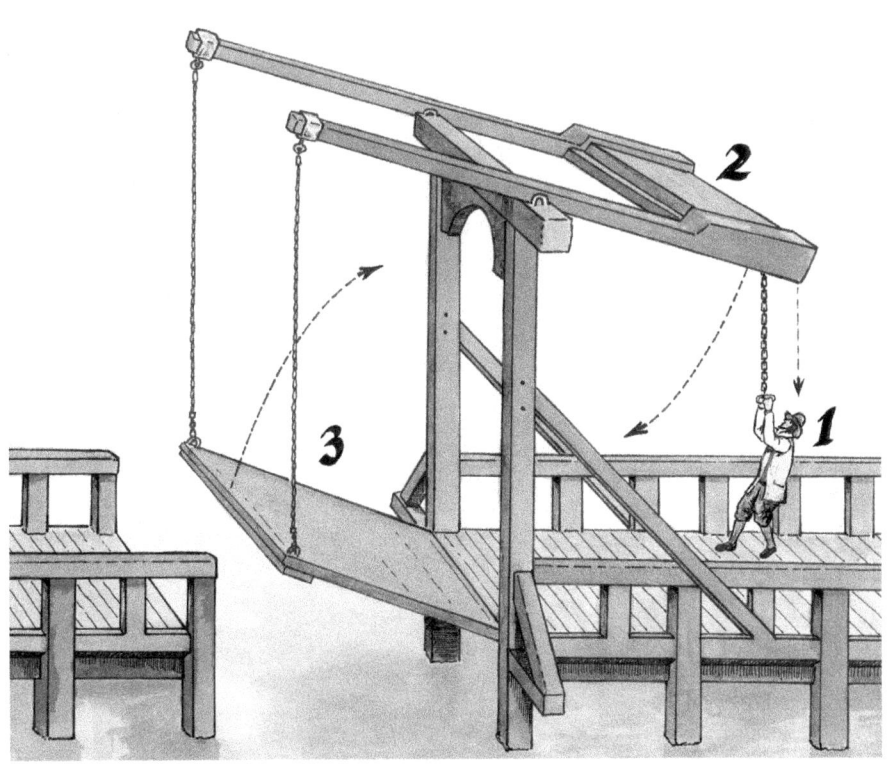

precautionally designed so that no blind spots could exist. The height of every outwork was calculated in order to command each other and be commanded from the main enceinte.

Outworks were furnished with parapets, ascents and surrounded by secondary ditches, which were less wide than the main ditch. The gorge (rear) of these works was always open which meant not protected by a parapet. So if an outwork was conquered it offered no protection to the besiegers. Communication between outworks and main body was effected through the wet ditches by means of wooden footbridges, which could be easily destroyed in an emergency situation. When there was no such footbridge, rowboats were used.

Bastions, ravelins and outworks were given names often with a patriotic overtone such as that of one of the seven provinces in rebellion against the Spaniards (Utrecht, Zeeland, Holland, Overijssel, Gelderland, Frisia, and Groningen), or the name of a famous noble leader linked to the Dutch revolt (e.g., Prins Maurits Bastion, Stadhouder Bastion). They could be named after a building they housed like *Kruittoren Dwinger* (powderhouse), *Bastion Meule* (windmill), *Jacobijnsdwinger* (near a monastery) or *Magazijn Bolwerk* (storeplace). Bastions could also be named from the neighborhood or the gate that they protected, for example *Oosterdwinger* (east bastion near the East Gate). Many other names were used.

Sconce

In addition to urban defenses, independent detached works were widely used. Called *schansen* (sconces) these military fortlets were small earth-made semi-permanent or permanent independent stronghold intended to defend a particular point or position, for example, passages, bridges or fords. Indeed the Low Lands included so many rivers and canals that it was impossible to conduct a campaign without making river crossings. Bridges and

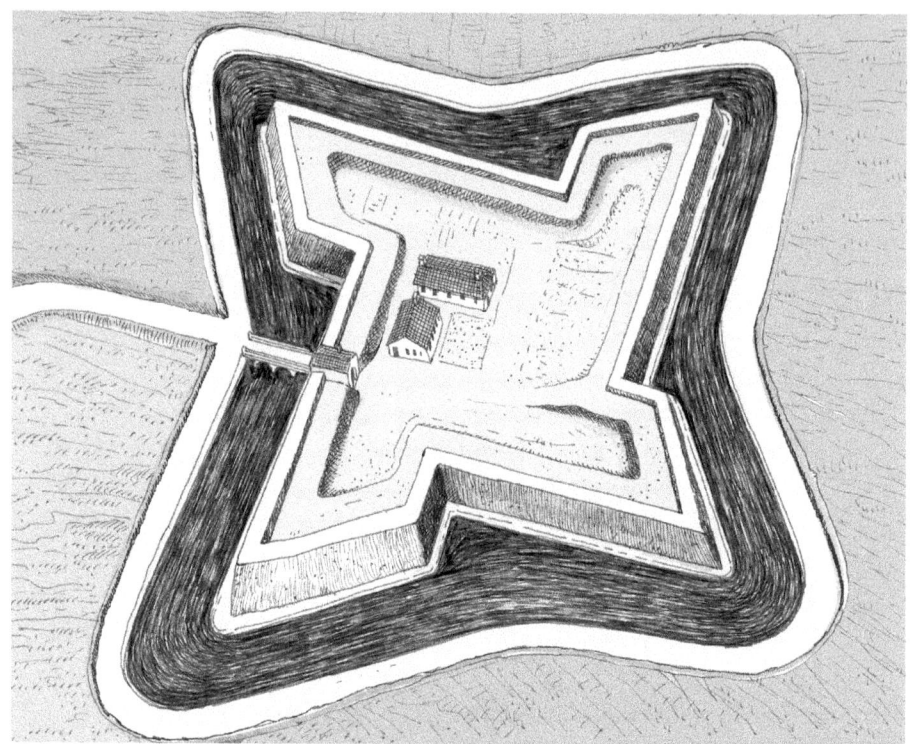

Ter Apel. The Ter Apel schans, built in 1593 on a passage in a swamp between Germany and Groningen, was a square sconce with half bastions on the corners.

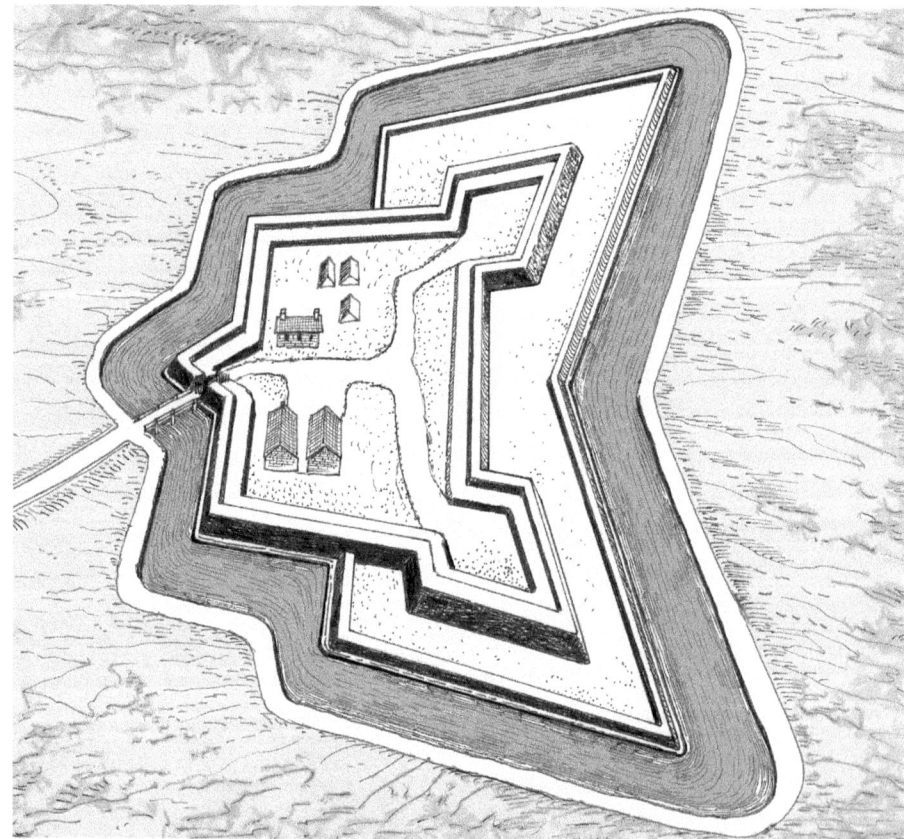

The Boner sconce, located in the province of Groningen, was built in 1589 in order to protect the access along a dike named the Hamdijk.

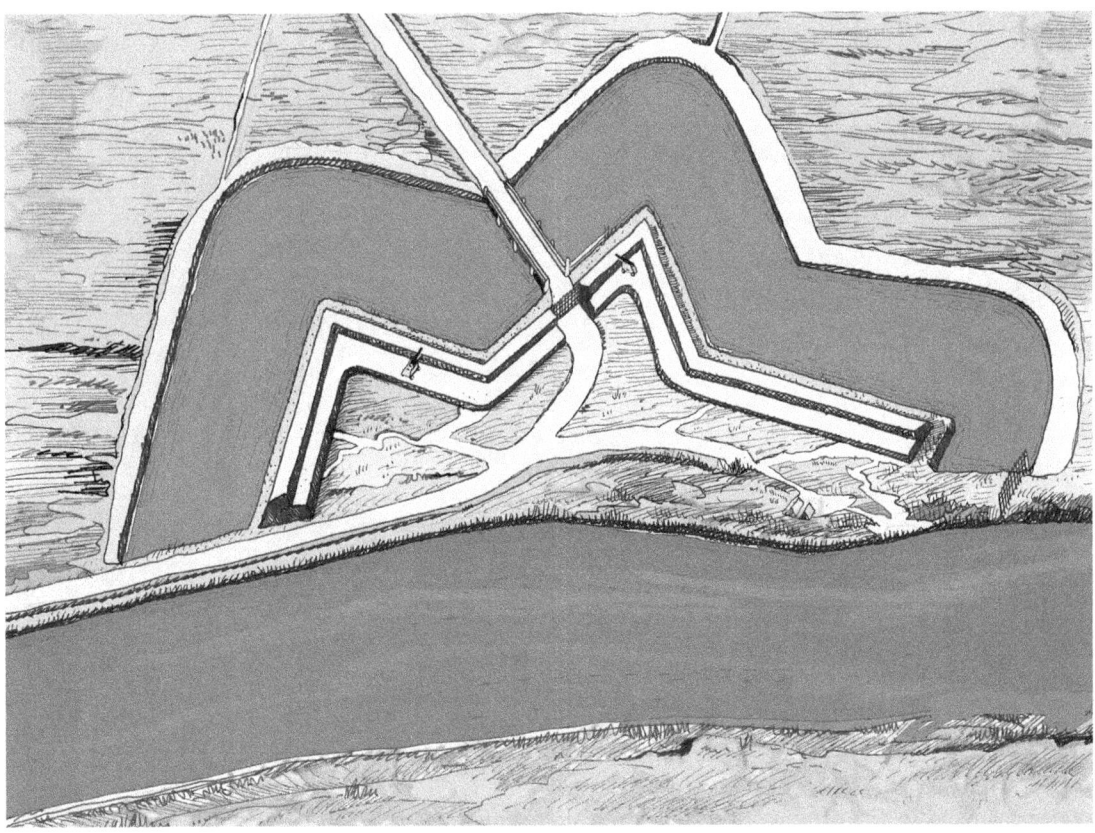

The simple Wolveschans included a double redan and a moat. It was built in 1593 to defend a passage in swamps between Groningen and Frisia.

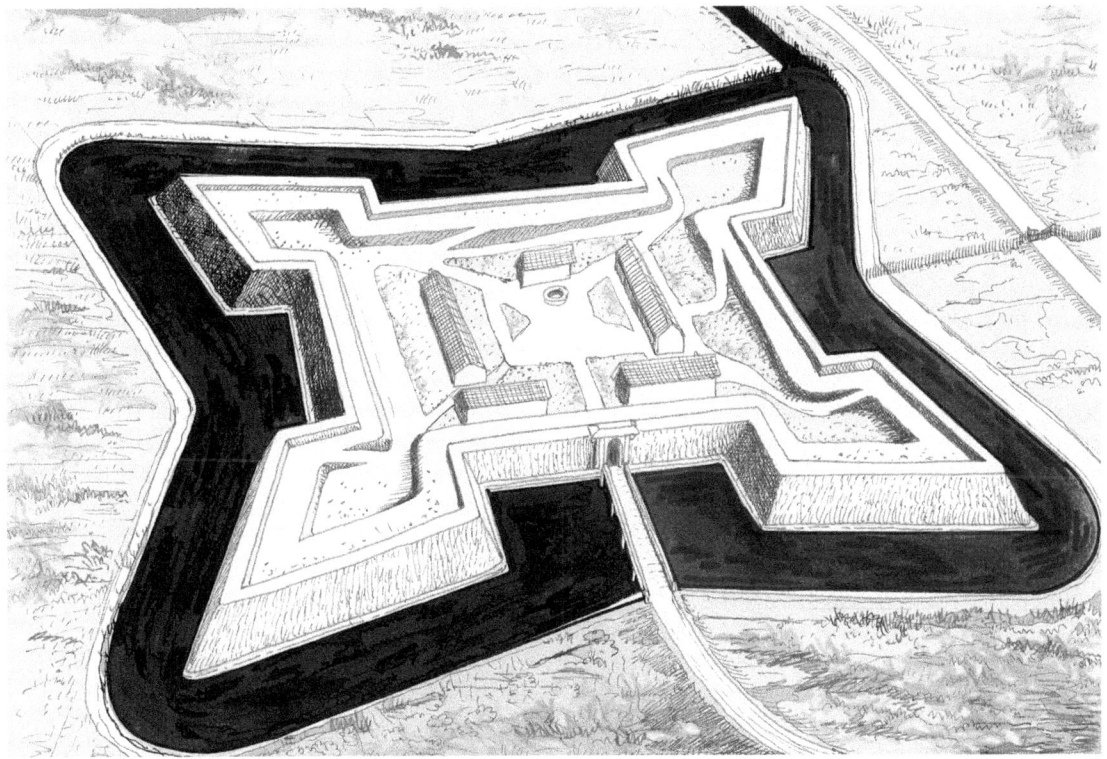

Breebergschans. The Breeberg sconce was built in May 1593 by order of count Willem Lodewijk. As a component of the Frisian Waterline, it was placed in a passage across marshes between Donkerbroek and Duurswolde in the western part of the province of Groningen.

Enumatil. Located west of Groningen city, the Enumatilschans was built in March 1582 by the Spanish army in order to control a bridge crossing a canal called the Hoendiep. The sconce was attacked and captured by the Dutch rebels in 1589. It was retaken by the Spaniards a year later, and re-captured by the Dutch in July 1591. The sconce remained in military use until the end of the 19th century.

fords were vital, and thus required to be held, controlled and defended with small fortlets. Sconces could also be temporarily constructed during a siege.

They had various shapes, but the most frequently encountered was a rectangle with a bastion at each of the corners. Other outlines also existed like a pentagon, a square or a triangle with bastions (or half bastions) at the corners or in the form of a hornwork close at the gorge for example. Irregular shapes were also used when local situations demanded it. Sconces were often built in waterlines between fortified towns, and along rivers and canals for defending fords, passes, dikes, sluices, and passages. When permanently occupied they included barracks for the small garrison, a house for the officers, and various ancillary buildings such as a powderhouse.

Citadel

Also a number of *dwangburchten* (citadels) were built. Indeed not all cities joined the rebellion willingly. At Groningen for example the population was majority Catholic and remained pro-Spanish. In 1594 Prince Maurice and Count Willem Lodewijk besieged and captured the city. When the citizens refused to pay their taxes in 1600, the General States ordered the construction of a pentagonal bastioned citadel to coerce them. In 1607 Groningen-city was incorporated into the Republic for good, the situation was checked, the taxes were accepted and paid, and finally the hated citadel was demolished.

Cityscape

In the late 16th century and first half of the 17th, the new program of fortifications included destruction of discarded elements, as well as design, construction and maintenance of new works and regular payment of garrisons. Although fortifications made of earth were much cheaper than masonry ones, the costs involved were extremely high. Traditionally the expense was to the municipalities, and financed by taxes levied on traders, merchants, and on the citizens themselves and on peasants from the vicinity. The expenses caused by the modernization were so considerable, however, that gradually the provincial authorities and General States had to intervene and arrange national funding for helping towns with limited budgets. Of course cities and towns located along the Republic's frontiers received a lot of such funds for improving their defenses.

While medieval defenses were vertical, bastioned fortifications were characterized by horizontality and depth. They spread out and occupied superficies that were sometimes larger than the land area reserved for inhabitation. As a result of the tremendous proliferation of bastioned fortifications, the shapes of cities and towns (at least the visual appearance on their outskirts) changed a lot with

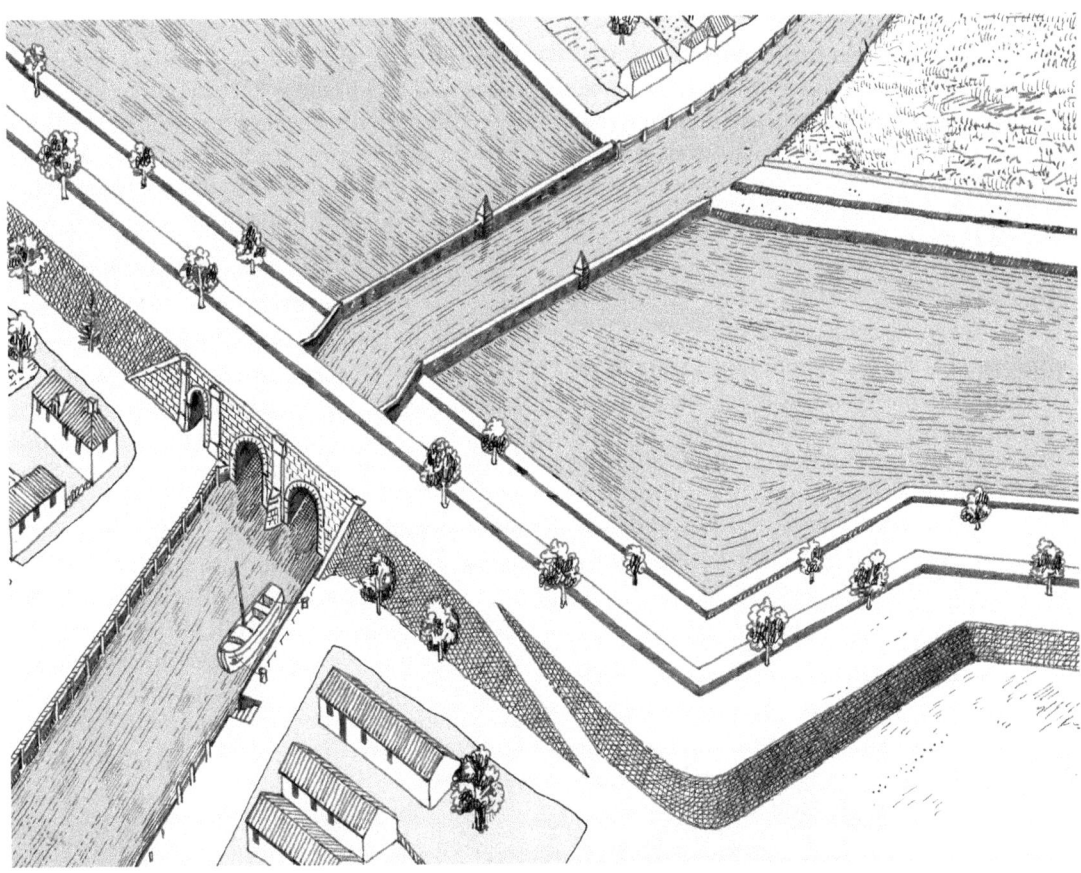

Batardeau. A batardeau was a cofferdam or an embankment or a small dike constructed across a ditch intended to retain water in a wet ditch or moat, or intended to separate dry and wet ditches or to isolate running sea or river-waters from standing moat waters. A batardeau could be hollow and fitted with a gallery in which sluices and watergates could be placed in order to regulate the amount of water in the moat.

Monk. The batardeau was built in masonry crosswise in the ditch and thus formed a dangerous weakness in the defensive system. To prevent its use as a means of crossing the ditch by enemy troop, it was usually made impassable by placing a masonry roundish obstacle called a monk.

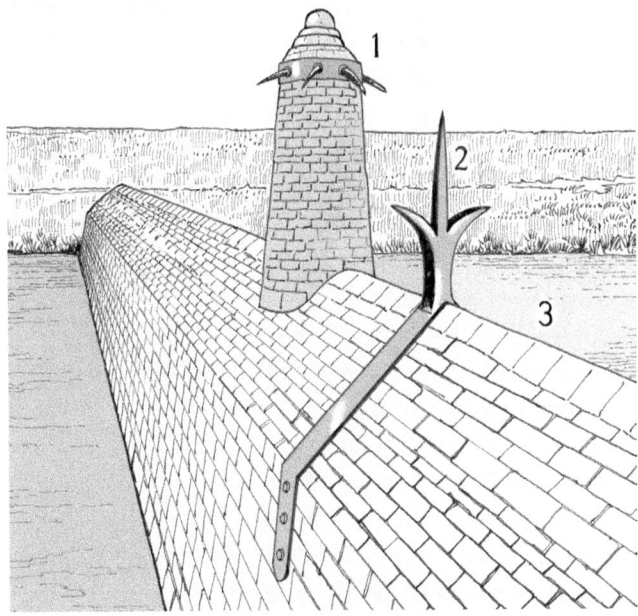

Obstacles on a batardeau included a monk (1), various forms of metal pointed spike-heads (2), and a sharp knife-edge or dos d'âne profile (3).

the establishment of floodable grounds, glacis, detached, advanced and outworks, and eventually citadel, wet ditch and bastioned works. For many Dutch towns these choking arrangements (i.e., excluding any possibility for urban growth) would remain so until 1874 when a royal and parliament law (Vestingwet) was promulgated allowing the dismantlement of the obsolete fortifications.

Dutch Engineers

Hastily improvised at the beginning of the Eighty Years' War, the (Old) Dutch System of fortifications was empiric, and grew up during the war. In the 17th century the ONS became internationally known, and indeed turned out to be a recognized and influential style of military architecture with its own typical rules, specific characteristics, and original features. It was studied, adapted, copied, commented, and codified by many military engineers, adepts, emulators and theorists.

Stevin

Originating from Bruges, Belgium, Simon Stevin (1548–1620) was an influential mathematician and engineer with a broad range of interests. He offered new insights and discoveries in the development of decimal numbers and the laws of inclines, gravity and hydrostatics. In 1577, Stevin occupied an administrative position in the financial department of the government of Flanders. Prior to that, he was employed as a bookkeeper and cashier for the city of Antwerp. Reportedly he travelled extensively through Poland, Prussia, and Norway from 1571 to 1577. Simon Stevin was a Protestant and by 1581 went into exile and joined the rebellion in the North, settling down in the Dutch city of Leiden. In 1594, Stevin developed a name for himself in yet another field, publishing the highly regarded work entitled *Stercktebouwingh* (Fortification).

Simon Stevin's method basically presented a style of fortification inspired by the Italian model, but which in addition offered an ingenious defense strategy using flooding the countryside, a tactic extremely well suited for the water-logged states of the Low Countries. Stevin's theoretical work had a great influence on the development of the Old Dutch system of fortification during the Eighty Years' War and was the inspiration to numerous engineers, notably Adriaan Anthonisz (discussed below). By the end of the 16th century, Stevin came to work for Prince Maurice of Nassau (the influential leader of the Dutch resistance movement) as a private tutor, counselor and surveyor. Stevin was appointed to an engineering position too, and in 1603 promoted to the rank of Quartermaster of the Army of the United Dutch Provinces.

In 1607, Stevin wrote a second treatise on fortification including inundation titled *Nieuwe maniere van Stercktebouw door Spilsluyzen* (New Manners for

Bastioned front by Simon Stevin, who advocated several kinds of bastioned fronts, notably this one featuring bastions with ears and cavaliers.

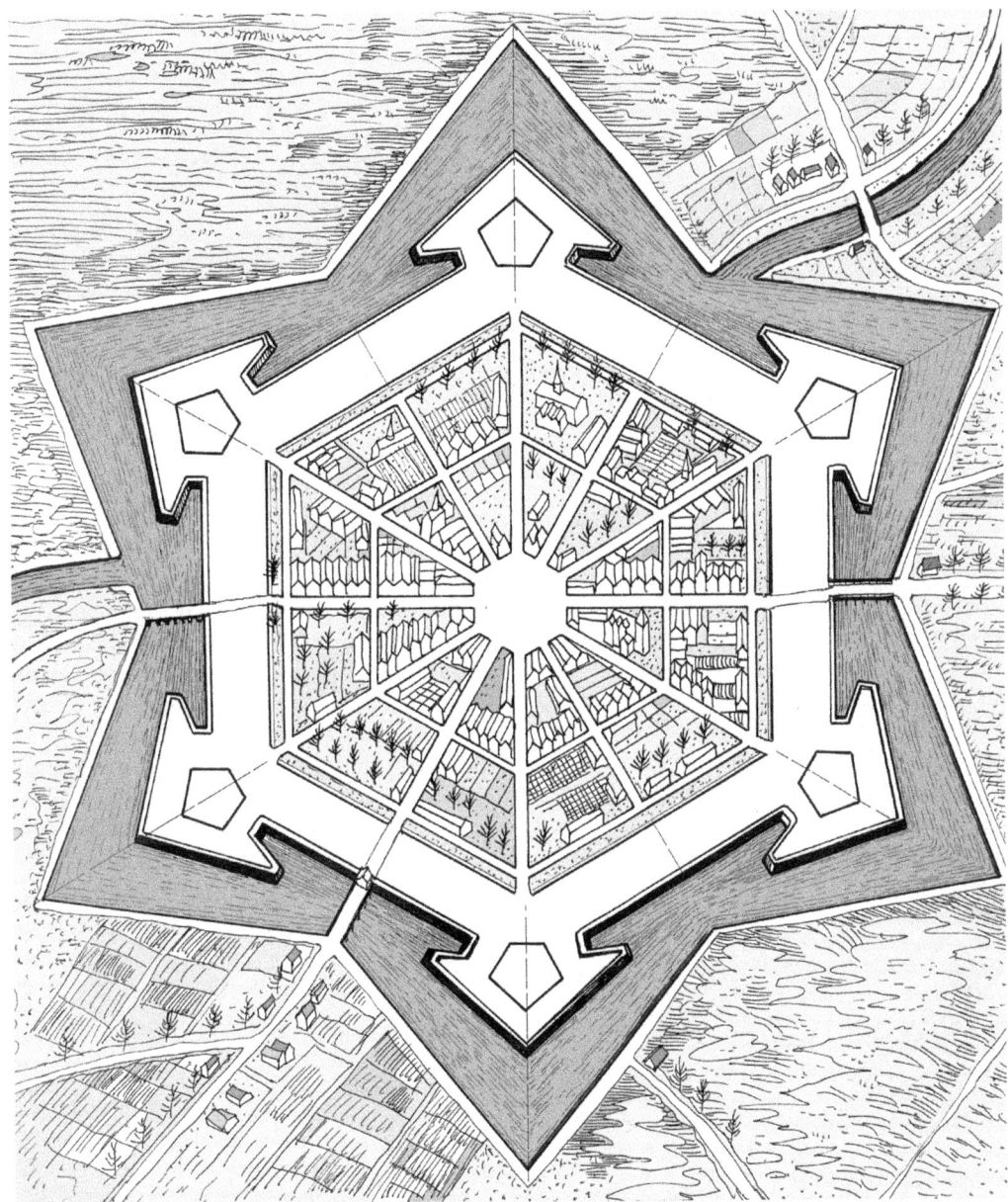

Ideal city designed by Simon Stevin. Among many civil and military engineers, architects, artists, and urbanists of the Renaissance arose the desire to create ideal cities, not focused on a cathedral or appended to a castle but offering an environment for civic life protected by bastioned fortifications. Many hypothetical and theoretical designs based on geometry and symmetry were proposed, but in fact only a few were ever implemented.

Fortifications and Sluices) in which he innovatively bound together siege strategy, tactics and fortifications, creating a scientific axis between them. For the first time Stevin placed the water defenses at the center of a new Dutch national method of fortification that would be used until the 1950s. Prince Maurice wanted to impose order and discipline in his army. He asked Simon Stevin to design a blueprint for siege works and military camps. In 1617 Stevin wrote a treatise called *Castrametario* (Military Camps). In this book, based on ancient Roman organization, Stevin defined where and how military camps should be located, established, organized and accommodated.

Schille

Hans van Schille (1515–1586) was a painter, engraver and cartographer. Originating from Antwerp, he worked for the Duke of Lorraine. In 1573 he published a treatise on fortification titled *Form und Weis zu Bauwen/Manière de bien bâtir* (methods for Good Building). In this book, re-published in 1580, Schilde presented various styles of bastioned fronts, as well as forms and plans of geometrically perfect "ideal cities" with radial urbanism enclosed by Italian styled fortifications. In 1576, Hans van Schille—together with Abraham Andriessen and Peter Franz collaborated to the design of the fortifications of Antwerp.

Bastioned front by Hans van Schille.

Specklin

Daniel Specklin or Speckle (1536–1589), born in Strasburg (today Strasbourg, France), was one of the first German-born theoricians of the bastioned fortification. At Vienna in 1561 he met the military engineer Hermann Schallantzer, and became his pupil. Specklin took part in the siege of Famagusta (Cyprus) in 1571 and travelled in the Low Countries in 1577. From 1564 to 1589, he designed bastioned work in Switzerland and Germany, notably at Ingolstadt in 1575, as well as in Colmar, Strasburg and Ulm.

Shortly before his death in 1589, he published a treatise titled *Architectura von Vestungen*, in which he expressed his conception of the ideal city, and proposed several methods of fortifications. Mainly inspired by Italian systems, Specklin's book became a standard reference, and must be placed in the forefront of the early treatises on military architecture. His work presented a wealth of information depicted with remarkably clear drawings and perspective illustrations, three-dimensional sketches, and breathtaking bird's-eye views combining plan, elevation and section on one sheet giving clarity to complex ideas. Specklin's book had a great influence on 17th century Dutch fortifications.

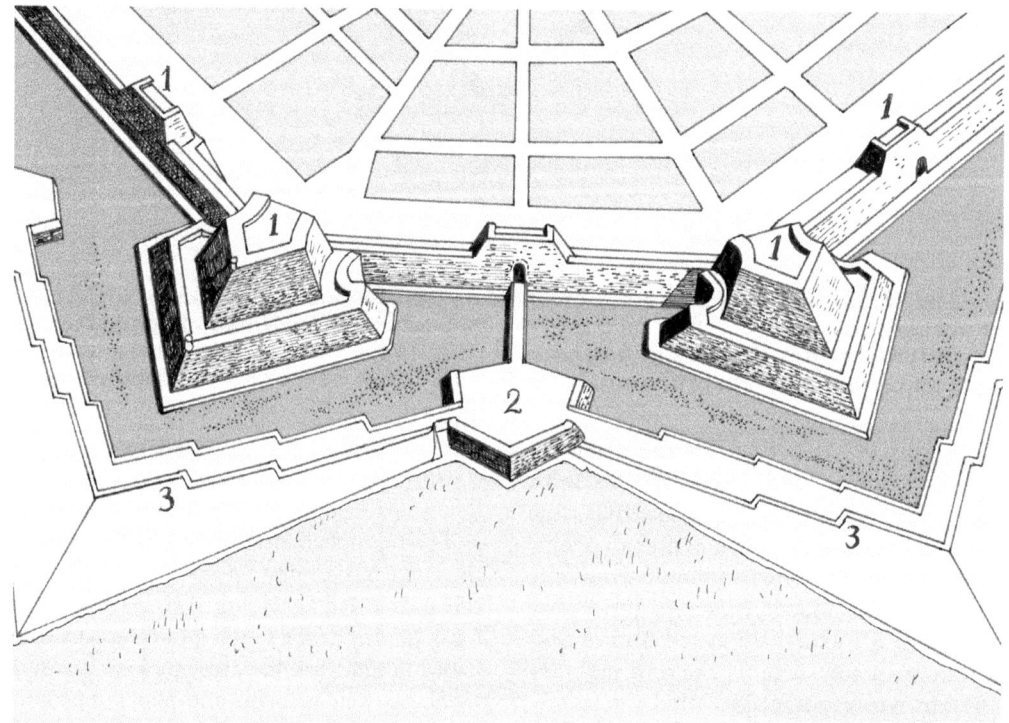

Specklin bastioned front 1589. Note the bulky bastions with ears and cavaliers (1) both placed within the bastions and in the middle of the curtains; the demi-lune (2); and the "saw-toothed" covered way en crémaillère (3) a tracé featuring receding or serrated steps enabling a greater development of flanking fire.

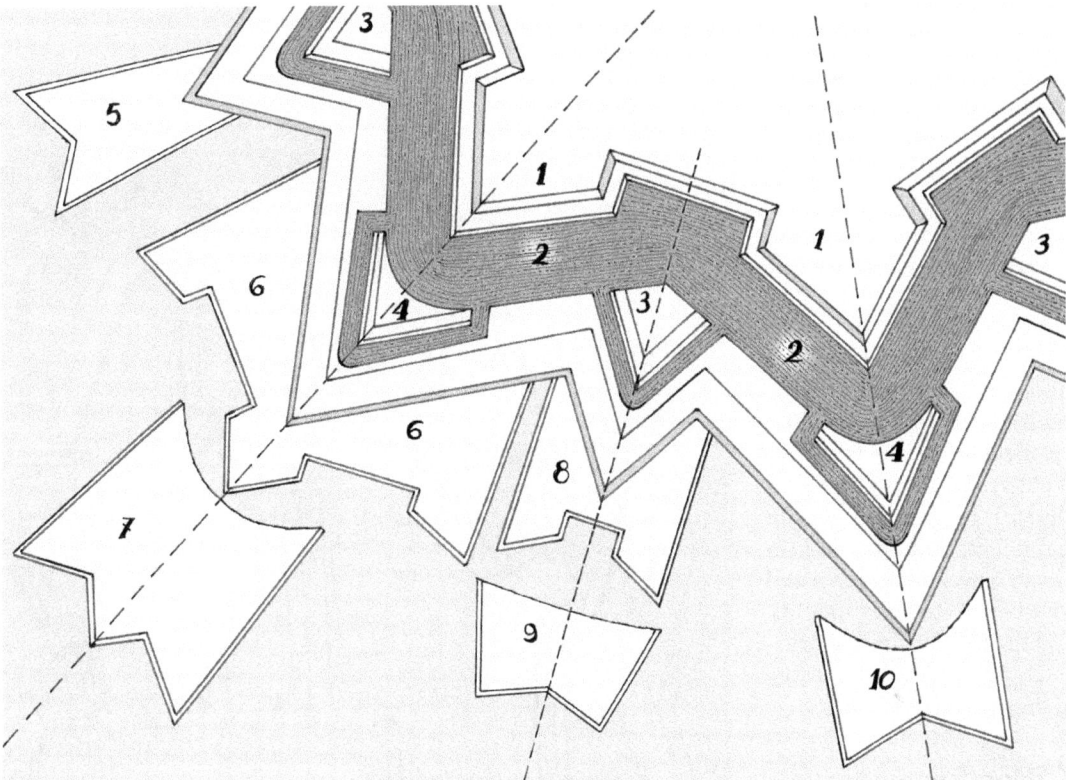

Freitag. Old Dutch fortification with outworks designed by Adam Freitag: 1: Bastioned front with fausse braie. 2: Wet ditch. 3: Ravelin aka demi-lune. 4: Halve maan "Half moon" (actually bonnet or small counterguard). 5: Swallow's tail. 6: Crownwork. 7: Double tenaille. 8: Hornwork. 9: Bishop's miter. 10: Simple tenaille.

Freitag

Adam Freitag (1602–1664), born at Torun in East Prussia, studied medicine at the university of Leiden in the Low Countries. As a surgeon and an engineer, he participated in the sieges of Den Bosch in 1629, and Maastricht in 1632. In 1630 he published *Architectura Militaris nova et aucta, oder Newe vermehrte Fortifikation* (Modern Military Architecture or New Augmented Fortifications) which enjoyed a great popularity, and was regarded as one of the most influential military engineering works of the 17th century.

The lavishly illustrated book was written in German, and eventually translated into French, English and Dutch. It focused on the Dutch manner of building fortification with ramparts and walls covered with earth rather than clad in stone, with a fausse-braie surrounding the scarp, and featuring wide, deep, and wet moats. The author advocated a strict geometrical correspondence between walls, bastions, gates, and the internal layout of the fortress. Freitag's work was mainly theoretical, pedagogic and particularly interesting since it constituted an excellent presentation of the so-called Old Dutch System of fortification, which was made available to a large public.

Marolois

Samuel Marolois (ca. 1572–1627) was a French Protestant mathematician who spent most of his life and career in the Netherlands where he cultivated strong ties with the architect, engineer and painter Vredeman de Vries and with the cartographer, engraver and publisher Hendrik Hondius. He served the Prince of Orange, William the Silent, in his struggle against Spain. Marolois was the author of several works, notably *Géométrie contenant la Théorie & Practique d'icelle nécessaire à la Fortification* (Geometry including theory and practice necessary to fortification) published in 1616, and *Fortification ou Architecture Militaire, tant Offensive que Défensive* (Fortification or military architecture, both offensive and defensive) published in 1615, and translated into Dutch in 1627.

Marolois, together with Adam Freitag was the principal commentator of the Old Dutch bastioned System. His book *Fortification ou Architecture Militaire* reflected a discourse internal to the discipline of military engineer. Marolois's work was a very technical book clearly intended for mature and experienced military engineers. Divided into two main parts (regular and irregular fortifications) it included few theoretical precepts, and few definitions of elements with only text and captions explaining the numerous figures and illustrations. The book was a great success with revised editions in 1628, 1633, 1634, and 1638.

Noyen

Sebastian van Noyen (1493–1557) was a Dutch military architect who entered the service of the King of Spain

Charles V in 1515. In 1550 he built the Brussels palace of cardinal Granvelle, an important edifice in true Italian Renaissance style. After a voyage to Italy, where he probably got acquainted with modern bastioned fortification, he participated in the siege of Metz in 1552, was appointed royal engineer, and designed fortifications for the city of Philippeville (now in Belgium) in 1555.

His nephew, Jacques van Noyen (born ca. 1523), succeeded him, and was appointed master-engineer by the Spanish King Philip II in 1561. Jacques van Noyen inspected fort Rammekens in 1564, and participated, together with Francesco de Marchi, in the design of fortifications at Thionville in 1567. In 1577, Jacques van Noyen turned coat and entered service of the Dutch insurgents, and worked on the fortifications of Brussels, Maastricht, Aalst and various fortified towns in the province of Artois.

Hondius

Hendrik de Hondt—latinized as Hondius (1573–1650)—was a Dutch cartographer and map publisher established at Amsterdam. In 1625 Hondius published a book titled *Description & Brève Déclaration des règles Générales de la Fortification*. Illustrated with beautiful plans and accurate maps, Hondius's treatise ranged from general matters and philosophy to detailed description and minutiae. It dealt with every aspect of fortification, siege warfare, and artillery with the focus on Dutch achievements. Like the works by Stevin, Freitag and Marolois, the theoretical designs advocated by Hondius were regarded as basic examples about the Dutch method of fortifying.

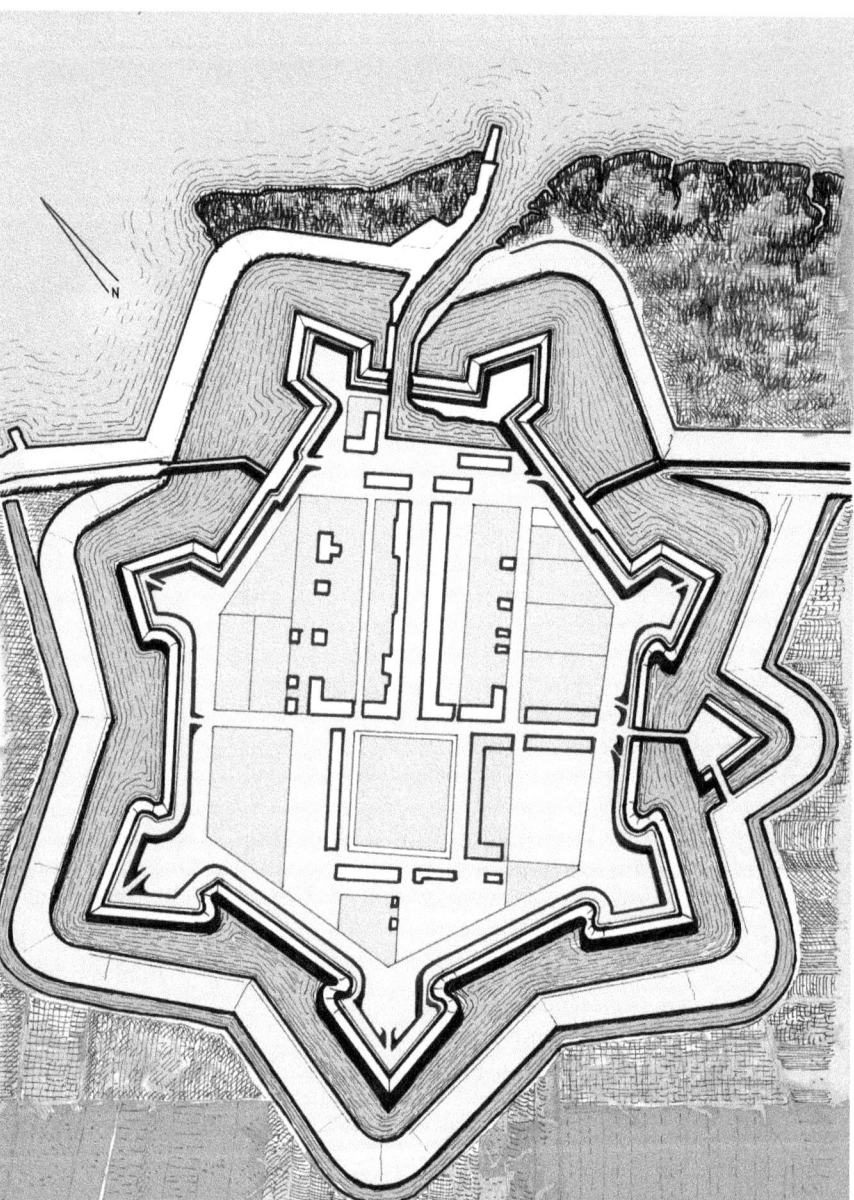

Willemstad. Located in present day Moerdijk on the Hollandse Diep in the province of North Brabant, Willemstad's ONS fortifications were designed by the engineer Abraham Andriesz by order of Prince William I of Orange. Built in the 1580s they are perfectly preserved today, and include seven bastions named after the Seven United Provinces (Holland, Zeeland, Drenthe, Groningen, Gelderland, Utrecht, and Overijssel).

Anthonisz

Little is known of the Dutch surveyor, cartographer and military engineer Adriaan Anthonisz (ca. 1541–1620) as a young man. In 1573, he successfully organized the resistance of Alkmaar then besieged by Spanish troops, and forced the ruthless Duke of Alva to withdraw. This marked the start of a brilliant military career in service of the rebellion and of the Republic of the United Provinces. Anthonisz was the elected burgomaster of Alkmaar from 1582 to 1601; was appointed Master-Engineer in 1578, and Superintendent of the Defenses of the provinces Holland and Utrecht in 1584; and commissioned as General Master of Fortifications of the United Provinces in 1586.

Whereas the works of Simon Stevin, Samuel Marolois and Adam Freitag were purely theoretical in designing and codifying the so-called Old Dutch bastioned system, Adriaan Anthonisz's work was totally practical. Between 1573 and 1607 during the first phase of the Eighty Years' War opposing the Dutch Protestant insurgents to Catholic Spain, Anthonisz was a very busy man. He created, designed, improved and modernized some 30 fortified places—notably Alkmaar (1573 and 1574), Hoorn (1576), Muiden (1577), Utrecht (1577 and 1584), Amsterdam (1578, 1586, 1593 and 1597), Naarden

Zwolle. The capital of the province Overijssel has been inhabited since the Bronze Age. It was a Salian Frankish settlement in Roman times, and a Frisian trading town during the Carolingian era. In August 1230 the bishop of Utrecht granted city rights to the inhabitants, and Zwolle became a member of the Hanseatic League from 1294. Medieval fortifications were established late in the 14th century. Modern ONS bastions and moats were added between 1586 and 1590. In the old city several fortified elements have been preserved until today notably the imposing medieval gatehouse Sassenpoort. The ONS bastioned fortifications are still clearly outlined with canals and moats.

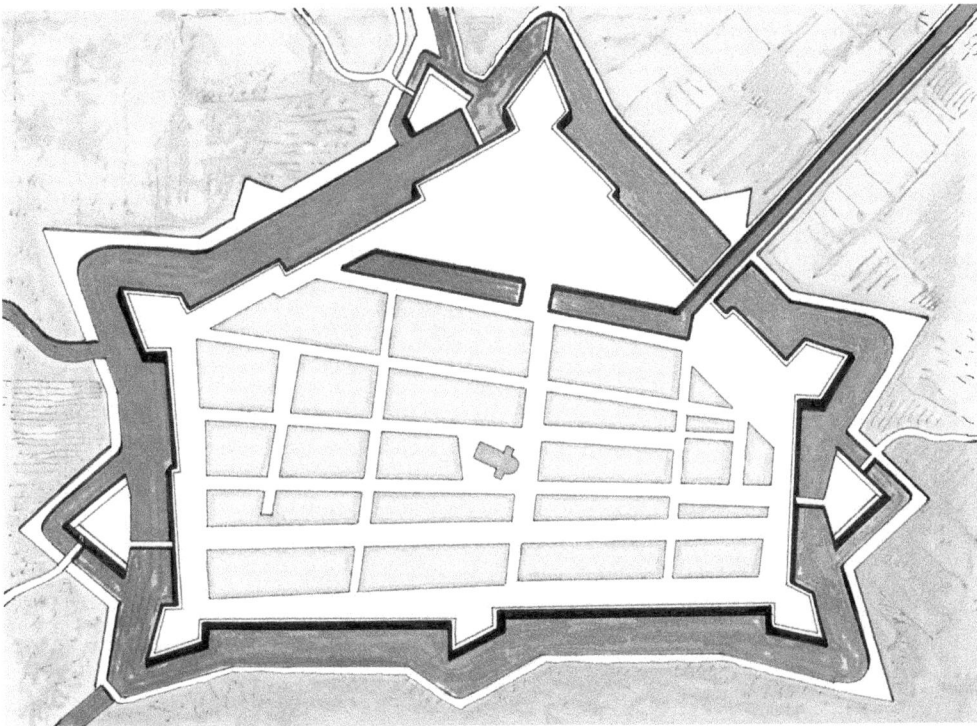

Naarden (province of North Holland) originated in ca. 996 from a small village. With the growth of the power of the Counts of Holland, it became a strategical place as it occupies a sandy ridge (called Utrechtse Heuvelrug) along the Zuiderzee, giving access to the important city and port of Amsterdam. From the Middle Ages the village was thus fortified as an advanced defense for Amsterdam. During the Eighty Years' War the small town was pillaged and most of its population was slaughtered by order of the Duke of Alva in December 1572. In the late 1570s Naarden was reconstructed and fortified by Adrian Anthonisz with six bastions, three ravelins and moats.

Enkhuizen. Situated at the Zuiderzee in the province of North Holland, the town was granted city rights in 1355. In the mid–17th century, Enkhuizen was at the peak of its power and was one of the most important harbor cities in the Netherlands. The bastioned fortifications were designed by Adrian Anthonisz and built in the 1580s.

Heusden. Located on the river Meuse (Maas) in North Brabant, Heusden received city rights in 1318. In 1577 during the Eighty Years' War, the citizens joined the revolt led by William, Prince of Orange. Because of the town's strategic position near the river Meuse, modern ONS fortifications (including bastions, walls, moats, ravelins, envelope and hornworks) were constructed between 1579 and 1597. Although badly damaged during World War II, Heusden has kept most of the 17th century fortifications.

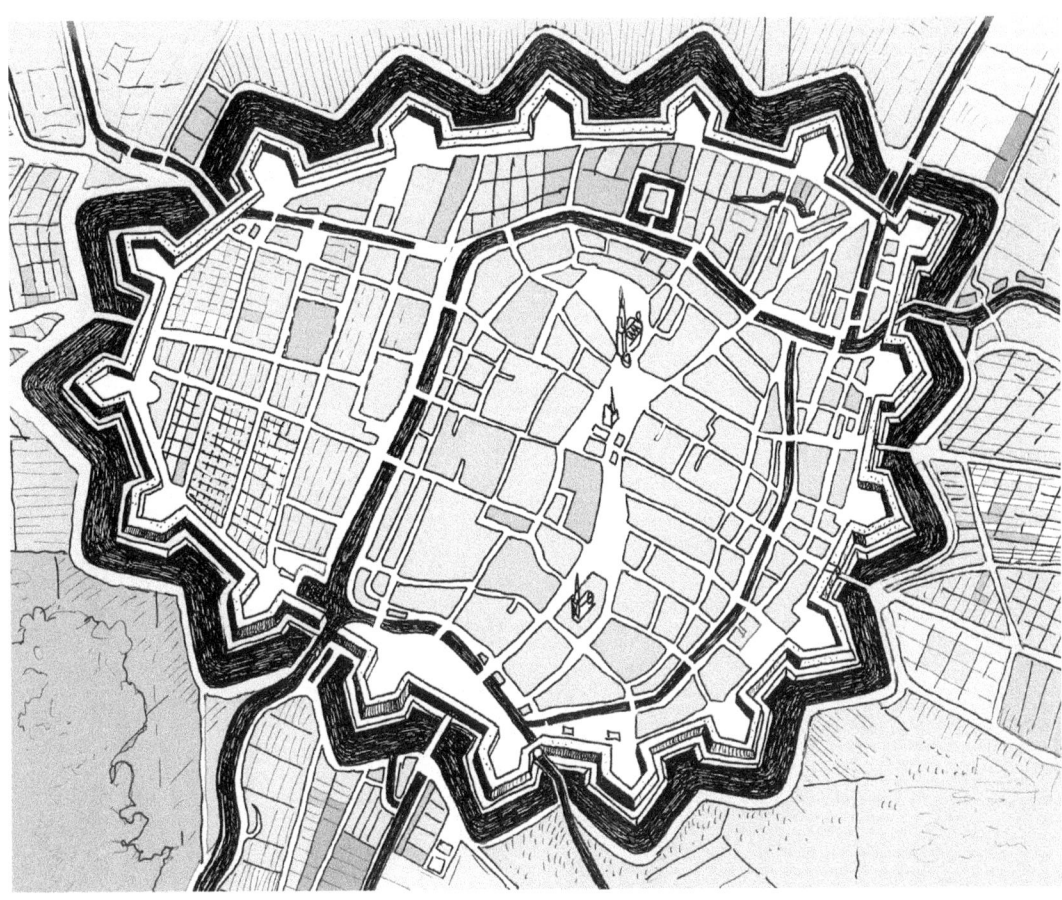

Groningen. The bastioned fortifications of Groningen (the capital of the province of the same name) were designed by master engineers Hillebrandt Smidt and Garwer Peters, and constructed between 1608 and 1624.

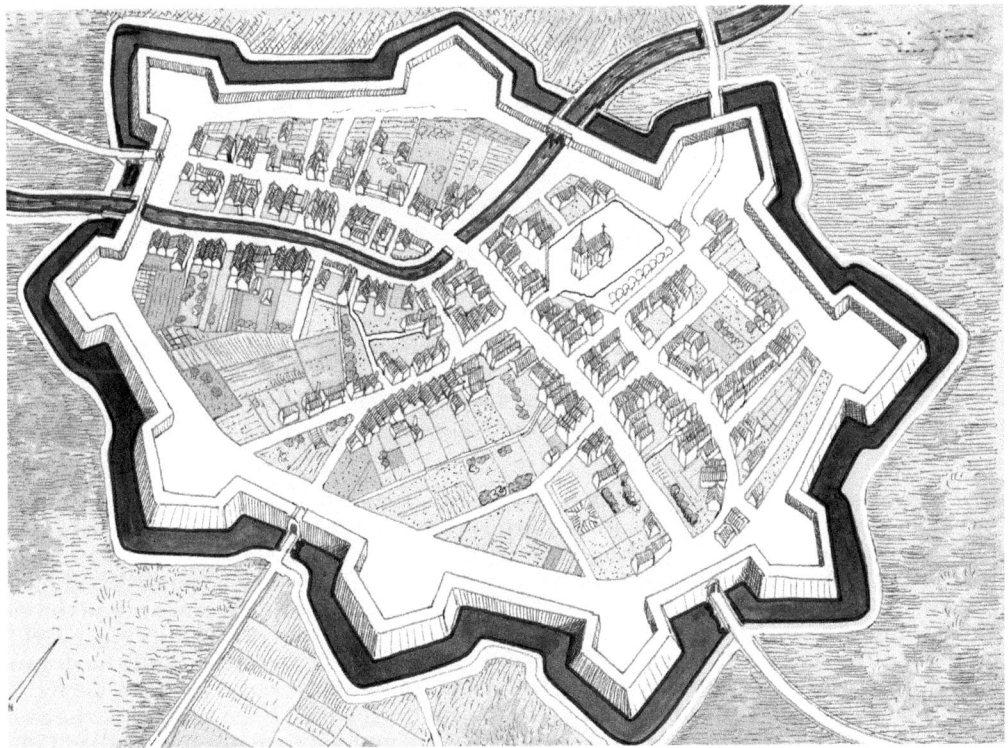

Winschoten. The small town of Winschoten in the northern province of Groningen was fortified in 1593.

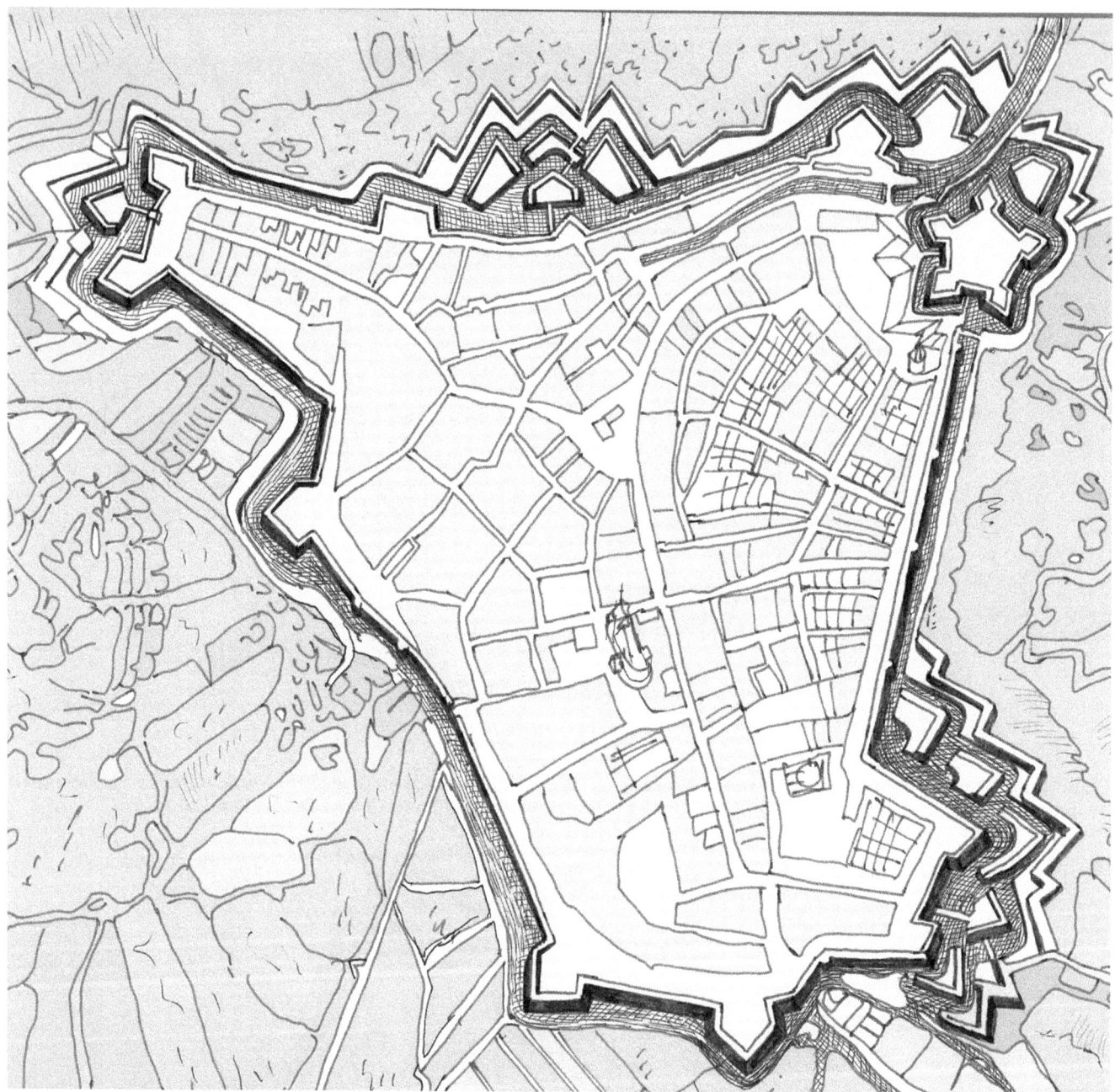

Den Bosch. 's-Hertogenbosch, meaning the Duke's Forest, and often shortened to Den Bosch, is the capital of the province of North Brabant. As an independent bishopric, the city was mainly Catholic and refused the rule of the rebel Calvinist United Provinces. It was vainly besieged several times by Prince Maurice of Orange. The city was finally conquered in 1629 by Prins Frederick Hendrik van Orange.

Opposite: Siege of Den Bosch. The Eighty Years' War saw many spectacular innovations in the domain of fortifications and siege warfare. Lasting from April 30, 1629, until September 14, 1629, the siege of 's-Hertogenbosch (Den Bosch) was an important victory for the Republican Army of the United Provinces led by Prins Frederick Hendrik van Orange (1584–1647). With a force of 24,000 infantry and 4,000 cavalry, Frederick Hendrik had hired 4,000 peasants to divert the two main streams feeding the swamps (the Dommel and the Aa Rivers) around the city by means of a double 40 km dike completely enclosing the city. Thus having created a polder, the Prins's engineers began to drain the grounds with horse-drawn mills. After the soil had sufficiently dried out, trenches could be dug to approach the city walls. Noble visitors from all over Europe visited the siege to admire the novel and spectacular attack method developed by the Prince of Orange. The capture of Den Bosch cut the town off from the rest of the duchy of Brabant and the area was treated by the Republic as an occupation zone without political liberties (called Generality of Brabant). Frederick Hendrik also successfully laid siege to Maastricht in 1632, Breda in 1637, Sas van Gent in 1644, and Hulst in 1645. For his tremendous achievements in siege warfare he gained the nickname of City Conqueror.

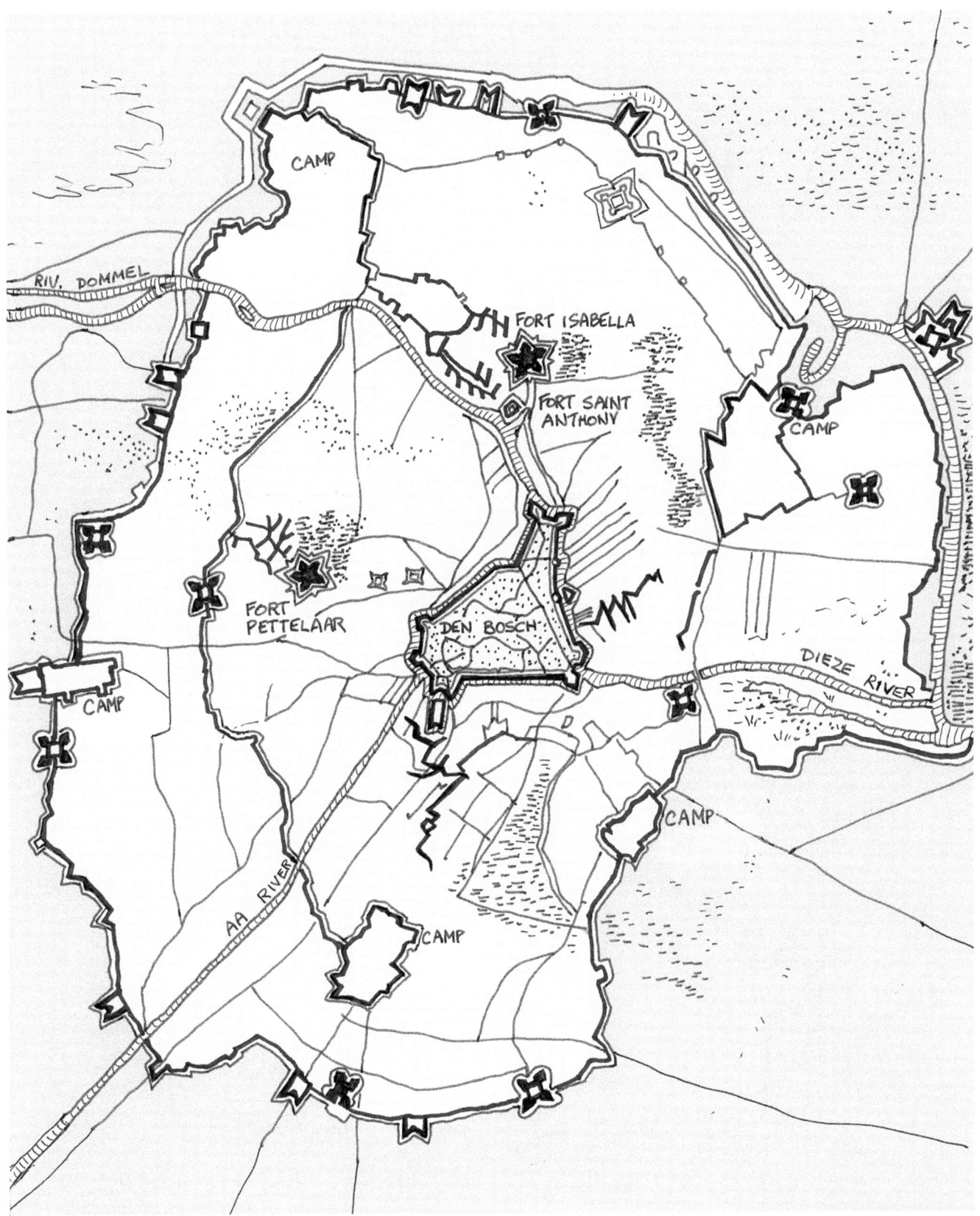

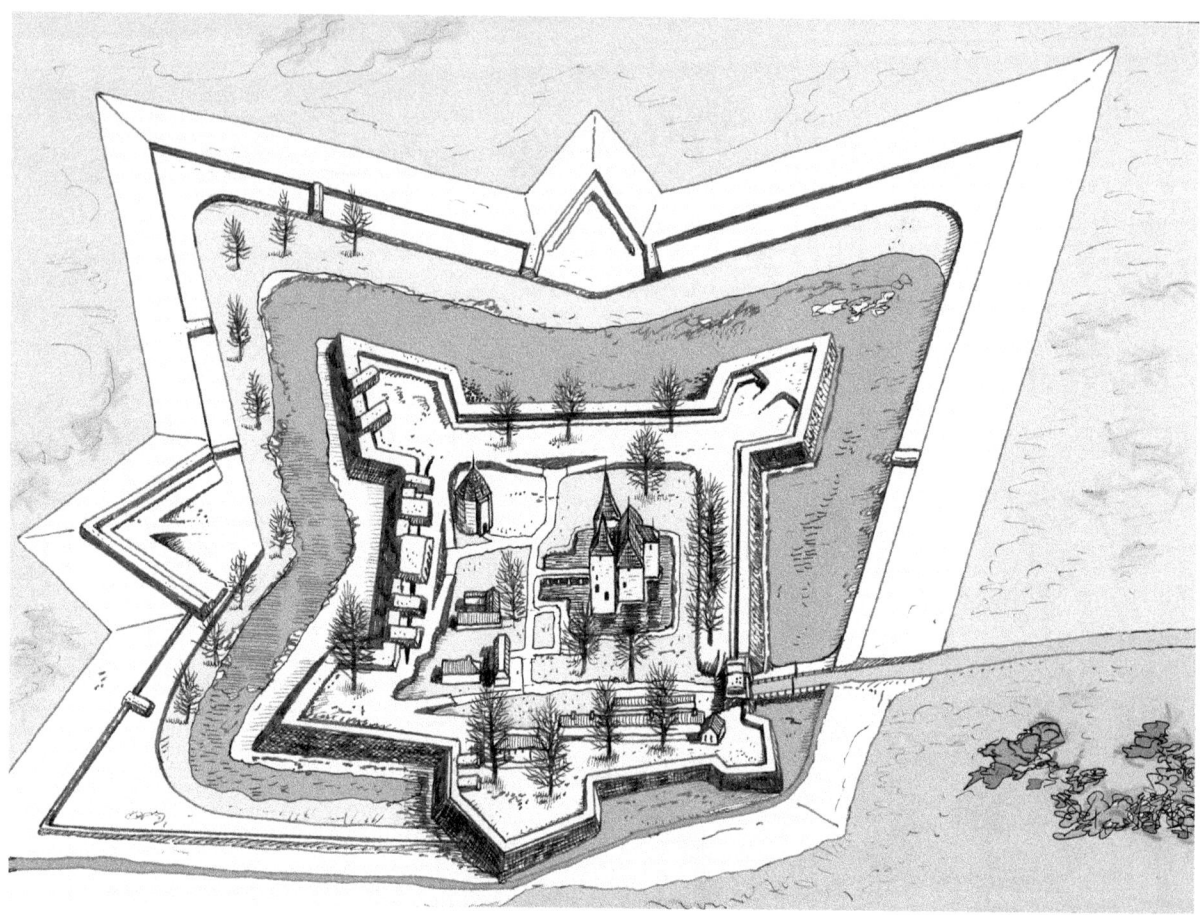

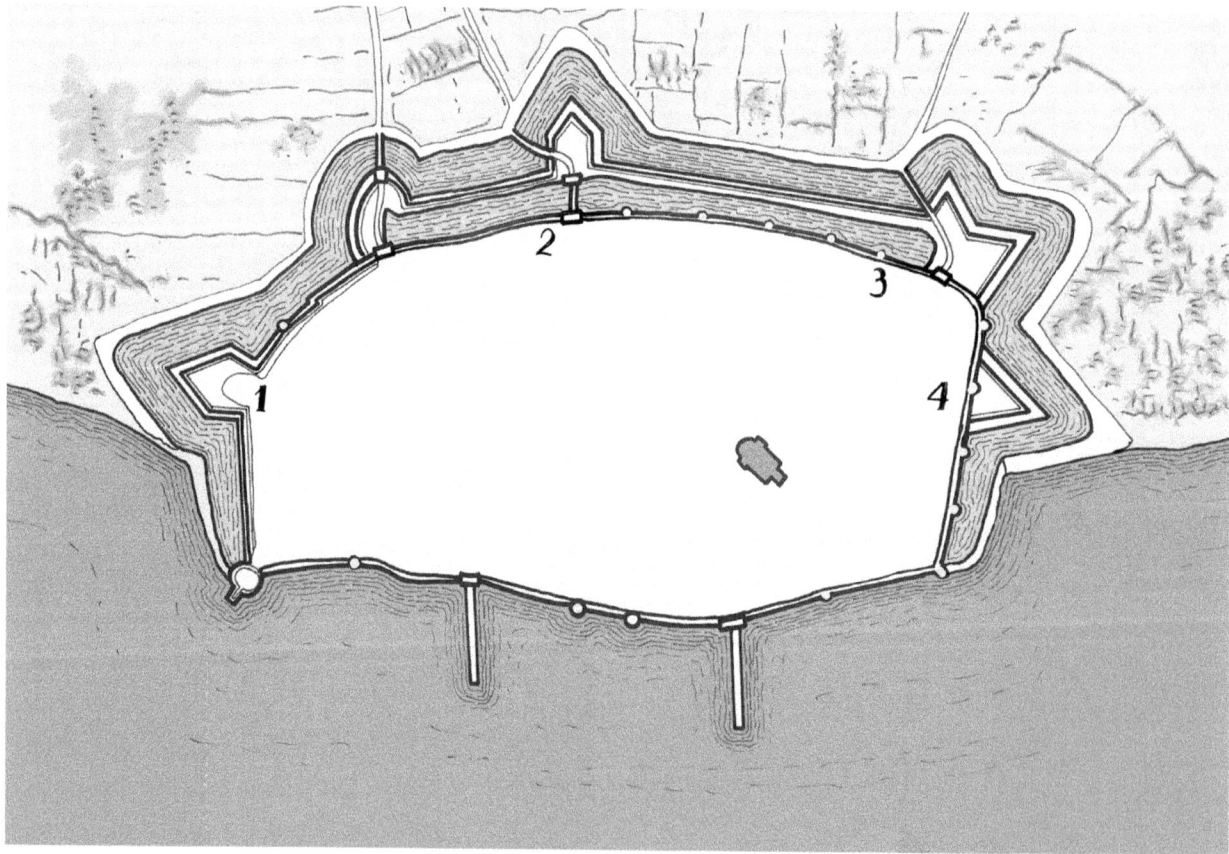

(1579), Kampen (1579), Gorkum (1579 and 1596), Harlingen (1579 and 1580), Coevorden (1580), Stavoren (1581), Hasselt (1583), Tiel (1584), Willemstad (1585), Woudrichem (1585), Goes (1586), Zwolle (1586 and 1590), Hardewijk (1586 and 1597), Enkhuizen (1588), Huisduinen (1588), Vlieland (1588), Schenkenschans (1589), Wijk bij Duurstede (1589), de Vaart bij Vresswijk (1589), Bellingwoude (1593), Bourtange (1593), Doesburg (1593), Amersfoort (1594), Deventer (1594 and 1597) and Zaltbommel (1596).

At first Anthonisz's work was a difficult task as the young Republic of the United Provinces was fighting for its very survival. Indeed Spain was then the greatest power in Europe, the Dutch medieval fortifications were outdated, funds were lacking and experienced engineers were few. Nevertheless, Adriaan Anthonisz proved a flexible and practical engineering man of war. Formed on the field "at the cannon's mouth," he managed to adapt the Italian bastioned system to the limited budget of the Republic, and the specific situation of the Low Countries. Without doubt he is the most important engineer who developed the Old Dutch System using broad ditches filled with water, as well as walls and bastions made of earth instead of expensive masonry.

Adriaan Anthonisz was also a renowned mathematician. He was the first Dutchman to calculate the numerical value of π (pi = circa 3.14159…). In 1589 he published a book on that subject. Since 1997 there is a statue of him by the sculptor John Bier in display near the shopping mall Noorder Arcade in Alkmaar.

Dalem Gate, ca. 1600 (Gorinchem)

Opposite, top: **Loevestein. Located on the Waal River near Zaltbommel in the province of Gelderland, Loevestein castle was built between 1357 and 1368. In the 1570s and the 1580s during the Eighty Years' War, the castle was reinforced with ONS earth walls, bastions, and ditches. By the end of the 18th century the stronghold became a part of the New Holland Waterline. Today the fortifications are still extant. They have been restored and are open to the public. They display a remarkable concentration of Dutch military architecture from the Middle Ages to the 19th century.**

Opposite, top: **Harderwijk. The fortifications of Harderwijk, situated in the province of Overijssel, were improved in 1590 by Adriaan Anthonisz. As the municipality had only a limited budget, Anthonisz made a simple design. On the northeast wall he had the existing artillery masonry roundel reinforced with a bastion (1) made of earth. A bastion (2) was built before the Saint Lucas gate. A tenaille (3) was placed in front of the Smee gate. The southwest wall was reinforced with a redan (4).**

Miscellaneous engineers

Of course, Adriaan Anthonisz was not the only engineer serving the Dutch Republic. He had many collaborators and colleagues, for example Jacob Kemp, who is known to have worked at Heusden, Gorkum, Arnhem, Breda, Deventer, Den Bosch.

Mattheus Joost, another colleague of Adriaan Anthonisz, is known to have worked at Steenwijk, Reide en Delfzijl.

Johan van Rijswijk (d. 1625) was a Dutch military engineer, and a colleague of Adriaan Anthonisz. During the period 1586 to 1600 he worked in Zeeland and Flanders notably at Grave, Bergen op Zoom, Woudrichem, Liefskenshoek, Oostende, Veere, and Hulst. Later he worked in Germany at Lippe, Bremen, Lübeck, Hamburg, and Ulm. Hildebrandt Smidt is known to have worked at the fortifications of Groningen in the north of the Netherlands in the 1620s. The military engineer Johan van Valckenburgh (1575–1625) entered the service of the Republiek der Zeven Verenigde Provincien in 1609. He is known to have worked in Germany at Lübeck, Emden, Bremen, Magdeburg, Braunschweig, Lüneburg, Rostock and Ulm. His main work was at the fortifications of Hamburg in the period 1619 to 1625. Johan van den Kornput (1542–1611) was a Dutch military engineer. Involved in the rebellion against Spain, he is known to have worked at the fortifications of Steenwijk and Delfzijl in 1580 and 1581. He died at Groningen.

Defensive Waterlines

After the proclamation of the Union of Utrecht that marked the foundation of the Republic of the Seven United Provinces, a national feeling appeared in the northern Low Countries. As the new Republic fought a war for survival and independence, an embryonic national system of defense was started. At first this system (based on flooding and fortifications) was adopted on a local and regional scale, but it developed as a distinctive Dutch national hallmark until the so-called Cold War in the 1950s.

As noted, Dutch engineers were quick to realize that flooding low lying areas formed an excellent defense against advancing enemy troops. Military flooding presented many advantages. On the larger scale it was very profitable because it necessitated fortifications and defensive troops only on high and dry grounds. On a smaller scale it made impossible the establishment and digging of entrenched approaches to lay siege to a place. It also made impossible the use of underground mines. To be effective, flooding had to be high enough in order to make impossible the progression of foot soldiers, horsemen, artillery trains and supply wagons. On the other hand it had to be low enough to prevent navigation with a maximum level of 60 cm to 1 meter (2 to 3 feet).

As can easily be imagined, the use of military tactical inundation

Veerpoort Gate, ca. 1601 (Schoonhoven)

required immense technical knowledge and important hydrologic installations in order to dominate and control such a dangerous element as water. The Eighty Years' War saw the establishment of the so-called defensive waterlines, combining military areas purposely flooded with fortifications (forts, sconces, and fortified towns) on the accesses, the passages through the inundations, as well as the high grounds and places that could not be flooded. Several regional waterlines were established during the long war against Spain. There is little point in describing them all; suffice it here to examine two of them.

Frisian waterline

In the north of the Republic, the building of the Friese Waterlinie (Frisian Waterline) began around 1580. The province of Frisia was by nature protected by three large bodies of water: the Zuiderzee in the West, the Waddenzee in the north, and the Lauwerszee (and many swamps) in the East. The Frisian Waterline was intended to form a barrier in the south and southeast. It included rivers, canals, lakes, and swamps, as well as purposely flooded zones, schansen (sconces), forts, and vestingen (fortified towns). It began at the Zuiderzee, running along the river Linde until Blessebrug, and that part was called the Tjonger-Lindelinie. The Frisian Waterline continued in a northerly direction to Kuinre along Wolvera, Heerenveen, Terband, Gorredijk, Donkerbroek, and Bakkeveen up to Frieschepalen. The line included inundations and many sconces, notably Leeksterschans, Sterrenschans (aka Bakkeveenschans), Zwartendijksterschans, Schans Frieschepalen, Makkingaasterschans, Bekhofschans, Kuinderschans, Sliekenborgschans, Blessebergeschans, Tolbrugschans, and Breebergschans.

In those strongholds soldiers could be garrisoned, and could patrol and watch the countryside, and defend the passages.

In the northeast, a part of the Frisian defense line was connected with another system: the Groninger Waterlinie. This ran from the city of Groningen up to the port of Delfzijl on the coast of the Wadden Sea. These defenses were used during the Eighty Years' War against Spain. The Frisian and Groningen Waterlines were also extremely useful during the 1672 Rampjaar (the so-called Disaster Year) when the German bishop of Münster Bernard von Galen (alias Bommend Berend) attacked the province of Groningen.

West Brabant waterline

In the south of the Republic, working on the West Brabant Waterline started in 1628. The line was intended to protect the island of Tholen and Schouwen-Duiveland in the province of Zeeland. It ran from the south at the *vesting* (heavily fortified city) Bergen op Zoom, and continued in northern direction to Steenbergen. It included swamps, canals, rivers, flooded zones, and the following strongholds: Redoubt Bieijenburg; Fort Moermont; Fort Pinssen; Fort De Roovere; and Fort Henricus. The southern part of that waterline was linked to another defensive system running along the island of Zuid-Beveland and the Hont—the large estuary and mouth of the Scheldt River leading to the port of Antwerp.

Flooding or inundation was the controlled submersion of more or less broad zones in order to create inaccessibility in moments of crisis.

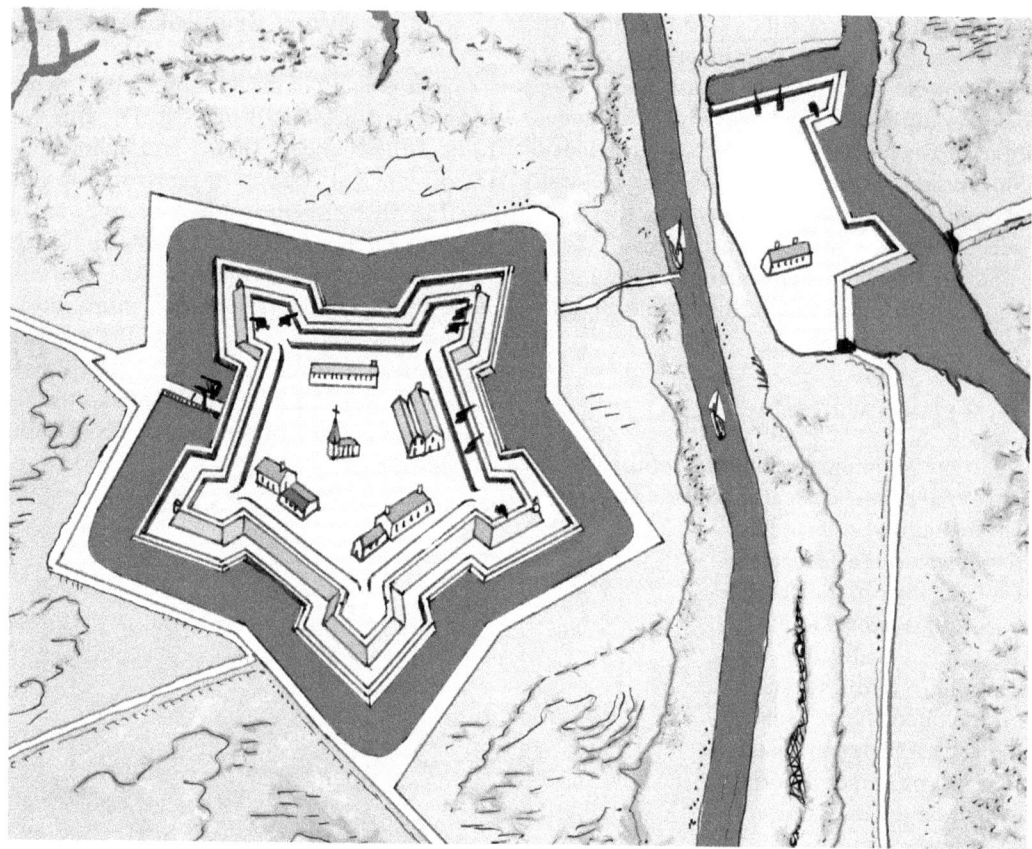

Fort Henricus. Located north of Steenbergen in Noord Brabant, Fort Henricus was a part of the West Brabant Waterline. Built in 1627 it was an important stronghold intended to protect the city and harbor of Steenbergen, the accesses on the dikes, and a system of sluices using the water of the Volkerak sea arm for regulating flooding in the neighboring polders. After 1816 the fort was abandoned and neglected until 2010, when it was restored for ecological, recreation and tourism purposes.

The Influence of Dutch ONS Fortifications in Northern Europe

The successful achievements of the Republic of the United Provinces attracted a great deal of attention from northern European lands including England, Scandinavia, and northern Germany. For many princes, kings and rulers the modern military reforms introduced by Prince Maurits and his successors served as a model. The engineering college at Leiden (founded in 1575) was regularly attended by foreign students. The Dutch Eighty Years' Wars against Spain was considered a kind of practical military academy for young European noblemen who wanted to make a career as army leader and military engineer.

The rather simple, and relatively cheap Old Dutch System of fortification acquired a good reputation. There were plenty of places to fortify in northern Europe where kings had many projects and huge budgets. The expertise of Dutch (Protestant) military engineers was welcome. The ONS was widely copied, and used until the 1670s.

Britain

There were close ties (notably strong anti–Spanish, and anti–Catholic feeling) between Anglican Elizabethan Tudor Britain and the Protestant Republic of the United Provinces during the Dutch War of Independence. This was reflected by flourishing trade, as well as mutual political and military aid. Actually the Dutch revolt was openly approved, encouraged and partly supported by Anglican England and the French Protestant Huguenot party.

In Britain the Dutch ONS was imported and applied

Opposite, top: **Fort Tilbury.** Designed by Bernard De Gomme in 1670, Fort Tilbury is located upon the Thames River near London. Although often modified and modernized, it is still today a fine example of Old Dutch fortification. It was constructed following the Old Dutch System including a regular pentagon with five bastions and ravelins, and a fausse-braie enlarged as a coastal gun battery directed toward the Thames River. The fort still exists, is made of earth works with very little masonry. It is surrounded by a wet ditch and a flooded area in typical Dutch manner. Fort Tilbury is now operated by the heritage agency English Heritage as a tourist attraction.
 Opposite, bottom: **Gosport** (*left*) **and Portsmouth** (*right*) **in the 1660s.** In 1660 Sir Bernard de Gomme became responsible for a major rebuilding of fortifications in Britain.

Chapter 4. The Eighty Years' War

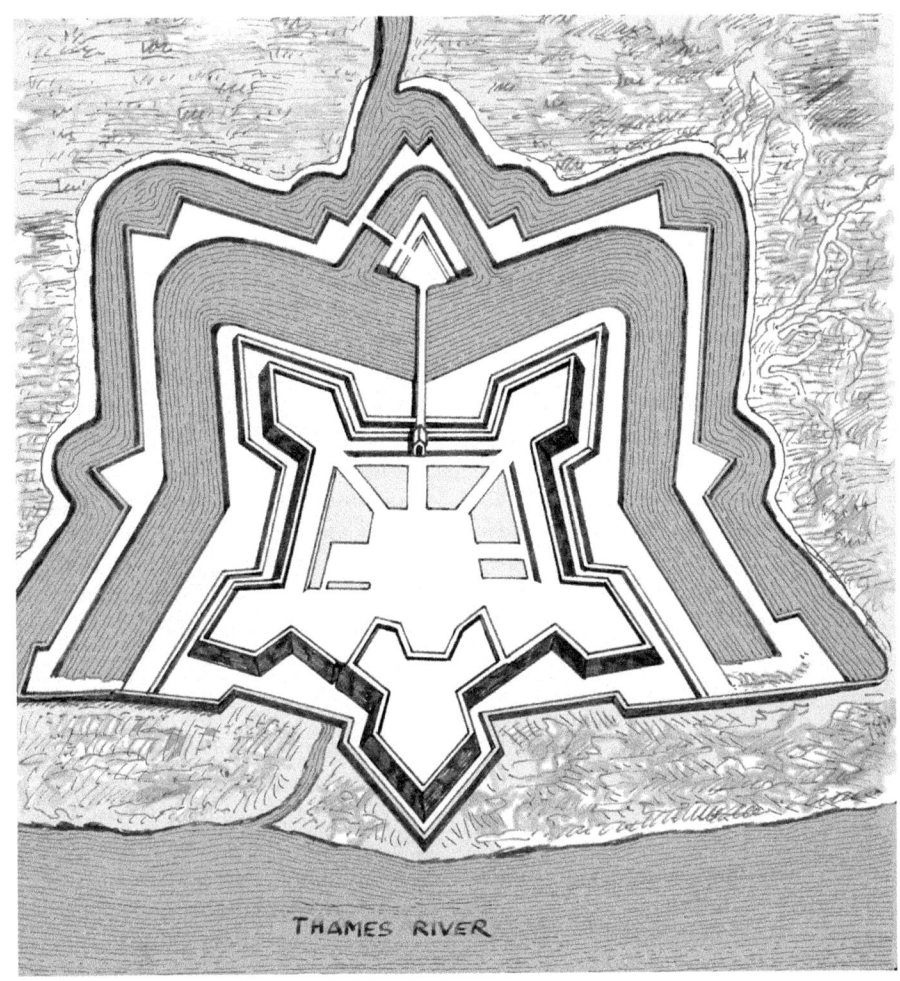

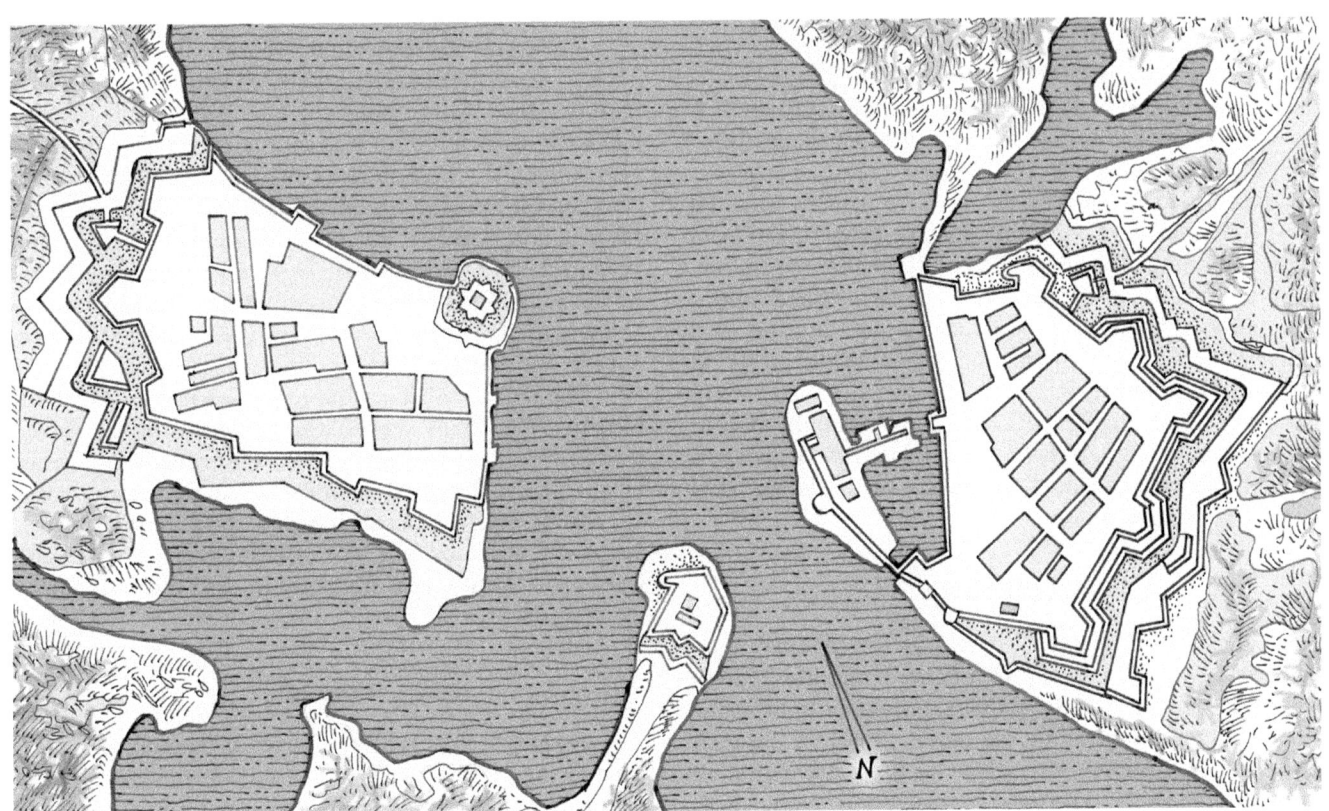

especially because of a Dutch military engineer named Bernard de Gomme (1620–1685). In his youth de Gomme had served in the campaigns of Frederik Hendrik, Prince of Orange notably in the Gennep operation of 1641. He accompanied Prince Rupert to England, and was knighted by Charles I. He served with conspicuous ability in the royalist army as engineer and quartermaster-general from June 1642 to May 1646, designing notably the fortifications of Liverpool. De Gomme left England after the 1646 defeats of the first English Civil War, and returned to the Netherlands, where he worked as one of several civil engineers at the construction of polders in Flanders. In June 1649, De Gomme received a commission from Charles II, then at Breda, to be quartermaster-general of all forces to be raised in England and Wales. He took part in the Battle of the Dunes near Dunkirk in 1658, and after the English Restoration returned to England, and was appointed Surveyor-General of Fortifications in 1660.

Among his first tasks were the repairs of Dover pier, the construction of fortifications at Dunkirk, and the surveying of Tilbury Fort. In 1665 he made designs for the fortifications at Portsmouth, and was commissioned to build a new citadel at Plymouth. In 1667, he gave advice for fortifying the Medway, Portsmouth, and Harwich. In 1673 and 1675 he was making surveys about Dublin. The highly regarded De Gomme died in November 1685, and was buried in the chapel of the Tower of London.

Of course, Bernard De Gomme was not the only engineer that brought and used the ONS in Britain.

John Rosworm (ca. 1630–1660) was originally a Dutch or German soldier and military engineer who served the Parliamentarian cause during the English Civil War. He fortified Manchester in 1642, Preston in 1643, Liverpool in 1643, and Yarmouth in 1651. After being appointed Army engineer-general in 1659, there is no further record of him. It is thought that he died in exile following the 1660 Restoration.

The British engineer Robert Ward published in 1639 a treatise entitled *Animadversions of Warre*, in many ways the most comprehensive book available to engineering officers at the beginning of the English Civil War (1642–51). Largely inspired by Dutch warfare, Ward's book included theory and design of fortification, with analysis on the attack and defense of places, and instructions on use of underground mines and countermines.

The British engineer Richard Norwood published in 1640 a work titled *Fortification or Architecture Military* in which he provided detailed instructions on the designing of Dutch-styled fortifications.

Scandinavia

Dutch engineers were also welcome in Scandinavia. As Sweden grew to be a great power in the 17th century, there were frequent wars in the Baltic region, and armed conflicts were common along the borders between Sweden and Denmark-Norway. Easy invasions routes from Sweden were fortified on the Danish-Norwegian border with new or upgraded fortresses during this period, effectively establishing the modern borders between the two countries.

An important fortification project was started in Kalmar (Sweden) in 1613–14 conducted by the Dutchman Andries Sersanders.

The military engineer Johan Sems worked in 1616 on the defenses of Christianhaven near Copenhagen, Denmark, and designed modern fortifications around the newly created city of Kristianstad in southern Norway in 1617.

Gothenburg in Sweden was fortified by several Dutch engineers (including Abraham de la Haye) in the 1610s. Later in 1653 the Dutchman Isaac van Geelkercken designed fortifications for Frederikstad (south of Oslo) in Norway.

The Dutch-born Anton Coucheron (1650–1689) played an important role in the history of Danish and Swedish fortifications. Anton Coucheron, together with his colleague Johan Caspar de Cicignon, played a major role in the fortification of the borders in both lands, notably Christiansen and Trontheim in 1681, Bornholm in 1684, Frederikshaven in 1687, and Kronborg in 1688.

Germany

While the Dutch were fighting for their independence, there was another conflict that raged in Germany and Central Europe: the so-called Thirty Years' War that lasted from 1618 to 1648. The causes, involvement of foreign nations, development, vicissitudes, and outcome of this war are beyond the limits of this study. However the war in Germany had great consequences in the Netherlands. One of these repercussions was a proliferation of fortifications, and therefore another fruitful market for Dutch military engineers.

A new port was founded in the 1610s at Glückstadt-upon-Elbe near Hamburg (north Germany) with fortifications designed by the Dutch engineer Abraham de la Haye. Johan van Rijswijk (d. in 1625) was a Dutch military engineer during the Eighty Years' War. As a collaborator of Adriaan Anthonisz in the period 1586 to 1600 he worked in Zealand and Flanders, notably at Grave, Bergen op Zoom, Woudrichem, Liefskenshoek, Oostende, Veere and Hulst. At the beginning of the 17th century he went to Germany and designed fortifications for the cities of Lippe, Bremen, Lübeck, and Hamburg among others.

The Dutch military engineer Johan van Valckenburgh (1575–1625) served the Republic of the Seven United Provinces during the War of Independence against Spain. He also designed fortifications in Germany notably at Lübeck, Emden, Bremen, Magdeburg, Braunschweig, Lüneburg, Rostock and Ulm. His most important work was the fortifications of Hamburg built in the period from 1619 to 1625.

Poland

The Dutch System was also used as far away as Poland, where King Wladyslaw IV (1595–1648) invited such Dutch engineers as Getkant, Tylman van Gameren and Beauplan, and encouraged Polish engineers Grodzicki and Arciszewski to study and improve their knowledge abroad. The King had a great interest in fortifications and the rather cheap and efficient Old Dutch System was quite popular. Between 1620 and 1648 the king ordered the construction of sconces and forts like Kudak, Pilica and Sluck; small fortresses, fortified military camps, border defenses like Dreszdenko, Wisnicz and Landcut; and fortified cities—some with a citadel, like Gdansk, Torun, Wroclaw, Szczecin, Grodziek, Brody, Brodnica, Wladislawowo, Kazimierzowo, and Czestochowa.

Dutch Fortifications in the Colonies

Dutch colonies and trading posts

As the European countries were starting to build their colonial empires, the conflicts between them extended to colonies as well.

Even before the end of the War of Independence against Spain in 1648, the Republic of the United Provinces started to create a colonial empire. Despite the tremendous costs, destructions and hardship caused at home, the Dutch continued expansion on the seas and discovery of new continents and new sea routes. Indeed the war with Spain had greatly disturbed the sea trading lanes, but the Dutch rapidly developed their own commercial transport fleet and launched daring exploration expeditions to find new sea routes and establish their own trading posts on a global scale.

Dutch shipbuilding skills were the most advanced of the time. This enabled them to gain mastery of the oceans, and to build up the largest trading empire until it was surpassed by the British in the 18th and 19th centuries. The Dutch had the world's biggest shipyards, and more money passing through Amsterdam than any other city in the world.

The Verenigde Oostindische Compagnie (VOC, United East India Company) was created in March 1602. The very wealthy company had the monopoly of trade between Holland and Asia during the 17th and 18th centuries until it was dismantled in March 1798.

By the mid–17th century, the tiny Dutch Republic had become Europe's economic powerhouse. With new, innovative sailing ship designs like the *fluyt*, new capitalist economic arrangements like the joint-stock company taking root, and a good understanding and management of military matters, Dutch commercial interests were expanding

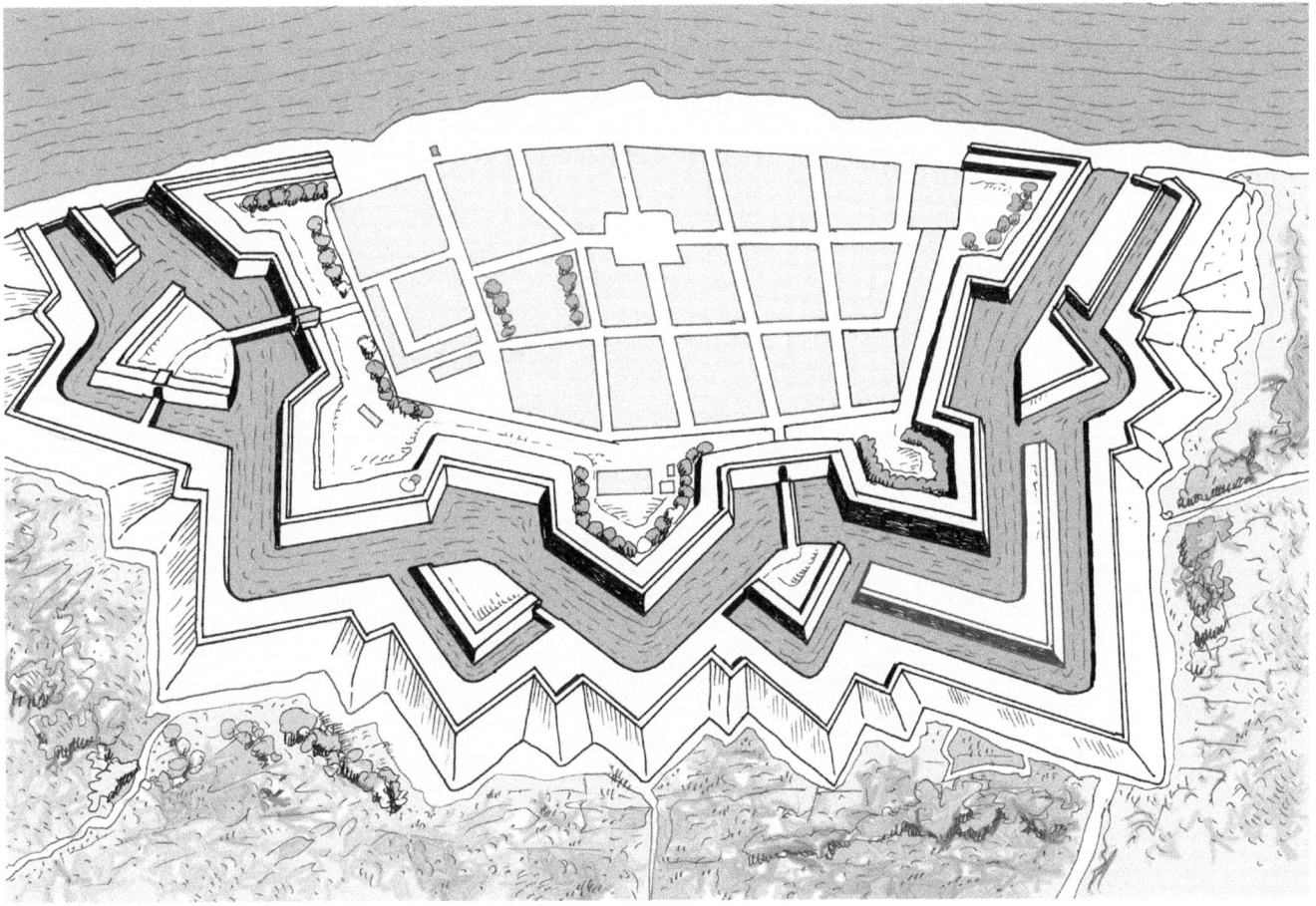

Fredrikstad. Situated near the Ford of Oslo (Norway) on the Glomma River, the city was founded in 1567 by order of King Fredrik II as a defense against neighboring Sweden. In the 1650s the ONS fortifications of the "Gamlebyen" (the Old City) were modernized by the engineers Coucheron and Cicignon. Today the fortifications are well preserved.

Copenhagen. The capital city of Denmark originated from a small Viking fishing village established in the 10th century. During the long reign of King Christian IV between 1588 and 1648, Copenhagen grew dramatically. Inspired by Dutch planning, the king sponsored the development of the district and port of Christianshavn with canals and fortifications. In 1658–59 the city was besieged by the Swedes under Charles X, and successfully repelled a major attack. After 1661, Copenhagen had asserted its position as capital of Denmark. The defenses were further enhanced with the completion of new bastioned ramparts and a citadel in the 1660s.

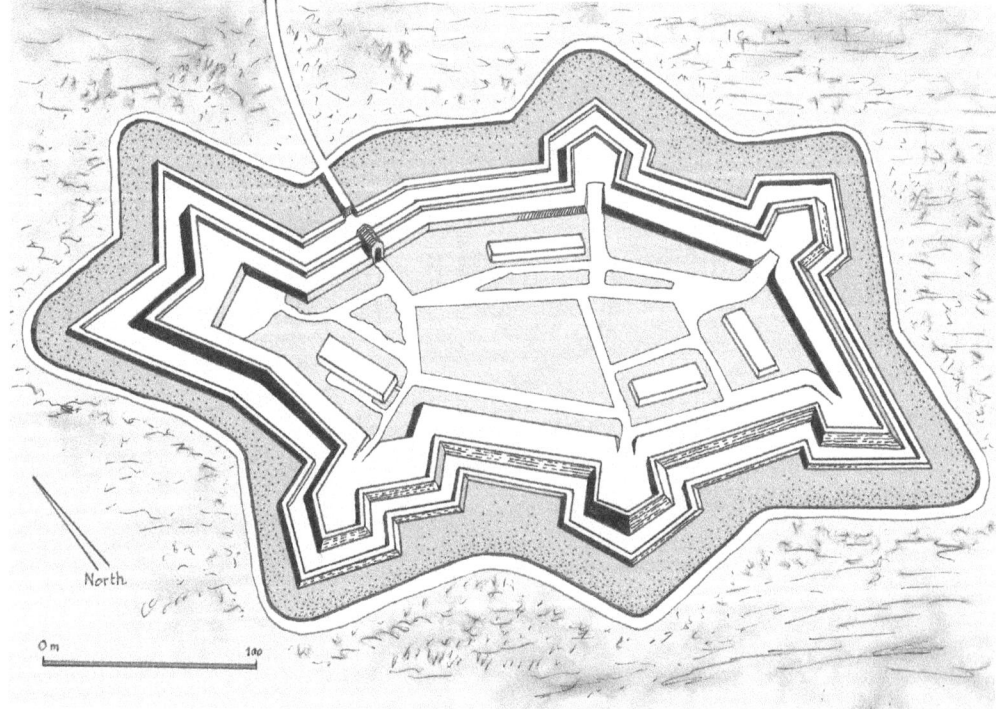

Fort at Wladyslawowo. This town and port is located on Poland's Baltic coast at the junction point on the road from Gdansk to the tip of the Hel Peninsula. The site was selected as a location for a naval base, and defended by an Old Dutch fort designed by the royal engineers Frederick Getkant, Jan Pleitner and Eliasz Arciszewski. The irregular fortifications were built in record time in 1634, and consisted of earthen walls, bastions, fausse-braie, moats, and covered way.

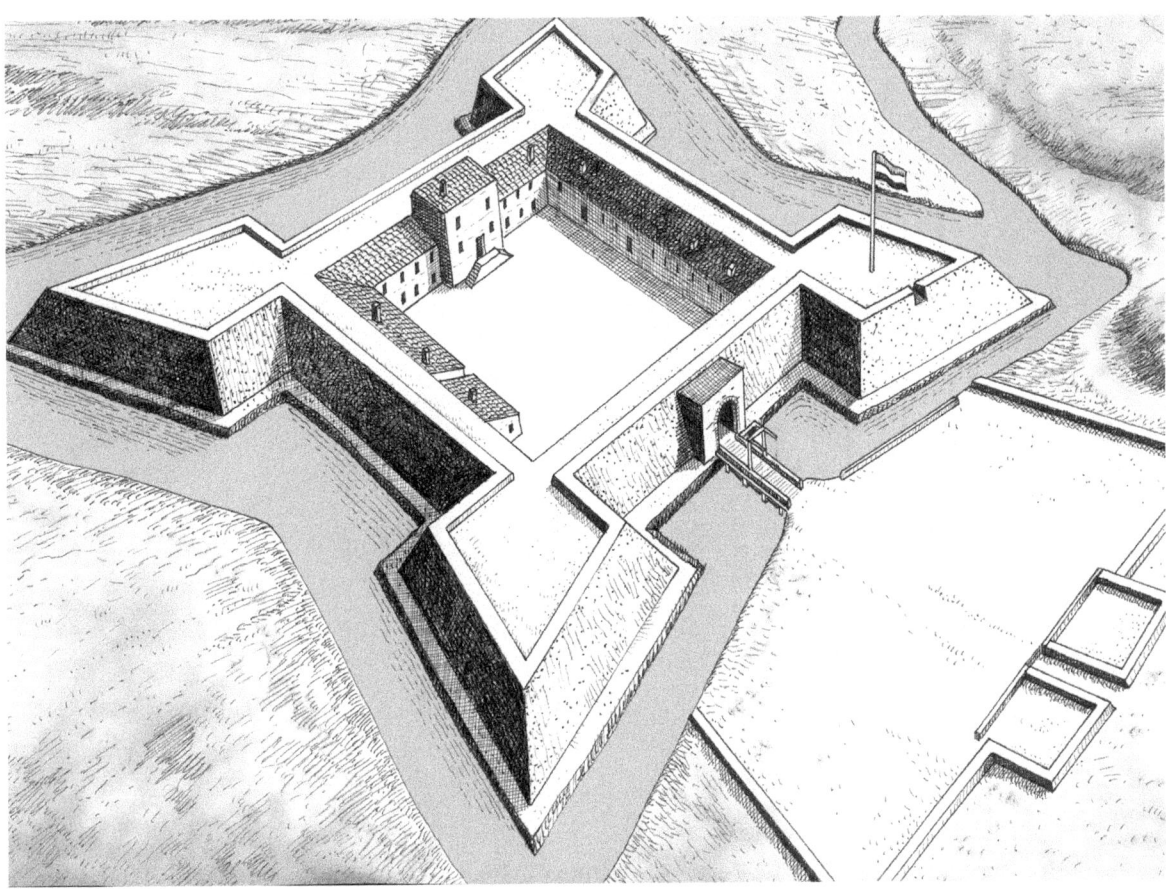

Fort at Cape of Good Hope. The area round the Cape of Good Hope in South Africa was first explored by the Portuguese Bartolomeu Dias in 1488. Called "Cape of Storms" (Cabo das Tormentas), and later renamed "Cape of Good Hope" (Cabo da Boa Esperança), the region became a trading post for ships en route to India and Asia. In 1652, the Dutch East India Company (Verenigde Oost-indische Compagnie, VOC) seized the place, made it a place of call for ships travelling to the Dutch East Indies, and constructed a square fort in ONS style. The fort of Good Hope was captured by Britain in 1795, returned to the Dutch in 1803, and retaken by Britain in 1806.

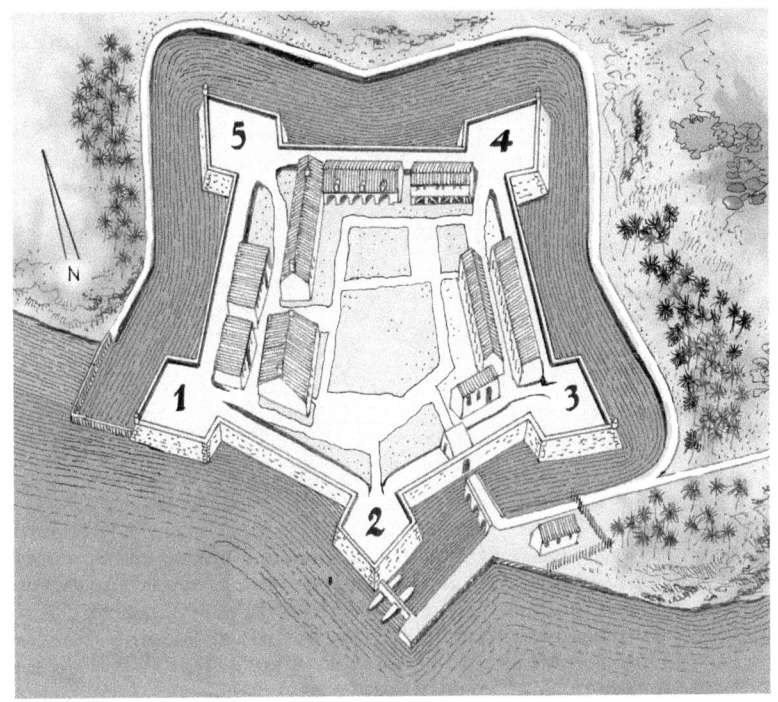

Jaffna. In order to defend the city and port of Jaffna situated on the peninsula at the northern tip of Ceylon (since 1972 named Sri Lanka), a fort was built by the Portuguese in 1618. The fortress was captured and enlarged by the Dutch under Rijcklof van Goens in 1658. In 1795, it was taken over by the British, and remained under the control of a British garrison till 1948. At the time of Dutch rule, the fort included five bastions and a moat: 1: Bastion Frisia. 2: Bastion Zeeland. 3: Bastion Holland. 4: Bastion Gelderland. 5: Bastion Utrecht.

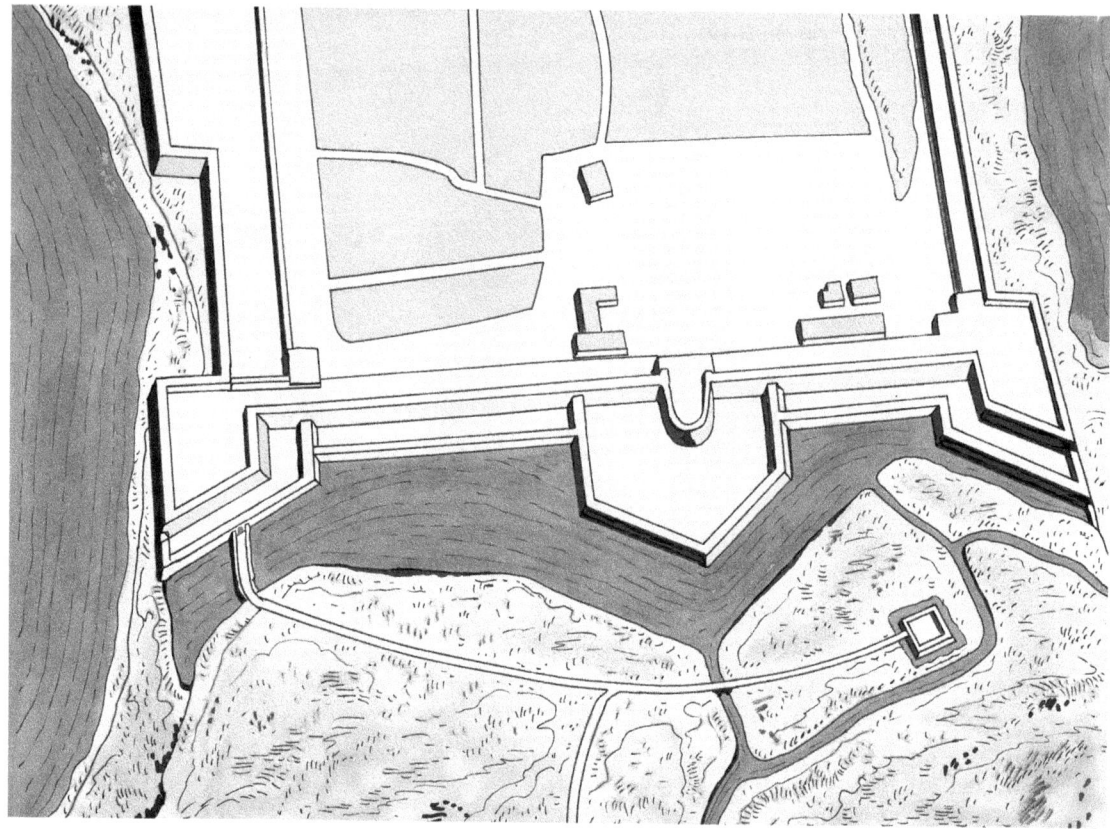

Galle (Sri Lanka). After having expelled the Portuguese from Ceylon (present day Sri Lanka), the Dutch endeavored to exploit the island and protect it against both European rivals and pirates. Under the rule of the VOC (East Indies Company), they fortified Colombo, the capital and several other key positions including Jaffna and Galle the chief town in the south of Ceylon. Built in the 1660s, the fortifications of Galle included a massive two-tiered bastioned line on the land front in the north, and a wall with bastions enclosing the peninsula dominating the harbor and the bay of Galle.

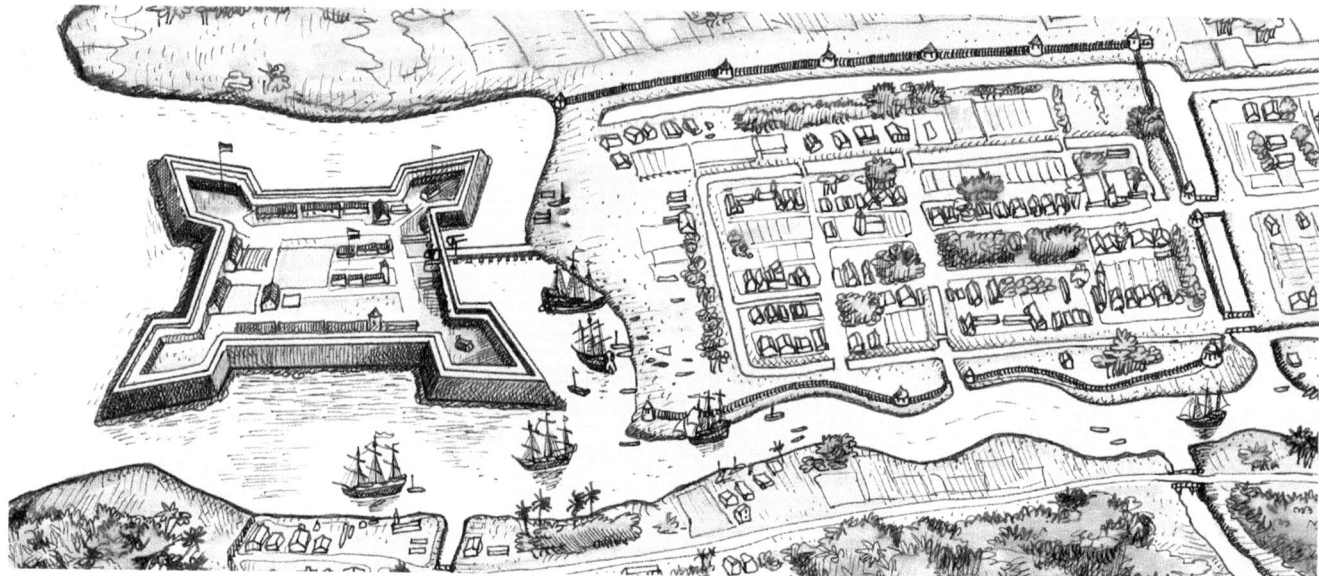

Jakarta. The capital city of Indonesia is located on the northwest coast of the island of Java. Originally occupied by the Portuguese, the Dutch took over control in 1619. They named the city Batavia, and the numerous islands were then known as Nederlandsche Oost Indië (Dutch East Asia). For much of 350 years they ruled the archipelago, first through the Dutch East India Company (VOC), and after 1800 as a nationalized colony. In the 17th century Batavia was defended by a bastioned fort and an enclosure. During the decolonization of Asia after World War II, Indonesia achieved independence in 1949 following an armed and diplomatic conflict with the Netherlands.

explosively across the globe. The United Provinces were the biggest maritime power of Europe, and Amsterdam was the most important financial center of the continent.

Colonial fortifications

The use of coastal artillery expanded from the Age of Discovery in the 16th century until decolonization in the second half of the 20th century. When a colonial power took over an overseas territory or when a trading post was created, one of the first tasks was to build some form of defense in order to subjugate the natives who could be hostile. Early settlements and trading posts were primitively fortified with palisades, and small coastal forts, towers, sconces, and blockhouses made of logs, mud or adobe. These early defenses were strong enough to repulse poorly armed local natives, marauders and pirates. But when rival European powers (equipped with guns and artillery) appeared, stronger fortifications were needed around the trading posts.

It was recognized that one land-based cannon equaled three guns of the same caliber mounted on a ship, due to the

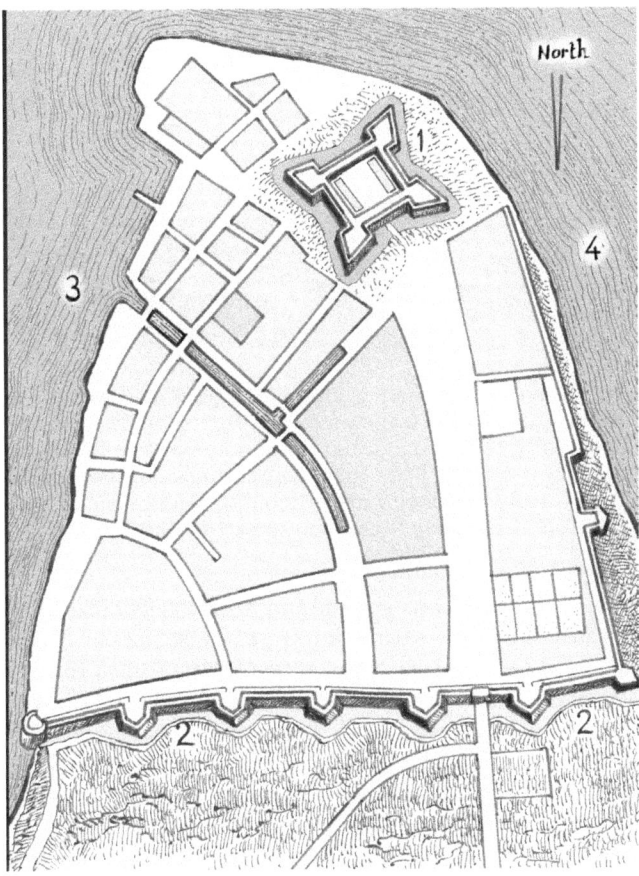

New Amsterdam (USA). The first native New Yorkers were the Lenapes, an Algonquin people who hunted, fished and farmed in the area between the Delaware and Hudson rivers. Europeans began to explore the region at the beginning of the 16th century. The cradle of New York, the island of Manhattan hemmed by the Hudson, East and Harlem rivers was first explored in 1524 by the Florentine explorer Giovanni da Verrazano, in French service. Verrazano called the site Angoulême after the king of France, François I of Angoulême. Dutch merchants, headed by Peter Minuit, purchased the island from the French in 1526. The Dutch colony was developed and fortified by a palisade and later a bastioned earth wall built in 1653 (today Wall Street). By that time a bastioned square fort was built at the place of the present Battery Park. In 1664, the Dutch governor Peter Stuyvesant had to yield the settlement to the English and the town was re-baptized New York in honor of the Duke of York who later became King James II. The map (simplified after the Castello Plan from 1660) shows Fort Amsterdam (1), the fortifications (2)—today Wall Street, the East River (3) and the Hudson River (4).

Fort Zelandia. Paramaribo, the capital of Suriname was founded in 1564 by French colonists who were ousted by the British. In February 1667 a Dutch expedition commanded by Abraham Crijnssen, took the colony. The territory became private property of the province of Zeeland. The new occupiers renamed the British pentagonal bastioned stronghold Fort Zeelandia. They enlarged the defensive perimeter and added a coastal battery. In 1782, they built another fort named Fort New Amsterdam and a redoubt called Leiden in order to defend the mouths of the River Suriname and its tributary, Commewine River. In the 19th century Fort Zeelandia lost all military value and from 1840 was used as a penitentiary.

Fort Orange Itamaraca Island, Brazil. For a while (between 1630 and 1654) the Dutch ruled a part of Portuguese Brazil notably the capital Bahia, and the rich sugar-producing region of Pernambuco. The Portuguese Catholic planters never accepted the Dutch Calvinist intruders and successfully revolted against them. The battle of Tobocas in August 1645 marked the beginning of the Dutch decline. In April 1648 and February 1649, others defeats sealed the fate of Dutch Brazil. The bastioned Fort Orange exists today and is situated on a beach of Itamaraca Island north of Recife (Pernambuco). It was designed by the Dutch military engineer Pieter van Bueren, and built in the early 1630s.

steadiness of the coastal gun. Obviously coastal guns had significantly higher accuracy than their ship-mounted counterparts. Land-based guns also benefited in most cases from the additional protection of coastal fortifications including forts, retrenchments, redoubts, breastwork, and coastal batteries. Those fortifications were built of more durable material—stone or brick (whenever available) holding thick masses of dirt. Logically fortifications in those overseas possessions were designed according to bastioned defenses, and built following theory and practice used in Europe. In this context, the ONS and later modernized land and coastal fortifications were widely exported to the overseas European colonies.

In the 17th and 18th centuries the Dutch Republic had created trading companies and started the development of a global network of settlements.

Africa

In West Africa, the Dutch were established since 1617 at Goree island west of Dakar (today Senegal), on the Gold Coast, on the Slave Coast, in Angola, Gambia and Arguin. From those bases they participated in the massive deportation of African people to be used as slaves on the American continent. In 1872 all Dutch western African possessions were sold to the British.

In 1652 the Dutch East India Company established a refueling station at the Cape of Good Hope situated in South Africa. Great Britain seized the colony in 1797 during the wars of the First Coalition (in which the Netherlands were allied with revolutionary France), and annexed it in 1805. The Dutch colonists in South Africa (known as *Afrikaners* or *Boers*) remained after the British took over. Later they made the trek across the country to Natal. By the end of the 19th century they tried without success to resist British invasion in what became known as the Boer Wars. In East Africa the Dutch had several settlements, notably in Mozambique (until 1721).

Middle East

In Persia, the Dutch East India Company established a trading station at Isfahan (the capital of the kingdom) that made commercial exchanges from 1623 to 1747. At Aden in Yemen they had a trading post from 1620 to 1739—with an interruption between 1623 and 1639.

India

In the Indian Ocean the Dutch were established on Mauritius Island east of Madagascar off Africa's southeast coast. This island was named after Prince Maurice of Nassau. It had been explored by Portuguese merchants in the early 16th century, and had been conquered by the Dutch East India Company. Mauritius Island was held from 1598 to 1710, then it was occupied by the French until 1810, and later became British.

The important port of Malacca on the western Malay Peninsula controlling the Strait of Malacca was seized from the Portuguese in 1641. During the Napoleonic Wars, it was yielded to Britain in 1806; it was later returned to the Dutch in 1816 and finally ceded again in 1824.

Dutch India included a number of trading settlements at Coromandel, Malabar, Suratte, and in Bengal administered by the *Verenigde Oostindische Compagnie* (VOC United East Indies Company).

The large island of Sri Lanka (then called Ceylon) located south of the Indian Peninsula was taken from Portugal by the Dutch East India Company (VOC) after a war that lasted from 1638 to 1658.

Once the Portuguese were driven out, the VOC re-used the existing fortifications. In many places the Company renovated them, and constructed new ones in order to meet the requirements of modern fort building. There were 22 Dutch forts on Ceylon, including several fortified ports like Colombo, Jaffna, and Galle. Placed along the shores of the island, strongholds were intended to keep rival Europeans at bay but also to call the local population and their rulers to order if necessary. Between 1665 and 1679 there was indeed a major rebellion led by Raja Sinha II. Another large-scale uprising took place between 1761 and 1765.

Ceylon was captured by the British in 1796 and remained a Crown colony until 1948.

Most Dutch possessions in India were yielded to Britain during and after the Napoleonic Wars in 1815.

In Pakistan the Dutch had a trading station in the city of Sindi (now Thatta) from 1652 to 1660. In Bangladesh they were established at Dhaka between 1664 and 1704.

Asia

Trading posts and settlements were created in the famous equatorial "Spice Islands" that soon became known as Dutch East India (today Indonesia). After having ousted the Portuguese in the early 17th century, the huge archipelago—including the large islands of Java, Sumatra, Borneo, Bali, Banka, Timor, Lombok, the Celebes, the Riouw, the Mollucas and many others—gradually became a permanent Dutch colony occupied until 1949 when Indonesia obtained her independence. Jakarta (then named Batavia) on Java island became the Asian headquarters of the United East India Company (VOC). Dutch New Guinea was retained separately until 1962, when it was yielded to Indonesia under pressure from the United States during the Vietnam War.

In Cambodia the VOC had trading posts at Phnom Penh during the period 1620–1667. In Vietnam, between 1636 and 1699, the Dutch had trading establishments at Hanoi in Tonkin, and at Hoi An in Annam. In China the Dutch had settlements on the island of Formosa (present-day Taiwan) from 1624 until 1681, and at Canton between 1749 and 1803.

In Japan the Dutch were allowed to run a trading post at Hirado in the period 1609–1641. Then the Japanese granted the Dutch a trade monopoly from 1641 to 1853, but solely on Deshima, an artificial island off the coast of Nagasaki. During this period the Dutch were the only Europeans allowed to trade with Japan.

Australia

Australia was discovered in 1606 by the Dutch explorer Willem Janszoon. The Indonesian islands, the northern coasts of Australia and the island of Tasmania were soon explored in the period 1642–1644 by the Dutch navigator Abel Tasman (1603–1659). In western Australia the Dutch established a small territory known as New Holland. As no serious attempt at exploration or formal colonization was ever made, the Dutch possession was simply taken over in 1770 by the British who renamed it New South Wales. In 1642 Abel Tasman also explored the Maori archipelago of Aotearoa and named it after the Southern-most province of the Netherlands: Nieuw Zeeland (later Anglicized as New Zealand).

North America

In what later became the United States of America, the Dutch West India Company had established a territory then known as Nieuw Nederland (New Netherlands) that included the areas of present day northeast Atlantic shores extending from the Delaware peninsula to Cape Cod. The settlements were initially situated on the Hudson River at Fort Nassau (in the period 1614–17) in present-day Albany, New York, and later relocated as Fort Orange in 1624. The trading post of New Amsterdam was founded in 1625 on Manhattan Island. The "New Netherlands" formally ended in 1674 after the Third Anglo-Dutch War when Dutch settlements were yielded to Britain. By then New Amsterdam was renamed New York. From its Dutch past, New York has kept the County of Nassau, Wall Street, the district of Harlem (named after the Dutch town Haarlem west of Amsterdam), the borough of Brooklyn (coming from the Dutch placename Breukelen), the sometimes pejorative term "yankee" (originating from the Dutch Christian name Jan-Kees) and the smokers a "Peter Stuyvesant" cigarette brand.

Central and South America

The so-called Dutch West Indies was a group of islands in the Caribbean Sea seized from Spain and colonized by the Dutch in the 17th century. Also known as *Nederlandsche Antillen* (Dutch Antilles) they included the islands of Saint Martin, the Virgin Islands, Tobago, Curaçao, Saint Eustatius, Aruba, Bonaire, and Saba.

The Dutch obtained from the British in 1667 the small territory known as Suriname (situated on the northeastern Atlantic coast of South America) after the Second Anglo-Dutch War, in exchange for New Netherlands (later known as New York) in North America. Suriname remained a Dutch colonial possession until 1975 when the country was granted full independence.

In northeastern Brazil the Dutch made an attempt to create a colony between 1630 and 1654. As a result there was a war with Portugal. Finally in 1661 by the Treaty of The Hague, the Portuguese agreed to pay a large sum of money to the Netherlands for abandoning all claims on the territory.

Chapter 5

The Golden 17th Century

A Blossoming Era

In Dutch history the 17th century is often referred to as the *Gouden Eeuw* (Golden Century). In spite of the war of Independence against Spain, it was a period of great fame and wealth for the Republic. Instead of being ruined by the war, the economy boomed, immigrants came in large numbers (notably Protestant refugees from Catholic Spanish-held Belgium), while many entrepreneurs started or extended their businesses. At the same time overseas trading settlements and colonies started to be established.

All aspects of science flourished with such persons as the lawyer Hugo Grotius (1583–1645), the philosophers Baruch Spinoza (1632–1677) and Pierre Bayle (1647–1706), the astronomer Christiaan Huygens (1629–1695), the scientist Anton van Leeuwenhoek (1632–1723), and the engineer Jan Leeghwater (1575–1650).

Art was particularly thriving in the paintings of the famous masters Rembrandt van Rijn (1606–1669), Frans Hals (1582–1666), Johannes Vermeer (1632–1675), Jacob Isaakszoon van Ruisdael (1628–1682) and Jan Havickszoon Steen (1625–1679). In the domain of poesy and literature the most prominent figures of this period were Gerbrand Bredero (1585–1618), Jacob Cats (1577–1660), Pieter Hooft (1581–1647), and Joost van den Vondel (1587–1679).

Wars against Britain

After the peace of Münster (treaty of Wesphalia in 1648), Dutch independence was recognized both by Spain and the German Empire. The Republic of the United Provinces reached then its political, cultural and commercial peak. Peacetime meant ground troops disbanded, mercenaries dismissed, and fortifications neglected. All public funds were used to finance the navy in order to increase trade and fight British commercial rivalry. Indeed a series of wars broke out between the Republic of the United Provinces and Britain. The First War (1652–1654), the Second War (1665–1667), and the Third War (1672–1674) were caused by competition in trade and mercantile interests. All these wars were fought at sea and in oversea colonies. The Fourth War (1780–1784) was caused by Dutch interference in supporting the American Revolution. Despite a few resounding successes, the wars against Britain spelled the decline of the Dutch Republic as a world power.

Dutch Fortifications in the Latter 17th Century

Hendrik Ruse

Successfully employed against Spain by the engineer Adriaan Anthonisz and his colleagues and collaborators, the Old Dutch fortification style tended toward obsolescence in the second half of the 17th century. The flank angle of 90 degrees was rather inappropriate to fire along the neighboring bastion and defenders placed in the flank could have the feeling that they fired at each other. There was also a number of disparagements about the advanced works. It was objected that they were useless, much too expensive to build and to maintain, and that they dangerously scattered the defender's forces.

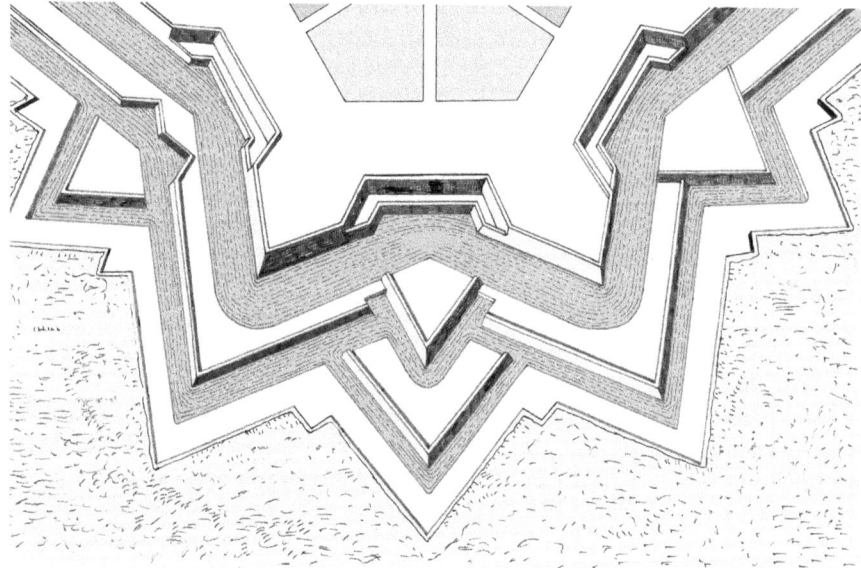

Fortification system designed by H. Ruse 1654. Hendrik Ruse pleaded for a better way to flank the ditch by opening the flank angle (meeting point of flank and curtain) to 120 degrees. Ruse also advocated connecting ravelins and counterguards to form a continuous envelope all around a place.

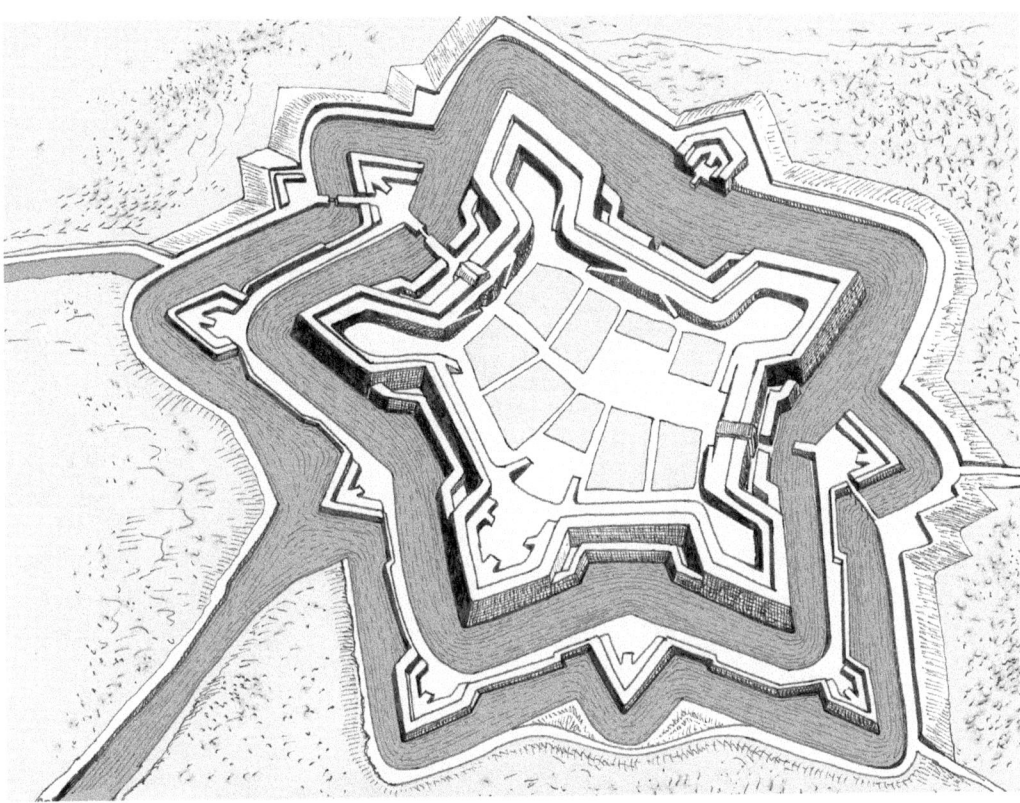

Kastellet, Copenhagen, Denmark. After the Swedish siege of Copenhagen in the late 1650s the Dutch engineer Hendrik Ruse was called in to improve the fortification. Ruse designed the pentagonal citadel named Citadellet Frederikshavn ("The Citadel Frederik's harbor") aka Kastellet ("the citadel"), constructed in the form of a pentagon with bastions at its corners.

Hornworks were criticized because of the limited number of guns their half-bastions could contain and because of the long vulnerable wings where guns and servants were dangerously exposed to enfilade fire. Critics observed that the purpose of horn- and crownworks was more to deter the enemy rather than actually strengthening the defense.

Among the soldiers, advanced works were unpopular because defenders posted there could have the feeling of being left alone or sent to sacrifice, thus tempted not to fight tooth and nail and soon retreating after a few rounds had been shot at them. Besides, once conquered, advanced works provided the besiegers with captured guns, space and material to carry on the siege.

Several engineers and military architects had noticed those facts, one of whom was a certain Hendrik Ruse (1624–1679) from Amsterdam. Born in Sauwert near Groningen, Ruse originated from a German-speaking Protestant family from Alsace who had migrated to the Low Countries. In 1654 Ruse designed an improved fortification system (strongly influenced by the French engineers Pagan and Vauban) that he published in a book called *Versterckte Vesting* (Fortified Fortress). Ruse notably pleaded for a better way to flank the ditch by widening the flank angle to 120 degrees. He also advocated connecting ravelins and counterguards to form a continuous envelope around the place as a replacement of the obsolete hornworks.

Ruse's book was translated into German and English in the 1670s, but his designs and propositions did not raise any interest in the Dutch Republic. In 1661 he got an opportunity to offer his service to the king of Denmark. For his new employer Ruse designed the fortress of Kalkar and the fortifications and the Kastellet (pentagonal citadel) at Copenhagen. As a reward for his service he was promoted to the rank of General and knighted to the rank of Baron Rüse von Rysenstein.

Rampjaar and the Franco-Dutch War (1672–1678)

Since its foundation, the Republic of the Seven United Provinces was supported by France. However the Bourbon king of France Louis XIV (born 1638, reigned from 1643 to 1715) considered the Dutch as trading rivals, seditious republicans and Protestant heretics. Tensions rose, hostility grew, and ultimately a war broke out. This conflict against the Dutch was wanted by both minister Colbert and Louis XIV. The main objective was to bring down the formidable economical strength of the Republic of the United Provinces, which stood in the way of Colbert's commercial development based on autarky and protectionism, called colbertism. The very Catholic "Sun-King" Louis XIV also wanted to eradicate Dutch Protestantism, and to silence the impertinent news-sheet writers from Amsterdam. Tolerant

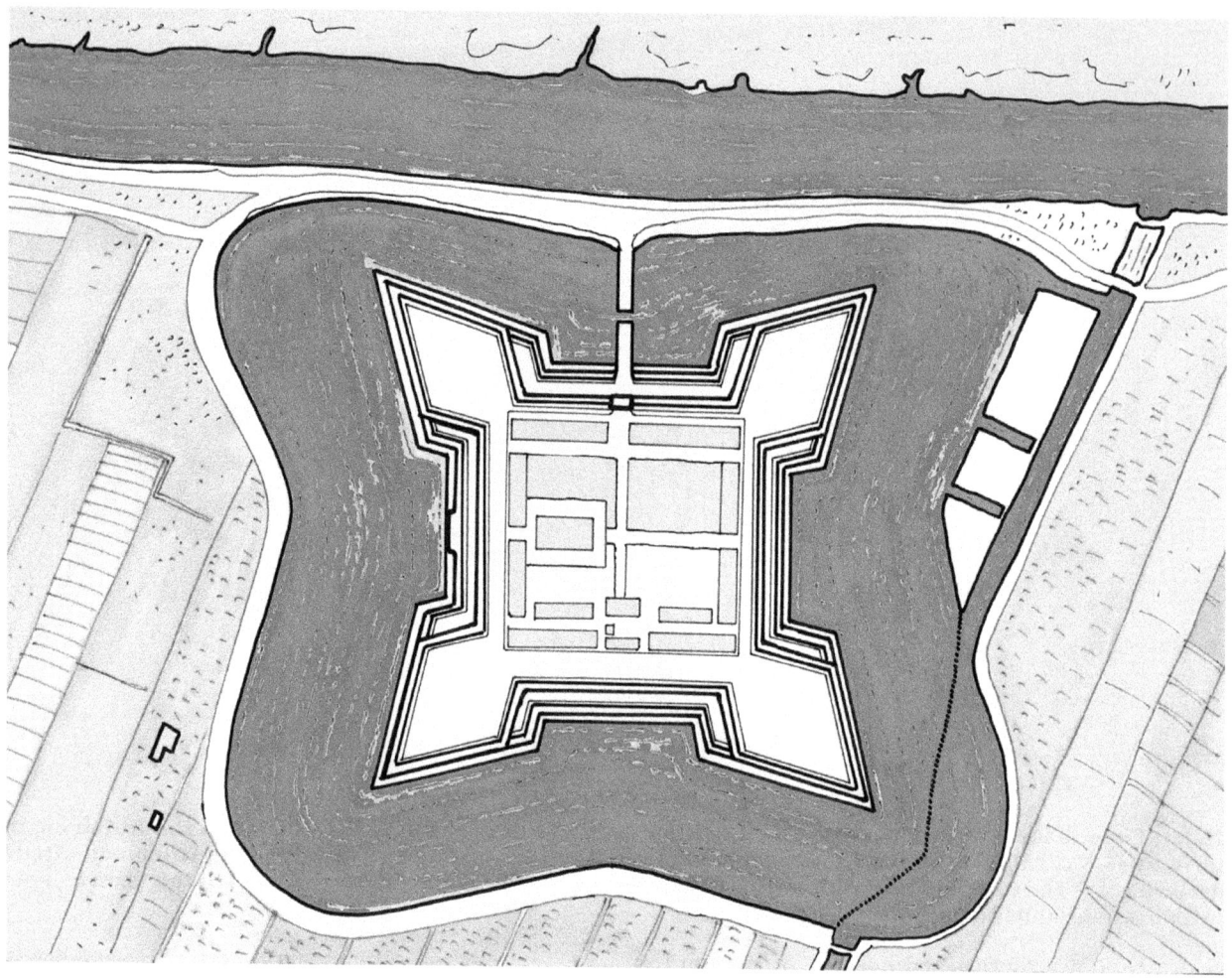

Fort of Wierick. Rebuilt in 1673, the Wiekickerschans is a fort situated between Woerden and Bodegraven in the province of South Holland. It was a part of the Old Holland Waterline, and displays features of the so-called Improved Dutch System.

Protestantism, commercial mercantilism, political liberalism, and the insolent bourgeoisie of this small but rich land exasperated the French Catholic absolute Sun-King.

With a large force of 100,000 well-equipped soldiers, led by the king himself, and commanded by skilled generals like Condé and Turenne, the French invaded Belgium, crossed the Rhine in April 1672, broke into Dutch territories and seized the cities of Orsoy, Doesburg, Arnhem, Deventer, Zutphen and Utrecht. Simultaneously, the north of the Republic was invaded by Louis XIV's allies, the German Archbishop of Cologne and the Bishop of Münster, while the English fleet attacked the North Sea coasts.

The well-prepared invasion did not, however, turn out to a decisive victory. Dutch Admiral Michiel De Ruyter defeated the British escaders at Sole Bay, the German prelates were stranded before Groningen and, in the south, the main French attack was stopped by vast and hastily made floodings in the region of Amsterdam. Louis XIV, although on a very strong position, made the mistake of refusing the peace offered by the Dutch. The young stadhouder Willem III of Orange-Nassau (1650–1702) came to power in 1673.

Willem restored the military, political and diplomatic situation by negotiating the neutrality of England and by setting a coalition against France, regrouping the German Empire, Austria, Spain and Lorraine. The Republic was saved in extremis, but had been dramatically close to total defeat. Therefore the year 1672 has remained in the Dutch collective memory as the *Rampjaar* (the Disaster Year). The war ended with the Treaty of Nimegue, signed on July 17, 1678. It represented the summit of Louis XIV's success and did reaffirm the superiority of the French Bourbons over the declining Spanish Habsburgs. The main loser was Spain, and France won respect by its victories, and Sun-King Louis XIV made himself the arbiter of the destiny of Europe.

Improved Dutch bastioned system

After the dramatic French invasion of 1672, the Republic of the United Provinces was forced to rethink and reshape its neglected and obsolete land defense system. Between 1678 and 1688, a new bastioned method was developed. Called *Improved Dutch bastioned fortification*, this system was directly inspired by Hendrik Ruse's design and also heavily

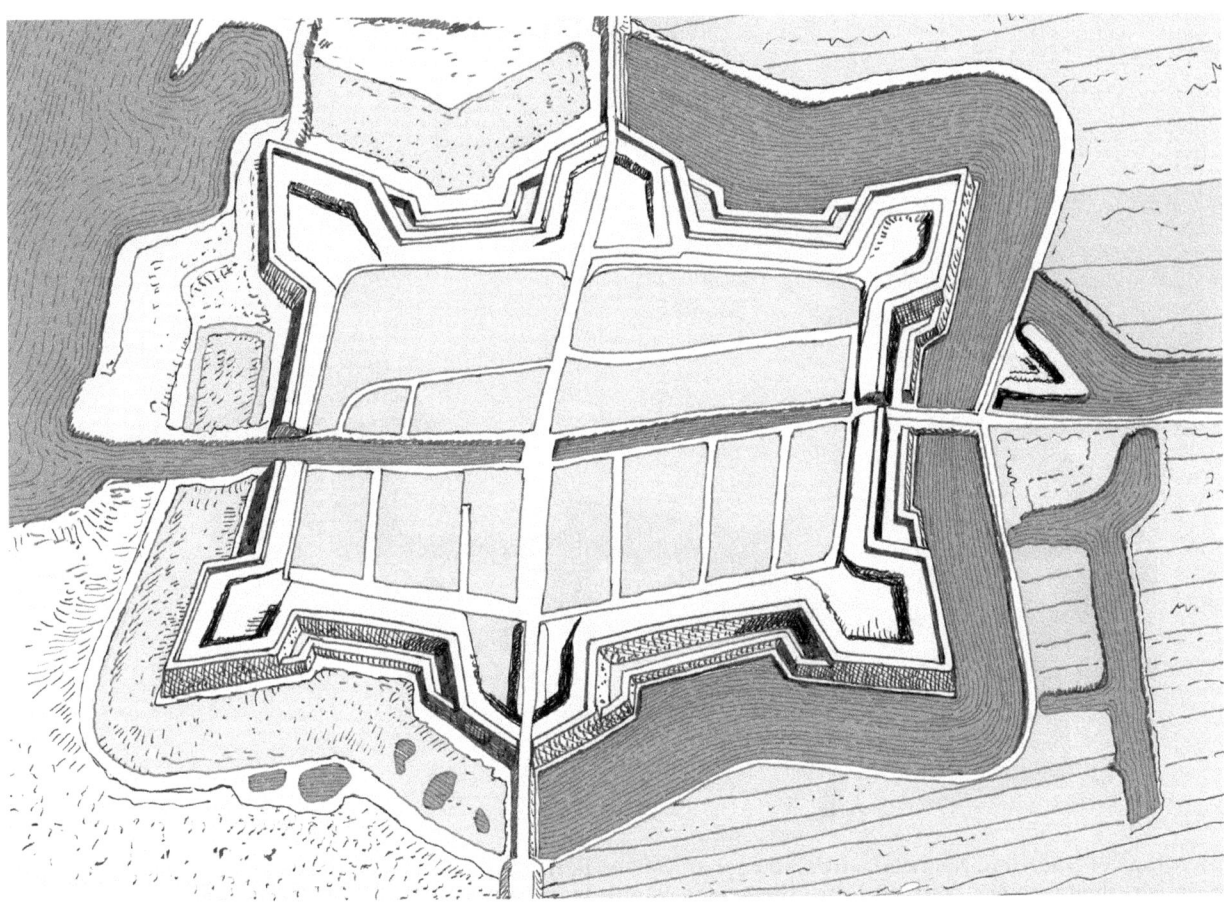

Plan of Nieuwpoort. Situated on the River Lek in front of Schoonhoven in the province of South Holland, the small town of Nieuwpoort was granted city rights in 1283. The stronghold featured a redoubt built in 1672 that protected a sluice allowing the flooding of the surrounding countryside. After the French invasion of 1672, Nieuwpoort was incorporated into the Old Holland Waterline and its fortifications were enlarged and refurbished in 1690 in the so-called Improved Old Dutch system. The fortifications of Nieuwpoort were abandoned in 1795, but left intact. They were restored between 1973 and 1998, and have become a nice tourist attraction.

influenced by the French method used by Louis XIV's engineer Sébastien Le Prestre de Vauban.

It seems that the so-called "improved method" was introduced in the Republic of the United Provinces by Paul Storff de Belleville. The German mercenary military engineer Storff de Belleville successively worked for the king of Sweden in the late 1650s, the Serenissime Republic of Venice, the Prince of Palatinate in 1666, and Spain in 1667–1668, before entering into the service of King Charles II of England for whom he designed several works, notably at Plymouth and Fort Charles at Kinsale in Ireland in 1672. A year later, the skilled Storff de Belleville joined the French army and took part to the siege of Maastricht in 1673. He then worked to the fortifications of northern France under the leadership of Vauban, e.g., at Courtrai, and took part to the sieges of Valenciennes, Condé, Dinant and Thuin.

In 1677 Storff was back in England, but again changed camp and joined the Dutch Republic of the United Provinces a year later. With the rank of General-Inspector, he probably introduced features of French military architecture in the Low Countries, and participated in the modernization of the fortifications of Naarden, Grave and Saas-van-Gent in the late 1670s, and 1680s. In 1684, Storff's career came to an abrupt end when he was involved into several financial scandals concerning the place of Grave and Den Bosch. The adventurer was then destitute, and condemned to death for embezzlement, but managed to escape abroad.

In 1685, Storff was in service of Venice again, with the rank of Sargente Maggiore di Battaglia and fought against the Turks in the Greek Peloponnese and at the fortress of Chiefala in 1686. After this date all traces of the mercenary Paul Storff vanish. The life of Storff de Belleville—like that of many other mercenary engineers was typical of a hired professional, who could get a job in any 17th century European army. Educated people and skilled professionals could become successful soldiers of fortune regardless of nationality, faith and conviction, in early modern Europe.

The main elements characterizing the short-lived transitional form known as Improved Dutch System were: the introduction of the use of masonry holding thick masses of earth; the tenaille replacing the fausse-braie; vast bastions often with curved multi-tiered flanks protected by massive orillons; the modification of the flank angle to 120 degrees allowing to give a perfect flanking as already advocated by

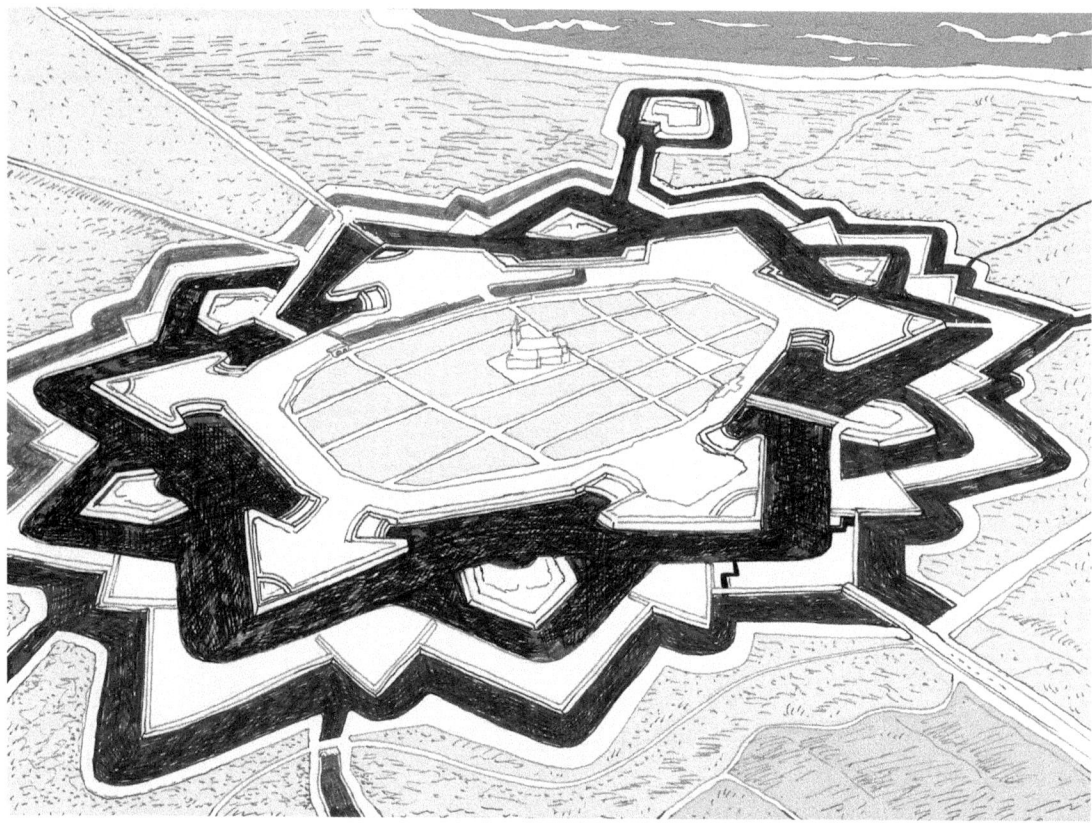

Fortifications of Naarden. Naarden, situated east of Amsterdam, grew from a small fishing village on the Zuiderzee. Naarden was destroyed in 1350, and the village was rebuilt by the count of Holland, Willem V. The simple earthen fortifications were replaced by a stone wall with towers in the 15th century. During the Eighty Years' War, Naarden was looted and the inhabitants slaughtered by the duke of Alva's son, Don Federico. Taken back by the Dutch Protestant insurgents, the city, key to the defense of Amsterdam, was fortified in 1580 by engineer Adriaan Anthonisz in Old Dutch style. Naarden was taken by Louis XIV in 1672. A year after, Stadhouder Willem III re-conquered the town and decided upon the erection of new modern defenses. Often wrongly attributed to Menno Van Coehoorn, the new fortifications of Naarden were designed by engineers Nicolaas Witsen, Jean-Baptiste de Bombel, Paul Storff de Belleville and Willem Paen. Completed in 1685 the fortifications of Naarden include walls, six elaborated bastions, six ravelins in the moat, an envelope, an outer moat, a covered way, and a glacis. Naarden was later incorporated into the Holland Waterline. Today Naarden's imposing and remarkable fortifications are perfectly preserved and house an interesting Dutch military architecture museum.

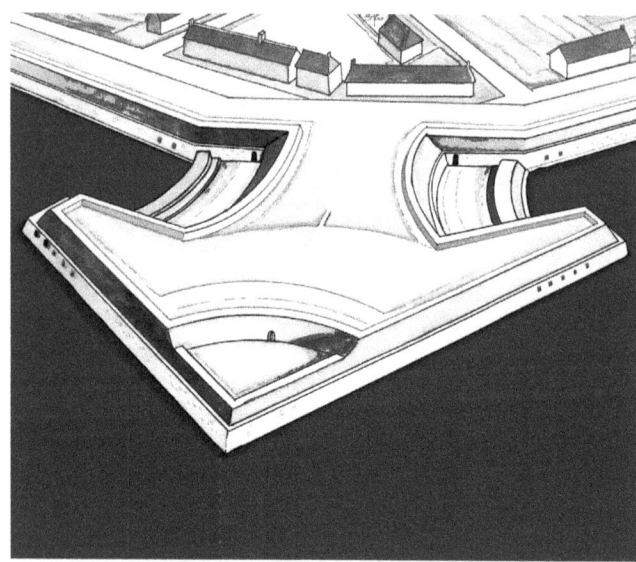

Bastion Nieuwe Molen at Naarden.

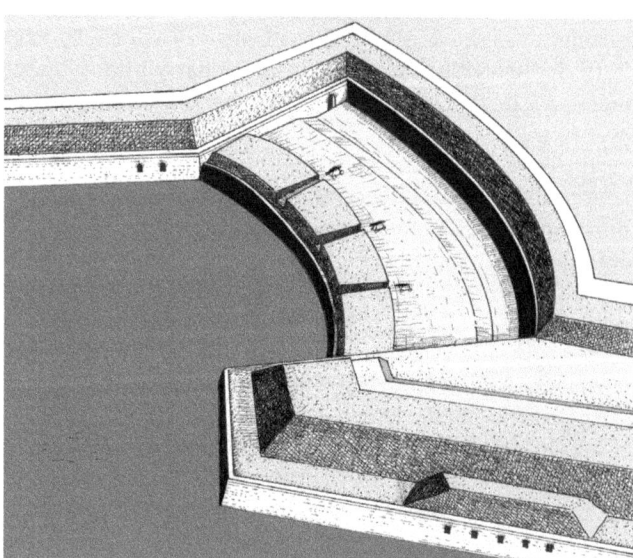

Concave bastion flank at Naarden.

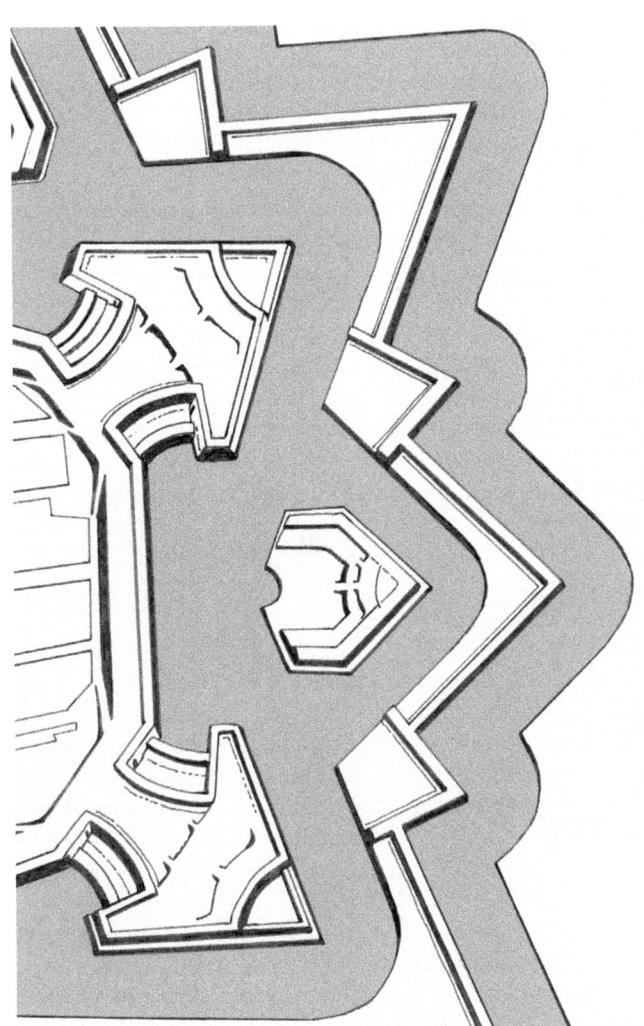

Bastioned front at Naarden include bastions and ravelins in the moat. Note that counterguards and lunettes were joined forming a continuous advanced envelope in front of which were an outer moat, a covered way, and a glacis.

Map of the Old Holland Waterline.

Pagan, Vauban and Ruse; and often an advanced moat ahead of the covered way.

The most noticeable realizations of this transitional period in the United Provinces were the fort of Wierick, the enceinte of Nieuwpoort on river Lek and, of course, the famous, impressive, still existing fortifications of Naarden.

Old Holland waterline

In 1672 the Republic had been saved by improvised floodings in the region of Amsterdam. So by pure sound reasoning the Dutch decided to base their national defense on inundations, but that would be on a larger scale, carefully prepared in peacetime, and carried out in moment of crisis or when a threat of invasion was feared.

The Oude Hollandsche Water-linie (Old Holland Waterline), intended to protect the western part of the United Provinces, and more particularly the rich and leading province of Holland, had already been planned during the Eighty Years' War by Maurits of Orange-Nassau and his half-brother Frederick Hendrick. It was called the Old Waterline because later a new line was created east of Utrecht.

As a weapon, water had proved its worth during the war against Spain on several occasions (notably at the siege of Leiden in 1574) and against the French in 1672. Work on the Holland Waterline started in 1629 and continued through the 17th, 18th and 19th centuries. The line ran from Muiden and Naarden near the Zuiderzee (today called the IJsselmeer) in the north, via Woerden, Ouderwater, Schoonhoven, Nieuwpoort, and down to Gorinchem, Woudrichem and Heusden on the rivers Waal and Rhine in the south. It would transform the province of Holland (the economic heartland of the Dutch Republic) into an almost impenetrable island.

Basically the waterline consisted of a combination of natural bodies of water, rivers, and canals, with judiciously placed sluices constructed in dikes allowing water to flow in low-lying grounds. The water level in the flooded areas was carefully maintained to a level of 60 cm (about two feet)— deep enough to make an advance on foot precarious and dangerous, and at the same time shallow enough to rule out effective use of boats—other than the flat bottomed *uitleggers* (patrol and gun barges) used by the Dutch defenders.

Under the water level additional (invisible) obstacles were dug including ditches, pits, small trenches, and wolf's pits (excavations fitted with a sharp wooden stake). In time of war, or when the threat of an invasion was feared, the trees lining the dikes that formed the only roads through the line, could be turned into obstacles called *abatis* (rows of felled trees lying lengthwise and parallel to each other with boughs and sharpened branches pointing outwards in the direction of the attackers' approach).

Although the Netherlands is a very flat country, the grounds between the Zuiderzee in the North and the Great Rivers in the South are not all exactly at the same level. So the flooded zones were divided into *kommen* (basins) separated from each other by dikes, and levees fitted with sluices and watergates. Of course the waterline was only a passive obstacle, and needed to be supported with active elements.

At strategic points along the line, strongholds, *schansen* (sconces), fortlets, forts and *vestingsteden* (fortified towns) were created, improved or reinforced with troops defending the so-called accesses (dikes, embankments, levees, and non-inundatable roads, later in the 19th century railway tracks) that ran across the waterline. This basic strategy became the standard defense of the Netherlands until the Cold War in the 1950s. The concept worked rather well but it had several weaknesses and drawbacks. During a cold winter the shallow flooded zones could freeze, and thus lose all efficiency as an enemy could simply advance on the ice. To counter this, running waterways were kept open and flowing, so that their movement could weaken, stop or delay the forming of ice. In extreme cold the ice would be broken up, for example by using maces, or even explosives.

A flooded area formed an impassable obstacle that exposed advancing enemy troops under the defenders' fire, but at the same time it was a passive and static obstacle that did not allow the defenders to launch a quick and large-scale counterattack on retreating enemies. As can be imagined, an inundation took some time to be carried out, and was ineffective in case of a sudden surprise attack.

The whole scheme was also a huge undertaking with enormous cost. It included the geographical re-shaping and modifying of a part of the country. It demanded gigantic structural investments, as well as advanced engineering installations for controlling, moving and conducting large quantities of water. It required sophisticated hydraulic techniques, and highly skilled and qualified personnel to manipulate on a large scale and dominate such a dangerous element as water. Fortunately the Dutch had a strong technical experience gathered in centuries of dike, dams and polders building in their eternal fighting against natural floodings.

As can be imagined the military use of inundation was extremely unpopular with the civilian population, villagers, and peasants involved to their dismay. Meadows, pastures, fields, orchards, cultivation grounds, country lanes, and land paths could be damaged or made useless for quite a long time especially when salt water of the Zuiderzee was used. Even a temporary and local inundation with fresh river water always meant disturbance, perturbation, forced evacuation, and ruinous loss of productivity. It took many years before the authorities would indemnify victimized and duped peasants.

Siege Warfare

In the 17th and 18th centuries, warfare was characterized by dynastic conflicts and a predominance of sieges, in which fortification and artillery played a central role. Strategy was dominated by carefulness and most risks were often calculated in advance. Eighteenth century royal rulers and their strategists frequently preferred the controllable and codified siege warfare rather than the hazardous chances of a bloody and uncertain battle in open field. Great battles of this period were seldom decisive in the sense that they brought the wars to an immediate end. They were often irrelevant unless they helped to determine the outcome of a siege. As a logical result, besieging a fortified place became a complicated military science elaborated and carried out by expert and specially trained engineering officers developing particular skills and technical terminology.

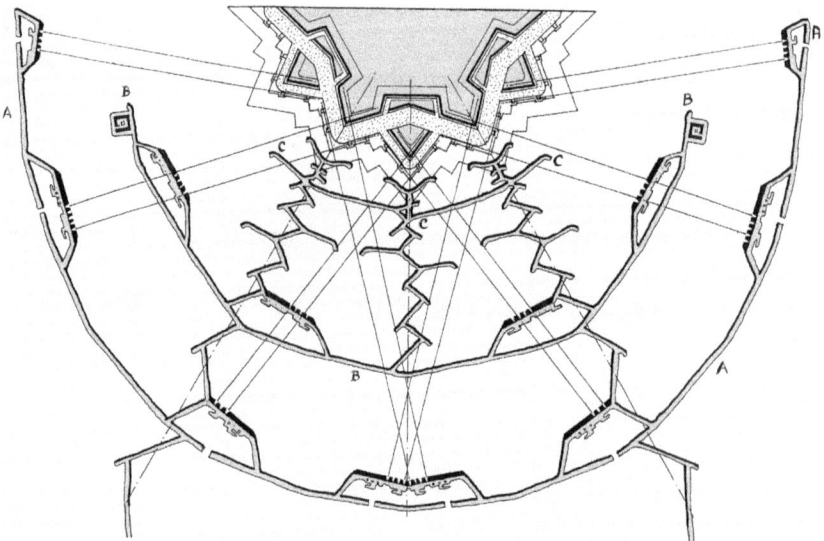

Vauban's siege theory: AA—first parallel with gun batteries. BB—second parallel with redoubts. CC—third parallel with cavaliers de tranchée.

In the 17th century, the French engineer Sébastien Le Prestre de Vauban (1633–1707) designed a systematic method of laying siege, using existing elements of field fortifications or *approaches* (parallels, saps and batteries). Parallels were trenches excavated by the besiegers alongside the attacked front, and enabling to get closer and closer to their objective in comparative safety. In theory, Vauban's siege method was characterized by the systematical use of four parallels.

The first parallel was established at the limit of defenders' guns range (about 600 m or 650 yards); it was used for the general communication and could also serve as countervallation line. The second parallel was excavated at approximately 350 meters (380 yards) from the defender's position. There, sheltered batteries were placed in the length of bastions faces to give enfilade and ricochet fire. The third parallel was dug at the foot of the glacis.

At that point Vauban advocated the construction of the so-called *cavaliers de tranchée* (raised structures, made of three or four tiers of gabions filled with earth) to attack the covered way. On the third parallel, mortars and pierriers batteries were also positioned to bombard at close range lateral outworks and to neutralize the defenders' fire.

The fourth parallel, also called *couronnement du chemin couvert* ("crowning" of the covered way) was established on the crest of the covered way. In these entrenchments, guns were deployed to carry out a breach, as sufficient weight of fire at close range could be mounted to batter a masonry wall into rubble. Vauban also advocated the use of mining, a method consisting of digging a gallery under the walls and exploding a charge of gunpowder that caused the destruction of the defenses.

The cannonade reached a climax and the walls cracked and crumbled; mines exploded, blowing the ramparts sky-high and opening a pathway into the heart of the besieged fortress. When the breach was made, the following step was the assault—a deadly hand-to-hand battle in the smoking ruins of the defenses. The assault was a crucial, bloody and decisive moment for both parties, unless negotiations were opened. A successful assault did not mean necessarily the end of the battle. The defenders could continue to resist and build an improvised barricade right behind the breach. They could also decide to retreat in the citadel or in the urban castle (when available). In that case, the attackers had to envisage a new siege.

Obviously not all sieges progressed with the clockwork

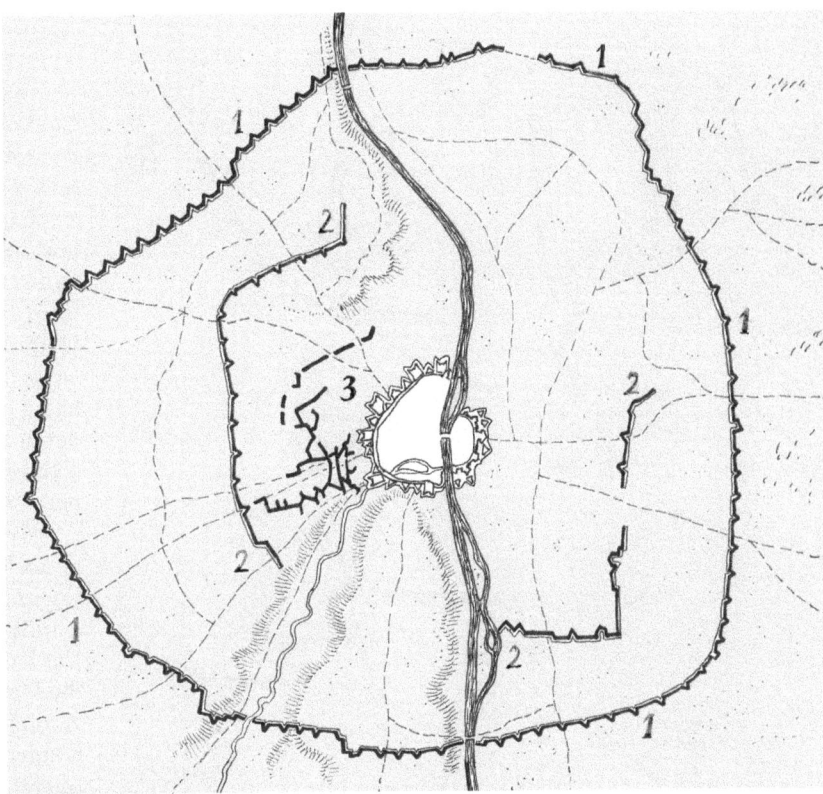

Circumvallation. Siege of Maastricht. The siege of Maastricht lasted from 15 to 30 June 1673 during the Franco-Dutch War of 1672–1678. The French troops were commanded by King Louis XIV and Vauban directed the engineering work. At this occasion Vauban introduced his new concept of parallel siege works: 1: The so-called circumvallation line was an external continuous entrenchment (generally including a ditch, and an earth wall with redans) dug by besiegers around a besieged place in order to invest it, blockade it, and isolate it from relief. 2: Another line of field fortification could also be established, the contravallation turned in the direction of the besieged place in order to resist sallies and counter attacks. Both lines protected the camps, artillery parks, and supply stores of the attackers. 3: Attack parallels, saps, and batteries.

precision and strict geometry as described above, because Vauban adapted his method to the natural particularities of the place to besiege. Nevertheless this method of laying siege constituted an undeniable progress. The business of conducting sieges became as formal and as pre-ordained as a game of chess. It considerably reduced the attackers' casualties, and frequently led to negotiations followed by the capitulation of the defenders after an honorable fight.

The method of laying siege designed by Vauban remained in use (with a few changed details) for a long time. It was still employed during the Napoleonic Wars (1802–1815), the siege of Sevastopol during the Crimean War (1853–56), during the American Civil War (1861–65), and during the First World War (1914–1918). The climatic siege of the entrenched camp of Dien Bien Phu, Tonkin, Vietnam (from March 13 to May 7, 1954) still featured elements of Vauban's systematic procedure. Indeed the Vietminh (Vietnamese communist independantists) made use of approaching parallels, trenches and saps, as well as underground mines resulting in a humiliating defeat for the French colonial army.

Menno van Coehoorn

Known as the "Dutch Vauban," Menno van Coehoorn (1641–1704) was the greatest Dutch military engineer of the second half of the 17th century. Often compared to the celebrated Frenchman Vauban, he was a distinguished combat engineer, both in attack and defense. Coehoorn and Vauban had several characteristics in common. Both came from the lesser rural nobility, both were great and talented workers, and both were patriots *avant la lettre*, largely devoted to their countries. They both had a large experience of war acquired by taking or defending fortresses and both were designers and creators of fortifications. But there stops the comparison between them.

While Vauban was concerned by social, economical, theological, colonial, or fiscal problems, Van Coehoorn was exclusively a commanding officer and a military engineer. Besides, the absolute King Louis XIV's Catholic France had nothing in common with the Dutch Republic of the United Provinces ruled by an oligarchy of rich Protestant merchants and Stadhouder-Prince Willem III. France was huge and continental, it counted a large rural Catholic population, and the arrogant king conducted an aggressive foreign policy. The Republic of the United Provinces was a small country whose wealth and power mainly rested

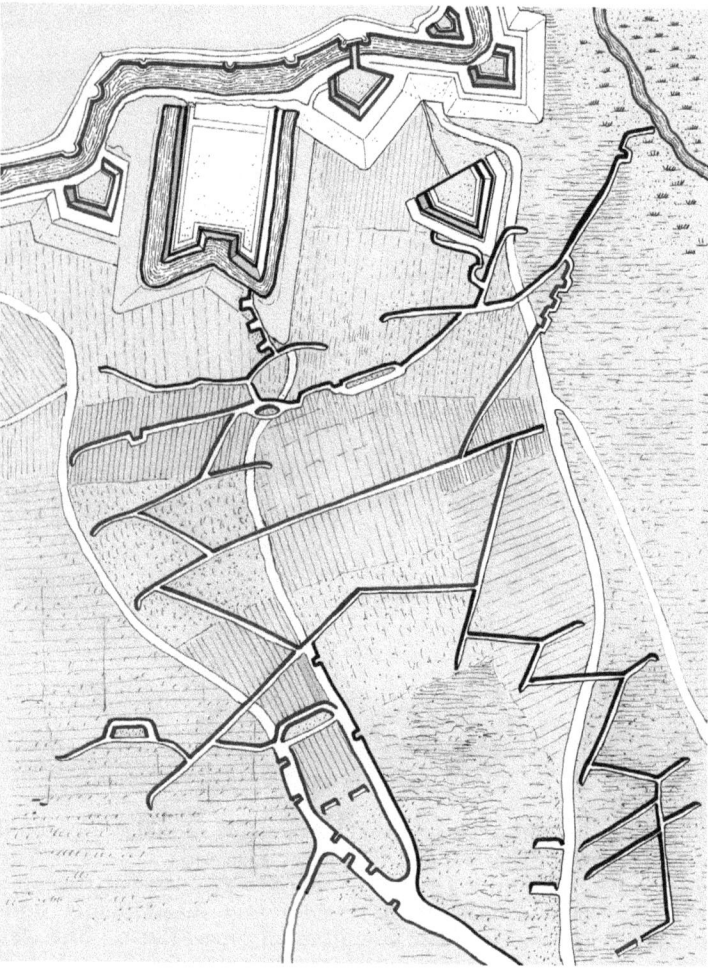

Parallels, saps, and siege batteries (Siege of Maastricht 1673).

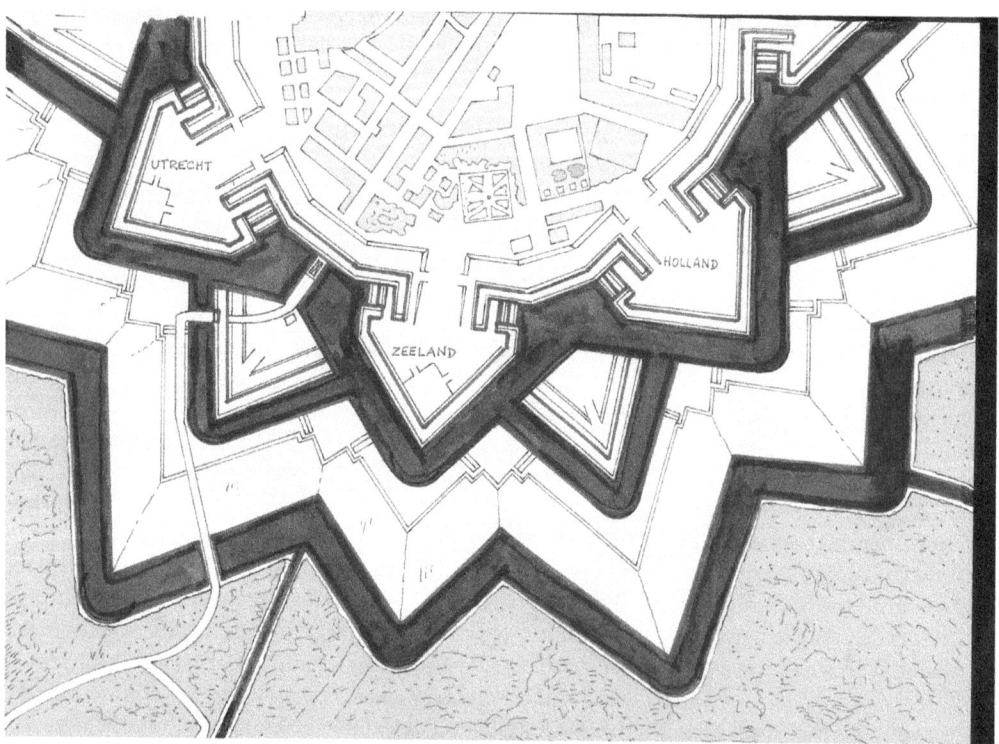

Below: **Project for Coevorden by Menno Van Coehoorn. Strategically situated on a narrow passage between marshes in the south of province Drenthe, Coevorden (meaning the ford of the cows) was developed by the bishops of Utrecht in the Middle Ages. The city was fortified by order of the King of Spain Charles V. It was taken by the Dutch independentists in 1592 and Prince Maurice of Nassau ordered the edification of new fortifications constructed between 1597 and 1607. The city displayed the model of an "ideal city" rather similar to Palmanova in Italy. The streets were laid out in a radial pattern within geometrical fortifications and extensive outer earthworks including a regular enceinte composed of seven bastions, a fausse-braie, seven ravelins, and a wet ditch. The depicted project by Menno van Coehoorn designed in the 1680s was not retained.**

NIEUWE VESTINGBOUW,

Op een natte of lage
HORISONT;
Welke op driederleye manieren getoont
wordt in 't Fortificeren der binnengroote

Van de

FRANSCHE
ROYALE SES-HOEK,
*Waar in de Sterkte der hedendaagsche
drooge- aan de natte-Grachten gevonden wordt:*

ALS MEDE

Hoe men tegenwoordig langs een Zee of Rivier
Fortificeert, en op wat manier men daar behoorde te Bouwen.

*Ider Methode geataqueert en vergeleken, so in
haar wederzijds Sterkten, als onkosten, met de Fransche
of hedendaagsche Vestingbouw.*

, DOOR

M. COEHOORN,

Collonel van 't Regement Infantery van de Heer
Prince van Nassau, Erfstadhouder van Vriesland, &c. &c. &c.

Te LEEUWARDEN,
By Hendrik Rintjes, Boekverkooper in de Peperstraat, in de
Zaadzaaijer. M DC LXXXV.

Title page of Menno van Coehoorn's book. It is titled: "New Fortress construction on a wet [ditch] of low horizon, which is demonstrated by three fashions of fortifying the royal French six-angle, which are found in contemporary dry and wet ditch, as well as how one should fortify along the sea or a river, each with method of attack and comparison with other strongholds and costs with the French way of modern fortification, by M. Coehoorn, Colonel in the Regiment of Infantry of his Highness the Prince of Nassau, Stadhouder of Frise, etc., etc., etc. At Leeuwarden by Hendrik Rintjes, Bookseller in the Pepper Street in the borough of Zaadzaaier, 1685."

Dutch Fortifications

Map of fortified border cities.

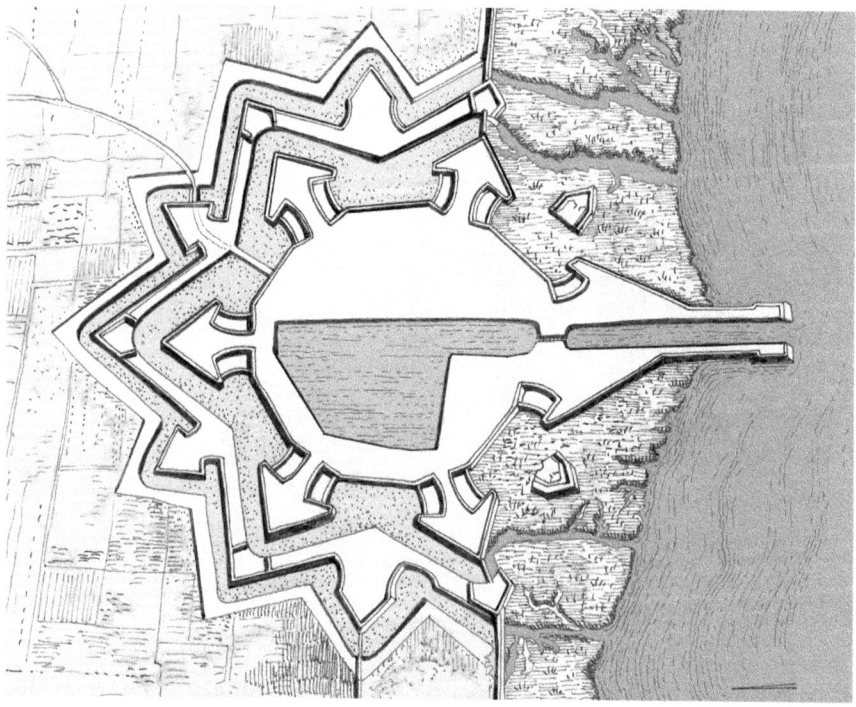

Project from 1688 by Menno van Coehoorn and his assistant Jan van Alberdingh for the military port of Hellevoetsluis. The influence of van Coehoorn's system is evident, but the project was not carried out, probably because of too high a cost.

on capitalism and international sea trade directed by a strong, austere, sober, hypocrite but rather tolerant and active Calvinist protestant community.

Menno Van Coehoorn was born in March 1641 at Britsum, a small village in the northern province Frisia. His ancestry was a modest country nobility probably originating from Sweden, and was traditionally military. The young Menno took service in the Dutch army in 1657 in his father's company, and quickly learned the trade of man-at-arms. In 1659 he was promoted to the rank of lieutenant and became an infantry captain in the Regiment Ernst van Aylua in 1660. His first combat occurred in 1665 when the German Catholic bishop of Münster launched a devastating raid through the northern province of Groningen.

During the war with France that began in 1672, van Coehoorn's company was engaged at Maastricht and he was wounded during the siege of July 1673. Menno participated to military operations during the Dutch War, notably at the battle of Seneffe near Namur in Belgium in August 1674. In November of the same year he was promoted to the rank of major. In April 1677 he took part in the battle of Mont Cassel, south of Dunkirk, and in 1678 to the battle of Saint-Denis, north of Maubeuge in France. His aggressive attitude and offensive spirit were rewarded by the rank of colonel commanding two battalions in the infantry regiment Nassau-Frisia in February 1679.

Van Coehoorn showed great interest in artillery. He introduced several improvements to guns for increasing mobility and firepower. He designed a small infantry portable grenade launcher that remained in service in the Dutch army until the 19th century known as Coehoon's mortar. His participation in sieges in the German Rhineland and Spanish Low Countries (Belgium) also aroused his interest for military architecture.

Menno Van Coehoorn had been shocked by the weaknesses of his country's defense during the Franco-Dutch War of 1672, and by the advanced siege warfare method developed by Vauban. In the 1680s, he was convinced that the time was ripe for a new system of fortification. By then, he designed new fortifications for the frontier city of Coevorden.

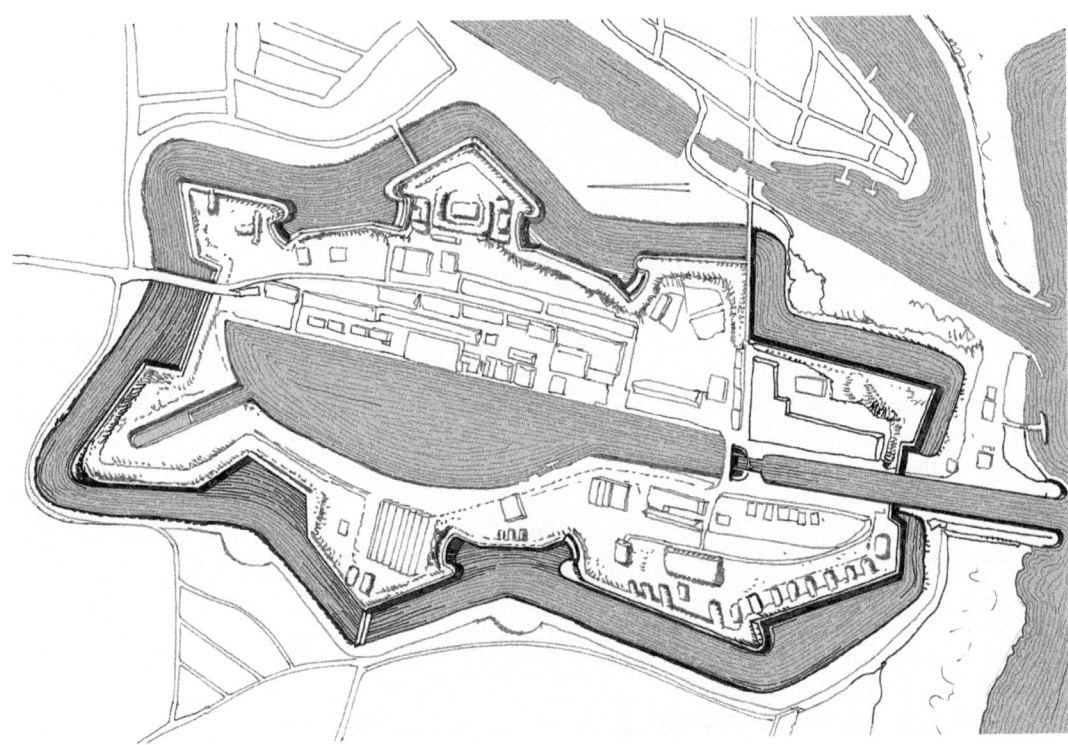

Hellevoetsluis. The military harbor Hellevoetsluis, situated in the province of South Holland, was created in 1598. The port included a dry dock, a sluice, stores, workshops, barracks, the admiralty house and a governor's residence. Hellevoetsluis was one of the most important harbors for the powerful Dutch navy. The naval installations were protected by a bastioned enceinte completed in 1695. Hellevoetsluis remained an important Dutch Navy base until 1957.

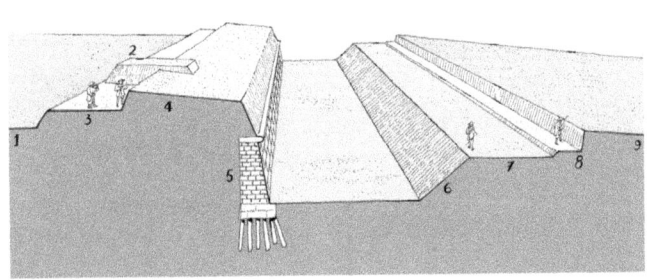

Cross-section of Menno van Coehoorn's fortifications: 1: Terre-plein. 2: Traverse. 3: Wallwalk. 4: Parapet. 5: Masonry scarp with foundations. 6: Sloping counterscarp. 7: Covered way. 8: Fire-step and breastwork. 9: Glacis.

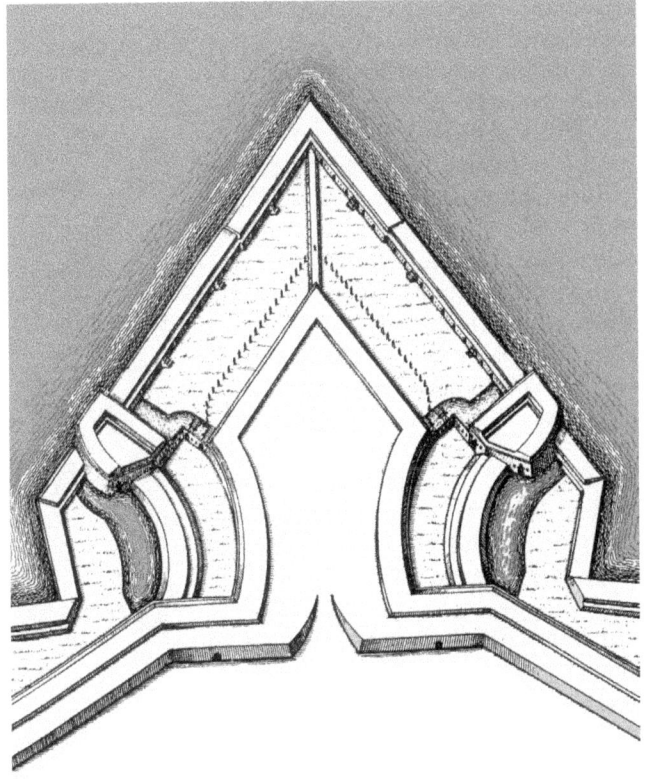

Bastion designed by Menno van Coehoorn.

His controversial project brought a quarrel with Naarden's designer, engineer Willem Paen. Van Coehoorn's design for Coevorden was refused because of its enormously high costs, but his conceptions were noticed and attracted attention. In 1682 he published a short polemical treaty entitled *Verstercking des Vijf-Hoecks met alle sijne Buyten-Werken* (Fortification of a Pentagon with all its Outworks). In 1684 he made a design for improving the defenses of the city of Grave (province of North Brabant). This project too was not retained but Menno van Coehoorn was not discouraged; on the contrary.

New Dutch System

Menno van Coehoorn's experience at war as well as the virulent quarrel with Willem Paen spurred his involvement in fortification. The result was a major technical book called *Nieuwe Vestingbouw op een natte of lage Horisont* (New Fortifications on a wet or low Horizon). In this treaty published in 1685, van Coehoorn presented a new defensive form, which eventually became known as Nieuw Nederlandsche Systeem (New Dutch System of bastioned fortification NNS) as replacement of the previously discussed Old Dutch System ONS.

Van Coehoorn's theories were, of course, not completely new. They were inspired by predecessors, e.g., Pagan, Ruse, Vauban and a mercenary engineer called Paul Storff de Belleville. Van Coehoorn's method took serious notice of Vauban's systematical siege warfare. Its most noticeable characteristic was the judicious adaptation to the marshy nature of the Low Countries. In *Nieuwe Vestingbouw*, van Coehoorn sharply criticized the obsolete Old Dutch system, his rival Willem Paen's view on fortification and Vauban's method, but he also expressed several new ideas regarding how to build fortifications.

Van Coehoorn's conceptions were characterized by the multiplicity and great saliency of the works, which were designed to suit the flat and often marshy sites of the Low Countries. His views were summed up in three theoretical and geometrical systems characterized by: the use of inundation; the combination of wet and dry ditches; the very broad main ditch with counterguard and large ravelins; the

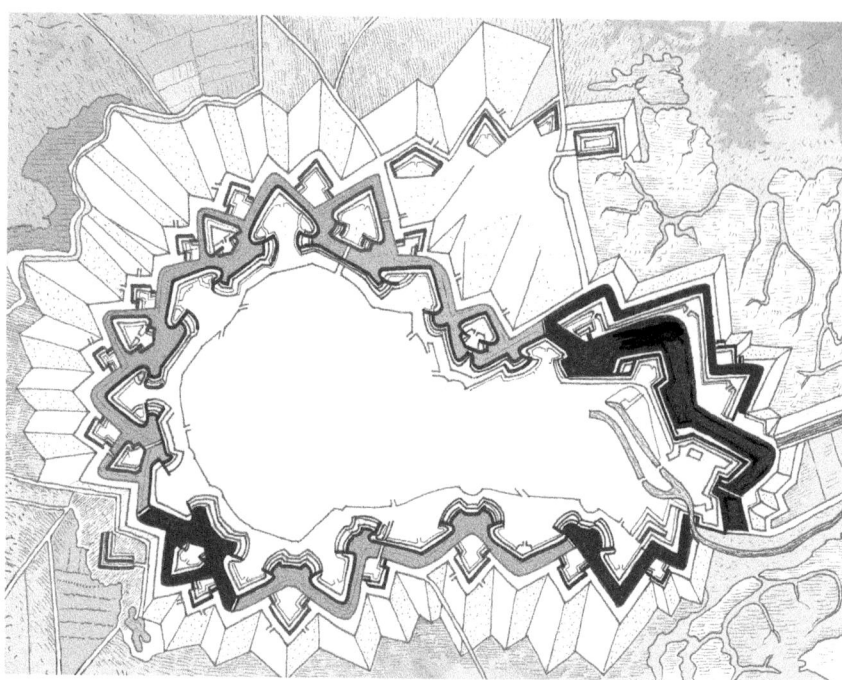

Fortifications of Bergen-op-Zoom. Menno Van Coehoorn's incontestable chef-d'œuvre was the fortifications of Bergen-op-Zoom situated in South-Brabant. Bergen-op-Zoom was one of the most important southern frontier cities. In 1700, Menno Van Coehoorn designed a powerful line of bastions, demi-lunes, ditches, a covered way with places of arms, a detached fort defending the harbor entrance and a vast flooding area. The fortifications were not completed until 1742. Unfortunately, the defenses of Bergen-op-Zoom were completely dismantled in 1869.

Boerenverdriet bastion at Bergen-op-Zoom.

roomy entrenchment (keep) of the ravelin; the replacement of the typical ONS fausse-braie with a tenaille (low parapet between two bastions); and the powerful enfilading fire from the bastions' flanks.

Van Coehoorn's book attracted some attention and

between 1701 and 1741 it was reprinted, re-published and translated into French, German, English and Russian. However his theories were not feasible because of their high costs. Also the area of ground taken by the defenses was five times the area within the body of the place, which leads one to wonder whether a fortress designed according to his theory could ever have accommodated sufficient garrison to man the main defenses and all the numerous outworks.

In the period between 1685 and 1691 Coehoorn served as a brigadier in the Dutch army. In June 1689 he took part in the sieges of Keizersweert and Bonn in Germany. He also fought at the battle of Fleurus near Namur (Belgium) in 1690. One might presume that he also worked as an engineer under Fortifications-General-Director François du Puy de Cambon and his successor, Charles du Puy de l'Espinasse. Indeed both fortification projects of Hellevoetsluis (1688) and Sas-van-Ghent (1694) present signs of strong influence by Menno van Coehoorn's theory.

Menno van Coehoorn's three systems

Van Coehoorn's systems were designed with specific principles in mind: to provide powerful flanking defense; to deprive any besieger of the means of digging approaches and artillery lodgments; and to facilitate possible sorties.

Menno van Coehoorn's first system served as the basis for several fortified cities, notably Bergen-op-Zoom, Breda, Hellevoetsluis, and Sas van Gent, as well as Namur in Belgium, and Mannheim in Germany.

Coehoorn's first system reintroduced Italian and French features: casemates, masonry and orillon. The lower part of the scarp (bastions and curtains) was made of masonry in order to resist erosion by water. The curtain angle was 120 degrees in order to ensure maximal flanking. The bastion flank was curved, fitted with two levels of gun emplacements to increase firepower, and protected by a large casemated orillon. The fausse-braie was replaced with a W-shaped tenaille and a counterguard (aka couvre-face) in the wet ditch. Demi-lunes and places of arms were fitted with a réduit (redoubt). The covered way was furnished with traverses. The counterscarp included flanking casemates and counterscarp galleries.

These were important contributions by Coehoorn to military architecture; indeed, placed in the reverse slope of a dry ditch, these elements enabled the defenders to bring fire to bear upon attackers who had got into the ditch; they were accessible from the place by underground tunnel. Advanced works (such as horn- and crownworks) were regarded as "useless furniture" and banned. Perhaps the most important feature was that Menno van Coehoorn had a clear understanding of the natural conditions of the Low Countries: the use of inundation was thus systematically encouraged and developed.

The second system, of which no example was ever built, included the same features as the first system but the emphasis was on increasing the defense in depth with two continuous external envelopes (joining ravelins and counterguards)

Van Coehoorn's first system.

Van Coehoorn's second system.

Van Coehoorn's ideal city.

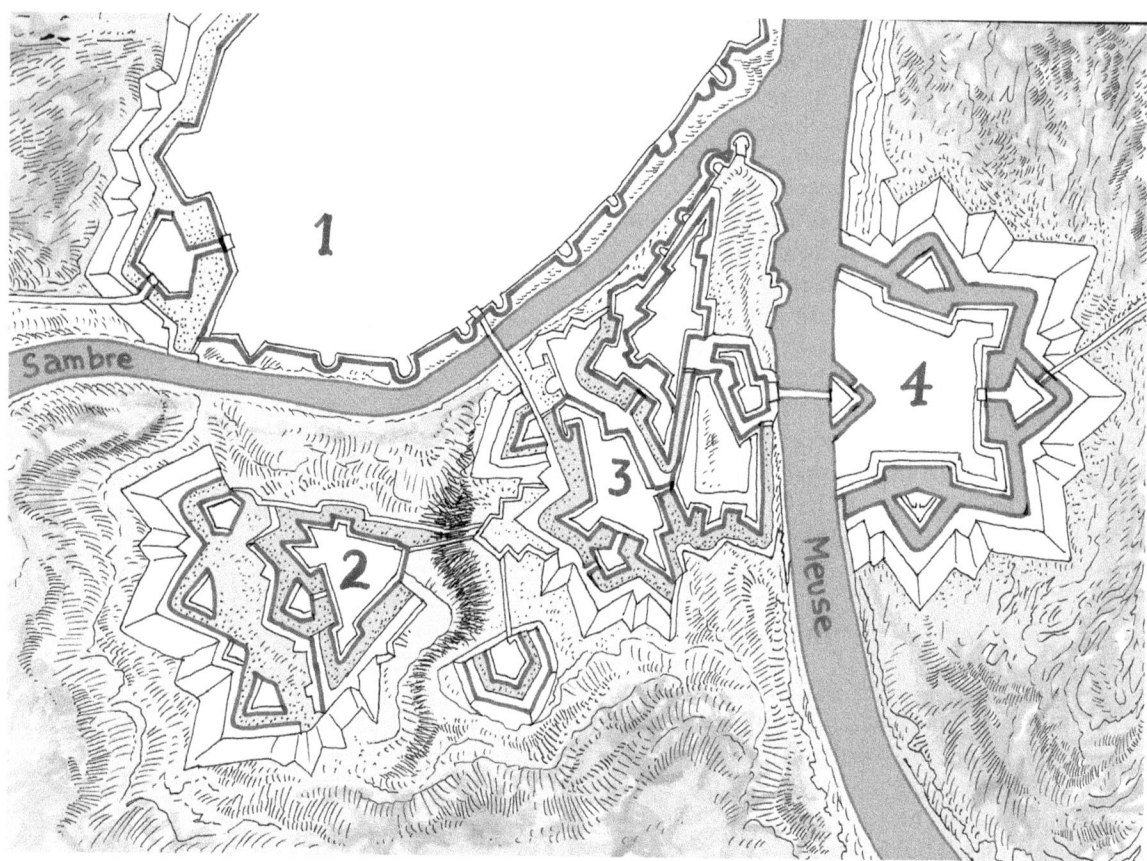

Fort Willem, Namur, Belgium. The town of Namur (1) lies in a strategic position at the confluence of two important rivers, the Sambre and the Meuse. Therefore the town was strongly fortified since the Middle Ages. In 1691 the Dutch fortification expert Menno van Coehoorn was sent to Namur to strengthen it amid fears of an imminent attack. He decided to add an outlying fort, known at first as Fort Coehoorn and later called Fort William (2). This self-contained stronghold was separated from the citadel called Terra Nova (3) by a small ravine. On the other bank of the Meuse River there was another fort called Bridgehead of Jambes (4) defending the Dinand Gate. Fort William was completed just in time to face the French siege of 1692 during the Nine Years' War. This siege was an epic battle between Coehoorn, who had been put in charge of the defense, and the skilled French engineer Vauban, who directed the attackers. After 37 days of siege, the French succeeding in capturing the hornwork in front of the citadel and the garrison surrendered.

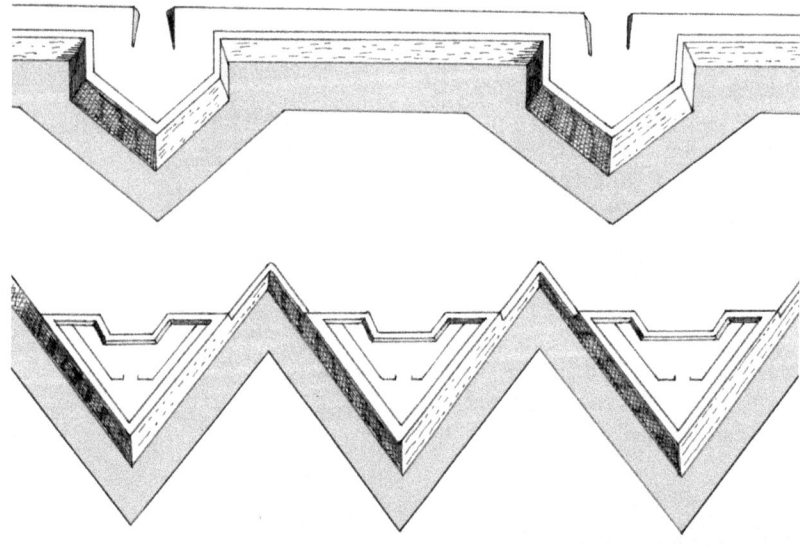

Bastioned front (*top*); Tenailled front (*bottom*).

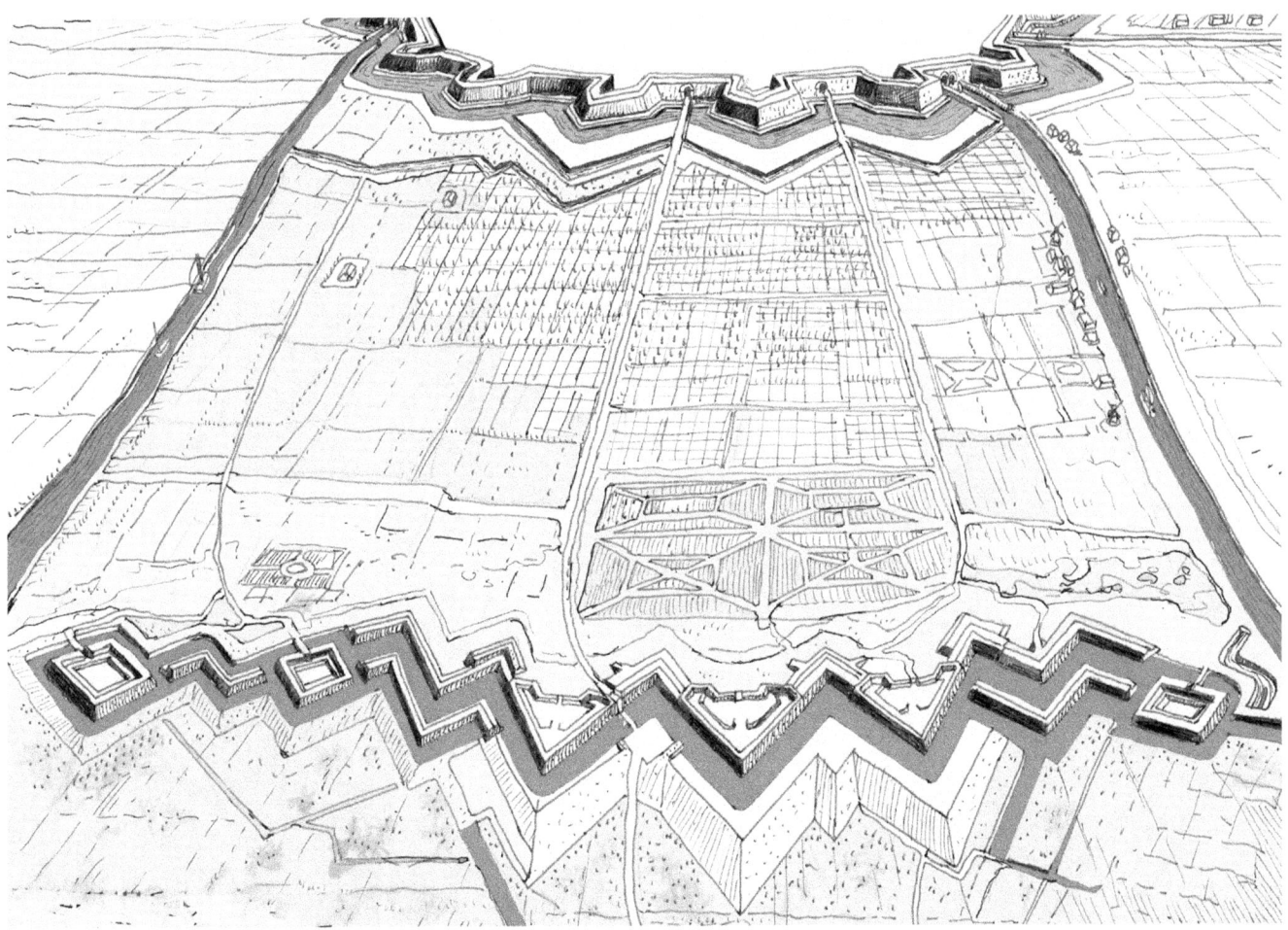

Tenaille line at Groningen 1700. The flanks of the line were protected by two existing deep canals forming a large entrenched camp.

separated by wet ditches. The third system (also totally theoretical) was an improvement to both first and second systems. Van Coehoorn abandoned the concept of the multiple envelope, and replaced it with vast self-defensible outworks (detached bastions, ravelins, and counterguards) placed in the broad wet ditch.

Menno Van Coehoorn's Works

General-director of fortifications

Menno Van Coehoorn was a theorist but above all a man of good sense, formed on the spot with a good war experience. His treaties (*Verstercking* and *Vestingbouw*) were published some years before he became much concerned with the actual construction of fortifications, and what he actually built was much simpler than what he had designed on paper. His realizations reflected his experience and his capacity for adaptation to local sites. His actual building was but little removed from Vauban's system.

In 1691, van Coehoorn's career reached a milestone when he worked in Belgium. He fortified Huy and Liège and designed a triangular bastioned fort (called Fort-Neuf of Fort Willem) at Namur. Louis XIV and Vauban besieged Namur in 1692. The city was obliged to capitulate and Fort Willem, commanded by van Coehoorn himself, was taken on June 30, 1692. Van Coehoorn had been wounded during the assault, and Vauban visited him on his hospital bed. The famous playwright Jean Racine reported on the battle and related the meeting of the two greatest 17th century military engineers. Though the gentlemen had a proper social conversation, they were rivals, enemies, and did not like each other. The resentful van Coehoorn never concealed his criticism of Vauban's way of fortifying, and the latter did not hide his contempt and found his opponent's methods of siege warfare brutal and unnecessarily cruel both for soldiers and civilians. Jealousy of trade?

After the siege, Vauban improved the defenses of Namur but could not complete the job. Namur was taken back by the Dutch army in 1695. Van Coehoorn actively participated to the reconquest, took his revenge and established a reputation as a city conqueror. During the combat, Fortifications-General-Director Charles du Puy de l'Espinasse was killed. Van Coehoorn was noticed by Willem III and obtained from the Stadhouder the title as a reward.

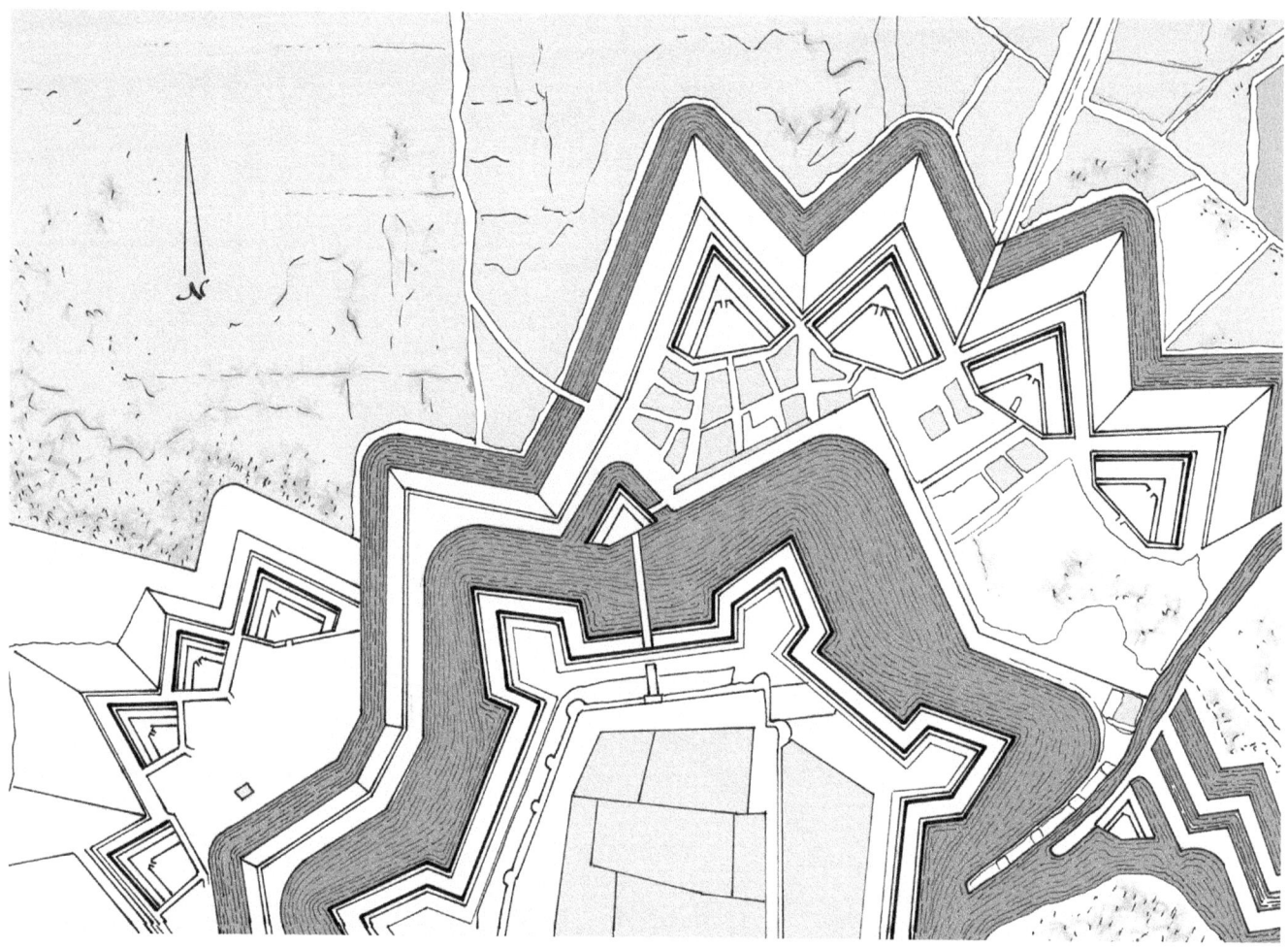

Tenailled line at Zutphen. Zutphen is situated at the confluence of rivers IJssel and Berkel, in the province of Gelderland. Taken by Prince Maurice of Nassau in 1591, the city was fortified in old Dutch style between 1592 and 1600. The defenses were modernized in 1701 by Menno Van Coehoorn with the addition of tenailled outworks. The fortifications of Zutphen were demolished in 1874.

From 1695 until his death in 1704, the career of Menno van Coehoorn reached its zenith. He accumulated the responsibilities of General-Director of Fortifications, Infantry-General-Lieutenant, and Artillery-General-Master.

In 1697 the Treaty of Ryswick allowed the United Provinces to place garrisons in cities in southern Belgium. In the 1700s Menno van Coehoorn organized a sort of Dutch "pré-carré," defending the borders of the United Provinces with fortresses, and inundations, including the following:

- The Line of Flanders in the south of province Zeeland with fortresses Sluis, Sas van Gent and Hulst.
- The Zuidwaterlinie (defenses south of the rivers Meuse, Waal and Rhine in province Dutch Brabant) with fortified cities of Bergen-op-Zoom, Steenbergen, Breda, Geertruidenberg, Heusden, Den Bosch and Grave.
- The IJssellinie (along the river IJssel covering the provinces of Gelderland and Overijssel) with the fortresses of Nijmegen, Arnhem, Doesburg, Zutphen, Deventer and Zwolle.
- The Noordoostenlinie (North-East Line protecting provinces Drenthe and Groningen) running from Zwolle to Coevorden, Bourtange and Nieuweschans.

Van Coehoorn also created a unit of 60 permanent professional architects and military engineers, which can be regarded as the origin of the Dutch Engineering Corps. With his professional collaborators he reorganized or designed new fortifications at Sluis, Hulst, IJzendijk, Cadzand, Geertruidenburg, Hulst, Vlissingen, Hellevoetsluis, Veere, Bergen-op-Zoom, Breda, Heusden, Den Bosch, Grave, Nijmegen, Arnhem, Doesburg, Zutphen, Deventer, Zwolle, Hasselt, Coevorden, Bourtange and Groningen. Wherever possible, Van Coehoorn made use of marshes, canals and rivers to create flooding zones between fortified cities in order to make borders impenetrable. It must be borne in mind that van Coehoorn operated in the Low Countries where excavating for more than four feet often came to ground water. In the Netherlands and the western part of Belgium, water and swampy ground were commonplace features that could be turned to useful account.

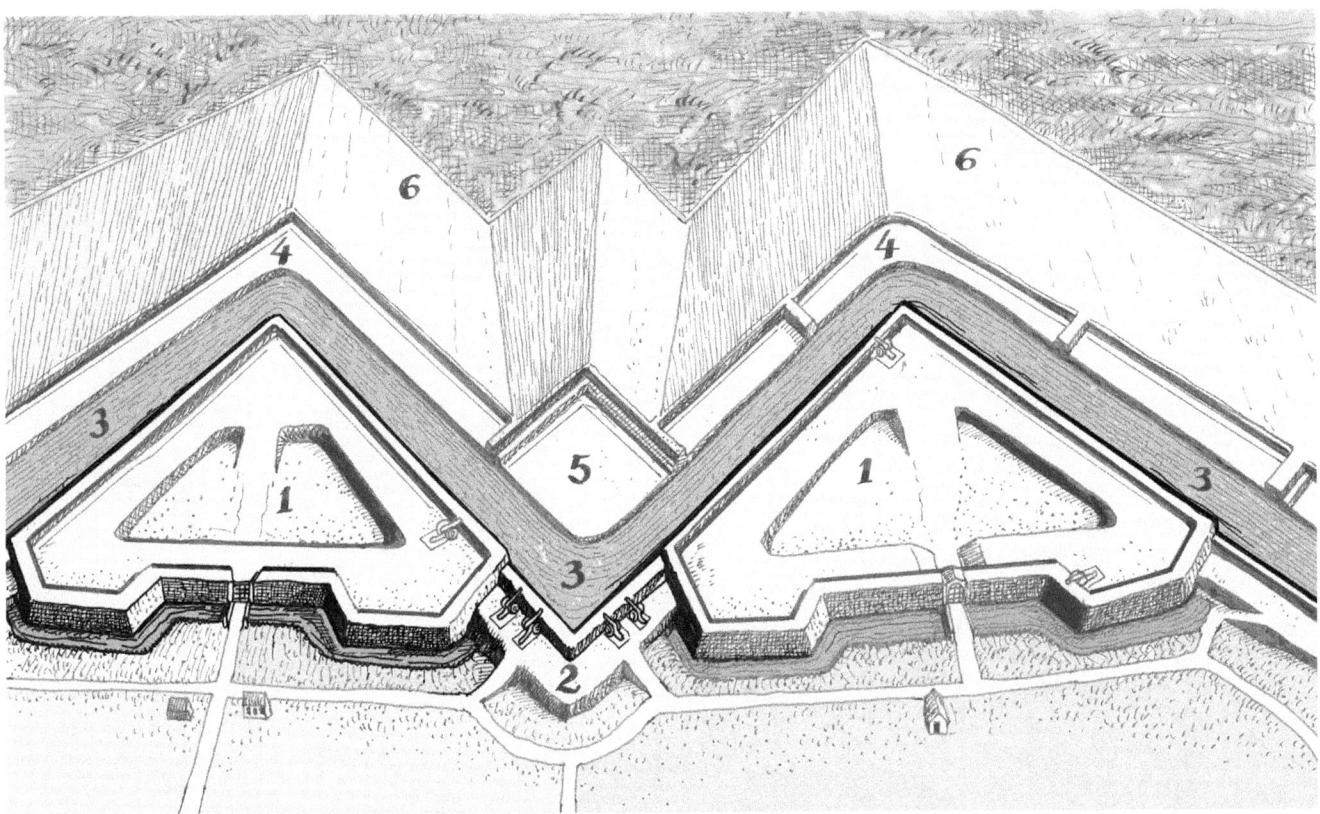

Tenailled lunette (Groningen 1700): 1: Enclosed lunette. 2: Gun battery. 3: Moat. 4: Covered way. 5: Place of arms. 6: Glacis.

At the same time as Vauban, Menno van Coehoorn pleaded for the constitution of vast *camps retranchés* (entrenched camps). Established outside cities, entrenched camps allowed for regrouping large armies. When it came to a siege they complicated the besiegers' approach.

Tenaille system

Van Coehoorn also reintroduced and modernized an existing concept of defense called a *tenailled system*. This method was characterized by the abandonment of curtains and bastions, to be replaced with redoubts (called lunettes) and batteries; the lunettes were closed at the gorge and therefore autonomous and difficult to take. Lunettes and batteries formed together a saw-tooth outline with right angles of 90 degrees. This system, also called the *perpendicular method*, was cheaper to build than a classical bastioned front, flanking was accurate and fewer soldiers were required for defending it. The main drawback was the long faces, which could easily be enfiladed by enemy fire. This weakness could be solved by *traverses* placed at regular intervals across the faces. That compartmentalizing confined the deadly effects of enemy projectiles to only a limited part of the position.

The tenailled system, however, was never really popular. It never supplanted the bastioned system, and only a few tenailled forts and lines were built in the 18th century—but

Tenailled line at Doesburg.

the idea was revived in the 19th century in the so-called Polygonal Fortification.

Van Coehoorn combined both methods (entrenched camp and tenailled enceinte) at Groningen (Helpman Line built in 1700), Nijmegen and Doesburg.

Menno van Coehoorn's Influence

Menno van Coehoorn's fame as a fortification builder and as an hydraulic engineer was so widely spread that several European kings courted him to serve them. Being hired as a mercenary was a normal and well paid practice in the 17th century, but van Coehoorn declined all solicitations, and faithfully stayed in service to the United Provinces. During the first two years of the Spanish Succession War (1702–1714), Menno van Coehoorn, aged 63, was infantry commanding officer in Flemish Zeeland and actively participated in action. As a besieger he took part in the

Coehoorn mortar. This small lightweight portable infantry grenade launcher, designed by Menno van Coehoorn, was adopted by the Dutch army in 1701 and remained in service until the 1860s.

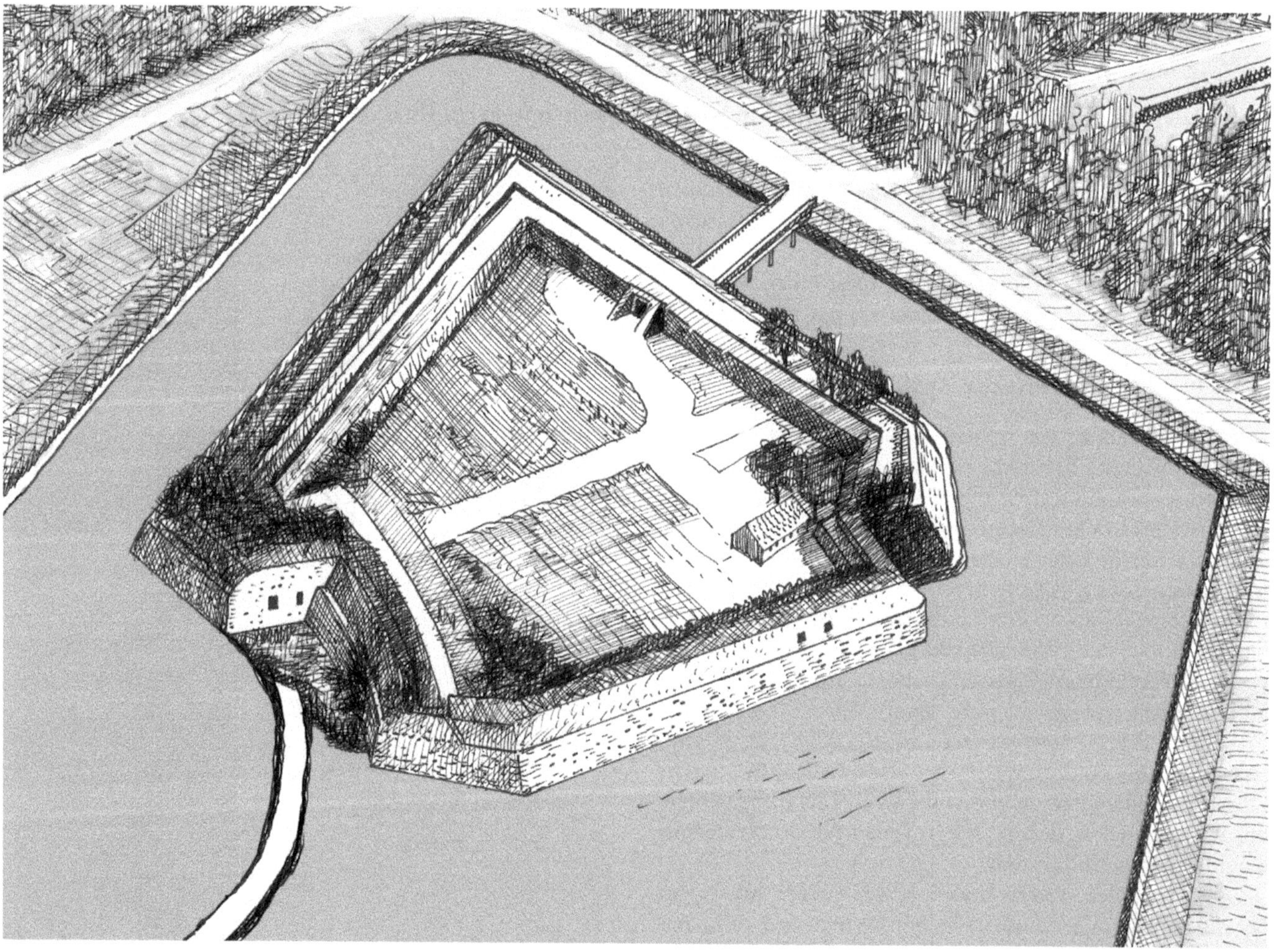

Demi-lune in Bergen-op-Zoom. This ravelin, located in the Pielekeswater near the Anton van Duinkerkenpark is one of the few extant examples of Menno van Coehoorn's work.

Hampoort Gate at Grave. Located in St. Elisabeth Street at Grave (province of North Brabant), the Ham gate was a part of the urban fortifications designed in 1688 by Menno van Coehoorn. The portico-shaped gate was constructed near the place of an old medieval gatehouse dating from 1309. The pediment carries the arms of the Republic of the United Provinces, and of William III Prince of Orange.

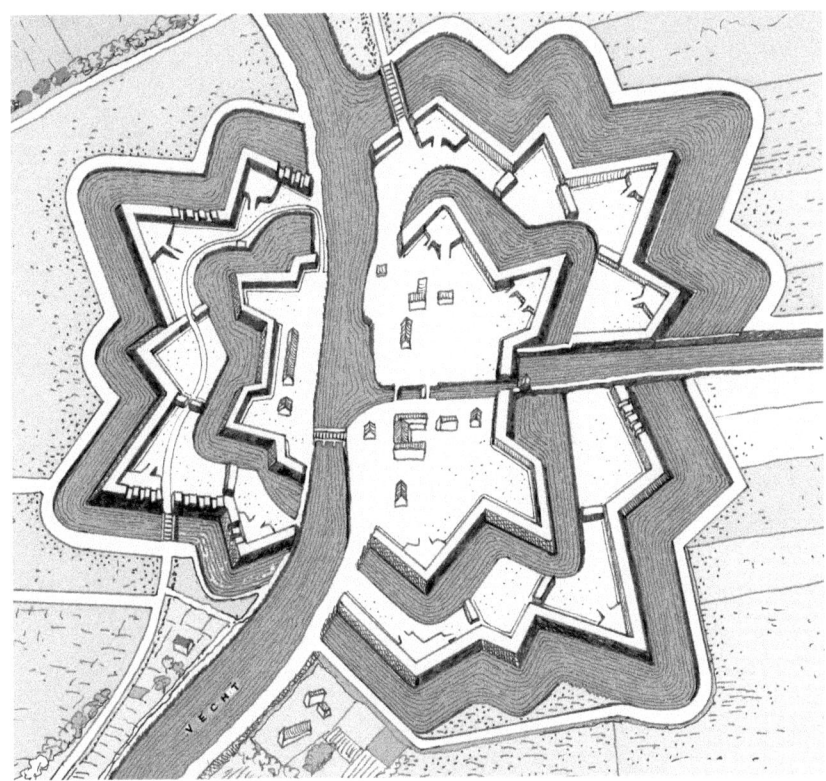

Tenailled Fort Nieuwersluis 1775. The concept of tenailled or perpendicular front pioneered by Menno van Coehoorn was further developed by the German military engineers Rimpler and Sturm, and by the French theoretician René de Montalembert. The tenailled system never supplanted the bastioned method, but several tenailled forts were built in the 18th century—e.g., Fort Nieuwersluis (depicted) from 1775 in the province of Utrecht.

capture of Huy in Belgium and Bonn in Germany.

Even though Menno Van Coehoorn's work is far less considerable than Vauban's, he was an important engineer with a great reputation. He is the creator of the Dutch engineering corps and practitioner of entrenched camp theory. His bastioned systems were applied in simpler fashion in Germany and in eastern Europe in the 18th century. For example, the fortifications of Belgrade and Temesvar (Timisoara in Romania) were strongly influenced by van Coehoorn's second system.

Van Coehoorn's book *Nieuwe Vestingbouw* was published in English in 1705, and according to the British academic Quentin Hughes (1920–2004), several of his proposed features were used by English engineers notably at places like Minorca and Gibraltar. The tenailled method and the introduction of the counterscarp gallery, without doubt, influenced Montalembert's perpendicular system, which gave birth to 19th century polygonal fortification.

If you ask in France who Vauban was, you may not expect a right answer but you are likely to find that the name has been heard of. This is not the case with van Coehoorn. Today in the Netherlands, Menno van Coehoorn is an unknown person to the public at large (except a few engineering officers, military buffs, and fortification historians) because Dutch people are less interested in history and much less nationalist than the French. But there is another reason. Compared to the huge legacy left by Vauban's, very little of van Coehoorn's realizations have been preserved: only a demi-lune near the Anton van Duinkerken public garden at Bergen-op-Zoom (Province of North Brabant); the Hampoort gatehouse at Grave (Province of North Brabant); and remnants of the tenailled line built at Doesburg (Province of Overijssel).

Menno Van Coehoorn died on March 17, 1704, in The Hague, attending a conference with the British General John Churchill, 1st Duke of Marlborough, commander of the Anglo-Dutch forces. Van Coehoorn is buried in the small church of Wijckel (municipality of Fryske Marren, province of Frisia) near Sloten where he had his residence in his native province Frisia. The Frisian States sponsored a monument over his grave that was decorated by the artists Daniel Marot and Pieter van der Plas. Van Coehoorn's son Gosewijn Theodoor wrote a biography, which was republished in 1860.

The name van Coehoorn, however, still lives on in Dutch army circles.

An infantry barracks was named after him at Arnhem from 1883 to 1967.

A private association was created on April 18, 1932, and called Stichting Menno van Coehoorn. Through the present its members devote their skills, talents and time to the preservation and promotion of Dutch historical fortifications.

In 1950 the Third Infantry Regiment of the Dutch Royal Army was redesignated Infantry Regiment Menno van Coehoorn. His family coat-of-arm (two standing black bears and two horns) has been given to the 47th Armored Infantry Battalion. Several towns have a street or a square bearing the name of this mostly forgotten and unsung hero of the 17th century.

Chapter 6

The 18th Century and the French Era

The 18th Century

Dutch decline

In the 18th century the domination of the great absolutist realms and empires of France, Austria, Russia, and Prussia started on the European continent, while British supremacy on oceans and seas gradually increased, encroaching on and finally outclassing the Dutch. The Republic of the Seven United Provinces often found itself quarrelling with surrounding countries and was also drawn into several major European wars. At the same time the dynamism and energy of the previous century dwindled. The Republic remained wealthy (at least its oligarchy), but it had no longer enough power to match the growing English and French competition.

In the long term, at sea the Netherlands were unable to keep up with Britain that had started a tremendous economic development (soon known as the "Industrial Revolution") and colonial growth, while France, and neighboring Prussia as well became stronger and more expansionistic than ever. So many enemies and competitors would have necessitated way higher military expenses than the Netherlands could afford. As a result the supremacy of the "Golden Age" of the tiny Dutch Republic came into decline and ultimately an end. Dutch political and economical interests became increasingly dependent on the goodwill of Britain, and Amsterdam lost to London her position as the trading center of Northern Europe.

The Barrier Treaty

The so-called Barrier Treaty referred to a series of agreements arranged and supervised by Britain and ratified between 1709 and 1715 that created a buffer zone between the Republic of the Seven United Provinces and France. The Barrier agreement granted the Dutch the right to occupy a number of fortresses in the south of the Austrian Netherlands—that was how Belgium was designated between 1714 and 1797. The Dutch were allowed to maintain garrisons in the following Barrier cities: Knokke, Ypres, Waasten, Veurne Dendemonde, Menen (aka Ménin), Doornik (aka Tournai), Mons, and Namen (aka Namur). In peacetime the Dutch were allowed to have 30,000 soldiers, and a maximum of 40,000 in time of war. Their maintenance was paid by Austria. As for the fortified cities of Aath, Bergen and Charleroi located between Doornik and Namur they were held by Austrian garrisons.

According to British interests the strategy behind that line of fortified cities was to create clusters of obstacles between the Flemish coast on the North Sea and River Maas (Meuse). It was meant to keep the French at bay and—in the eventually of aggression—fight on foreign Belgian soil. In practice, in peacetime the Barrier scheme was an unlikely solution that only generated a lot of financial problems, irritations and political tensions between the populations of Belgium, Austria and the Republic of the United Provinces.

During the War of Austrian Succession (1740–1748) the military worth of defensive lines proved wanting. Ultimately the plan worked ineffectively as a means of defense, particularly after 1756 when Austria and France entered into an alliance. Then there was in fact no purpose in the Barrier Treaty any more. The whole scheme was cancelled by Austria in 1781.

The Barrier represented a first advanced line of defense. As early as 1700 General-Director of Fortifications Menno van Coehoorn, and during the 18th century his followers, advocated the establishment of a continuous system of inundations. Together with the Holland Waterline, there were inundations planned around marshes in eastern Groningen, in the region of River Vecht, in the province of Overijssel, at the southern border in North Brabant, and in Zeeland. However, in the 18th century, the defenses were poorly coordinated, all too often neglected, and not properly carried out at a national level with centralized command.

War of Austrian succession

For its defense against France the Dutch Republic relied upon the Barrier cities in Belgium. However the General-Director of Fortifications between 1709 and 1730, Guillaume le Vasseur des Rocques, tried to alarm the political authorities about the weakness of the defenses in several reports he wrote in the 1720s. Following Menno van Coehoorn, Le Vasseur des Rocques advocated the establishment of inundations on the eastern bank of the River IJssel and between Grave and Bergen op Zoom, as well as the reinforcement of the defense lines between Arnhem, Nijmegen and Grave, but to no avail. Nothing (or very little) was done to improve the defense of the actual borders of the Republic.

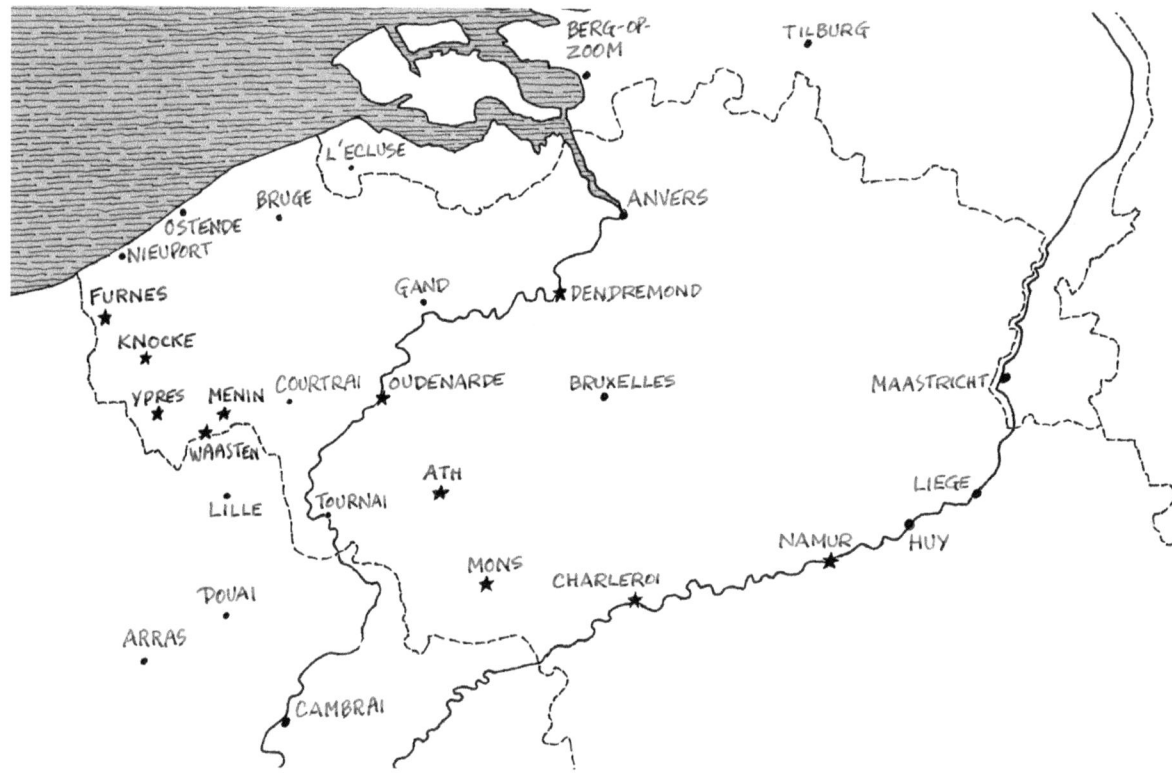

Map of the Austrian Netherlands with the main Barrier fortresses.

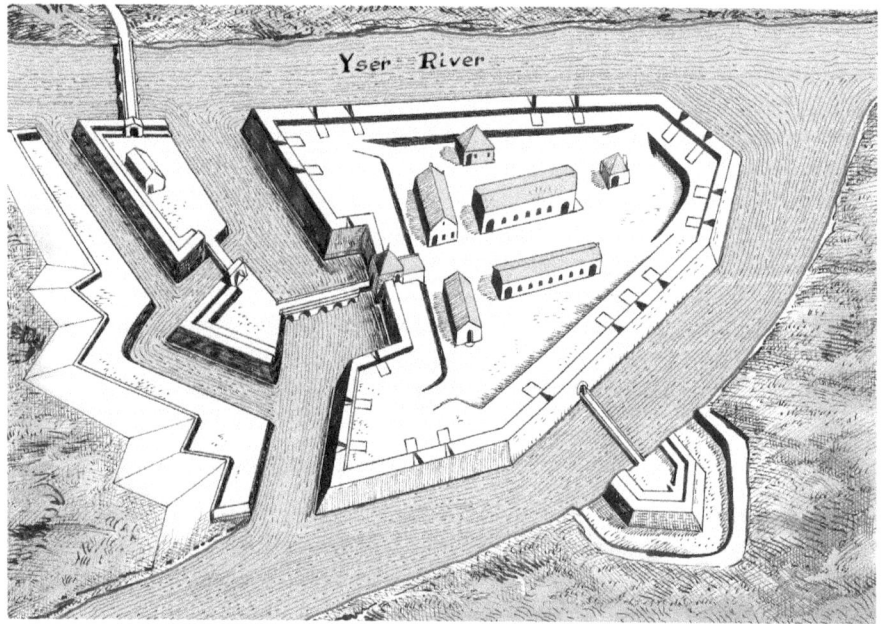

Fort Knokke (aka Fort de la Kennocque or Fort de Knocke) was located at the confluence of the Yser River and the Ypres Canal about 8 kilometers (5 mi) southwest of Diksmuide, western Flanders, Belgium. An important stronghold built in the 1580s, it was successively occupied by the Spanish Empire, the Kingdom of France, Habsburg Austria and the Dutch Republic. It was modernized in 1678 by Louis XIV's celebrated military engineer Sébastien Le Prestre de Vauban, and held by the French until 1712. Fort Knokke was one of the strongholds in the Austrian Netherlands occupied by the Dutch according to the Barrier Treaty in 1713. The French captured the fort after a two-month siege in 1744 during the War of Austrian Succession. In 1781, the Austrian Emperor Joseph II ordered its demolition.

The Grebbeline was established in 1745, however, in order to reinforce the defense of Amsterdam. Designed by the military engineer Bernard de Roij Jr., it comprised inundations and fortifications spreading in the flat valley between the Utrecht hills and the forested Veluwe region. It ran in a southeastern direction from the Zuiderzee (IJssel Lake) in the North down to Amersfoort, along Veenendal, and down to Rhenen on the Rhine River in the South. The Grebbe Line was neglected in the 19th century but was reactivated in 1939.

The Dutch Republic was involved into the War of Austrian Succession (1740–1748), a series of conflicts breaking out after the death of Emperor Charles VI of Habsburg in 1740. In 1744 and 1745 the French laid siege to Dutch-held Barrier fortresses at Menen and Tournai. This persuaded the Dutch Republic in 1745 to join the Quadruple Alliance, but this coalition was badly crushed at the Battle of Fontenoy in May 1745. In 1746 the French occupied most of the Barrier fortresses, and many large cities in the Austrian Netherlands.

Chapter 6. The 18th Century and the French Era

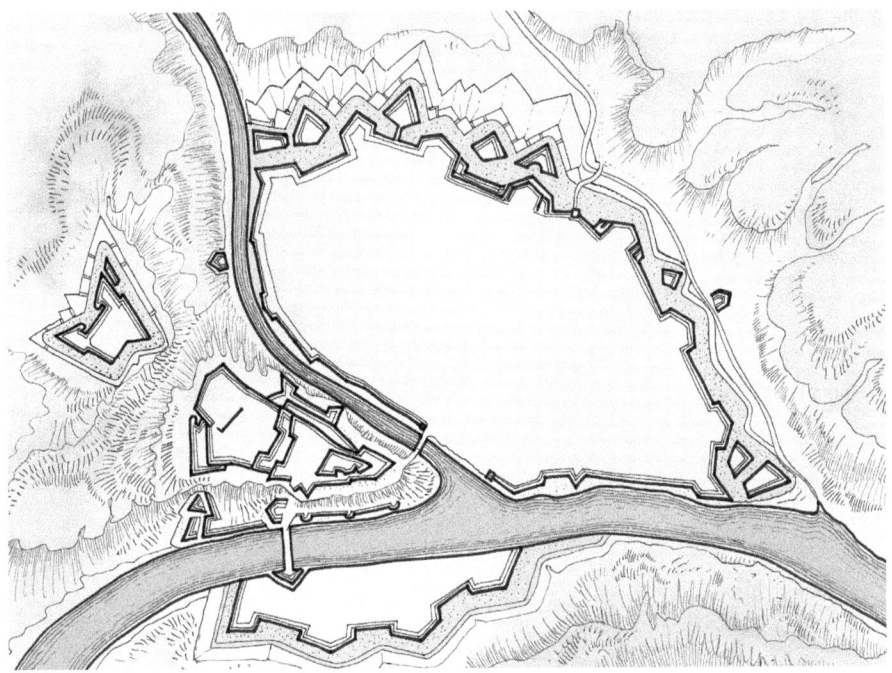

Namur (Namen in Flemish) is located at the confluence of the Sambre and Meuse rivers. The city was founded in Celtic times and developed during the Roman era. The fortifications of the city followed the evolutions of warfare and were reconstructed, adapted and modernized several times, notably by Menno van Coehoorn, and Vauban after the siege of 1692. In the 17th and 18th century various subsidiary positions were built, including a strong citadel. In the 19th century the urban bastioned enceinte and outworks were obsolete, and replaced with a ring of independent polygonal forts. Together with Dinant, Huy and Liège, Namur formed part of the so-called Meuse Citadels.

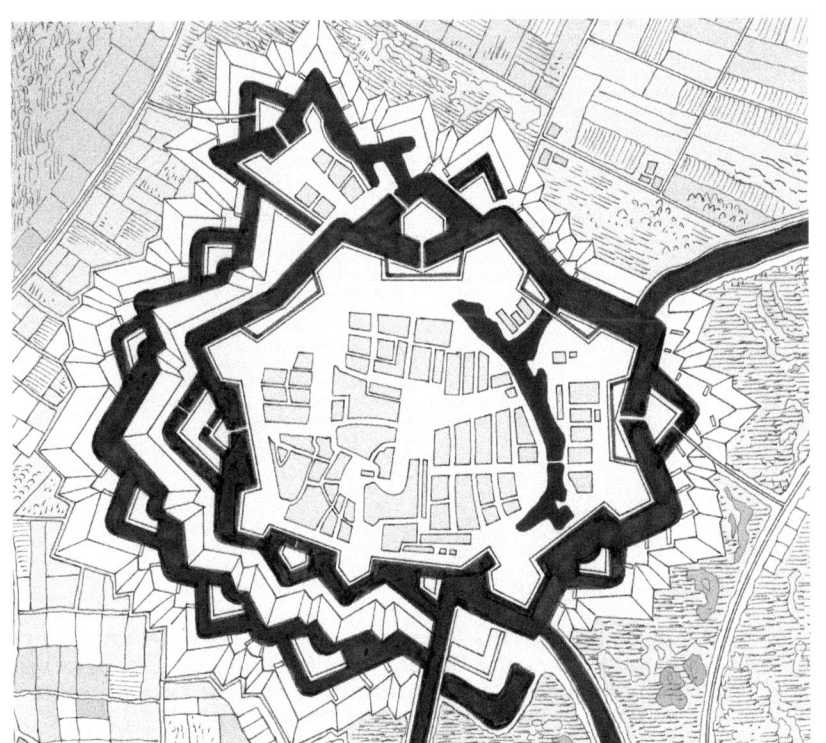

This reminder of the French invasion in the "Year of Disaster" 1672 caused panic and the hasty restoration of the stadtholderate in the person of Prince of Orange William IV. From 1746 to 1748 the war dragged on indecisively, and ultimately was brought to an end with the Treaty of Aix-la-Chapelle (1748). The War of the Austrian Succession saw the appearance of a new and strong power on the European stage: Prussia.

The period 1752–1784

This period was marked by a long regency stained by corruption, disorder, and misrule. All power was concentrated in the hands of a few oligarch families while William V of Oranje assumed the function of stadtholder.

Very little improvements were brought to the national fortifications during this time, even though Carel Diderik Du Moulin (General-Director of Dutch Fortifications from 1730 to 1774) had warned on numerous occasions about the neglect and weakness of the existing defenses. Du Moulin,

Furnes. The city of Furnes (Veurne in Flamish) originated from a small village named Furnae in 877. The city developed around a motte-castle built about 1040 by the counts of Flanders. Fortifications comprising a broad moat and an earth wall were built during the French-Flemish war of 1213–1214. These were replaced with a stone wall with gates and 33 towers between 1388 and 1414. By 1578, the four gates were adapted to the increasing power of firearms by the addition of artillery bulwarks. During the Nine Years' War (1688–1697) Furnes was incorporated into Vauban's Pré-Carré to help protect the strategically important port of Dunkirk. New modern bastioned defenses, designed by Vauban, included a bastioned enceinte, outworks and wet ditches. Pressures of time and limited finances meant that the new works were built in several stages between 1693 and 1713. When the fortifications were hardly completed, Furnes came under the control of the Austrian Empire as a consequence of the Treaty of Utrecht in 1713, and became a border fortress against the French. The fortifications were totally demolished in 1781 by order of Emperor Joseph II of Austria.

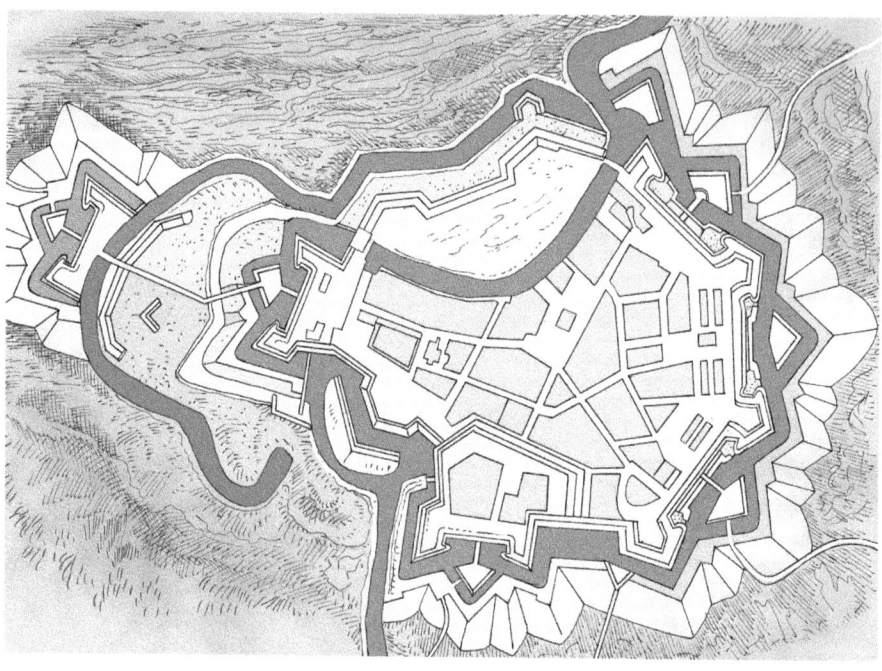

Menin. Situated between Tourcoing (now in France) and Kortrijk (Courtai, now in Belgium), grew from 1087 as a crossroads over the River Leie. Because of its strategic position on the roads connecting Lille to Bruges, and Yper to Kortrijk, Menin was besieged no fewer than 22 times between 1579 and 1830. In the 12th century, the lords of Menin constructed a fortified residence and the town developed into an important medieval center of the cloth industry. The small city was further fortified in 1578 during the Wars of Religion with six Italian-styled bastions. In 1678, Menin was captured by the French and included into the Pré-Carré when Vauban rebuilt the fortifications which had been pulled down some twenty years before. Completed by 1689, the defenses included 11 bastions, four gates, a wet moat, tenailles, ravelins, two advanced hornworks, a glacis, and a flooding area along the Leie river plain. For political reasons, the fortifications were demolished in 1774 under the reign of Louis XV. After the fall of Napoléon I, in 1817, the defense was rebuilt by the Dutch, following the French layout, but with significant changes including casemates and bomb-proof shelters. Completed in 1830, the fortifications were again dismantled in 1852. Fortunately the demolition was only partly carried out and Menin has preserved interesting vestiges of Dutch-styled fortifications.

Siege of Bergen-op-Zoom (July 12–September 16, 1747). The map shows the French siege works (parallels and saps) for attacking the Coehoorn Bastion (*left*), the Pucelle Bastion (*right*), the Dedem Ravelin (*middle*), as well as the covered way and the places of arms. The siege of Bergen op Zoom attracted the attention of all of Europe and the capture of the city forced the Anglo-Dutch allies to enter negotiations leading to the Treaty of Aix-la-Chapelle in 1748, putting an end to the Austrian War of Succession (December 1740–October 1748).

Chapter 6. The 18th Century and the French Era

Siege Fort de Roovere 1747. Located near Halsteren (province of North Brabant), Fort De Roovere was an earthen sconce built by order of Prince Maurice of Nassau in the early 17th century as a part of the Dutch Waterline (Hollandse Waterlinie). In 1747, during the Austrian War of Succession (1740–1748) the fort was besieged by the French. When the nearby fortified city of Bergen op Zoom was captured, the siege was lifted. In the 2010s extensive renovation of the fort have taken place for recreational functions, notably preservation of nature, cycling and hiking.

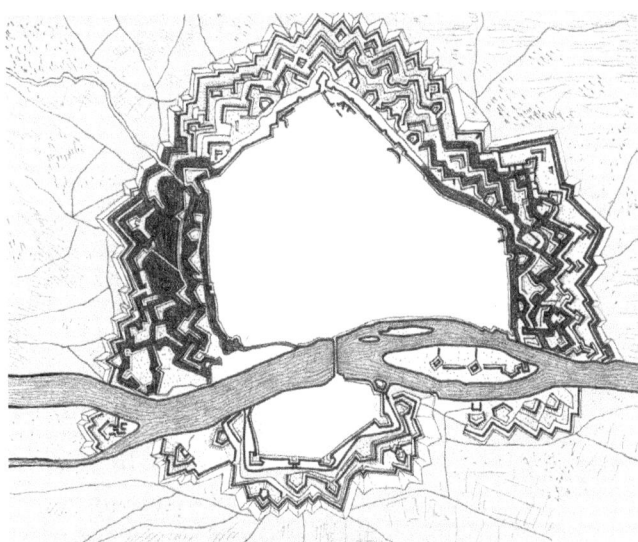

Maastricht ca. 1784. Strategically situated on the river Meuse in the south of the Netherlands (province of Limbourg), Maastricht (called Traiectum ad Mosam in Roman time) has always been an important crossroads on the way from Liège to Cologne. Therefore the city was always fortified. Two main walls with towers and gatehouses were successively erected in the Middle Ages, as well as bastions in the Renaissance. During the Franco-Dutch War in 1673, the city was captured by the French. The French occupation of Maastricht, from 1673 to 1678, was marked by repair and improvement of the fortifications with the construction of many out- and advanced works. Maastricht was yielded back to the Dutch after the treaty of Nimegue in 1678. The fortifications were significantly reinforced in the late 18th century by the Dutch. They were dismantled between 1871 and 1878 but many vestiges fortunately can be found today.

just like his predecessor Le Vasseur des Rocques repeatedly demanded new modern guns, improved inundations, new forts in key positions and better and enlarged urban fortifications around important border cities like Breda, Bergen op Zoom, Hulst, Sas van Gent, and Venlo.

In Maastricht, Du Moulin managed to obtain significant improvement. In the 1770s modern fortifications, inundations and countermine galleries were added to Maastricht—the capital of the southernmost province.

Decline of the Republic

During the Seven Years' War from 1756 and 1763, the Republic was allied to Britain by a defensive alliance, but also bound to France by an advantageous commercial treaty. So the Dutch were in a perilous and delicate situation, but managed to keep out of the war, and made considerable profits.

Things then went badly wrong during the American War of Independence (1775–1783). The Republic of the United Provinces was still allied to Britain but refused to enter the war. Furthermore, because of trading disagreements with Britain, the Dutch established diplomatic (and fruitful economical) contacts with the American insurgents (notably selling weapons and ammunitions). As a result the disastrous Fourth Anglo-Dutch War (1780–1784) broke out. This conflict was marked by a series of successful British attacks against Dutch colonial economic targets. The war ended tragically for the Dutch and increased their decline. The Treaty of Paris (1784) consolidated the British hegemony, and sounded the knell of Dutch greatness.

In the late 18th century, economic and social crises raised tensions in the Republic to new heights. Political movements and liberal factions questioned the ability of the oligarchy and Stadhouder William V to govern. After 1784, the declining Republic grew restless under William's rule. There was turmoil and agitation, and an open rebellion was started by the Patriots' movement. Influenced by the American Revolution, the anti–Orangist and pro–French Patriots sought a more democratic form of government. The Republic and the Orange stadhouder reacted with severity, relying on Prussian forces and a small contingent of British troops to suppress the rebellion.

Batavian Revolution

Although still a strong colonial power by the last decades of the 18th century, the Netherlands had lost her position of leading world power. The Republic of the United Provinces was internally collapsing due to the aged government structure and lack of innovation. Meanwhile France had only become stronger, and warfare itself had changed, too. The old Republic finally broke down after an era of sustained political decline, and economic decrease. By then the 1789 Revolution had broken out in France. The U.S. Declaration of Independence (1776) and the French Revolution (1789–1799) introduced new notions of universalistic and progressive institutions like individualism, administrative rationality (notably standardized measurements), nationalism and patriotism, centralized government, parliamentarianism, equality, and many other modern principles. These ideas were popular in the Netherlands, and underground clubs and secret factions (notably the Patriots' party) were promoting it.

In early 1793 the French revolutionary army invaded the Low Countries. General Dumouriez's army captured Breda, Geertruidenberg, and Klundert. Actually the Netherlands was not really conquered by Revolutionary France. It was a popular uprising that gave the Netherlands to the French. A part of the Dutch population revolted and started a new republic with the support of French forces. Soon the Dutch underground Patriots rose up, overthrew the municipal and provincial governments, and proclaimed

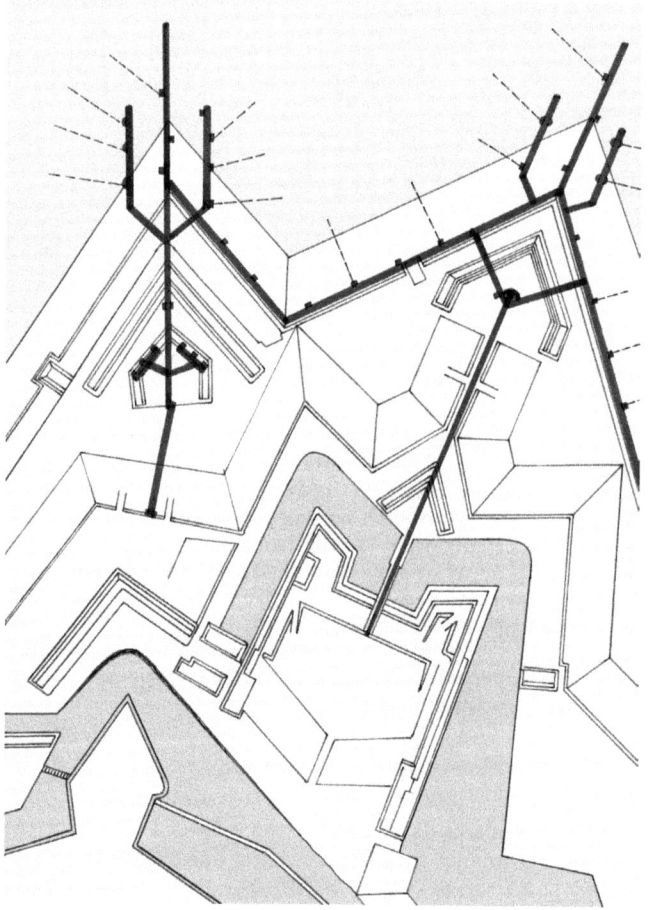

Hoge Fronten at Maastricht. Between 1753 and 1777, the fortifications of the city of Maastricht (particularly the northwest part) were greatly modernized under the supervision of the following General Directors of Fortifications Peter de la Rive, Carel D. Dumoulin and F.S. de Veye. The modern fortifications were influenced by the work of Menno van Coehoorn. Note that the black lines on the plan indicate the main underground countermine galleries. Today a significant part of the Hoge Fronten has been preserved and turned into an oasis of calm and green in the busy city.

Underground countermine galleries at Maastricht. The black lines indicate permanent masonry tunnels, underground listening-posts (aka écoutes), and subterranean ramifications. Arranged in a symmetrical, regular pattern, they were designed and built together with the fortress. They had their entrances on the counterscarp under the covered way or under a ravelin. From the main galleries, secondary branches could be dug when enemy activities had been detected and located. Then gunpowder was placed and ignited with the purpose of blasting enemy mine galleries, killing attacking miners and exploding defense works taken by the besiegers.

the *Bataafse Republiek* (Batavian Republic) in Amsterdam. In 1795 Stadthouder William V fled to England, the States General was dissolved, and the oligarchic, sclerosed, and corrupt Republic of the Seven United Provinces was abrogated.

The French Era

From 1795 to 1814 the Netherlands were ruled by the French, a period known in the Dutch history as the *Franse Tijd* (French Era). Upon the ruins of the Republic of the United Seven Provinces, the newly created Batavian Republic adopted many positive French Revolutionary institutions, notably the useful metric system, the stable and united system of minting, the liberty of religious worship, the abolition of privileges, and the French equalitarian Code of Laws. However, to the Patriots' dismay, the new Republic, instead of bringing a democratic regime with more freedom, gradually became virtually a puppet of France, particularly after the seizure of power by Napoleon Bonaparte.

The occupation of the Low Countries with their rich, hard-working trading citizens, their good fleet, the important ports of Amsterdam, Rotterdam, Vlissingen, and Antwerp (in Belgium) allowed the French to control an important part of the North Sea, a significant and major asset for the war and the blockade aiming to starve Britain. In 1806, as Napoleon grew impatient and dissatisfied with the Dutch Republicans, the country was instituted as the *Royaume de Hollande* (Realm of Holland) with Louis-Napoléon (1778–1846)—the Emperor's youngest brother as king. Ironically, the first king of the kingdom of the Netherlands was thus a Corsican Frenchman. Louis (called Lodewijk in Dutch) was also the father of the future Emperor Charles-Louis-Napoléon III (1808–1873, reign 1852–1870).

Louis made his official entry to The Hague with his wife Hortense de Beauharnais and children in June 1806. The affable, rather romantic and generous Lodewijk took his role of King of Holland rather seriously, and was determined to bring a genuine national cohesion to the loose federation of Dutch provinces. King Lodewijk had many ambitious plans for *his* country. He refused to impose conscription, relying instead on German Hessian mercenaries. He also tried to arrange a mild version of the Continental Blockade System. Louis—who had come to enjoy his new position and perhaps to liking the Dutch more than the French—soon exasperated his elder brother. Indeed the French Emperor had no interest in developing the Netherlands. All he wanted was money to finance his wars, and conscripts to fill the ranks of his armies. In 1810, the dictatorial Napoléon was extremely irritated by his brother's attitude who preferred to serve the interests of his subjects rather than satisfy the demands of France. In 1810 King Louis was fired, and the realm of Holland was simply absorbed into imperial France. In September and October 1811 Napoléon visited the Netherlands for a tour of inspection as he intended to transform the Netherlands into a fortress against hated Britain. The Emperor and his wife Marie-Louise briefly sojourned at Vlissingen, Dordrecht, Utrecht, Haarlem and Amsterdam.

Being a part of Napoleonic Imperial France was a terrible ordeal for the Dutch. The continental blockade imposed by Napoleon ruined the economy of the Low Countries traditionally based on trading with Scandinavia, southern Europe, Germany and England. Many Dutch colonial possessions were seized by Britain—notably South Africa, Guyana and Ceylon. Moreover the conscription and the heavy casualties resulting from Napoléon's ceaseless wars added to the resentment of the population. Some 15,000 Dutch draftees were forced to accompany Napoléon on his ill-fated Russian campaign in 1812, and only a few of them survived this terrible and foolish war. In 1813, the anti–French feeling running high, the Dutch entered into rebellion and joined the European anti–French coalition. Gradually the French troops of occupation were harassed and forced to withdraw. The turbulent years of French occupation came to an end when Napoléon was finally defeated in June 1815 at Waterloo in Belgium.

The French Era laid the foundations for a fresh start for the country, with a more centralized state and a new monarchy. In December 1815, the Dutch recovered their freedom, and the Prince of Orange became King of the Netherlands.

Maastricht. Close up of a musket loop.

French Napoleonic Fortifications

Napoleonic fortifications were severe and martial, complying with grandeur, robustness, solidity and economy. There was comparatively little creativity, being basically classical and making a general use of the traditional bastioned system. In certain aspects it displayed a curious return to some medieval forms like the standardized *tour modèle* with machicoulis balconies. These traditional and backwards elements were mixed with innovations influenced by Montalembert's new concepts, like the massive artillery tower, the general use of stone and masonry for vaulted casemates, and an increased use of caponiers, subterranean passages and riflemen's galleries concealing the defenders from enemy sight and fire.

Napoleonic fortifications also adopted the so-called Haxo casemate. This was a firing chamber, fitted with a gun port, and arched with strong masonry covered with earth. This innovation opened the rear of the casemate to the terre-plein and greatly helped to solve the crucial smoke problem by enabling ample natural ventilation. Other advantages were that it allowed mobile guns to be run in and out easily and that daylight could enter inside the dark casemate.

The French imperial military architecture rejected classical obsolete outworks like the horn- and crownworks, but it made a general use of advanced works like the demi-lune, the lunette and the counterguard to develop defense in depth, thereby taking account of the significant artillery improvements occurred during the second half of the 18th century.

The Napoleonic system of fortification is possibly best illustrated by the bastioned design by the imperial engineer Count François de Chasseloup-Laubat (1754–1833). This was basically a classical bastioned system à la Vauban-Cormontaigne mixed with modern features and adaptations influenced by Montalembert's systems and innovations introduced by the Napoleonic engineers François Haxo (1774–1838), and Jean Claude Elénore Le Michaud d'Arçon (1733–1800).

French Realizations in the Netherlands

French Napoleonic fortifications in the Netherlands and Belgium were mainly focused on the protection of the important ports and arsenals of Antwerp, Amsterdam and Den Helder. In early 1805 tensions grew between France and Prussia, and there arose the possibility of a war in which the small Batavian Republic would be involved. Prime-Minister Schimmelpenninck then gave orders to the Minister of War, Cornelis Rudolphus Theodorus, baron of Kraijenhoff, and to the Colonel-Director of the Dutch engineering corps, Croiset, to plan some kind of protection against the dangerous and threatening neighbor.

Map of the Netherlands 1810–1814. The ancient Dutch provinces were renamed and re-organized into French administrative departments.

Chapter 6. The 18th Century and the French Era

This policy was continued by the first king of Holland, Louis-Napoléon Bonaparte. Existing fortified points were modernized and several others were created at the border between the Netherlands and Germany.

Napoleonic Imperial France had many more projects for the Netherlands and Belgium. However, the rapid decline after 1812 and ultimately the fall of the Napoleonic Empire in 1815 did not allow enough time for their realization.

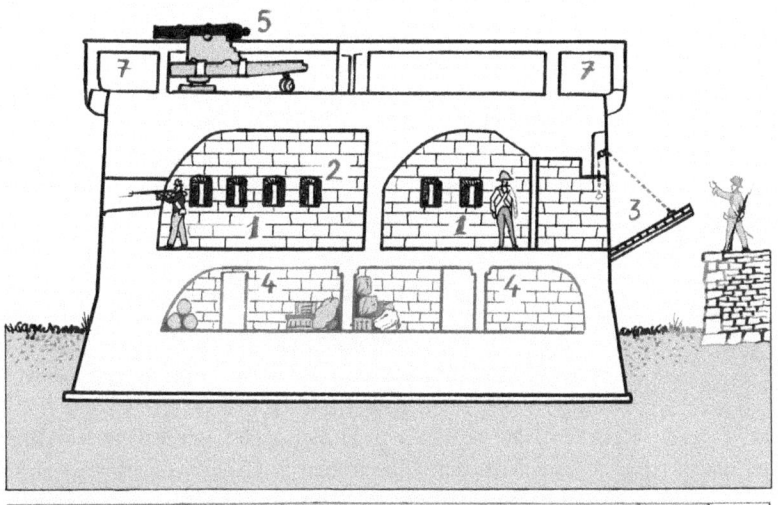

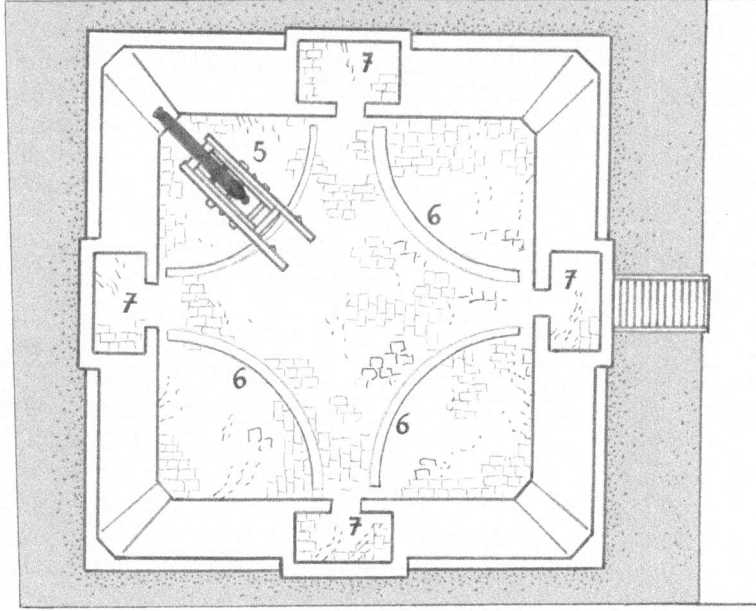

French tour modèle No.1. Designed and approved in 1811, the French tour modèle No.1 was a standardized square fortified work. The cross-section (*top*) shows the garrison's quarter (1) pierced with musket loopholes (2); and the drawbridge (3) giving access to the tower across the ditch. The basement (4) housed a food store, a magazine, and a cistern. The plan of the top platform (*bottom*) shows: one of the four 36-pound guns (5) on a Gribeauval fortress mount pivoting on a rail (6); and brattices with machicoulis (7) built on the faces of the wall. The brattice was a small projecting balcony fitted with an opening permitting defenders to throw missiles down upon assailants. This medieval element was probably more decorative than suited for active defense.

Kraijenhoff's posts

Since the 17th century Amsterdam was the most important city of the province of Holland and of the Netherlands. Consequently the city had to be well protected, and together with the so-called Old Holland Waterline, the French decided to establish a line of defenses all around the city. The so-called Kraijenhoff's posts were a number of temporary defensive positions built during the French Era in order to control the accesses to the capital of the Netherlands.

Krayenhoff's posts are often regarded as the origin of the later built "Stelling van Amsterdam" (Position of Amsterdam). The construction of those posts was started in 1797 when Prussia threatened to invade the Batavian Republic. In 1805 all accesses to Amsterdam were fortified, and reorganized in three large areas: South Front (between Pampus and the Lake of Haarlem) with 25 posts; West Front (facing Haarlem) with 13 posts; and North Front (North Holland) with 12 posts. The Kraijenhoff posts were not standardized. They had a semi-permanent character, were made of earth, often featured a ditch or a moat, and were often coupled with local inundations at least wherever possible.

They included posts (surveillance and control positions), coastal batteries (intended to defend shores), land batteries, schansen (sconces), redoubts and fortlets (designed to protect accesses, passes, and inundation installations) as well as numerous breastworks, embankments and other small earth-made defensive positions. The posts were not continually manned but only hastily refurbished and occupied by troops in time of crisis. They usually included temporary barracks, an ammunitions store and various supply sheds and huts, all made of wood. Today many perishable earth-made posts with their wooden shacks have disappeared.

Improvements in the New Holland waterline 1796–1814

The New Holland Waterline (see Chapter 7) was planned in the period 1796–97 by the General Director of Fortifications Cornelis R.T. Kraijenhoff. It was actually a substantial enlargement of the existing waterline. Kraijenhoff advocated the incorporation of the important city of Utrecht within the defended perimeter. Therefore he proposed to displace the inundations

French tour modèle.

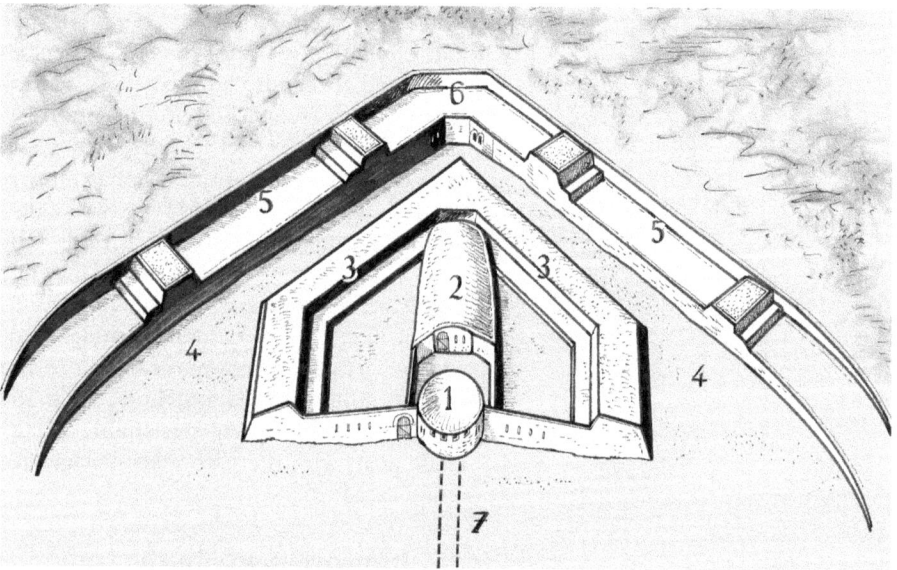

Lunette d'Arçon. The lunet designed by engineer Le Michaud d'Arçon was a self-support fortlet including: a round fortified tower in the gorge (1) used as quarter for the garrison and as flanking defense at the rear of the work; a bomb-proof vaulted ammunition store forming traverse (2) on the terre-plein; two faces forming a scarp wall with parapet and barbette gun emplacements (3); a ditch (4); a counterscarp with a covered way (5) fitted with traverses; and, under the covered way, a flanking coffer (6) consisting of two masonry casemates for enfilading fire. Placed in the reverse slope of the ditch, these elements enabled the defenders to bring fire to bear upon attackers who had got into the ditch; both reverse casemates were accessible from the lunette via an underground passage. The lunette was often detached and placed ahead of a fortress. It could be connected to the main body via a subterranean gallery (7).

and all strongholds defending the accesses to the east. This new defense scheme was later designated the New Holland Waterline.

In 1811 Kraijenhoff's plans were approved by Emperor Napoleon Bonaparte who then declared: "*La ligne de Naarden à Gorcum doit donc être considérée comme la vraie frontière de l'Empire*" (the line from Naarden to Gorcum must thus be considered as the true border of the [French] Empire). Because of the fall of the Napoleonic rule, the construction of the New Holland Waterline was not carried out.

After the defeat of France in 1815, Kraijenhoff's plans were reconsidered and re-actualized. By then, the new King Willem I acknowledged the importance of a new line of defense. The New Holland Waterline was a major undertaking that completely reorganized the defense of the Kingdom of the Netherlands in the 19th century, as we shall further discuss in the following part.

The Beverwijk line

As a satellite of France, the puppet République Batave was unwillingly involved in the Napoleonic Wars. In August 1799, an Anglo-Russian task force landed in the north of the province of Holland and marched in the direction of Amsterdam. The franco-batave forces, after a series of bloody battles at Bergen, Alkmaar and Castricum, managed to force the raiders to re-embark and leave. The incursion, however, had clearly demonstrated the vulnerability of the defensive system around the important city and port of Amsterdam. To palliate this weakness, the Kraijenhoff's posts were established.

In addition, the so-called Line of Beverwijk was created at the instigation of General Guillaume Brune (1763–1815). It was designed by Lieutenant-Colonel Claude Gillet of the French Engineer Corps, and by his Dutch colleague Lieutenant-Colonel Kraijenhoff. Both designers were influenced by the theoretical book *Tracé d'un système de lunettes disposées en lignes* (Outline of a system of lunets placed in line), written by a certain Simon François Gay de Vernon (1760–1822)—a teacher in mathematics and fortification at the Paris Polytechnique School.

Established between February and July 1800, the line was composed of 26 detached *lunettes* (redan-shaped triangular fortlets) and several ammunition dumps, and engineering places. They were spaced out in two rows in the dunes in the north-west of Amsterdam and spread from Wijk-aan-Zee in the West near the coast up to Beverwijk landwards in the East. Each lunette was independent, and most were triangular in plan (a few were pentagonal). Each work was defended by a glacis, and a ditch. Each was fitted with a thick earth breastwork protecting infantry and artillery, and had a palisaded gorge.

No attack ever occurred again, and after the departure of the French in 1814, the provisory and perishable lunettes were neglected, and abandoned. Soon the grounds were sold to local farmers and the works were progressively swallowed up in the landscape. Today only a few small earth mounds in the flat North-Holland countryside indicate the remnants of lunettes.

Haxo casemate. The French military engineer François Haxo (1774–1838) was rather conservative in matters of fortification, and heavily relied upon classical bastioned works, but the gun casemate he designed was quite innovative. Laid inside the rampart (1) of a permanent fortified work, the Haxo casemate was made of vaulted masonry. Protected from enemy artillery fire by a thick layer of earth (2) it absorbed the shock of shot and shell impacts and prevented the roof from damaging and collapsing. The casemate was, of course, fitted with an embrasure (3). Its originality was that it was open at the rear, allowing light inside and smoke and other toxic fumes to escape out. The Haxo casemate was widely used all over Europe during the 19th century.

Den Helder

Den Helder, situated in a marshy peninsula in the Dutch province of North-Holland, originated from a simple medieval fishers' village named Huisduinen. After centuries of hard toil the region north of Amsterdam was protected by dikes and lands reclaimed from the sea by polderization. In the 15th century Den Helder became a port, and coastal redoubts were built in the 1580s. When the United Provinces became independent from Spain in 1648, Den Helder—which with the island of Texel controlled the passage to the port of Amsterdam via the then open Zuiderzee—was fitted with coastal batteries.

The port and the Dutch fleet were taken in January 1795 by 400 hussards led by Major Lahure, a part of General Pichegru's army. This episode has become famous as the French horsemen attacked the ships by charging on the ice of the frozen port. Den Helder was attacked and held for a while by the British in August 1799. Den Helder was radically transformed by order of Napoléon in 1811 to constitute an important naval arsenal, dockyard and a military port for the French navy.

The project for Den Helder was quite ambitious. On the land front the arsenal was to be defended by a citadel and a bastioned enceinte; both these works were never built. Three large advanced bastioned forts (named forts Lasalle,

Haxo casemat (rear view).

l'Ecluse and Dugommier) were actually constructed. They were linked together by continuous walls reinforced with redans and lunets. In addition the marshy grounds south of the arsenal could be flooded, thus making Den Helder a formidable camp retranché (entrenched camp).

In the dunes facing the North Sea several coastal forts and batteries were planned (Morland and Dufalga). In 1814, after the departure of the French all fortification works

Fort de Ruijter. Built in 1810–1811 by the French, Fort de Ruijter is located near Willemstad in the province of North Brabant. It originally consisted of a Napoleonic tour modèle no. 2 (standardized square tower) made of brick masonry to which a shell-proof barrack and various service buildings covered with dirt were later added. The fort was surrounded by a pentagonal enceinte featuring an earth wall, and a moat. It had a garrison of 69 men. After the departure of the French in 1813 the stronghold was renamed Fort Sabina Henrica. It was greatly enlarged and modernized in the early 1880s as a part of the national defense line in the so-called Stelling van het Volkerak en het Hollandsch Diep (North Brabant Position). The fort was decommissioned in 1981. Purchased by Staatsbosbeheer (National Forest Management Department), it has been open to the public since 2012.

Post Nr. 8. Another of the Kraijenhoff posts was Nr. 8 Abcoude, intended to defend the access to Amsterdam. It was actually composed of two gun batteries placed on each bank of the Gein River. The post was located in the South Front near the Oostzijdse windmill between Geindijk-Nigtevecht and Abcoude (Province of Utrecht). Made of earth it was built between 1805 and 1810. The double entrenchment was not permanently garrisoned, and later was incorporated into the Stelling of Amsterdam. Armed with two 10 cm guns, it was then an interval battery covering the grounds between Fort Abcoude and Fort Winkel. It remained in military use until 1959. The post has been reshaped in the 1980s, and is now managed by the Vereniging Natuurmonumenten (Dutch Association for the Preservation of Nature).

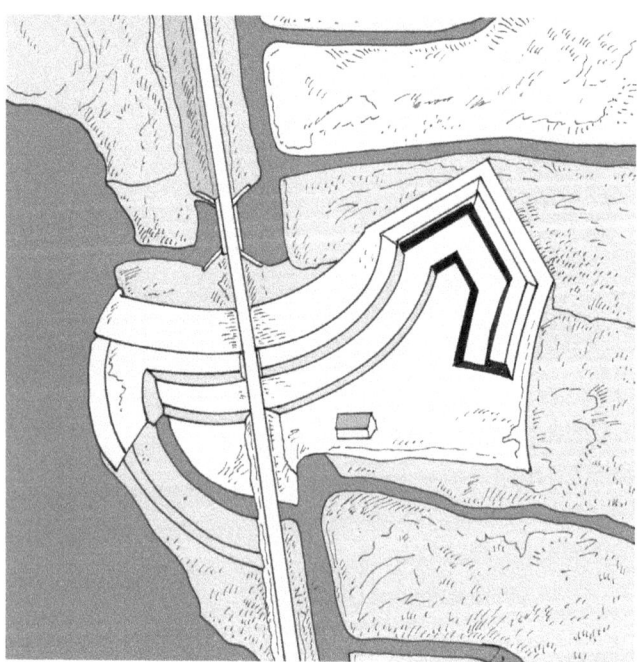

Post Nr. 1 Kouhorner Braak. Located in the North Front of the Line of Amsterdam, the Post Nr.1 was one of the so-called Kraijenhoff posts intended to defend the access to Amsterdam. Built in 1806–1810 by order of Minister of War Kraijenhoff, it included a curved breastwork placed across the main road. The post remained in military use until 1926.

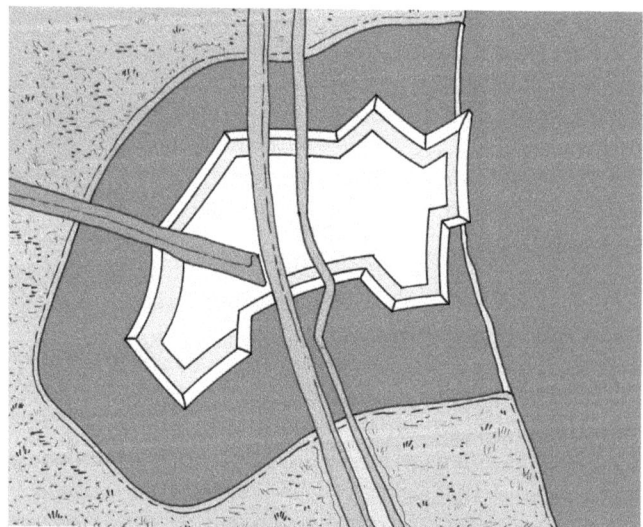

Post Nr. 6 Werk aan de Zwet. The Post Nr. 6 was located in the West Front on the dike named Spaarndammerdijk across the highway and the canal (later the railway track) between Amsterdam and Haarlem. It remained in military use until 1927.

Chapter 6. The 18th Century and the French Era

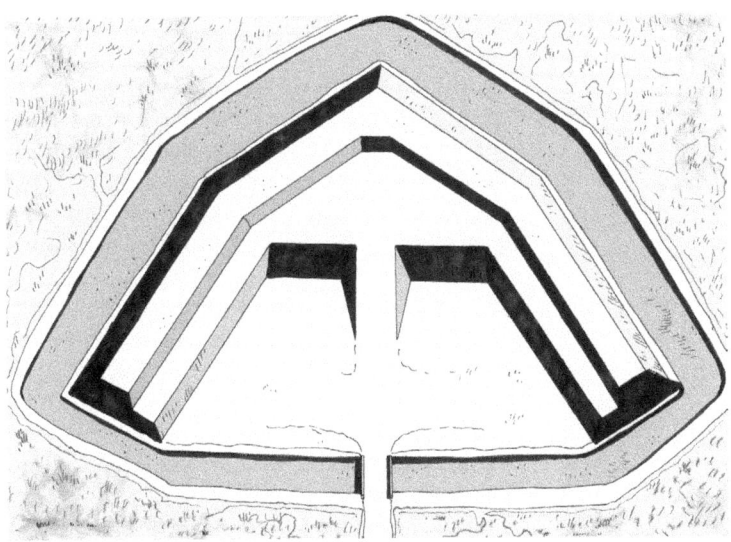

Lunette in the line of Beverwijk northwest of Amsterdam.

were taken over by the Dutch navy, and given Dutch names. The installations were completed in the 19th and 20th centuries. During World War II (1940–1945) the base was used by Nazi Germany and heavily bombarded by the Allies. After the war, the town and the navy installations were re-built and enlarged. Den Helder remains today an important Dutch navy base.

Northern Netherlands

The French also built or reactivated a number of strong points in the North of the Netherlands in the provinces of Drenthe and Groningen (then designated *Département de l'Ems Occidental*, and *Département des Bouches de l'IJssel*). These included improvements to major border defenses like Groningen and Bourtange, and small frontier works intended for surveillance roles, namely repression of smuggling between Germany and the Netherlands.

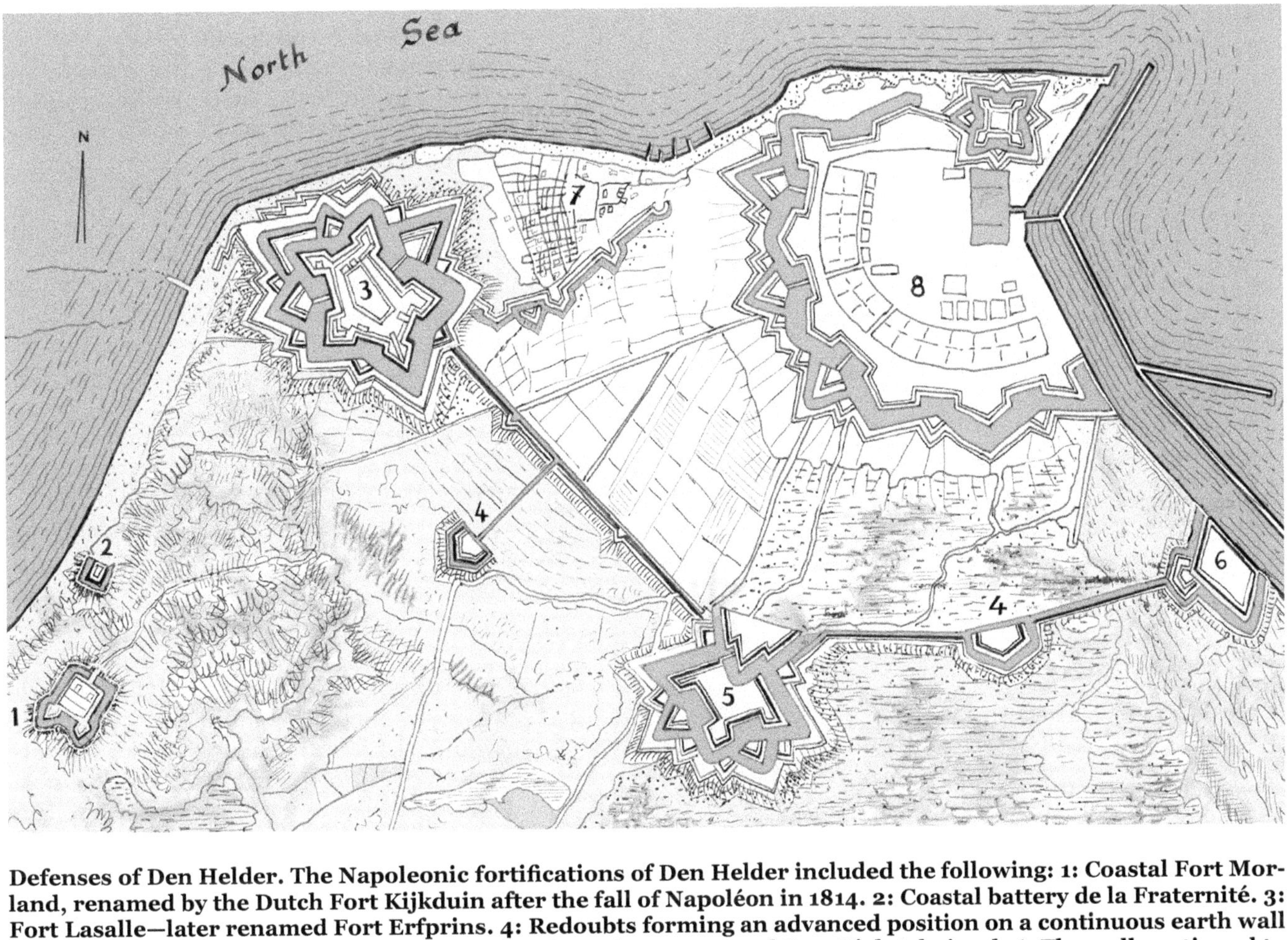

Defenses of Den Helder. The Napoleonic fortifications of Den Helder included the following: **1:** Coastal Fort Morland, renamed by the Dutch Fort Kijkduin after the fall of Napoléon in 1814. **2:** Coastal battery de la Fraternité. **3:** Fort Lasalle—later renamed Fort Erfprins. **4:** Redoubts forming an advanced position on a continuous earth wall joining Fort Lasalle to Fort l'Ecluse. **5:** Fort l'Ecluse—later renamed Fort Dirk Admiraal. **6:** The wall continued to Fort Dugommier, later renamed Fort Oost-Oever. **7:** There was a project to fortify the village of Den Helder, the naval base and the arsenal (**8**) but this was never completed. Several additional coastal batteries were temporary established between Fort Lasalle and the arsenal including Batterie de la Révolution, Batterie du Roi de Rome, Batterie Le Réparateur, Batterie de l'Indivisibilité, and Batterie de l'Union.

Fort Lasalle Den Helder. Today named Fort Erfprins, Fort Lasalle was one of the three main forts built by the French to protect Den Helder. The fort was a classical bastioned pentagon with outworks, envelope, two wet ditches and a traversed coastal gun battery facing the North Sea.

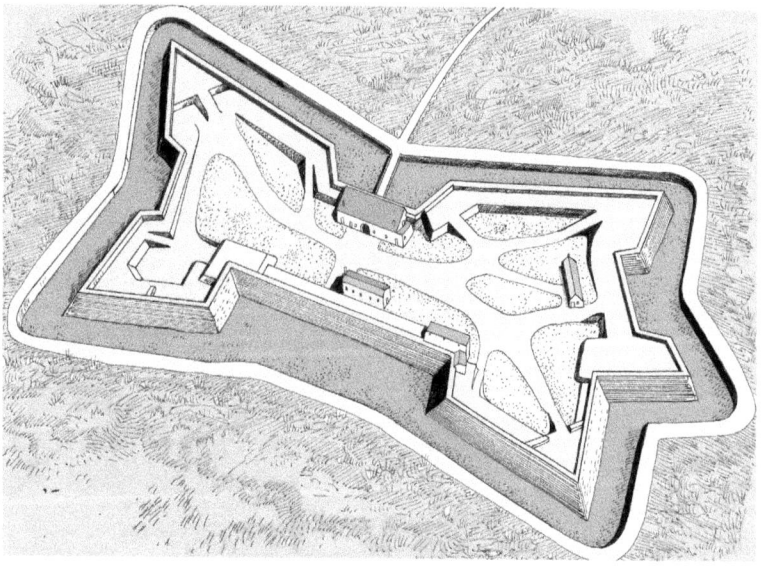

Fort Dufalga (1810 project). The construction of Fort Dufalga started in 1781 as a battery in the dunes south of Den Helder at a place known as Kleine Keeten. The fort was eventually renamed after Louis Marie Joseph Maximillien de Caffarelli Dufalga (1756–1799), a French military engineer who was killed during the campaign of Egypt at the battle of Saint John of Accro in 1799. The fort was to be made of earth and designed according to the bastioned system. It would have had a length of 380 m, and was intended to house twenty-two 36-pound guns and 16 mortars. In the gorge of the work it was planned to edify a standardized "tour modèle" for the garrison. However, in 1812 the fort was not yet completed, having only a bastioned front facing the sea, a palisade defending the rear, and a temporary barrack made of wood. The fort, never completed, was abandoned in February 1814. Today it has vanished in the dunes and nothing remains of it.

Antwerp

Although Antwerp is located in Belgium, not Netherlands, a few words must be said about the Napoleonic realizations. Napoléon's main preoccupation in Belgium was the protection of the access to the strategic port of Antwerp. Antwerp appeared in the 3rd century CE and developed in the Middle Age as an important commercial port, notably for trading with England. Medieval fortifications were modernized by the Italian Donato Buoni di Pellezuoli in the 1540s. A citadel designed by Francesco Pacciotto and Bartholomeo Scampi was built in 1567. In 1794, Belgium was annexed by France and divided into nine administrative departments.

In 1803 the First Consul Bonaparte decided to make of Antwerp (Anvers in French) a strong navy arsenal, a "loaded pistol aiming at England" in his own words. Soon, work started for the construction of a large military port and shipyard on the right bank of the Scheldt River. In 1809 a sudden and brutal British raid clearly showed the weakness of the defenses: the raiders occupied the port and town of Vlissingen and made important destructions before withdrawing. After this attack, it was decided to increase the defenses (notably the access to the estuary of the Schelde River leading to Antwerp).

Then ambitious plans were made to transform the city into a formidable fortress with bastioned enceintes and detached forts. Paciotto's citadel was reinforced by a continuous envelop, demi-lunes, counterguards and covered way *en crémaillère* (sawtooth-shaped). The much neglected Spanish bastioned enceinte from 1540 was modernized by the addition of various outworks. The huge undertaking was however not completed at the time of Napoléon's fall. Between 1815 and 1830 Belgium was incorporated into the newly created United Kingdom of the Netherlands.

Vlissingen and Breskens

The port of Vlissingen (*Flessingue* in French, *Flushing* in English) is situated on the southern part of Walcheren Island in the Dutch province of Zeeland. Because of its excellent strategic site on the north side of the Eastern Scheldt mouth, the French Napoleonic occupiers decided to make of Vlissingen a sort of dependence of the port of Antwerp. From 1796 to the end of the Napoleonic Empire, Vlissingen was the object of numerous projects including enlargement and modernization of its maritime installations: shipyards, dry docks, arsenal, quays, sluices, magazines and stores, workshops and barracks.

Fort De Schans Texel island. Strategically situated north of Den Helder on the island of Texel, the Fort De Oude Schans ("the Old Sconce") was designed by Dutch military engineers Jan Crab and Adriaan Dooren in 1572 to defend the passage leading to the then still open Zuiderzee and to Amsterdam. Completed about 1574, the fort was constantly improved and enlarged, notably in 1665 during the Second Anglo-Dutch War (1665–1667). During the War of American Independence (1775–1783), the fort was used as a hospital for sick and wounded British sailors who had been captured. The Fort Oude Schans (re-named Fort Central by the French), occupied by some 300 men, both French and Dutch was attacked in 1799 by a strong party of Anglo-Russians who had landed at Callandsoog. The fort was taken for a short while  by the raiders. During his visit in 1811, Napoleon, anxious to protect the waterway leading to Amsterdam, ordered the construction of a demi-lune, and the reinforcement of the covered way which was transformed into an envelope with large places-of-arms and a second outer wet ditch. Two detached works known as Redoute and Lunette were also built about 700 m west of the fort as additional defenses. The digging works were carried out by Spanish prisoners. The fort, which was also used as a prison for British prisoners of war was occupied by the French until November 1813. The Fort Oude Schans on Texel Island was several times modified and adapted to modern artillery, and used by the Dutch as observation post and coastal battery until 1922. In 1931, following the reinforcement of the dike protecting the island against sea erosion, the fort was partly dismantled. Today about half of the work has been restored and forms a cultural historical object in the wild nature of the island.

Bourtange. Fort Bourtange is situated in the Dutch northern province of Groningen. The fort, controlling a small sandy passage in large impassible marshes that marked the natural frontier with Germany, was built in the 1590s by order of Prins William of Orange. Fort Bourtange, originally intended as a simple temporary bastioned pentagonal sconce made of earth, was later incorporated into the defense of the Dutch United Provinces. The fort was enlarged in 1607 and 1645 by the addition of outworks, wet ditches, and a crownwork. Advanced defenses were added by the French, comprising the line of Abeljeshuis, and the redoubt Bakoven. Fort Bourtange lost its military value in the 19th century. It was dismantled and the grounds were sold to local peasants in the 1850s. Fortunately, in the 1960s, the municipality decided to revive 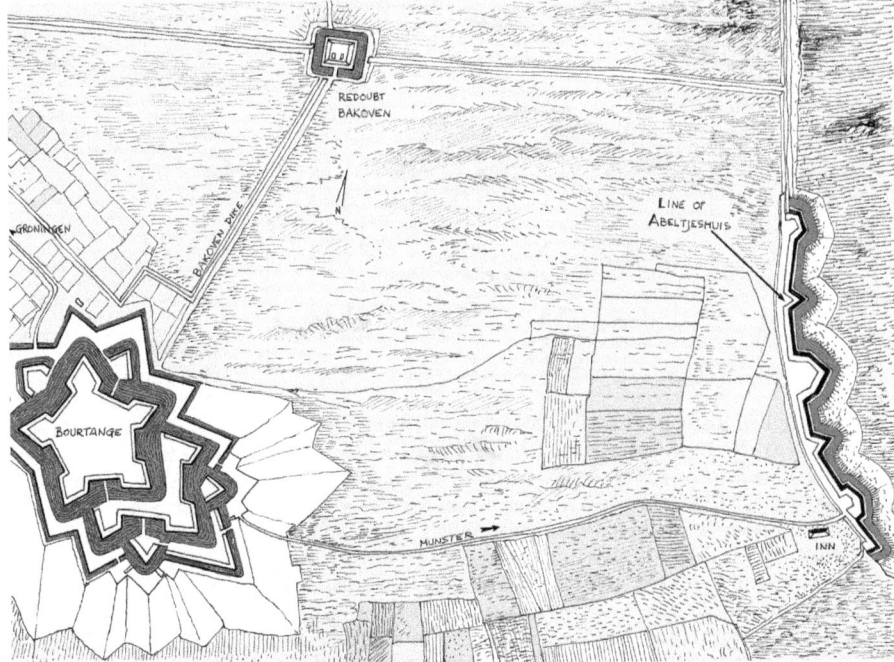 the old fort. It was completely and exactly rebuilt as it was in 1742. The fort is now a major tourist attraction where the past is cleverly and skillfully reenacted.

Above: Line of Abeltjeshuis at Fort Bourtange. The strategically important passage between Bourtange and Germany was reinforced by the French occupiers. Between what today are border poles 181 and 182, they established a line, called the Abeltjeshuis Linie (named after a small inn "the House of Little Abel") composed of a central bastion (1), two redans (2) and two batteries (3) linked together by an earth wall defended by a wet ditch.

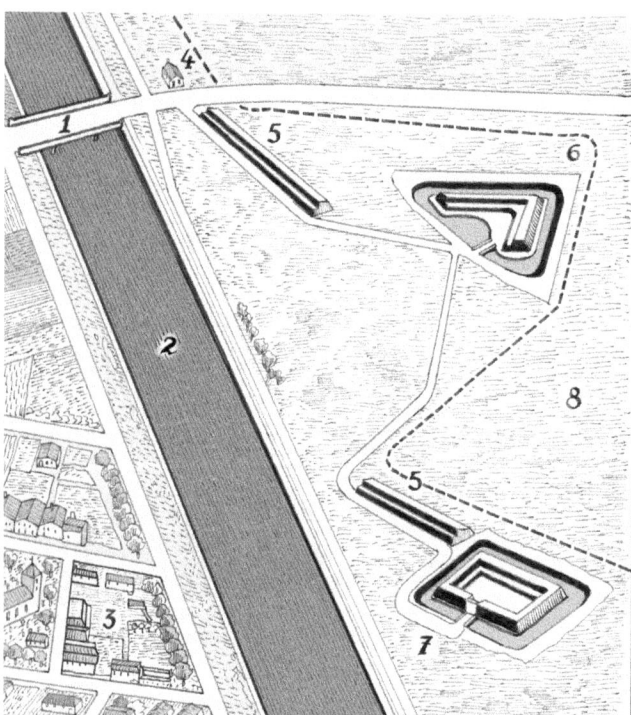

De Leethe border checkpoint. Near the village of Bellingwolde at the frontier with Germany at border markstone No. 188, the French occupiers buit a defensive post in 1797. This included a flèche (an arrow-shaped redan), a redoubt and two infantry positions. The post was also intended to thwart smuggling, which had always been a fruitful activity between Germany and the Netherlands. The Leethe control post was abandoned in 1828. Today both works have been restored and acts as tourist sights. The sketch shows the following: 1: Wijmer Bridge. 2: B.J. Tijdens Canal. 3: Village of Bellingwolde. 4: Border post. 5: Infantry breastwork. 6: Flèche or arrow. 7: Redoubt. 8: Germany.

Delfzijl in 1812. The village of Delfzijl in the north of the province of Groningen was probably established in 1272 by monks who built a zijl (sluice) on the river Delf. By 1500, Delfzijl developed into a port, which played a role in the Eighty Years' War (1568–1648) opposing the Dutch Protestant independentists to Spain. The city was fortified during that period in Old Dutch Bastioned System. During the Napoleonic occupation, the French established a coastal battery facing the sea (Batterie du Nord, 1 on the plan); added a demi-lune (2) in front of the Land Gate; arranged a vast central place-of-arms (3); and refurbished the old Kostverloren hornwork (4) protecting the southern Farmsum Gate. The French garrison was besieged by the Cossacks and capitulated on May 23, 1814. Delfzijl remained a part of the Dutch defense until 1874.

Chapter 6. The 18th Century and the French Era

Katshaarschans. The Katshaarschans ("Sconce of the cat's hair") was built by the Dutch in 1672, during Louis XIV's invasion of the Low Countries. It was a simple rectangular earthen redoubt intended to control the Katshaar pass, a narrow passage between Dalerveen and Vlieghuis in the marshes east of the town of Coevorden in the province of Drenthe. In 1797 the neglected redoubt was re-activated by the French, and a new advanced fortified point in the form of an open hornwork was added. This had a front length of about 150 m. The redoubt and the advanced work formed what was called the Post of Katshaar.

Emmerschans. The Emmerschans (Sconce of Emmen) in the province of Drenthe was built by the French in 1800. It was a simple redoubt made of earth, surrounded by a wet ditch, with a front length of about 60 m and a width of about 45 m. Together with the Katshaarschans, it was intended to defend the border of the province of Drenthe against an attack from the East. The Katshaarschans and the Emmerschans remained a part of the Dutch defenses until 1851. Since 1935 both redoubts are owned by the Stichting Oude Drenthe (Old Drenthe Association) and today still stand in the flat and green landscape.

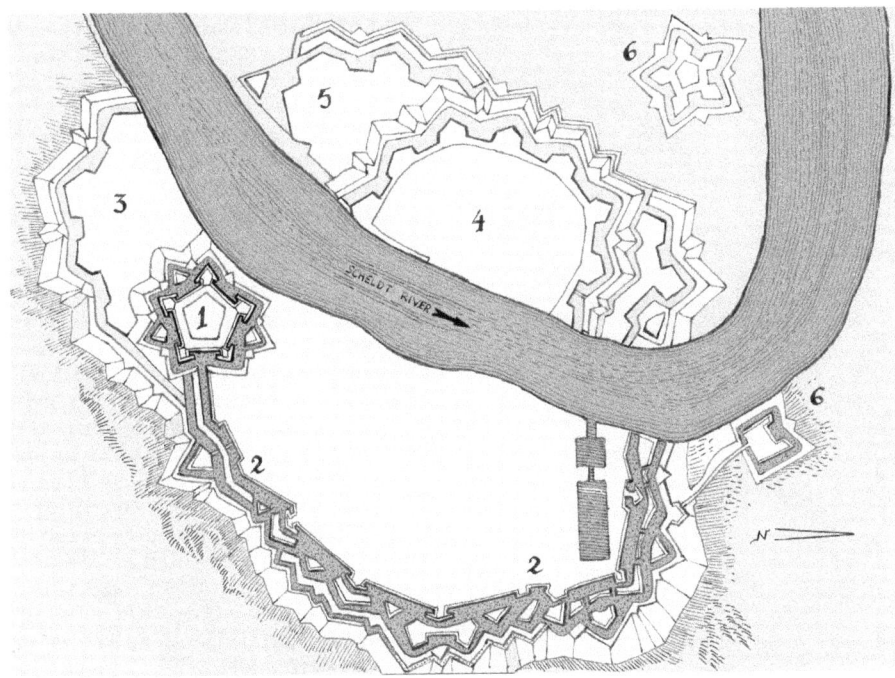

Project for Antwerp 1810. The French Napoleonic scheme was quite ambitious. It included the modernization of the citadel (1), and the old Spanish enceinte (2), as well as the creation of a vast entrenched camp (3) south of the citadel, and a city expansion (4) on the left bank of the Scheldt, with another entrenched camp (5) and detached forts (6).

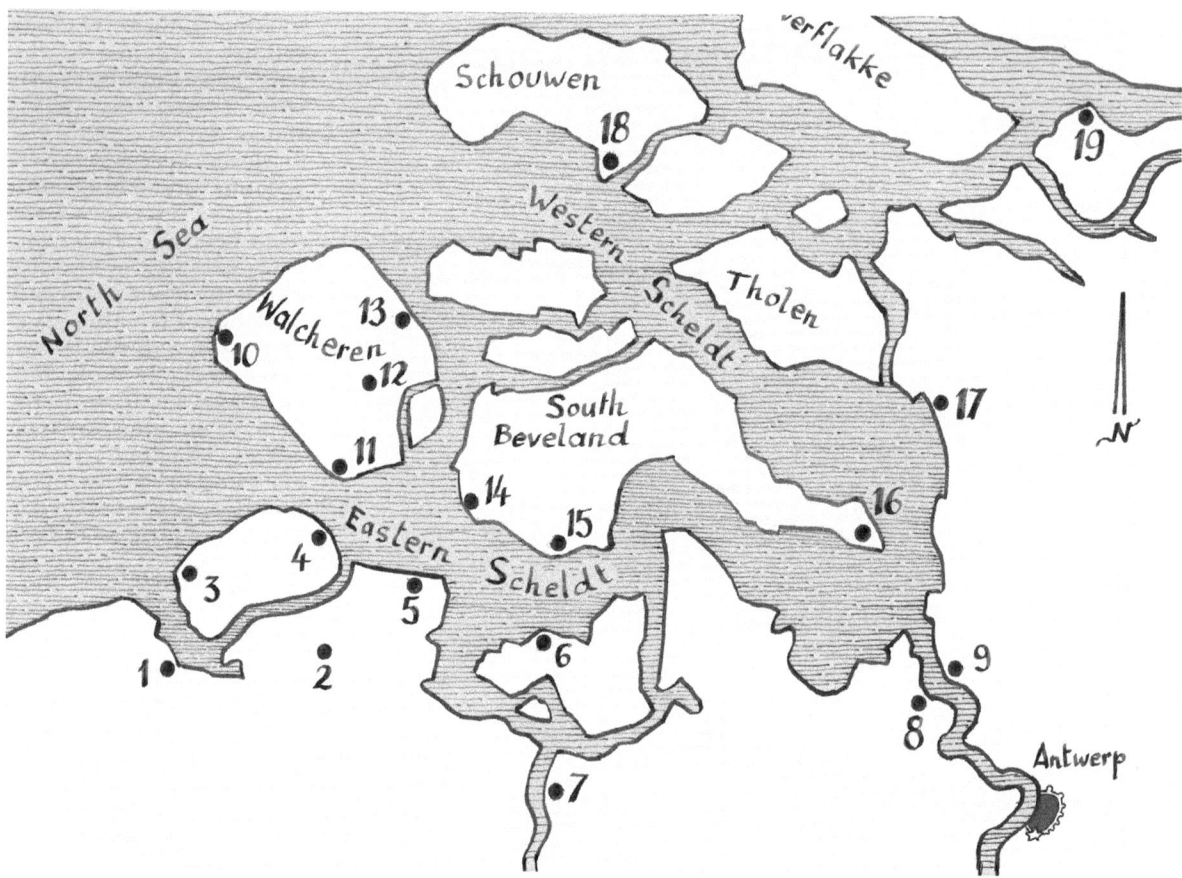

Defenses of Antwerp. Following the British attack of 1809, the defenses of the port and arsenal of Antwerp were reinforced and enlarged by order of Napoléon. They involved the whole province of Zeeland and included the fortified cities, strongholds and forts of l'Ecluse or Sluis (1), IJsendijk (2), Cadzand (3), Breskens (4), Hoogplaat (5), Terneuzen (6), Sas-de-Gand (7), fort Liefskenhoek (8), fort Lillo (9), Westkapelle (10), Vlissingen (11), Middelburg (12), Veere (13), Borselen (14), Ellenwoutswijk (15), Batz (16), Bergen-op-Zoom (17), Zierickzee (18) and Willemstadt (19).

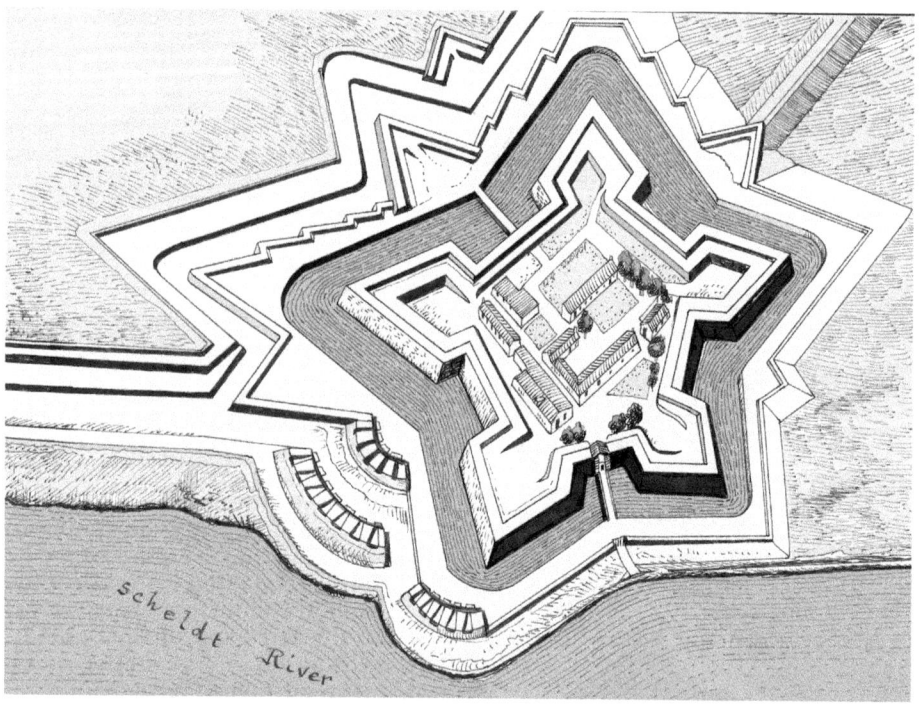

Fort Lillo (Antwerp) ca. 1810. Fort Lillo was built between 1579 and 1582 by order of stadhouder William of Orange on the right bank of River Scheldt north of Antwerp. Together with Fort Liefkenhoek (placed on the left bank), they were intended to defend the access to the port of Antwerp. After the British attack of 1809, Napoléon ordered a refurbishment of the fort. A new powderhouse was built, a gun battery and two fortified shelters were established, as well as a covered way "en crémaillère" (sawtooth-shaped). Fort Liefskenhoek, on the opposite bank was also modernized by the French occupiers. Declassed as a military place in 1894, Fort Lillo still exists and forms a green oasis, an island of calm and nature in the heavily industrialized suburbs of Antwerp.

Chapter 6. The 18th Century and the French Era

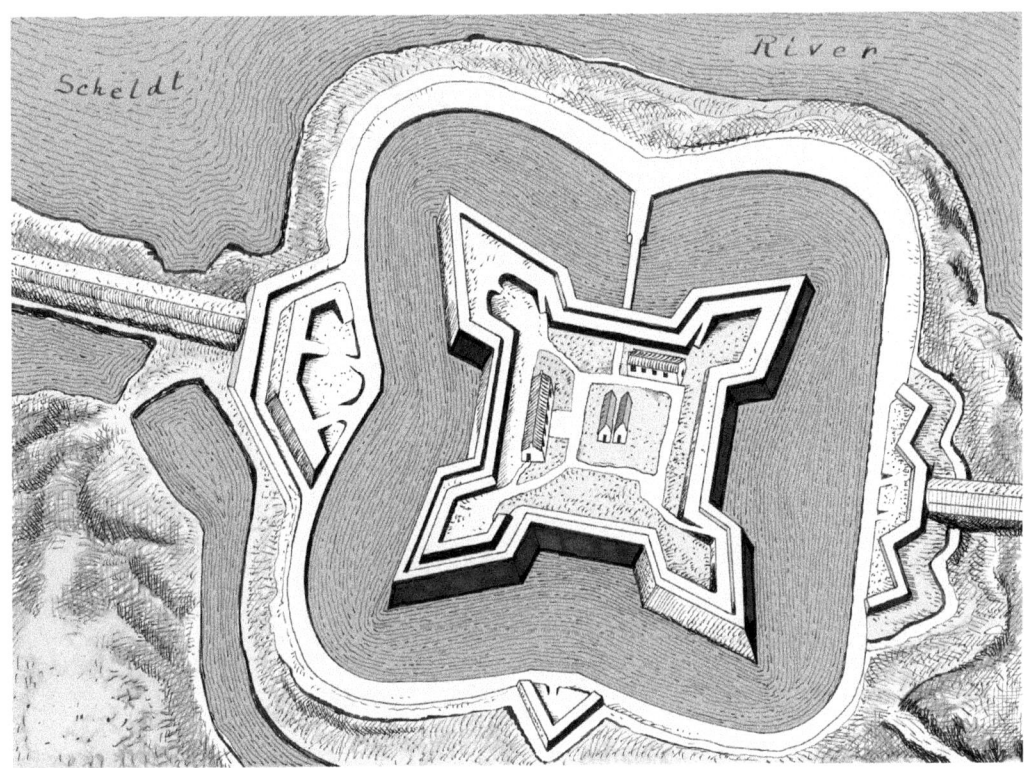

Fort Liefkenshoek (Antwerp) ca. 1810. Smaller sister located on the left bank, Fort Liefkenshoek was constructed and discarded at the same time as Fort Lillo.

Project for Vlissingen ca. 1810. The projected fortifications of Vlissingen were designed by the French military engineer Lamy. The city and the port (1) were defended by a bastioned enceinte and floodings (2). Three large detached bastioned forts (3), as well as redoubts and coastal batteries (4) were planned. The Renaissance bastioned fort Rammekens (5) was incorporated into the defenses of the town.

Fort Rammekens. Situated east of Vlissingen, Fort Rammekens was built in 1547 by order of Mary of Hungary (sister of Emperor of Germany and King of Spain Charles V, and then governess of the Low Countries). Designed by the Italian engineer Donato de Boni and constructed by the Dutch master-builder Peter Fransz, the fort includes a diamond-shaped enceinte, one casemated bastion turned towards the Scheldt and two flanking half-bastions turned towards the rear. During the French Era, Fort Rammekens (1) was incorporated into the defenses of Vlissingen, modernized, and reinforced with a large crownwork (2) and a line en crémaillère (3) on the land front.

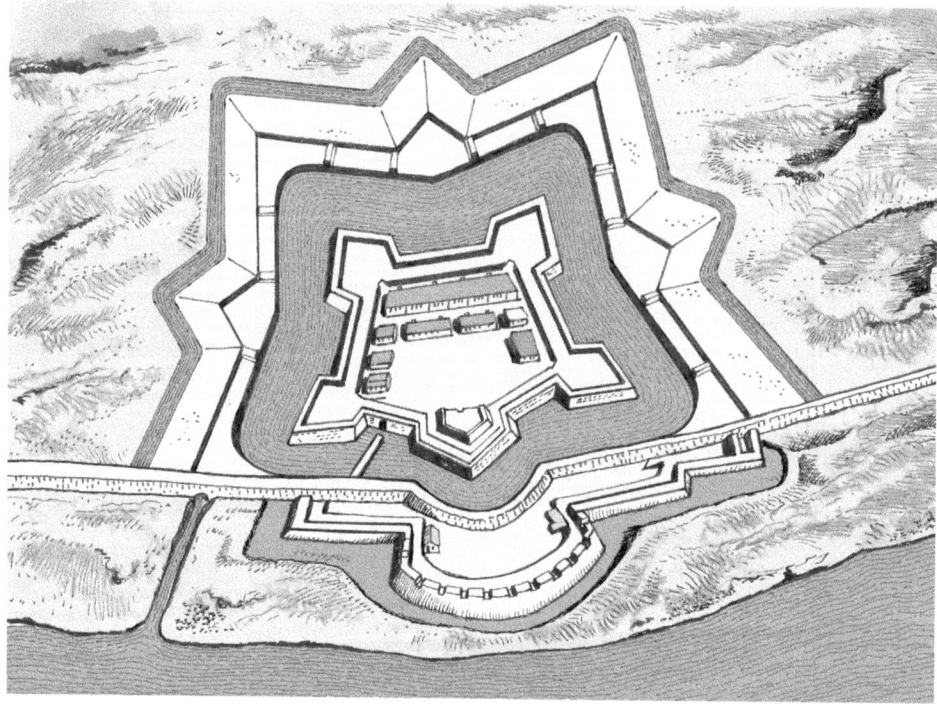

Fort Impérial 1811 Breskens. Situated near Breskens on the left bank of the Eastern Scheldt, Fort Impérial was built in 1811. It was a classical bastioned pentagon with wet ditch, covered way, glacis and additional outer wet ditch. The fort also included a gun battery placed on the opposite side of the dike; this was armed with twelve 36-pound guns, twelve 24-pound guns, six 16-pound guns and twelve 12-pound mortars. After the French withdrawal in 1814, the Dutch renamed the stronghold Fort Frederik-Hendrik.

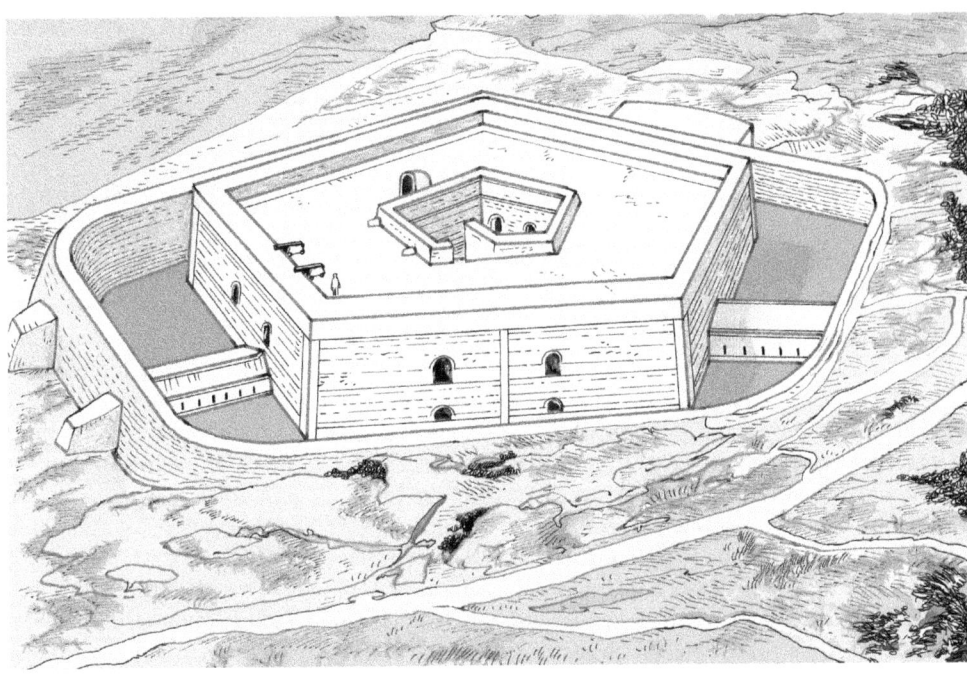

Fort Napoléon at Oostende (Belgium), ca. 1811. The construction of this fort was ordered to protect the port of Oostende from a British attack. The attack never happened, and the fort was completed only upon the fall of the French Emperor in 1814. The fort, quite modern for its time, showed some Montalembert's influence. It is a massive casemated pentagon with a gun battery on its open top terrace, and its deep dry ditches were defended by caponiers. During the two world wars of the 20th century, the fort was used by the Germans. After 1945, it fell into neglect. Since 1995 it has been managed and restored by the Flanders Heritage Association. Today Fort Napoléon has withstood the test of time, and houses a museum and a restaurant. Worthy of mention is the German Navy coastal battery (MKB Hundius) that stands in the dunes close to Fort Napoléon and which gives a typical example of what the defenses of the World War II Atlantic Wall actually were.

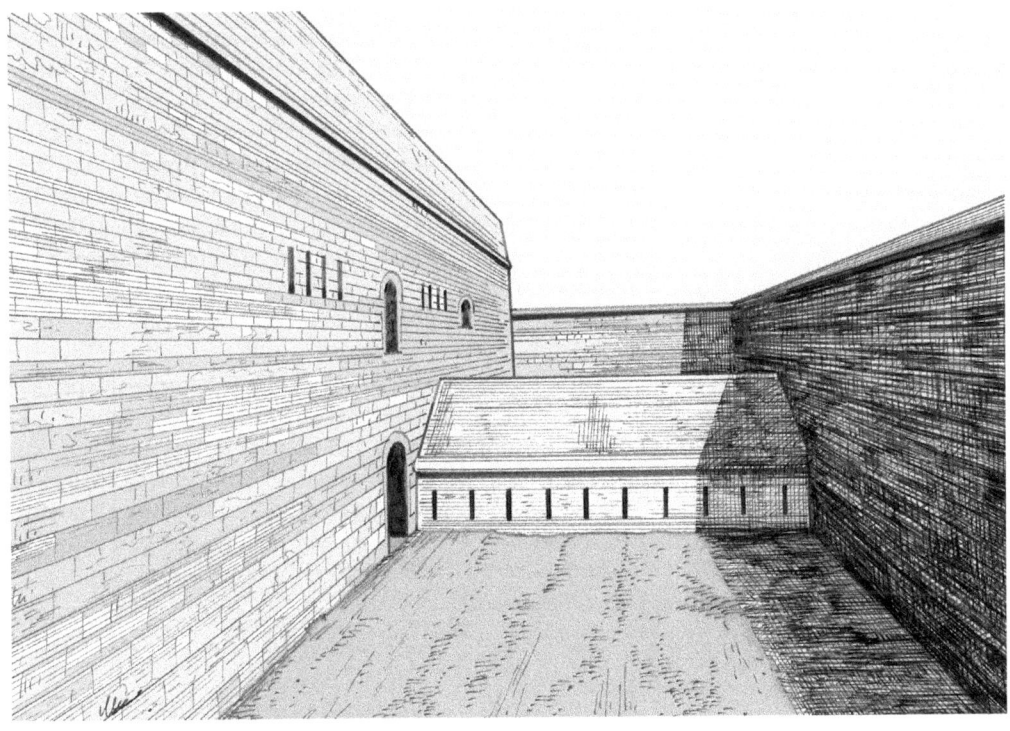

Caponier Fort Napoléon (Oostende).

After the devastating British raid of 1809, Napoléon ordered the reinforcement of the fortifications. The obsolete Dutch bastioned urban enceinte from the 17th century, and the old Renaissance Fort Rammekens were modernized while a belt of detached forts, and coastal batteries were planned both on the land and sea fronts of Vlissingen.

For the defense of the southern bank of the Eastern Scheldt estuary, the French planned a series of fortified positions at Cadzand, Breskens, Hoogplaat and Terneuzen. At Breskens, for example, they built in 1811 the fort Napoléon, the fort du Centre, the fort Impérial and the lunette Caffarelli. It was also planned to establish an entrenched camp around the village of Breskens including a bastioned wall and a pentagonal fort, as well as a canal connecting to Brugge. The fall of the Empire prevented these works from being carried out.

Chapter 7

The 19th Century

The Industrial Revolution

After the French dictator Napoleon I had been defeated in 1815 and sent away for good on the small island of Saint Helena, where he died in May 1821, Europe did breathe freely again and turned to other matters. A great congress of the victorious Allies met at Vienna, Austria, in 1815 to restore the state of affair that the Napoleonic storm had greatly disturbed. From 1815 until 1870, an uneasy peace was maintained in Europe. The continent settled down back to "normal business." Trade picked up again, and remote places like Asia and Africa took on added importance. There were good reasons for reviving attention to faraway countries.

Western Europe entered into a new industrial, economic and social organization. Even in its infant stage, the new society had appetites and capacities that got bigger and bigger every year. This new upheaval was the so-called *Industrial Revolution*—an economic transformation that gradually but inexorably took the place of the ancient and obsolete social organization. Under the impact of 18th century "enlightenment," a steady growth of knowledge and a general clearing up of men's ideas about the world in which they lived was in progress. In the 19th century, medieval remnants of feudalism with privileged noble lords and common serfs deprived of rights, fiefs and manors, and almighty clergy, were being supplanted by a new social and production system: capitalism.

Mainstays were still monarchs, courts, noblemen and senior ecclesiastics, but they had to share their power with a new enriched lay bourgeoisie of enterprising businessmen, resourceful capital holders, greedy investors and speculators, ambitious bankers, gifted financiers, daring merchants, audacious adventurers and imaginative inventors. Like all historical movements industrialism and capitalism developed slowly. Actually the Industrial Revolution had started in the 18th century in England, and since the Renaissance the European bourgeoisie of financiers, traders and merchants had developed an early form of liberal capitalism.

In the 19th century this evolution blossomed when new processes of production, new technology and new machinery appeared. Steampower for example provided energy for factories, railways, and steamships. Significant improvements in metallurgy, development of electricity, telegraph, agricultural fertilizers, and spectacular advances in medicine and science radically changed life. Later on by the end of the 19th century and beginning of the 20th, the invention of the internal combustion engine enabled the development of automobile vehicles and aeroplanes. All these innovations coupled with new methods of working like the division of labor, assembly line and chain-work made mass production possible.

Many peasants were pushed off their farms and forced to become city workers, where they made a thin living in enormous building projects, large workshops, and huge factories. Towns grew and spread, they became cities announcing present day metropoles, conurbations and megalopolis. Growing industrialism needed cheap raw materials to produce goods, and compliant markets to sell them. This, of course, was not new, but in the 19th century the various sorts and the enormous quantity of raw materials needed could simply no longer be supplied by pre-industrial handicraft methods. Before the Industrial Revolution, no European powers wished to subject Asia and Africa to their political control. One was generally satisfied with a few influential coastal trading stations and market ports. From the 16th to the 18th century, the older empires of the pre-industrial era were maritime and mercantile. If with a few exceptions (like the United States of America), the Europeans usually had no territorial ambitions beyond the protection of ports, way stations, posts, and trading centers, and—on the whole—one did not venture far inside the unknown and dangerous hinterland.

In the 19th century, on the contrary, under the principle of protected trade caused by ruthless competition, it was deemed necessary to rule and exercise political influence in areas in which one did business. Asia and Africa as sources of raw materials and as markets were particularly attractive. As a consequence most European powers took part in a wide movement of racist and aggressive imperialism and predatory colonialism.

The United Kingdom of the Netherlands

Union with Belgium 1815–1830

After Napoleon's defeat, European ambassadors met between November 1814 and June 1815 at the Congress of Vienna in Austria. The purposes of this conference were numerous. Among other things it was intended to reshape

Europe and provide a long-term peace and stability on a continental scale. The conference of Vienna formed the framework for international politics until the outbreak of the First World War in 1914.

For the Netherlands its provisions included the following. The national kingdom of the Netherlands was created and comprised an enlarged configuration combining with the defunct Dutch United Republic, the former Austrian Netherlands (present day Belgium), the Great Duchy of Luxembourg, and the Principality/Bishopric of Liège. It was then known as the United Kingdom of the Netherlands. It was a constitutional monarchy ruled by King William I of the House of Orange-Nassau.

William I was thus the second king in the history of the Netherlands. The first monarch was the Frenchman Louis-Napoléon Bonaparte. Until then the Orange-Nassau had been *stadhouders* of the Dutch Republic—actually lieutenants (place holders) representing the Crown of Spain (at least in theory). With the United Kingdom of the Netherland it was expected that the Dutch commerce and agriculture combined with the Belgian industry would make a prosperous association for a cohesive state.

Wellington Barrier

The new enlarged United Kingdom of the Netherlands had then a long border with France, and needed a modernized system of defense. In a convention held in August 1814 it was agreed between Britain and the Netherlands that the great Barrier scheme based on fortified cities in Southern Belgium would be reactivated. The system was partly financed by Britain and supervised by Arthur Wellesley, 1st duke of Wellington, called the Iron Duke (1769–1852)—the Irish-born commander of the British army during the Napoleonic Wars and later prime minister of Great Britain (1828–30). Therefore the line of fortified cities was called Wellington Barrier.

The southern part of the new Dutch state would become a buffer zone that would prevent future French hegemony on the continent. Although many urban fortifications in south Belgium had crumbled into a useless state, a series of 25 city-fortresses was established, modernized, altered, restored, repaired, and expanded between the Scheldt and the Meuse Rivers. The main strong points of this defensive concept included the cities of Antwerp, Gent (Ghent), Liège, Huy, and Maastricht. The large-scale strategic program was not yet fully completed in the late 1820s when it was suddenly shattered by unsuspected political events that radically changed the situation.

Belgian independence 1830

From the start the (northern) Dutch rule was unpopular in the French-speaking southern part of the Kingdom. Notably the provinces in today's Wallonnia did not favor the policies of William I. The two parts of the Low Countries (The Netherlands and Belgium) had been separated for two-and-a-half centuries, and the union scheme did not work smoothly. Following political, linguistic, religious, and economical tensions, the United Kingdom of the Netherlands broke down when the Belgians revolted and seceded in August 1830.

The Dutch tried but failed to suppress the rebellious secession during the so-called Ten-Days' Campaign. With the support of France and Britain, an independent and constitutional monarchy was established in Belgium under the leadership of King Leopold I from Saxe-Coburg und Gotha. After a period of tension, an agreement was signed in 1839 (Treaty of London) making official the separation of the two sovereign kingdoms. Luxembourg finally separated from the Kingdom of the Netherlands in 1890, when the Dutch King William III died without a male heir, which was a condition to rule the Grand Duchy of Luxembourg.

Peace and prosperity

In spite of (or perhaps because of) the secession and independence of Belgium, the 19th century after 1830 was a time of peace and prosperity at least for the Dutch bourgeoisie. On the international scene the great political affair in the 19th century was neutrality. After the costly 18th century wars and the loss of many colonies to Britain, the traumatic Napoleonic occupation, and the short-lived and unfortunate Belgian venture, the Dutch had come to the conclusion that they no longer could compete with their now powerful and overwhelming neighbors.

It became quite clear that the tiny Netherlands had really nothing to gain in taking part of any alliance whatsoever. The wisest stance to follow was thus the observance of prudent and strict neutrality keeping the Netherlands out of trouble while trading with everybody.

In the second half of the 19th century, through slow but constant economic growth, and important constitutional reforms (reluctantly granted by the rich oligarchic bourgeoisie and the monarchs to the lower social classes), the Netherlands became a parliamentary monarchy, and a liberal, capitalist, and modern industrial state.

With the *de facto* independence of Belgium in 1830, the Dutch borders were brought back to what they were before the French Era. However in the meantime many things had changed; notably, artillery had made significant advances. As a result the systems of defense of the Netherlands needed modernization.

Until deep in the 17th century the defense of the Netherlands was concentrated on a network of important fortified cities, which played a military, strategic, administrative and economical role. In the 18th century the emphasis lay in waterlines established between these key fortresses. In the 19th century the focus on waterlines was continued, while new and adapted fortifications were designed and built in order to block the accesses.

Dutch Fortifications in the Period 1830–1860

New fortifications

One of the most obvious effects of the Industrial Revolution in the 19th century was the simultaneous development of new attack weapons and, in response, new defensive methods of fortification. After the Napoleonic Wars new principles of fortifications appeared, many of which were based on theories developed by the French officer René de Montalembert (1714–1800) and several German military engineers.

Montalembert's leading idea was that for a successful defense it was necessary for the artillery to be superior to that of the attacker. New artillery fortresses were needed, as well as new forts of the kind he advocated in several books published in the years 1776–1784. Montalembert's concept, called *Fortification Perpendiculaire*, led to the adoption of casemates in several tiers in preference to open parapets. These new fortifications were characterized by abandoning the use of bastioned fortifications, which had demonstrated their obsolescence during the Napoleonic period.

According to Montalembert and other military engineers a modern fortress had to be a simple-shaped artillery stronghold bristling with weapons placed in casemates, huge caponniers, and massive cylindrical artillery towers connected by underground communications. The central flanking work resulting from this arrangement was a kind of 16th century Italian caponier that was re-introduced and greatly developed. For a square fort of about 400 yards a side, Montalembert proposed over 1,000 casemate guns; and one of his caponier sections showed 10 tiers of masonry gun-casemates, one above the other.

This was of course much too large, and much too expensive. Montalembert's theoretical designs must be considered in some ways unworkable but the basic idea was sound. The essence of his system was its simplicity, which could be applied to any sort of ground, level or broken, and to long or short fronts.

Montalembert was also one of the first to foresee the coming necessity for detached forts for the defense of towns and frontiers. He advocated the establishment of one or more rings of detached and autonomous forts. These detached works armed with heavy artillery covering a wide radius of fire were actually powerful bases of fire. Each fort could defend a large area and could cover its neighbor with overlapping fields of fire. A belt of such forts could defend a whole city and could replace a continuous enceinte. A line of them could protect a port, a whole frontier or a strategical passage.

Defense in depth, made up of a series of parallel lines of fortresses, strengthened by fortified complexes around manufacturing cities and communications centers, could turn the whole state into one impregnable stronghold. This was the theoretical reasoning behind the American, British, German, Belgian and Dutch military construction in the 19th century. Indeed, every industrialized European power subscribed to this strategic thinking in some way. Besides, forts,

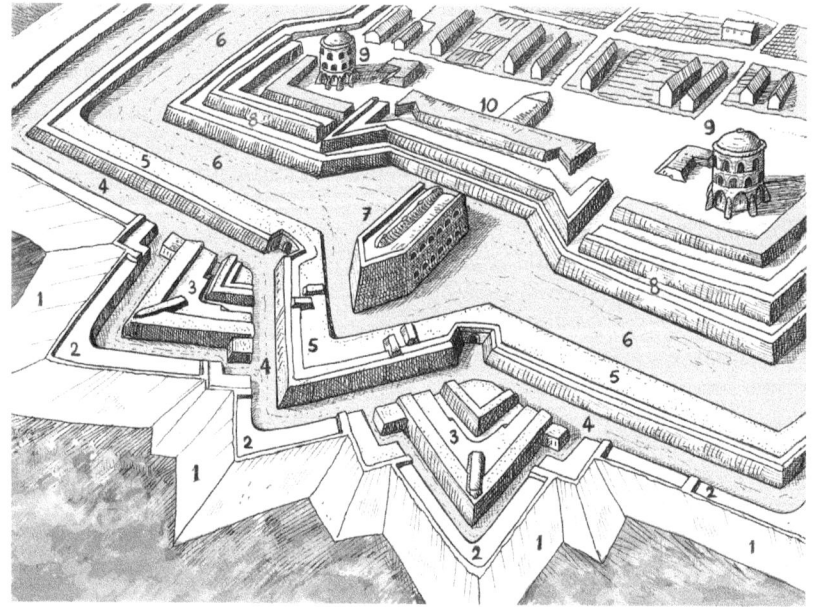

Montalembert's fortification. According to Montalembert, the defenders' best asset was to direct as many guns as possible on the enemy besieging force from a simple work comprising the following elements: 1: Glacis. 2: Covered way. 3: Advanced lunette. 4: Advanced ditch. 5: Envelope. 6: Main ditch. 7: Caponier (a large jutting out casemated structure projecting from the main body of a place into the ditch in order to provide flanking fire). 8: Main wall. 9: Large, cylindrical artillery tower. 10: Bombproof barrack for garrison. It is doubtful whether a Montalembert fortress would have been able to hold a sufficient garrison to man all guns, together with the necessary infantry, cavalry for sorties, and service troops and their stores and ammunitions. However, the concept was sound and influenced 19th century military architecture.

Montalembert's artillery tower. This multi-layered artillery tower designed by Montalembert—a close-up of (9) in the previous illustration, was a large circular casemated work bristling with guns. It was to be built with extremely strong and thick masonry, and included quarters for the gunners, as well as ammunitions, supply and water stores. The top of the tower was arranged as an observatory and command post. The ground floor included riflemen loopholes for close-range defense.

together with armored steam battleships, size of army, heavy artillery and (later) railway mileage, were symbols of national status.

Montalembert is often regarded as the precursor of the so-called polygonal system that dominated the second half of the 19th century military architecture.

The Germans were prompt to design and build new fortifications based on Montalembert's theory. Modern works appeared in 1834 at Germersheim (south of Mannheim in Rhineland-Palatinate). A new system called "Old Prussian Method" was employed at Koblenz, Mainz, Ulm and Ingolstadt.

It should be noted that the bastioned system was never totally abandoned in Dutch (and French) fortifications. Until the 1880s bastions and walls with thick breastworks (often constructed in earth and almost always in combination with wide moats) were still widely used.

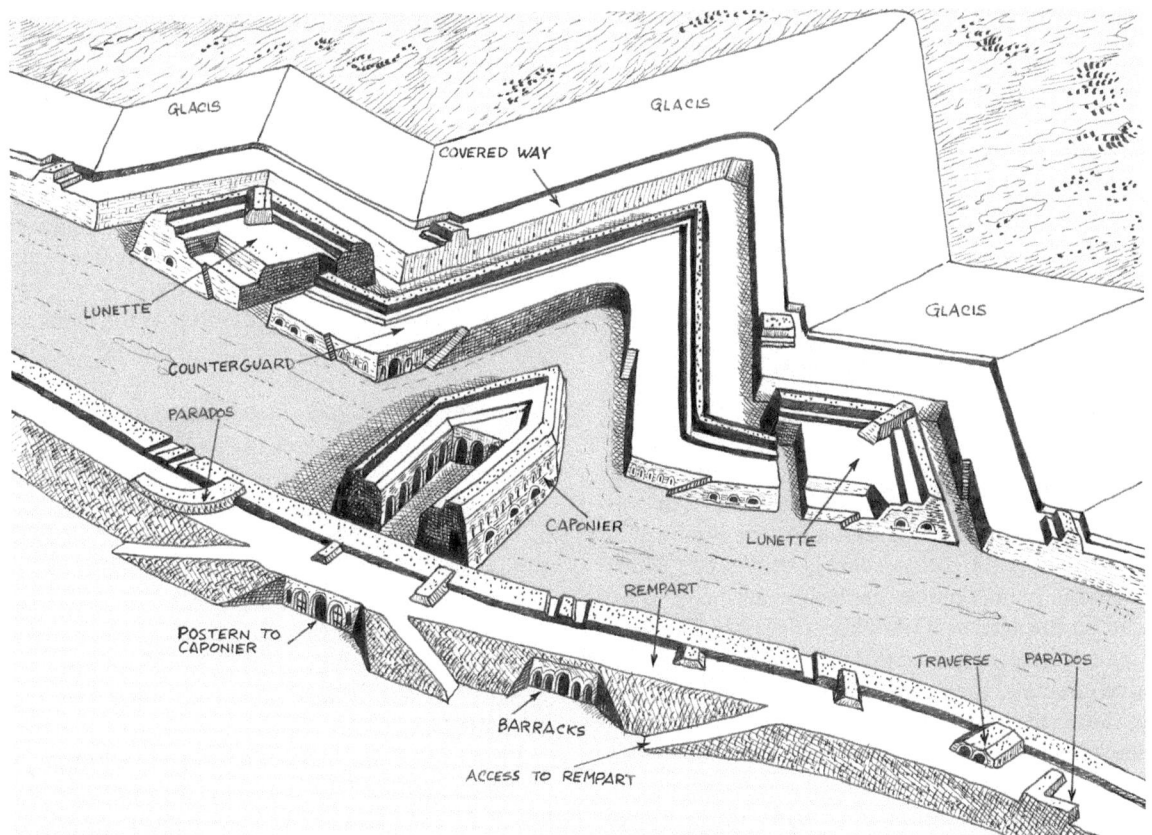

Ingolstadt. Located on the River Danube North of Munich in Bavaria (Germany), Ingolstadt was fortified in modern fashion between 1826 and 1855 after designs made by Colonel Michael von Streiter and Colonel Peter von Becker.

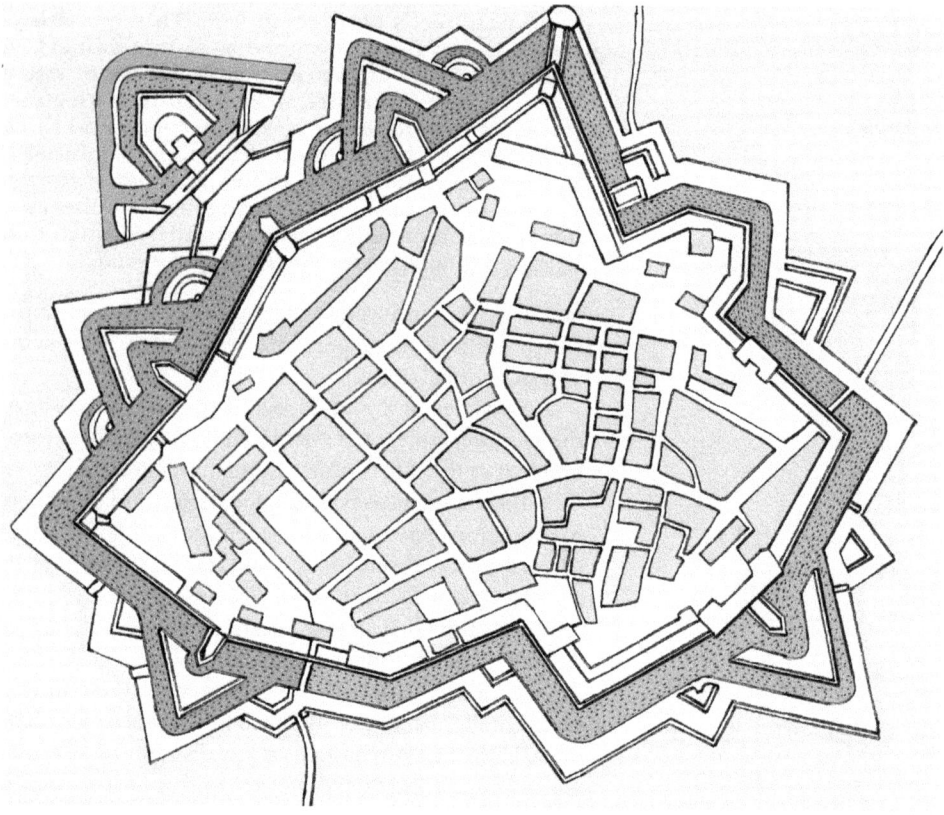

Groundplan of Germersheim. Situated on the river Rhine in the province of Rhineland-Palatinate (Germany), Germersheim was fortified between 1834 and 1855 by the engineer von Schmauss following the so-called Old Prussian System, based on Montalembert's theory. The fortifications included fronts defended by large caponiers as well as a ring of detached forts.

The New Holland Waterline

After the French Era, the national defense of the Netherlands was reorganized. The scheme of a new waterline designed by Cornelis Kraijenhoff was re-actualized, and King Willem gave his approval. It was then decided to establish a new waterline that would include the important city of Utrecht. Therefore the New Holland Waterline (as it became known, NHW in short) was partly shifted east of Utrecht.

The basic principle remained the same as defined by Maurits and Frederick Hendrik van Nassau in the 17th century: protect the heart of the Netherlands (the province of Holland and the capital Amsterdam), fend off invaders behind inundations and repulse them with new modernized forts placed on the accesses. However this strategical view was not totally accepted. Indeed this scheme left many provinces undefended and potentially abandoned many Dutch citizens to the mercy of aggressors.

The new waterline ran from the Zuiderzee (today called IJsselmeer) in the North at Muiden, Naarden and Weesp. Then it slightly bent to the east in order to include the city of Utrecht. It continued in a southern direction to Vianen and Culemborg on the River Lek. It ran further down in a southwest direction to Asperen, then down to Gorinchem to meet the Great Rivers: Boven Merwede, Waal and Rhine.

This line of inundation was an ambitious, enormously expensive, and impressive feat of engineering, turning the threat of water into an efficient ally. It included an ingenious system of sluices, dikes, floodgates, and canals. It was divided into 10 *kommen* (basins). The line had a length of 85 km (50 miles), and a width varying from 3 to 5 km (about 2.5 miles). It covered an area of ca. 50,000 hectares.

At strategic points where the land was too high to be flooded (the so-called accesses) modern strongholds were positioned. These fortifications comprised two modernized medieval castles (Muiderslot and Slot Loevestein), several fortified towns (Naarden, Weesp, Utrecht, Nieuwegein, Leerdam, Gorinchem, and Werkendam), and more than 60 major defensive permanent works including forts, batteries, fortlets, fortified sluices, lunettes, and redoubts. As can be easily imagined such an ambitious program took years to put together. The whole scheme was designed and constructed from 1815 to 1886 with many interruptions and reorganizations caused by political and economic events, but also by the tremendous development of artillery all through the 19th century.

In the first phase from 1815 to 1826 the focus was on the establishment of NHW inundations and the construction of new forts around Utrecht.

In the period 1825–1830, funds and attention focused on fortifying the South of the Kingdom in Belgium. After the secession of Belgium, a second phase of construction started. Then, from 1841 to 1864 new fortifications were built (notably the so-called *torenforten* or tower-forts) along the river accesses. Between 1867 and 1872 many forts became suddenly obsolete when rifled artillery (see below) was introduced, and therefore improvements were carried out in the design of all existing fortifications. In 1879 and 1883 the NHW was once more reorganized with improved inundations.

Finally in the 1880s a new crisis without precedent emerged when extremely powerful projectiles were introduced. By then all forts made of masonry and earth layer had became obsolete. By the last decade of the 19th century the art of fortifications entered a new era dominated by reinforced concrete and armor.

Map of the New Holland Waterline. From 1815 to 1940 the Dutch Waterline protected the Dutch National Réduit (the province of Holland). When enemy forces were advancing, whole swaths of polders land were flooded by an ingenious system of inundation combined with forts blocking the accesses.

Utrecht forts

The forts defending the north, east and south of Utrecht were a part of the new waterline. As just discussed the New Holland Waterline (NHW) running from Naarden in the north to Gorinchem in the south included the city of Utrecht within the defended area. In the period 1816–1824, new inundation zones were established, and new strongholds were built for the defense of accesses leading to or passing through the region of Utrecht, notably forts De Klop, De Gagel, Blauwkapel, De Bilt, Vossegat, the four Houtensepad Lunettes, and Jutphaas.

The materials used to construct them remained brick masonry covered with sturdy layers of stamped earth. In style, the Utrecht forts were traditional designs inherited from the 18th century and French Napoleonic influence. They consisted of or were often based on a double couronne (crownwork). Fort Jutphaas included tenaille works on its sides. As for the row of four independent pentagonal lunettes in the Houten plain in the South west of Utrecht, they were typically French style. Some forts (like De Bild or Blauwkapel) were mixed structures with innovative features including a *reduit* (keep) placed in the gorge where the garrison could withdraw for a last stand when the rest of the position was captured.

Dutch Fortifications

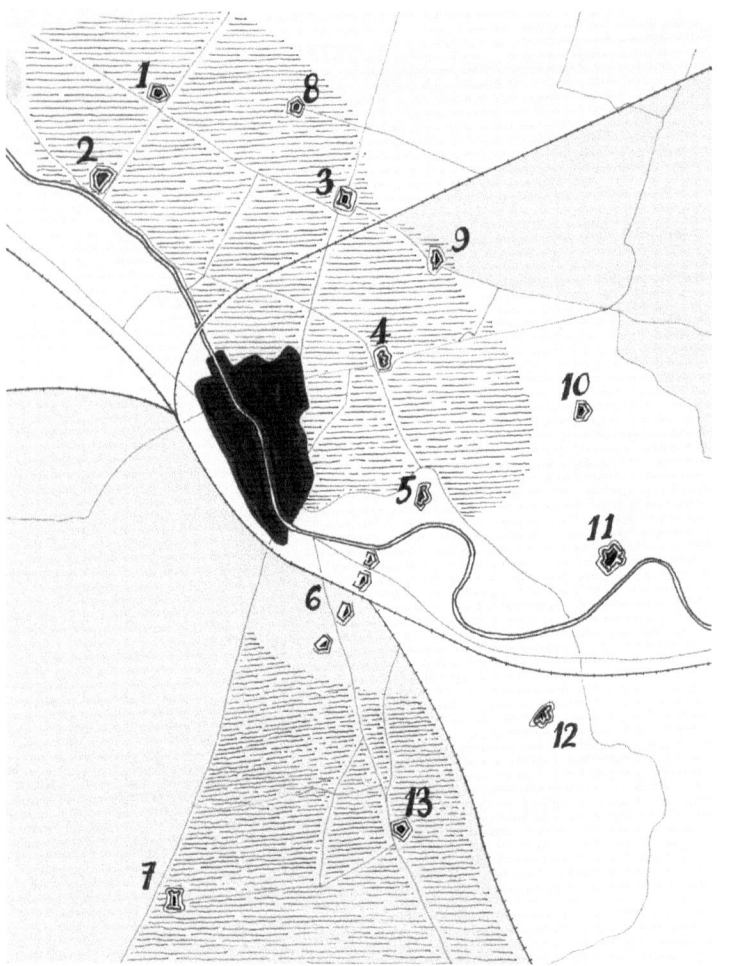

Map of Utrecht forts. Forts built in the first belt between 1816 and 1824: 1: Gagel. 2: De Klop. 3: Blauwkapel. 4: De Bilt. 5: Vossegat. 6: Lunets (I, II, III and IV). 7: Jutphaas. Forts buit in the second belt in the period 1867–1881: 8: Ruigenhoek. 9: Voordorp. 10: Hoofddijk. 11: Rijnauwen. 12: Vechten. 13: Het Hemeltje.

Fort De Bilt (aka Fort op de Biltstraat) was constructed in the years 1816–1819 across one of the main highway leading to Utrecht. Like most early NHW forts it included a wide ditch filled with water, bastioned earth walls with gun batteries, and a reduit (a kind of keep) in the rear housing quarters and stores where the garrison could withdraw.

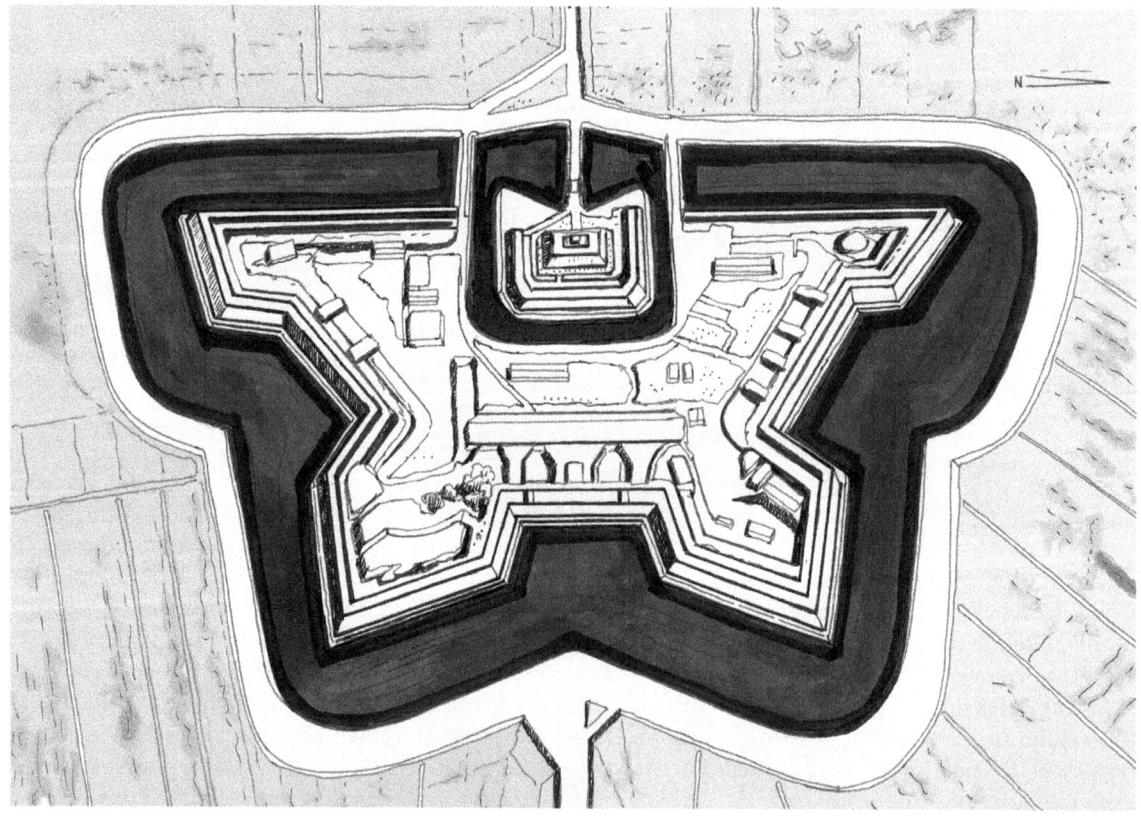

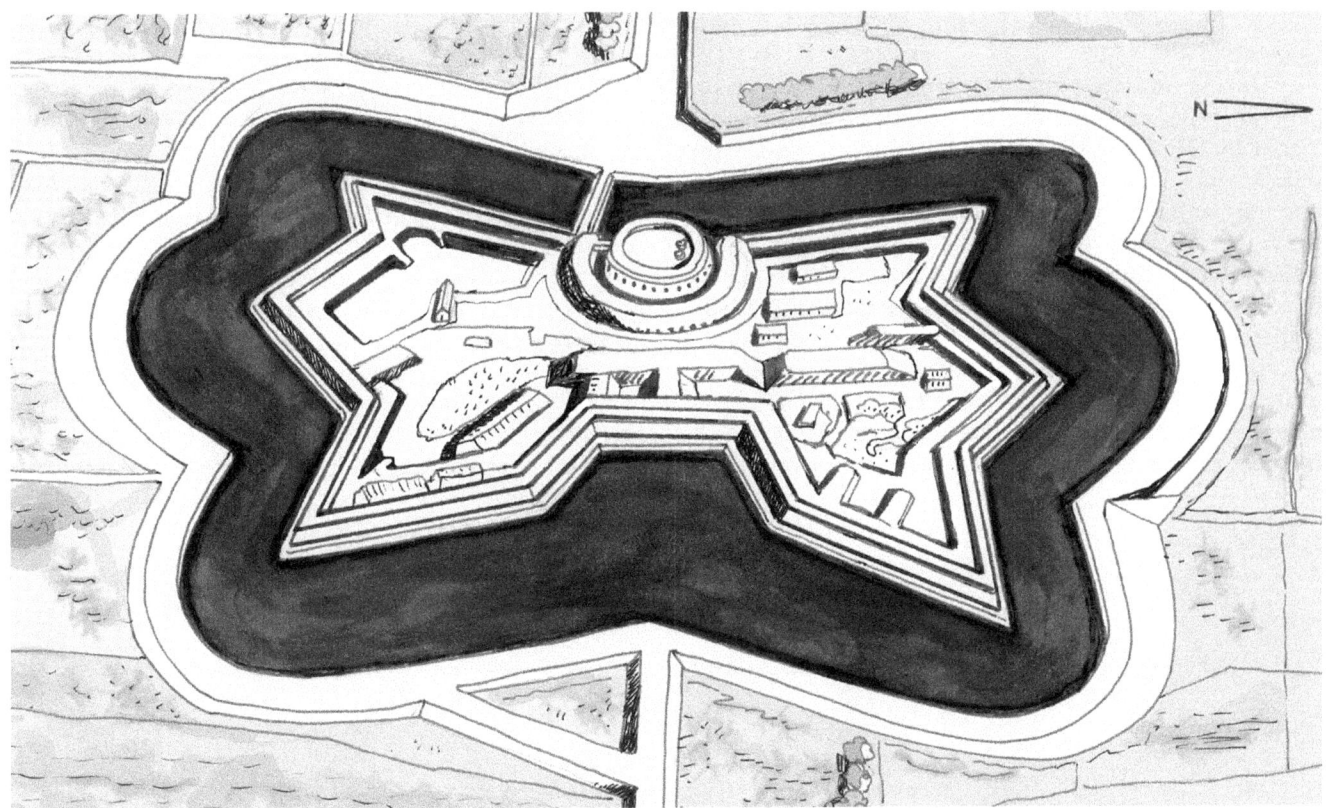

Above: Fort Jutphaas. Located south of Utrecht near the village of Nieuwegein, Fort Jutphaas was built in 1819–1820 across a main access to Utrecht called the Overeindseweg. The stronghold included a wide moat, a tenailled enceinte made of earth, and a roundish two-story high defensive wachthuis (guard house) made of strong masonry constructed in 1846–1848 at the rear of the fort. The fort was armed with five 12 cm guns, and two 8 cm guns served by a garrison of four officers and 133 soldiers. Later the armament was modernized with 15 cm howitzers.

Right: Lunetten Utrecht. Built in the flat valley of Houten in the southeast of Utrecht between 1819 and 1829 the four lunettes (designated I, II, III and IV) were part of the New Holland Waterline. Together with Fort Vossegat they were intended to defend that part of the Utrecht high grounds that could not be flooded. Quite similar in shape and structure, the lunettes were pentagonal strongholds with two faces meeting at a salient facing the enemy, two small flanks on the sides and a closed gorge at the rear. They were made of brick masonry and thick earth breastworks, and surrounded by a wet ditch.

Torenforten

During and after the Napoleonic Era there was a general revival of the large circular artillery tower, illustrated by the French standardized Tour, British Martello towers, enormous Maximilian towers in Austria and Italy, and the Dutch torenforten. In the period 1830–1850, the Dutch forts constructed or improved in the NHW were typical artillery towers. The new forts developed in the Netherlands were loosely based on Montalembert's conceptions.

The Dutch *torenforten* (tower-forts) consisted of large oval or cylindrical structures—some could be as large as 40 meters (43 yards) in diameter. They were made of thick brick masonry in order to be shellproof. They were generally three or four stories high, and housed quarters for the garrison, as well as supplies, and ammunition stores. The numerous artillery pieces were emplaced into casemates (firing chambers). The towers were practically always surrounded with a wet ditch.

The artillery placed in the tower was intended for long range, and close range defense was carried out from an enclosure made of low earth walls. These low works encircling the tower were of various designs including bastioned or tenaille enceintes. They included thick breastworks, masonry ammunitions stores covered with thick layer of dirt, and numerous compartmented open guns emplacements. In peacetime the guns were parked in underground garages called *remises*. The whole fort (including tower and enclosure) was bordered with a large wet moat, and a glacis.

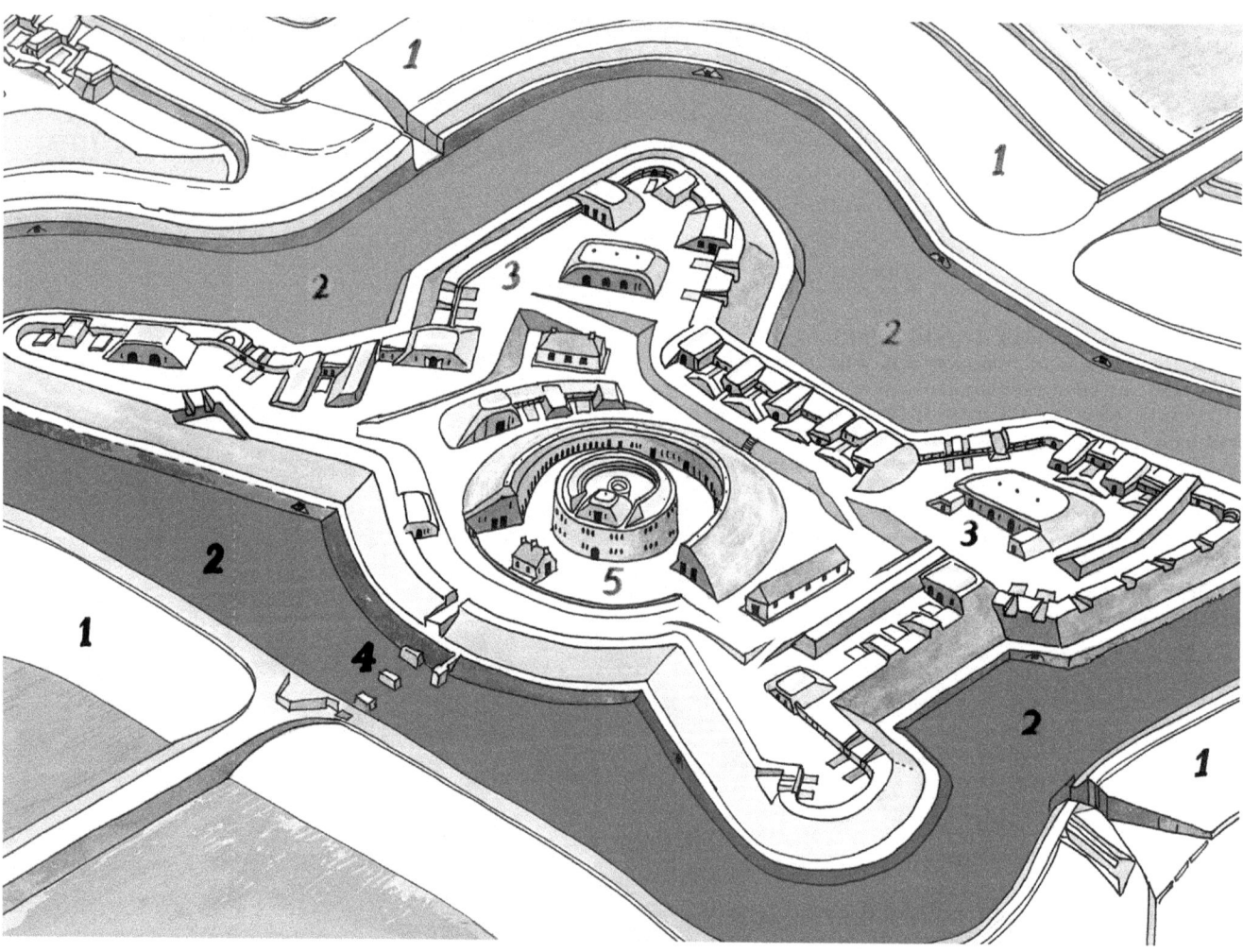

Hagendorp Torenfort. Although Fort Hagendorp did not exist (it was a school example of a typical 1840s Dutch torenfort), it presented many similarities with Fort Everdingen built between 1844 and 1849. Hagendorp displayed for engineering students the basic elements: the glacis (1), the moat (2), the compartmented artillery emplacements placed in bastions and curtains (3), the entrance in the gorge (4), and the tower (5) with its protective counterscarp gallery added in the 1860s after the introduction of rifled artillery.

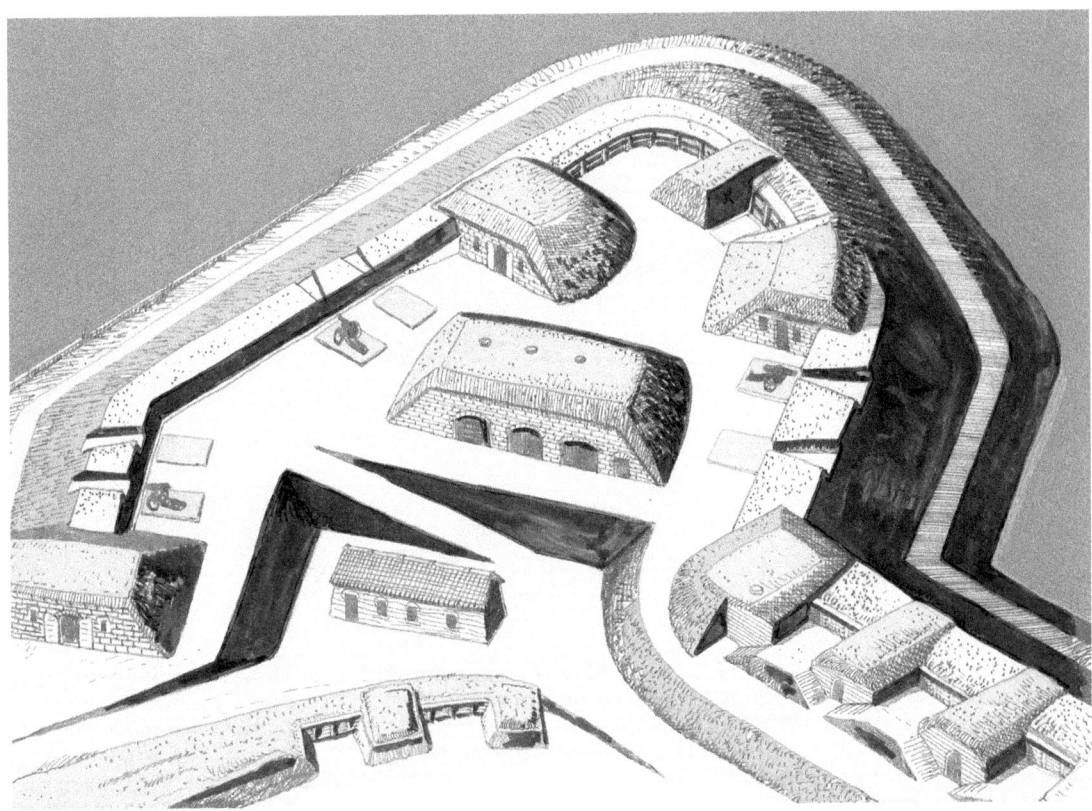

Artillery emplacements. The artillery emplacements were compartmented with traverses and parados. Note that traverses and parados could be hollow and arranged as vaulted masonry storeplaces in order to shelter the gunners and store ammunition.

Emplacement: 1: Embrasure. 2: Parapet. 3: Traverse. 4: Parados. 5: Gun platform. 6: Access ramp.

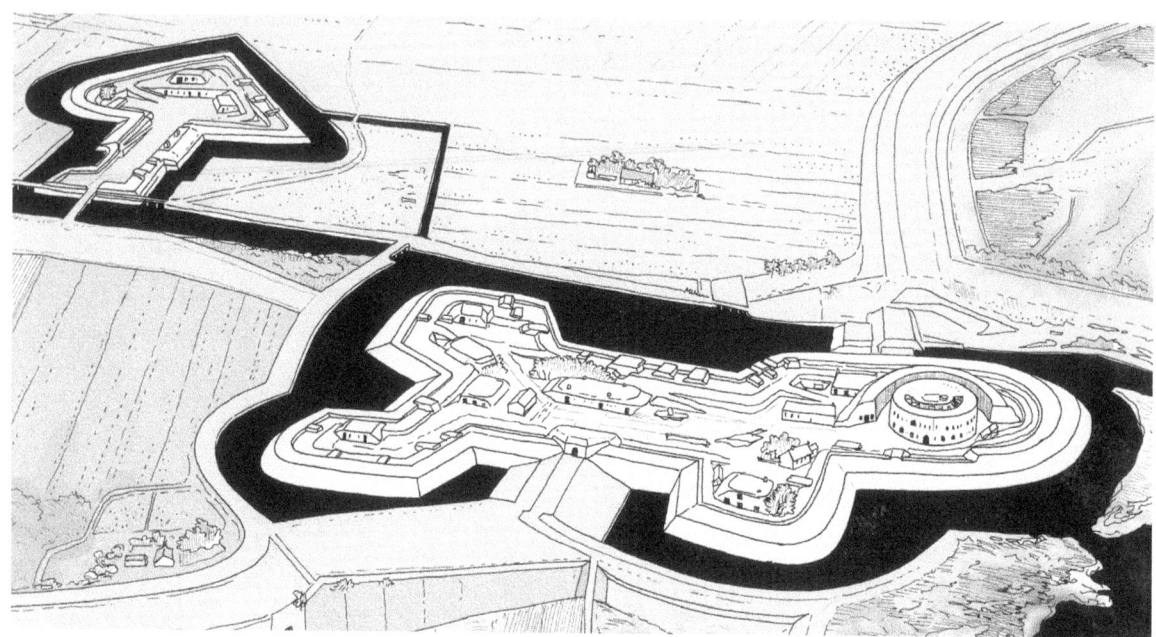

Honswijk. The torenfort Honswijk located on the Lek River near the village of Houten in the province of Utrecht was built between 1842 and 1848. Together with the neighboring Fort Everdingen, the torenfort Honswijk was intended to defend a dike running along the Lek River, an important inundation canal (called De Snel) and a sluice allowing for the regulating and flooding of that part of the New Holland Waterline. The tower was quite large, and its top featured neo–Gothic crenellation and machicolation as decorative elements. The garrison of the fort included some 550 to 650 men serving 34 canons. About 400 m north of the fort, a triangular lunette called Lunet aan de Snel (*top left*) was constructed in the years 1845–46. Fort Honswijk was greatly improved in the period 1878–1881. The tower was lowered and protected with a counterscarp gallery. Various shellproof buildings were added and the protection of the sluice was significantly enhanced. The fort was garrisoned during World War I and occupied by the Germans during World War II. After the war it was briefly used as a prison, and served until 2013 as ammunition store for the Dutch armed forces. Since then the fort has been managed by the municipality of Houten.

Asperen. Fort Asperen located in the village of Lingewaal in the province of Gelderland was built in 1845 to defend a passage on a dijk (called the Zuider Lingedijk) in the New Holland Waterline. The torenfort was surrounded by earth walls and a wide moat. It included three stories, and its walls were made of brick masonry with a thickness of 1.50 m. It could house a garrison of 219 soldiers and officers serving 14 guns. After decades of neglect the torenfort was renovated in 2012, and has become a tourist attraction.

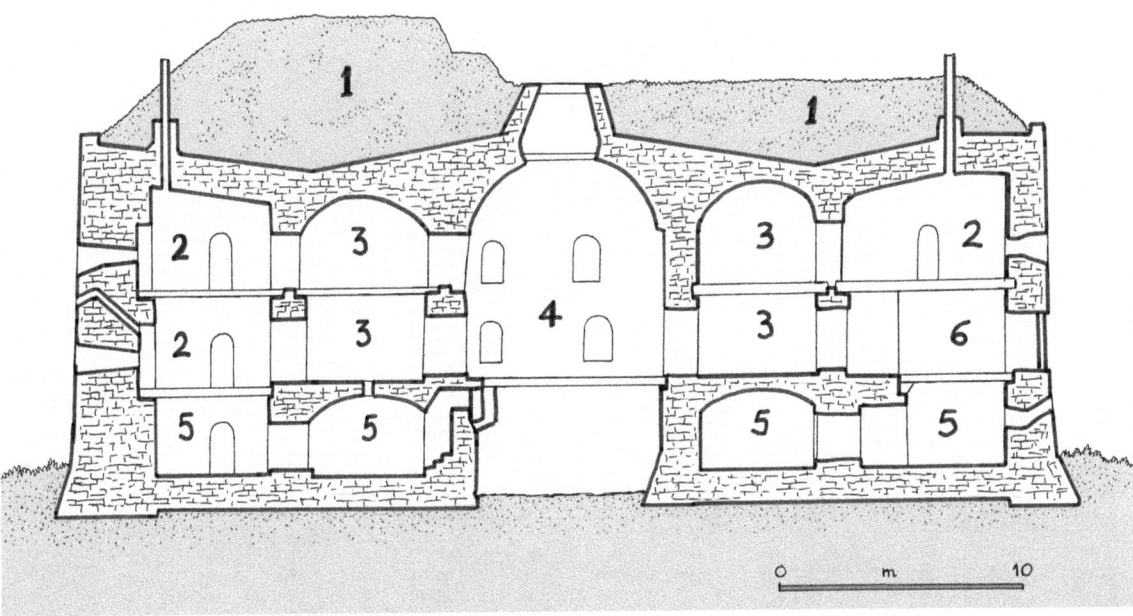

Fort Vuren (cross-section). Located in the municipality of Lingewaal near the medieval castle of Loevestein on the Wall River (Gelderland), the torenfort Vuren was built in 1847. It was intended to defend a dike called the Waaldijk that led to the city of Gorinchem. It had a garrison of 420 soldiers serving 30 guns. The cross-section shows the roof protected with earth (1), the artillery casemates (2), the garrison quarters (3), the central shaft (4) allowing for light and fresh air, the storeplaces (5) for ammunitions, supplies, food and water, and the entrance hall (6). The torenfort is now a hotel and restaurant.

Torenfort Muiden. This torenfort was part of the Vesting Muiden in the New Holland Waterline, and was built between 1850 and 1852 to cover the mouth of the harbor and flank the Zeedijk (Sea side dike). It was oval in shape, made of brick masonry, and included two stories housing quarters for the ca. 80-men garrison, a kitchen, several ammunition stores and an underground tank collecting rainwater. Canons were placed inside casemates, and on top of the building. The tower was surrounded by an 8 m wide moat, and included a drawbridge. Similar designs were also built at Fort de Klop near Utrecht and at Fort Ossemaarkt at Weesp.

Rifled gun. The rifled gun was a major advance in gunnery. It had spiral grooves in its bore that made the shell spin and thereby greatly increased accuracy and range.

The Rifled Artillery Crisis, ca. 1860

The second half of the 19th century was marked by tremendous advances in metallurgy, chemistry, technology and weapons systems, making fortifications obsolete before they were completed. In the 1860s appeared a new kind of artillery pieces. Their barrel was "rifled," meaning that they were grooved—fitted with a narrow spiral cut in the bore. This innovation made the shot spin thereby allowing for greater accuracy, and much longer range of fire than previous smooth-bore guns.

During the Napoleonic Wars, the maximum range of artillery was about 2 km. After 1860 the rifled cannons had a range between 3.5 km and 4 km (about 6 miles). The rate of fire too was significantly increased as breech loading started to appear. At the same time, new shells were introduced, including propellant, explosive and projectile in one single round. Rapidly it became clear that existing fortifications basically made of brick masonry and thick layers of earth no longer offered a sufficient protection. Given the long range of the new artillery and the destructive power it had, the problem as ever was twofold: how to keep out the projectiles launched by the new weaponry; and how to install and operate that very same equipment.

Improvements in the New Holland Waterline

The Dutch, untroubled since the Napoleonic wars, seriously overhauled their defenses in the 1860s when rifled artillery was introduced. This new and unpredicted factor greatly disturbed all planning by the military.

With the introduction of rifled guns the era of the tower forts came to an end. Again, neutral Netherlands was obliged to rethink her system of national defense. The first counter-measures taken were improvements in the already discussed *Nieuwe Waterlinie* (New Holland Waterline or NHW), notably new fortifications in the form of detached forts placed on the main accesses.

The impressive Dutch torenforten built in the 1840s and 1850 on strategical passages proved vulnerable when rifled artillery and new shells were introduced in the 1860s. The first reaction was to lower them by removing one or more stories in order to diminish the surface exposed to enemy fire. Another measure was the construction of a screen wall known as a *counterscarp gallery*. This new element was a large roundish shield made of brick masonry filled with earth. A torenfort fitted with a counterscarp gallery was thus well protected but it could no longer be used as an artillery emplacement, as most casemates had become useless with their embrasures obstructed. The tower could still be used as quarters for officers and troops, and as storehouses. As for the guns they were then placed in open batteries compartmented by traverses in the surrounding earth walls.

Certain parts of the NHW required new fortifications. Around Utrecht for example, the usefulness of the first line of forts constructed in the period 1817–1821 was considerably reduced. Utrecht had always been an important hub of road traffic, and in the 1860s it had also become a central junction for railroad tracks. In order to protect Utrecht and stop any enemy attack towards the capital Amsterdam, a new advanced line of detached forts was built in the period 1867–1881 including forts Ruigenhoek, Voordorp, Hoofddijk, Rijnauwen, Vechten, and Het Hemeltje.

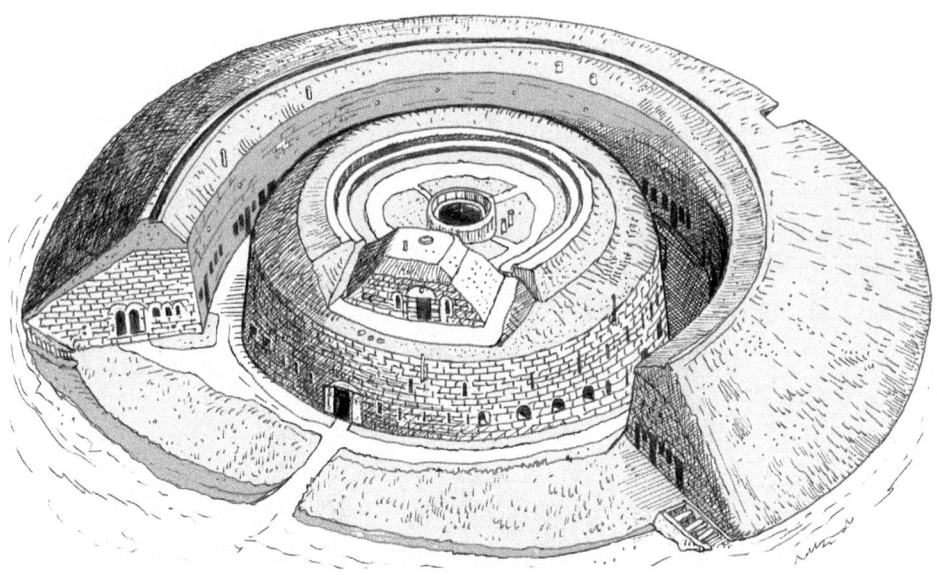

Torenfort with counterscarp gallery.

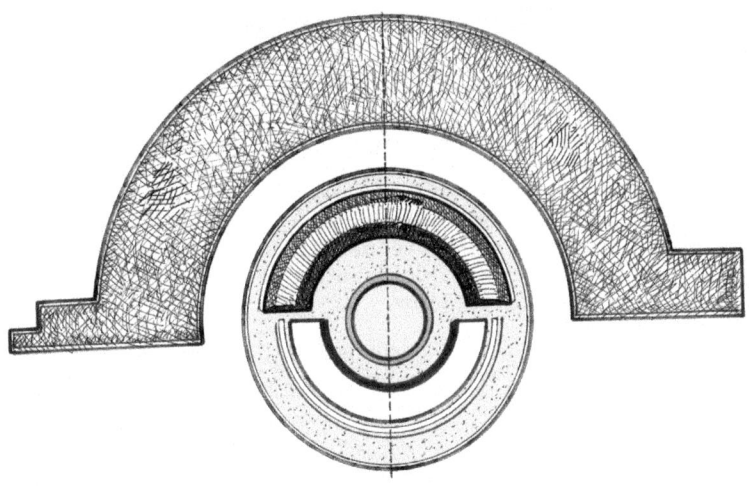

Counterscarp gallery (groundplan).

Counterscarp gallery (cross-section).

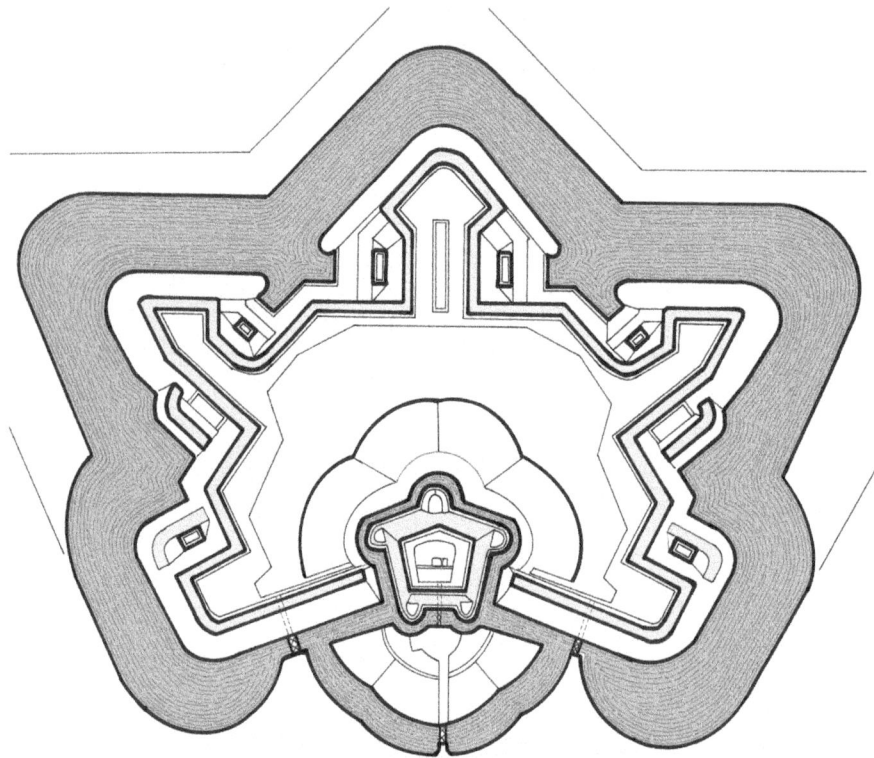

Vechten (plan). Fort Vechten in the Position of Utrecht was built in 1867 to 1870 near the village of Bunnink. A part of the New Holland Waterline it was an additional advanced fortress designed to cover the flat Houten plain that could not be flooded. It was also intended to protect the access of the railroad track Arnhem-Utrecht constructed in the 1860s. Fort Vechten was a huge stronghold covering 23 hectares, and designed according to a curious hybrid style, mixing both traditional bastioned tracé and polygonal features such as a reduit (a small self-containing fort inside the fort) for a last ditch defense, reverse flanking batteries, shellproof barracks, shellproof ammunition stores, and underground remises (garages) for the guns.

Vechten entrance to the reduit. A reduit (aka redoubt or keep) was a small fully enclosed and independent fortlet generally placed within or in the gorge of a polygonal fort. In this self-sufficient stronghold within the fort the garrison could retire, and continue to fight even when the rest of the work was taken. The reduit was often used as quarters for the garrison, command post, and administration place for the commander.

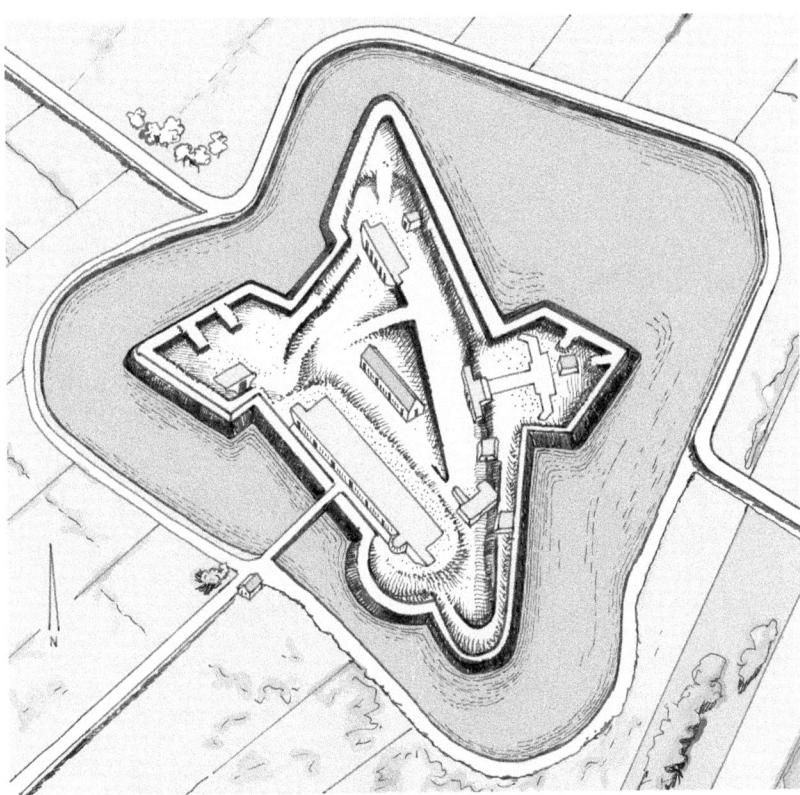

Fort Voordorp (Utrecht). Located north of Utrecht, Fort Voordorp was built in 1869–1870 as an additional advanced stronghold protecting a dike called Voordorpsedijk, and the railroad line Utrecht-Amersfoort (completed in 1863). The fort was still constructed according to the earth-made bastioned system, and included a moat, a half subterranean shellproof barrack, and open traversed artillery emplacements. In 1885 the fort was garrisoned with 242 men serving 34 field guns. In peacetime the guns were stored inside masonry garages called remises.

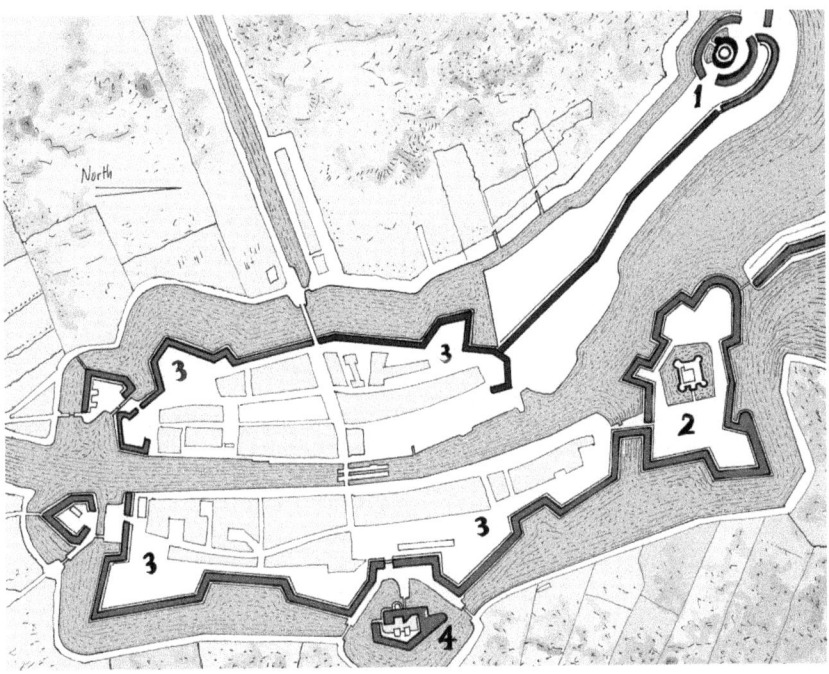

Muiden (plan). Developments in warfare during the Franco-Prussian War of 1870 (notably the introduction of rifled artillery) prompted programs of upgrade and the construction of new fortifications in the NHW. The position of Muiden improved between 1874 and 1877 included a torenfort (1), a bastioned enclosure around the old medieval Muiderslot castle (2), earthen bastioned walls (3) defending the village, and an independent triangular masonry casemated stronghold called Gebouw C or Muizenfort (4).

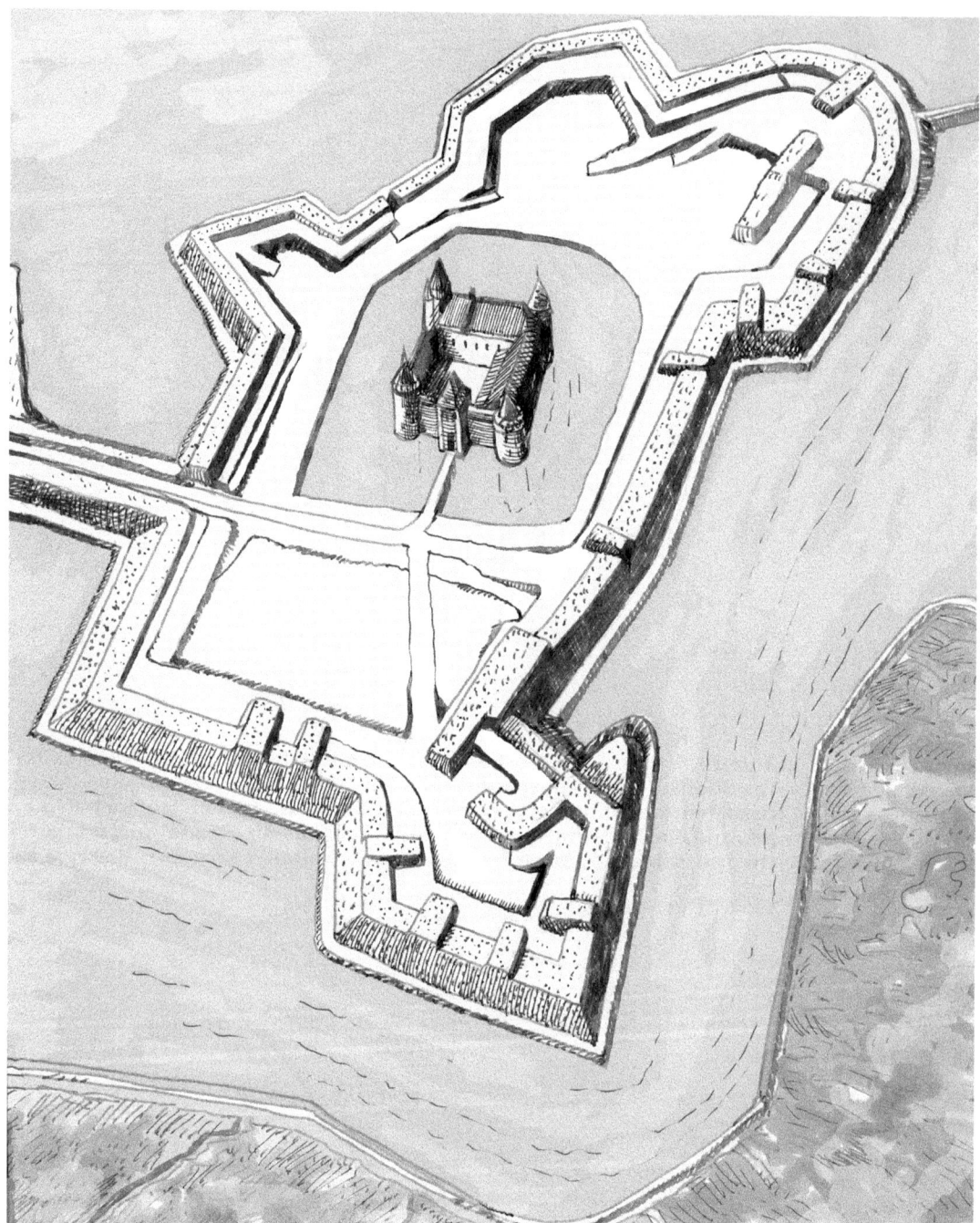

Muiden castle. The 13th century Muiderslot was incorporated into a fortified group called the Stelling of Muiden within the sector Ouderkerk in the NHW, and surrounded by earthen bastioned fortifications.

New waterlines

The New Holland Waterline was the main line of resistance, but it was reinforced with several other lines set up near the borders.

For example, the Grebbelinie, established after 1869 started at Spakenburg, and ran from the Zuiderzee in the north to Rhenen near the Waal and Rhine Rivers in the South. This advanced line was intended to take and delay the first enemy attack, in order to win time and to carry out inundations and manning the forts in the NHW. The Grebbe line just like the NHW included flooding and forts on the accesses.

The Zuiderfrontier (Southern Border defense line) was another waterline running from Willemstad in the west to Grave and Cuijk in the east leaning on the Meuse, Waal and Rhine Rivers. A part of the Southern Border, the Stelling of the Hollandsch Diep and Volkerrak included the fortified cities of Willemstad, and Klundert, as well as a number of detached forts, including Fort De Hel, Fort Sabina Henrica, Fort Duquesnes, Fort Bovensluis, and Fort Buitensluis.

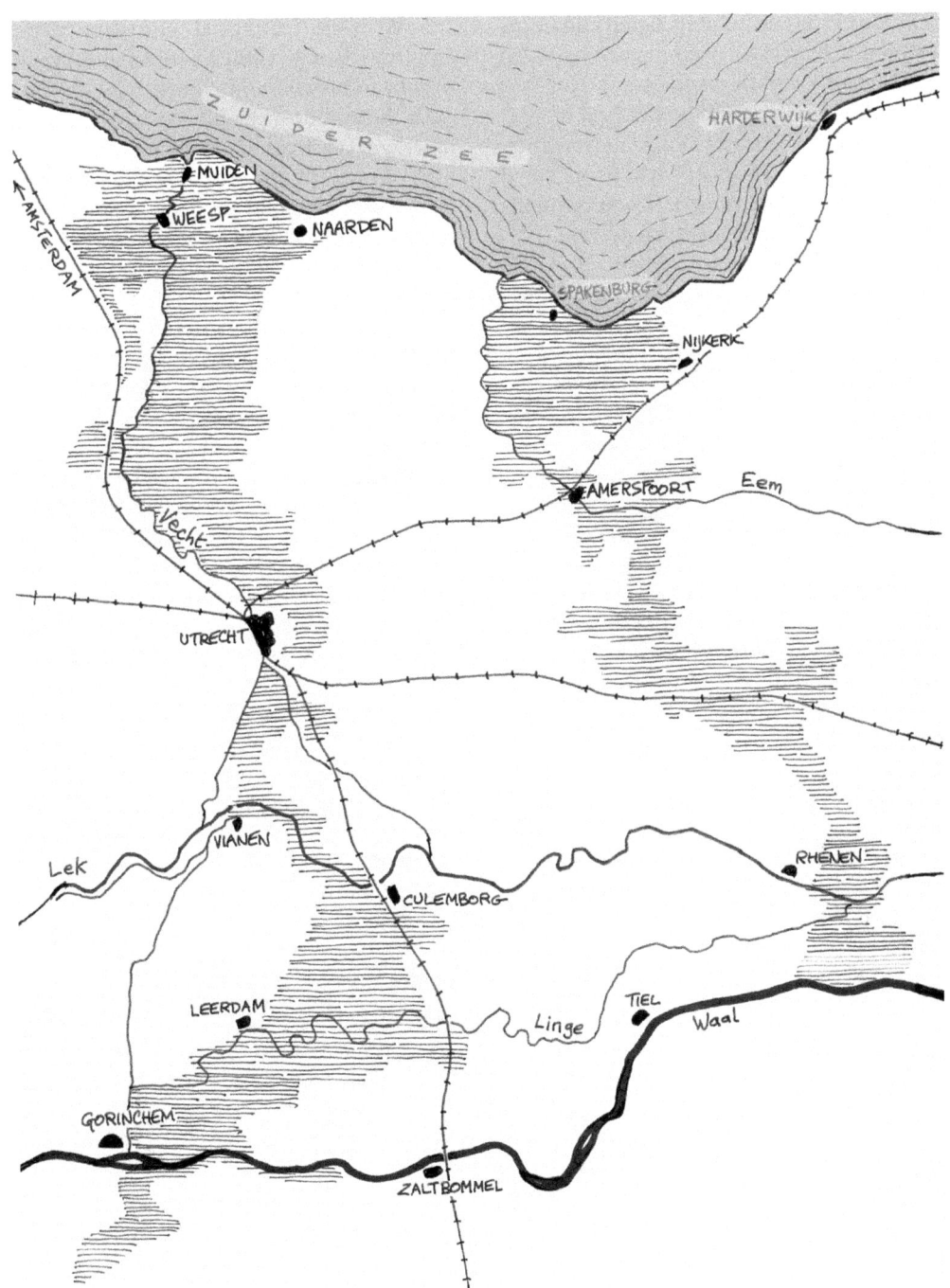

Left: NHW; *right:* Grebbeline.

Vestingwet 1874

New defense lines

The crisis caused by the advances in means of attack created a period of uncertainty and a crisis in the fortifications. There were many reasons to improve the defense of the Netherlands after 1870. Together with the tremendous advances in military technology—notably the rifled artillery, there were also political changes: the Kingdom of the Netherlands was confronted with a new danger. Until 1870, the main enemy was France (and Britain at sea), but after the Franco-Prussian War (1870–1871) the Unification of Germany (called the Second Reich) placed a dangerous new great industrial power close on the eastern border of the Netherlands.

From 1870, although the Netherlands proclaimed strict neutrality in European affairs, all Dutch fortifications were oriented against an enemy coming from the East, while France and Britain were considered as potential allies in case of extreme emergency. So in the 1870s, it was clear that the time was ripe for a thorough re-structuring.

The so-called *Wet tot Regeling en Voltooiing van het*

Vestingstelsel [aka *Vestingwet*] (Law for Regulation and Completion of the Fortified System) was promulgated on April 18, 1874, by King Willem III and the then Minister of War August Weitzel. The law was a deep and methodical reorganization of the Dutch national defenses along modern lines, and mainly facing the German Second Reich. It indicated the following defensive works and *stellingen* (linear positions of defense) that were to constitute the national defense.

(1) The Nieuwe Hollandsche Waterlinie (NHW New Holland Waterline) from the Zuiderzee along Utrecht and the Lek River, along the Merwede River all through the Land of Altena, and down to the New Merwede River;

(2) The stelling of the Geldersche valley in the Lower-Betuwe working as an advanced defense line to the NHW;

(3) The stelling along the Hollandsch Diep (broad waterway connecting the Rhine to the Meuse River) and Volkerak (waterway south east of the island of Goeree-Overflakkee);

(4) The stelling of the mouth of the River Meuse (aka Maas) and Haringvliet (waterway between Voorne-Putten, Hoeksche Waard and Goeree-Overflakkee);

(5) The stelling of the military port of Den Helder (North of Amsterdam);

(6) Works covering the crossing points on Rivers IJssel, Waal and Maas;

(7) The stelling of Amsterdam (a ring of forts around the capital);

(8) The Zuider Waterlinie (Southern Waterline) from the Maas River at Saint Andries to the Amer River near Geertruidenberg;

(9) Works defending the Wester-Schelde in Zeeland.

Urban fortifications

In addition the Vestingwet allowed the dismantlement of the useless urban fortifications of many Dutch cities such as Deventer, Zutphen, Elden (South of Utrecht), Grave, Nijmegen, Den Bosch, Groningen, Leeuwarden, Harlingen, Breda, Bergen op Zoom, Breskens, and several others. The urban fortifications of Arnhem, and Utrecht had already been demolished respectively in 1829 and 1830.

This decision was particularly welcome and enthusiastically acclaimed by the public at large. Indeed urban fortifications (most of which had been established in the 17th century and even earlier) were profoundly resented as choking, obsolete, pointless and insufferable limitations in space at the time when the Industrial Revolution drastically demanded expansion and growth. The carcan of fortification had no longer any military value and only stood in the way of development.

With the Vestingwet, most Dutch cities felt liberated, and from then on started a tremendous growth allowing new urban, industrial and commercial development, together with the addition of numerous suburbs, and new communications (particularly large boulevards, as well as intercity highways and railways). Ancient bastions, walls, and outworks were demolished, wide ditches were filled or transformed into ponds or canals, covered ways and advanced works were razed, and glacis were leveled making wide spaces available for new practical and purposeful constructions.

These included factories and production centers, dwelling neighborhoods, cemeteries, railway installations (for example marshaling yards, workshops and stations), universities, schools, hospitals, exhibition halls, large boulevards and avenues, and (later) peripheral highways around cities, and all the facilities needed by a modern efficacious city. Public gardens and parks were established too, for example the Noorderplantsoen (north public Park) at Groningen—an elegant public garden that was completed and open to public in 1880. The design and layout resemble a romantic English garden style, characterized by sandy paths meandering between lawns, groups of trees and bushes, gentle mounds, and serpentine ponds, inspired by wild nature.

Most of the Dutch cities started their modern development after the 1874 Vestingwet. However, modernism had a negative side too. In the feverish enthusiasm of demolishing "useless" old objects and reconstructing only new productive ones, many witnesses of the past, irreplaceable edifices, and significant historical buildings were irreversibly destroyed and lost forever.

Plan of Zutphen (Gelderland). The dotted line indicates the tracé of the dismantled bastioned fortifications.

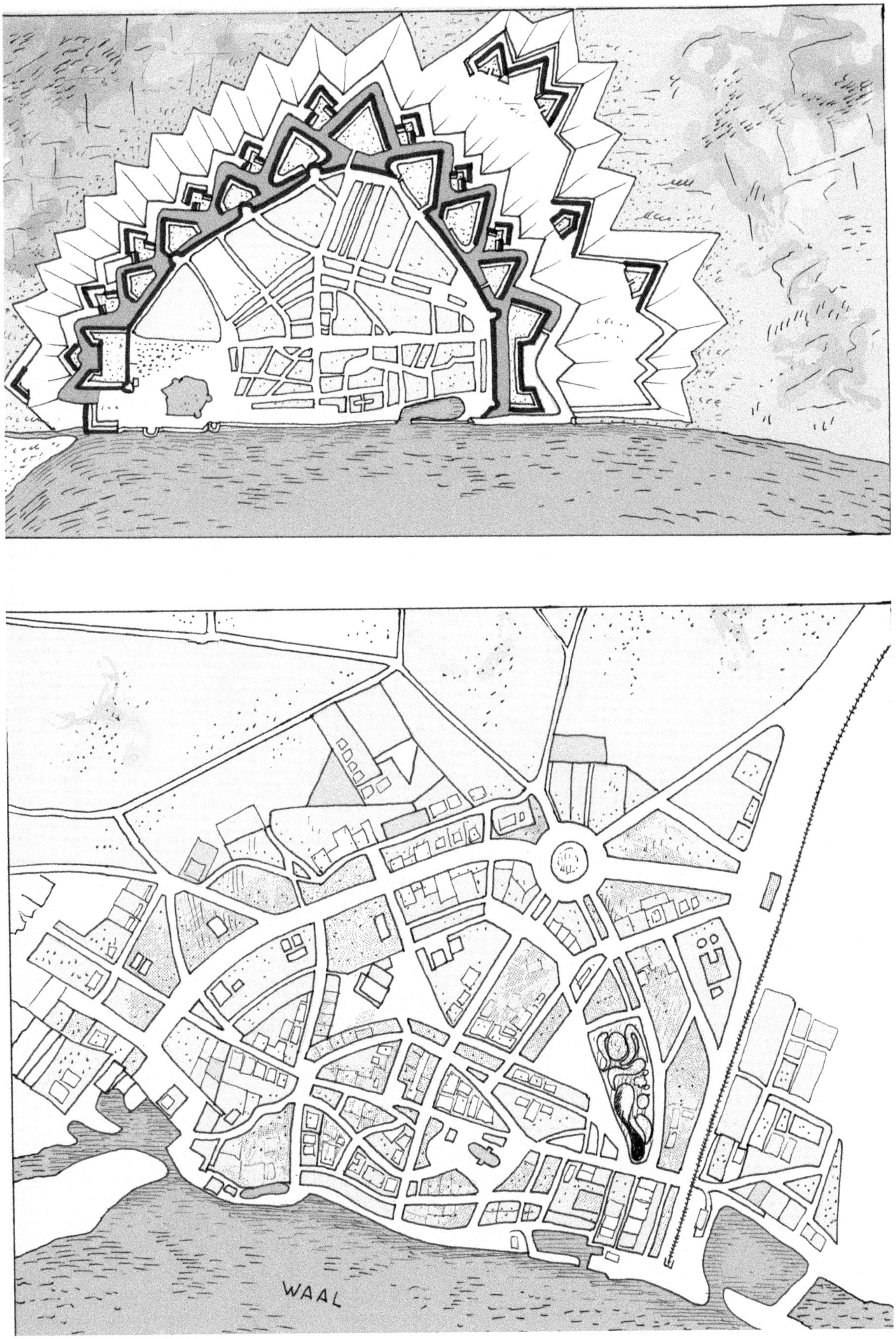

Nijmegen. Plan of the city of Nijmegen before (*top*) and after (*bottom*) the Vestingwet.

Kruittoren at Wijk near Maastricht. Built between 1397 and 1400 the Grote Toren (Big Tower) (later named Kruittoren [Powder Tower]) formed the northernmost point of defense in the medieval fortification of Wijk, the suburb of Maastricht located on the right bank of the Meuse River. The tower was about 20 m high, and divided into four stories. In May 1867 King Willem III gave permission to dismantle the old city defense, and this important and remarkable example of Middle Age military architecture was destroyed a year later. (After a photograph by Th. Weijnen from 1868.)

The Polygonal System

After the Napoleonic Era various German states, but more particularly Austria, Bavaria and Prussia, developed new systems of fortification that were referred to under the generic term of the *Polygonal System* because of their pentagonal shapes. Based on the theories of Montalembert, polygonal fortifications included many artillery pieces well protected behind thick earth walls, traverses and parados (thick screens) and installed in casemates covering all possible attack zones. Low caponiers protruding across the ditches replaced the old angled bastions for flanking and close defense. An independent casemated reduit in the shape of a fortlet was often constructed in the gorge of the main work.

The Germans also made extensive use of detached forts, which were strongly armed artillery units keeping the enemy at a safe distance. Each fort was located within gunshot of its neighbor to allow overlapping fields of fire and mutual support.

The polygonal fortification was a revolutionary change from the prevailing orthodoxy of forts designed with angled bastions for defense. Polygonal forts could be more easily adapted to the terrain and allowed a greatly increased number of heavy guns mounted on the ramparts in order to repulse attackers. These forts were built entirely of masonry, using large quantities of shaped stone, bricks and thick layers of earth. They were always provided with a dry ditch or a wet moat, six to twelve meters wide (similar in yards) and bounded by the main wall of the fort on one side and a counterscarp on the opposing side. Entries were typically by drawbridges.

Inside the fort's perimeter were multistory barracks with facades facing interior courtyards. Barracks were semi-recessed into the walls and included mess halls, kitchens and

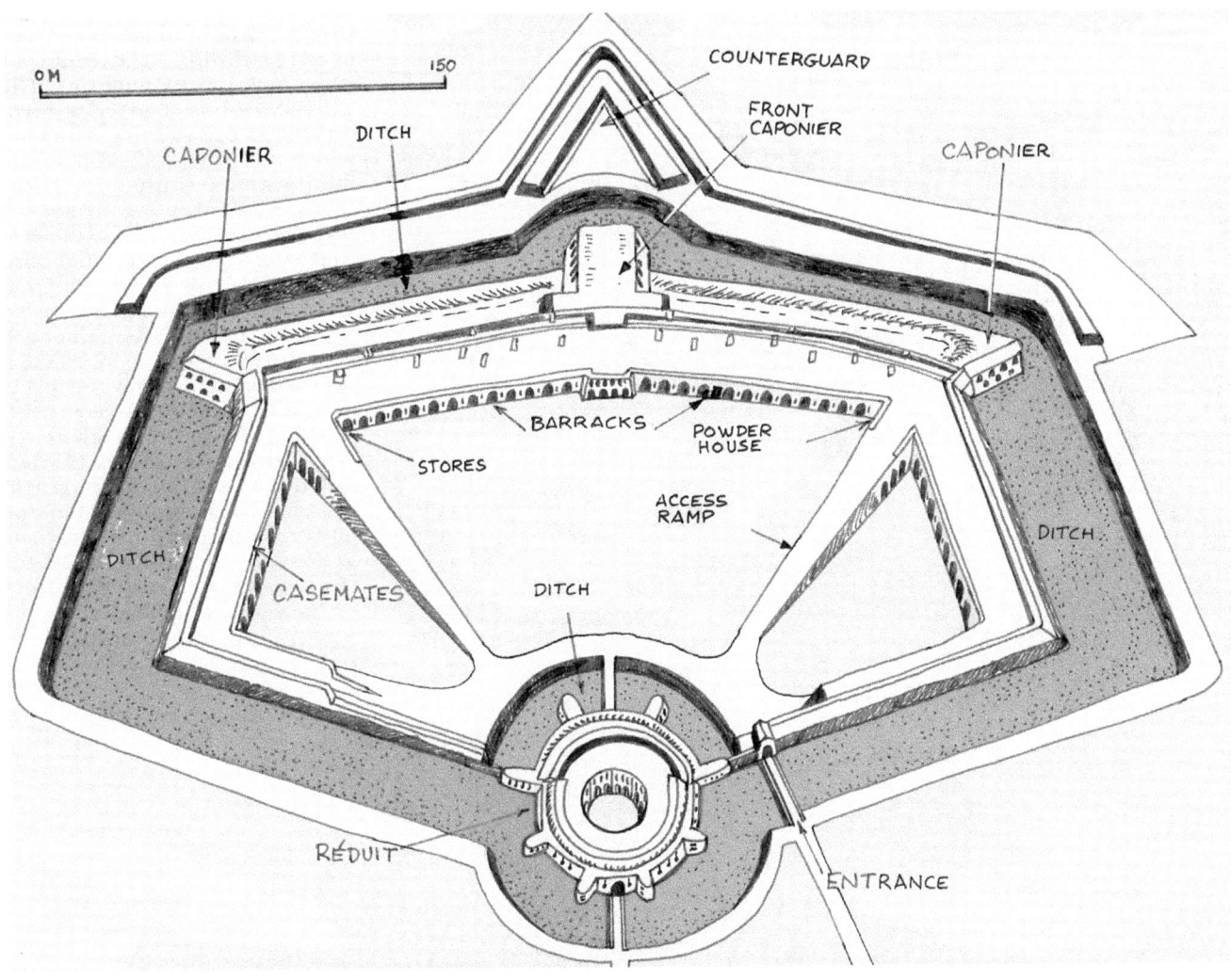

Fort Brockhurst. Located at Gosport, in Hampshire (Great Britain), Fort Brockhurst was a typical polygonal fort. Designed by engineer William Crossman, it was built between 1858 and 1862. It was one of a chain of five similar forts (Elson, Grange, Rowner and Gomer) intended to defend the access to Portsmouth. Brockhurst was surrounded by a moat and featured a circular reduit (keep). The artillery was deployed in casemates placed under the ramparts, and on top of the walls reached by two long ramps on the enclosed parade ground in the middle of the fort. There were flanking caponiers, one at the front, and two on the shoulders (at the angles of the ramparts) to allow riflemen to cover the ditch.

cisterns. Powder magazines were buried for protection from artillery, located behind triple-locked double doors, and illuminated indirectly from lamp rooms to prevent accidental explosion.

The artillery was placed in masonry casemates or deployed in the open air on top of the ramparts in compartments (called traverses and parados) sheltered by thick screens made of masonry and earth. Shelters were provided for ready ammunition. In many cases during peacetime, artillery was located in masonry covered remises (garages). Special infantry positions in the form of thick earth breastworks and caponiers were provided for defense of the ditch by riflemen.

Soon the polygonal system was adopted by all industrialized nations. Polygonal forts were widely constructed in the 1850s, 1860s and 1870s in Britain (e.g., Chatham, Plymouth, and Portsmouth). About 1860, the Prussian system was improved and applied in Belgium by General Henri Brialmont, particularly at Antwerp. After 1874 the French finally abandoned the bastioned system inherited from King Louis XIV and engineer Vauban. Then General Séré de Rivières started a huge reorganization of French fortifications, notably at Paris, Lyon, Langres, Epinal, and Verdun. In the USA, during and after the Civil War (1861–1865), many coastal forts (e.g., Fort Pulaski, Georgia) displayed European polygonal features.

In the Netherlands only one fort was designed and built according to the polygonal principle: Fort Rijnauwen constructed between 1867 and 1881 to defend the city of Utrecht. Why the Netherlands did not follow the new trend can be explained by economical reasons. It was much too expensive to rebuild all fortifications on the accesses in the various waterlines. Besides the Dutch kingdom was officially neutral, and its political leadership gambled that this would be respected.

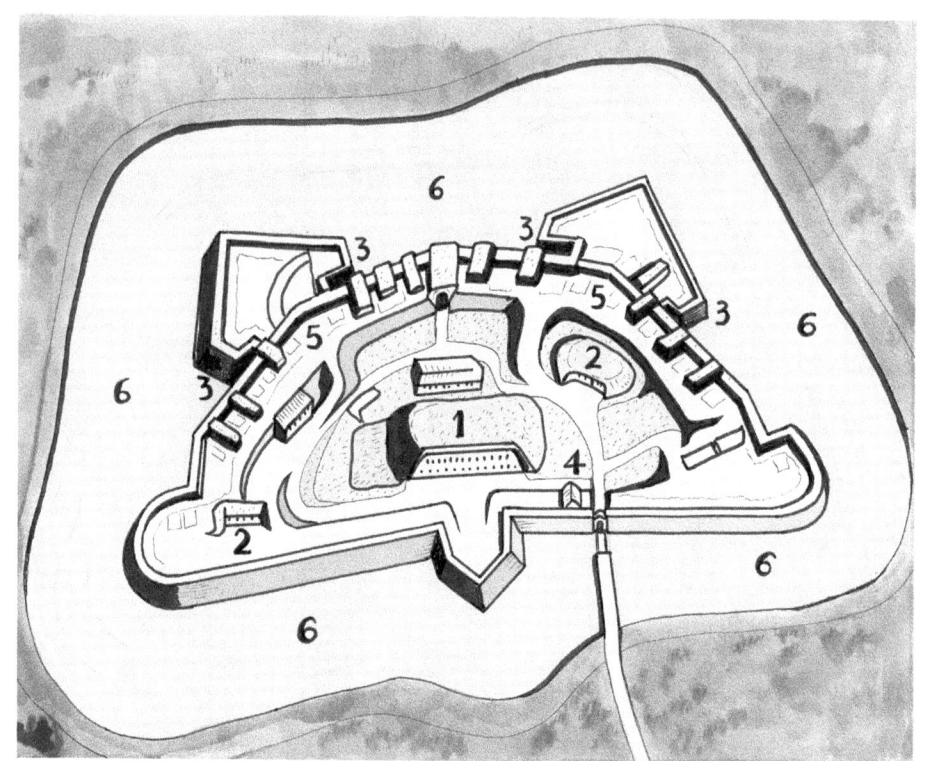

Fort Het Hemeltje. Built between 1878 and 1881 Fort Het Hemeltje is the last and southern-most fort of the second advanced line of strongholds at Utrecht. It was named Het Hemeltje ("The Tiny Paradise") after a farm located in the vicinity. The fort, which still featured evident elements of bastioned fortifications, was intended to defend the high ground around Houten that could not be flooded, and protect the railroad track from Utrecht to Den Bosch. It covered 7.5 hectares, had a length of 275 m and a width of 140 m. It had a garrison of 285. Today the fort is owned by Staatsbosbeheer (SBB National Forest Management). The illustration shows the following: 1: Shellproof barrack. 2: Remise (garage). 3: Flanking battery with ammunition store. 4: House for the fortwachter (fort guard). 5: Artillery battery on the rampart. 6: Moat.

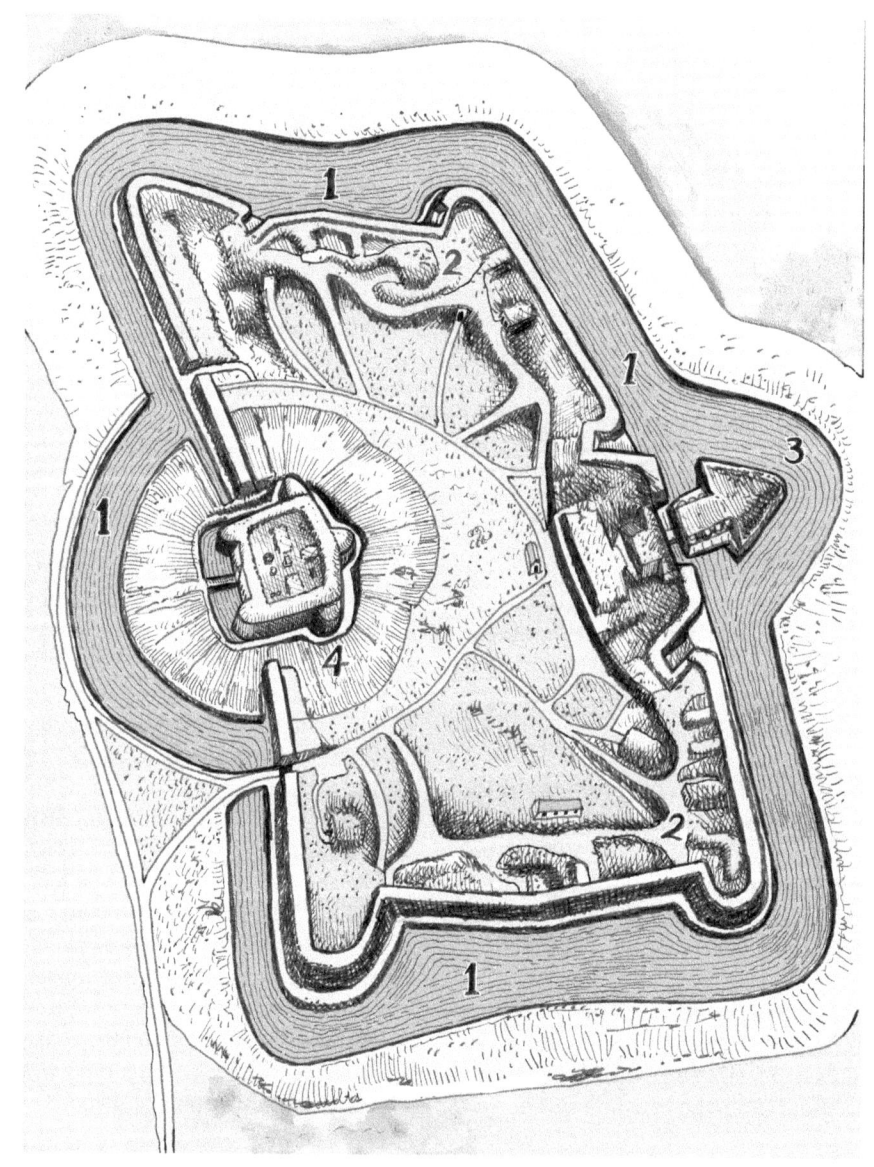

Chapter 7. The 19th Century

Rijnauwen reduit entrance. The masonry reduit was a fort in reduction within the fortress. It was self-containing with its own glacis, moat, entrance, and drawbridge, and small flanking caponiers.

Bottom, facing page: Fort Rijnauwen was part of the second ring of advanced works built between 1867 and 1881 to defend the Plain of Houten—the eastern access to the city of Utrecht. The other forts were Ruigenhoek, Voordorp, Hoofddijk, Vechten and Het Hemeltje. Fort Rijnauwen covers some 32 hectares, and is the largest fort in the New Dutch Waterline. The front has a length of 240 meters, and the sides are 195 m long. It is an unique and impressive structure displaying the characteristics of a polygonal fortress, based on the Belgian General Brialmont's theory. It was made of sturdy brick masonry covered with thick layer of earth. It included a 30 m broad wet ditch (1); ramparts bristling with gun batteries (2); a front caponier (3); and a powerful réduit or keep (4) placed in its gorge. In 1877 shellproof barracks, remises and depots were added. The fort could accommodate 19 officers, 32 NCOs, 515 soldiers and one washing woman. In 1872 armament included 105 cannons, howitzers and mortars of various caliber from 8 cm to 29 cm. Fort Rijnauwen was garrisoned during World War I, and hasty refurbished, modernized, and reinforced (notably with the construction of concrete MG pillboxes) in 1939. During World War II, Fort Rijnauwen was used by the German army as an ammunitions depot and place of execution by firing squads of Dutch and Belgian resistance fighters. After the war it was used as an ammunition depot until 1975 when it passed to the management of Staatsbosbeheer (SBB National Association for the Preservation of Nature). For years the fort was closed to the public, and as a result is now covered with thick vegetation and is home to many plants and animals, including rare varieties and species. Kingfishers, roe deer, grass snakes and weasels feel right at home here, and hundreds of bats spend the winter in this green oasis in the most populated part of the Netherlands. To ensure that the natural habitat is disturbed as little as possible, the fort can only be visited as part of SBB guided group tours.

Fort Ruigenhoek. Built between 1869 and 1870, Fort Ruigenhoek belongs to the second belt of detached fortresses defending Utrecht. The illustration shows the entrance of an underground shellproof ammunition store.

Crisis in the 1880s

High explosive projectiles

After the introduction of rifled barrels in the 1860s, tremendous advances were made in the quality of projectiles in the 1880s. Then another revolution occurred in artillery, notably the development of much more powerful explosives. These metallurgic and chemical developments multiplied the power of artillery against fortifications. Tests against existing fortifications indicated that forts built previously were obsolete. New high explosive shells could penetrate deep in earth walls and cause complete destruction of brickwork and stonewalls. Masonry and thick layers of earth were no longer resistant.

Artillery and gunners deployed in the open on the fort superstructures had become extremely vulnerable to new developed anti-personnel shells exploding in flight in the air projecting with violence down on them a rain of lethal bullets, fragments and splinters—originally invented by the British General Henry Shrapnel (1761–1842). The rifled breech-loading gun and the new high explosives had finally ended the long reign of bastioned fortifications. A new solution was required, and military engineers were compelled to revise their designs for new defensive works and adapt existing ones to the new predicament

Concrete and armor

The answer was found in the use of high-strength concrete, which was more resistant than masonry to explosives. Coming from the Latin word "concretus" (meaning compact or condensed), concrete is a construction material composed of cement and other cementitious materials such as fly ash and slag cement; aggregate (generally a coarse aggregate made of gravel or crushed rocks such as limestone, or granite, plus a fine aggregate such as sand) bound with water. Reinforced concrete, which was widely used to replace masonry, included steel rods (or rebars) incorporated into the concrete mass.

Concrete offers a good resistance to compression while the metal framework opposes to traction. The combined complementary of concrete and metal rebars forms a solid, compact, massive and monolitic material. The development of reinforced concrete allowed new fortifications to deal with the extremely destructiveness of new explosives.

Rebuilding all existing forts in concrete of course was an impossible an immensely expensive task if not an impossible one. As a result many forts already existing were left as

Opposite, top: **Fort Pannerden.** Built in the period 1869–1872, Fort Pannerden was a key position in the New Holland Waterline. It was an advanced sperfort (barrier fort) placed at the joining point of the Rhine and Waal rivers in the province of Gelderland in the southeast of the Netherlands. It covered 2.6 ha and occupied an important strategic significance in that it also controlled the Pannerden Canal, which supplied a part of the water for the inundations of the New Dutch Waterline and could potentially be used as a route towards the main line of defense. The entrance of the fort (1) included a bridge crossing the 6 m wide dry ditch (2). The ditch was defended by a continuous escarp gallery, a double caponier (3) also known as "cat's ears," and three other single caponiers (4). The long-range artillery was placed in open emplacements (5) on the top platform. In the middle of the fort there was a massive central shellproof partly underground core (6) divided into rooms for officers, quarters for the troops, water reserve, supply and ammunition stores, a kitchen with a bakery, infirmary, washing facilities, and other services. Fort Pannerden was upgraded during the period 1885–1895. The open batteries were completely reconstructed and covered with armor and concrete while the roofs of the buildings were reinforced with concrete. During World War II on May 10, 1940, during the German invasion of the Netherlands, the then obsolete fort saw a short combat when it was surrounded and attacked. After the war the fort was dismissed, and fell into disuse. In November 2006, it briefly became the focus of national news stories when a group of squatters were dislodged by a spectacular police operation.

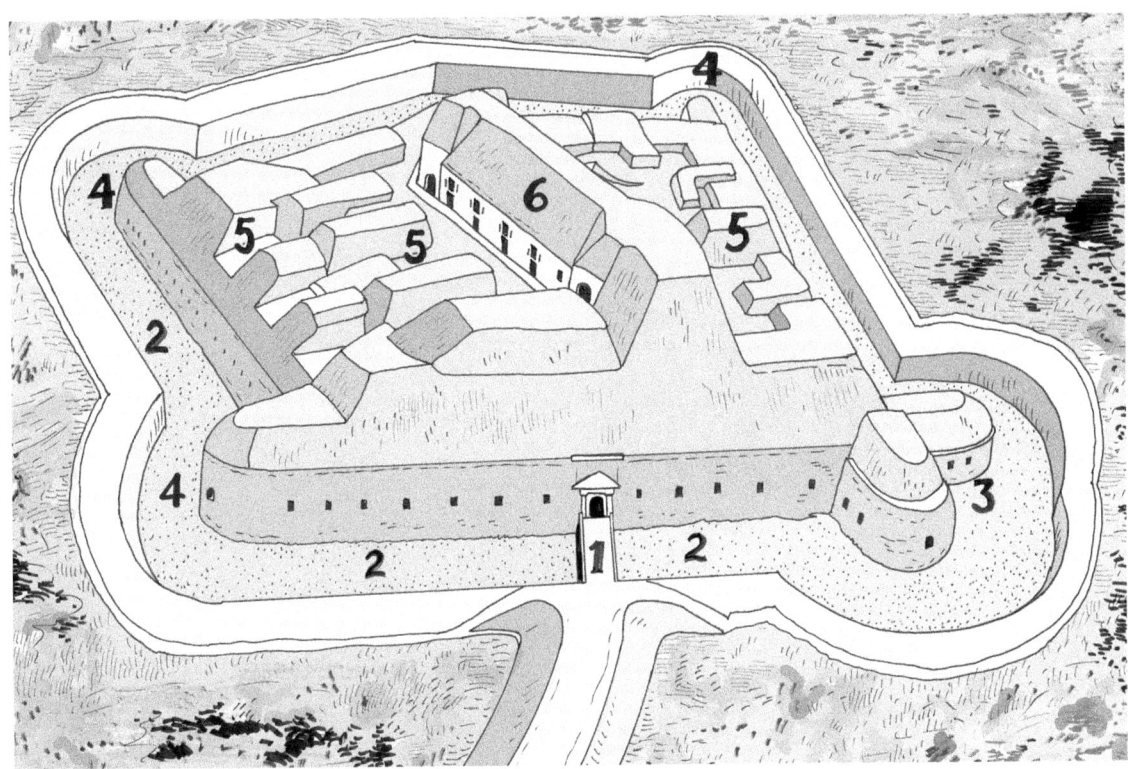

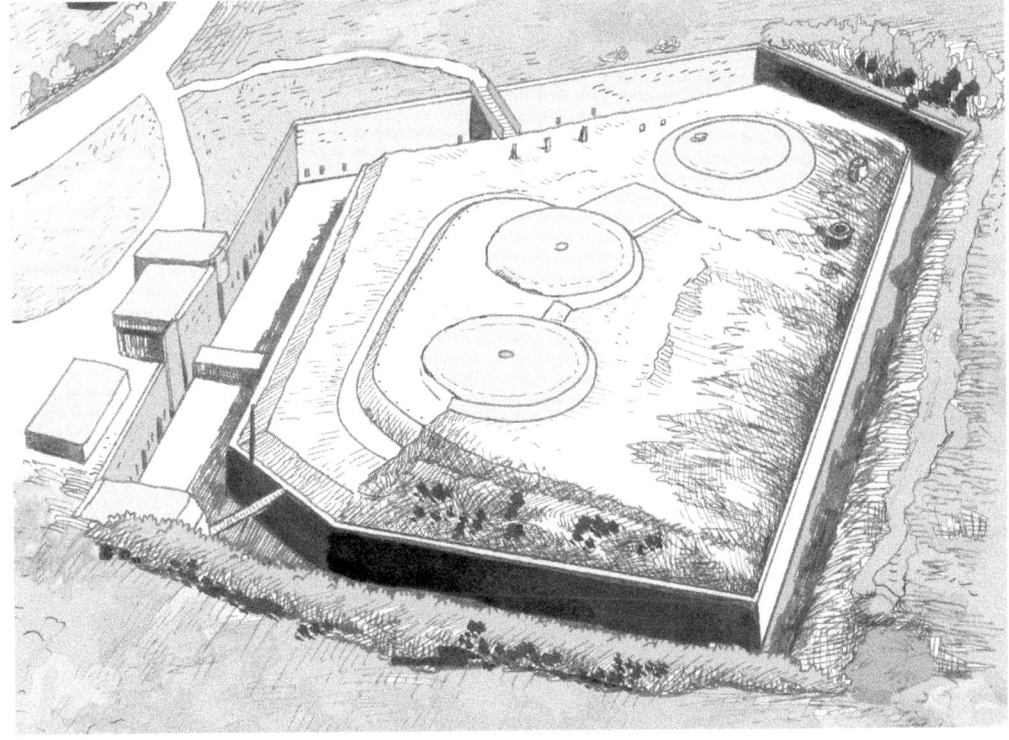

Fort Maasmond Hoek van Holland. The coastal pantser (armored) Fort Hoek van Holland, also known as Fort Maasmond, is located near the port of Hoek van Holland on the North Sea. Built between 1881 and 1890 it was intended to defend the strategically important mouth of the canal Nieuwe Waterweg (New Waterway), making the city of Rotterdam easily accessible to large ocean vessels but also to enemy warships. Placed on the north bank of the canal, the fort is made of masonry and concrete. It is enclosed by a dry moat defended by counterscarp galleries. It featured a gepantserde geweergalerij (an armored riflemen gallery), living quarters for a garrison of about 350 soldiers, an access road, magazines for ammunition, three rotating armored gun turrets, and various ancillary and service rooms. During World War I, the fort was manned and refurbished but soon proved to be woefully outdated. In 1939, a modern battery, including three 15 cm guns, was built in the sand dunes further to the west. The original fort was then assigned a supportive task. During the occupation of the Netherlands between 1940 and 1945 the fort was incorporated into the German Atlantic Wall and housed a telephone exchange and a radio jamming station. By then all metal parts (including the armored cupolas) and all useful equipment were removed and recycled for the German war production. Only the riflemen gallery, shielded by heavy armor plating, was spared. After the war and until 1975, the fort was used as a military storage depot. Since 1987, it has been renovated and become a permanent museum.

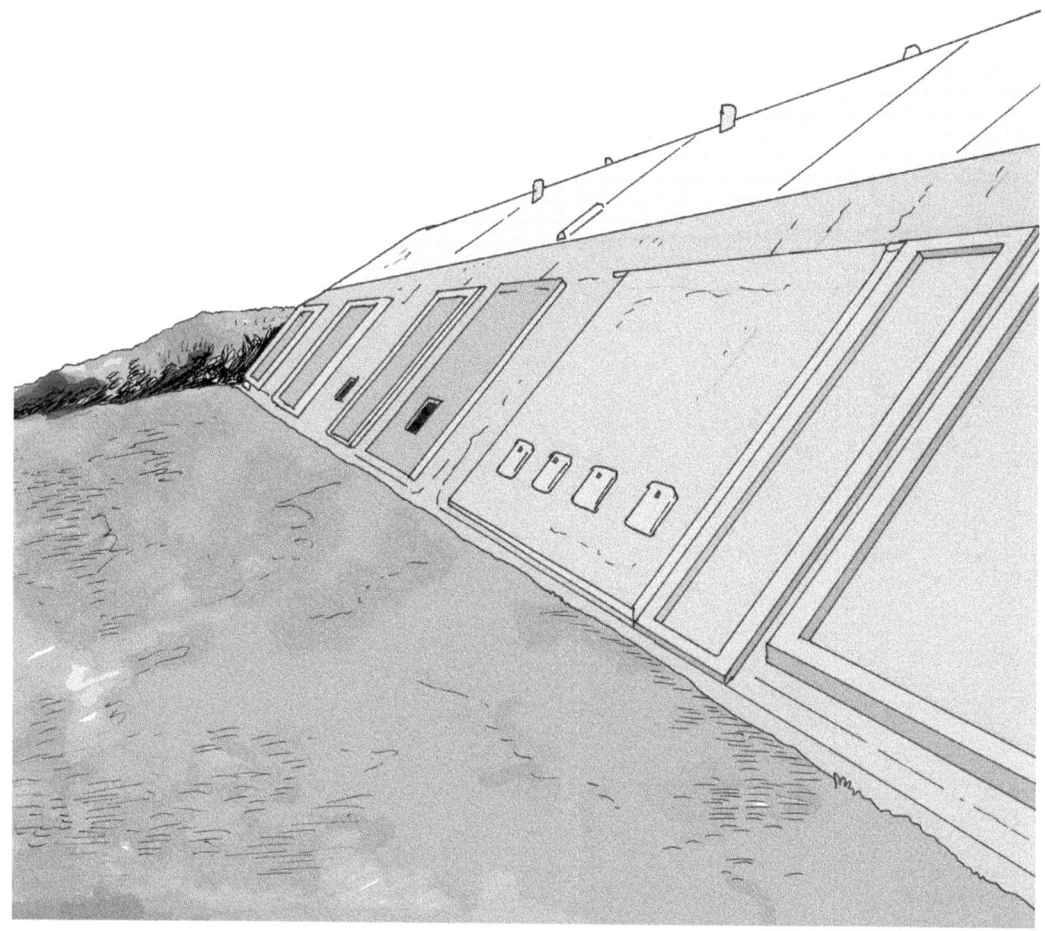

Armored infantry gallery Fort Maasmond.

Gruson turret Fort Hoek van Holland. The main armament at Fort Hoek van Holland consisted of three rotating armored turrets, each armed with two short 15 cm guns. The turrets were purchased from the German company Gruson. Today concrete replicas have been installed to replace the steel gun cupolas that were looted by the German occupiers during World War II.

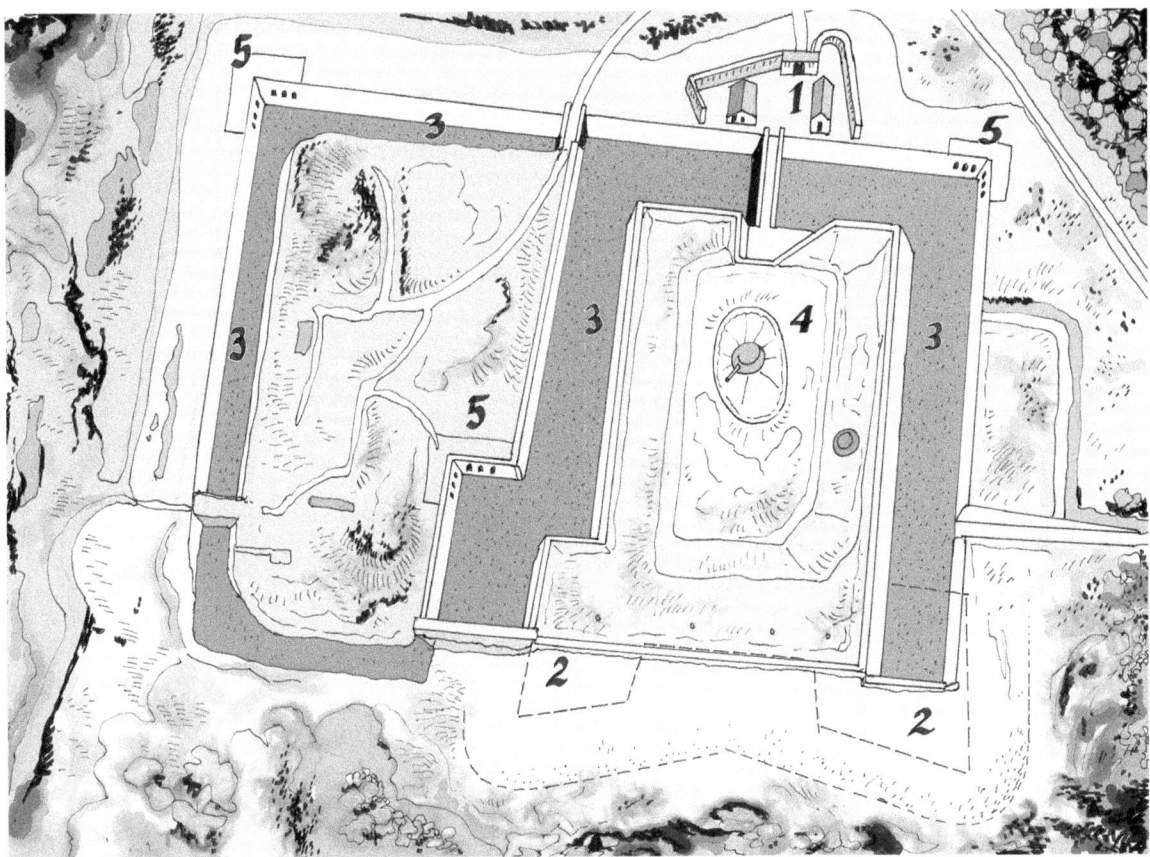

Fort Kijkduin. Located in the dunes near Den Helder, Fort Kijkduin was a coastal stronghold defending the approach of the important military harbor of Den Helder in the province of North Holland. Built by the French as an enclosed hornwork with two full bastions, it was completed in 1813. It could accommodate about 700 soldiers. It was originally named Fort Morland after Colonel François Louis Morland (1771–1805), who was killed at the Battle of Austerlitz in December 1805. After the departure of the French in 1815, it was re-baptized Fort Kijkduin ("dune sight"). In 1897, the old Napoleonic fort was modernized. The entrance was defended by an advanced tambour (1). The two original bastions (2) were demolished. The dry ditch (3) was expanded. Walls and roofs were reinforced with concrete. An oval concrete structure with an armored gun turret (4) was built. Concrete coffers (5), counterscarp galleries and a system of underground passages replaced the obsolete and vulnerable caponiers for the defense of the dry ditch. Today Fort Kijkduin has the status of Rijksmonument (national monument) and since 1996 houses the Noordzeeaquarium (North Sea Aquarium) and a restaurant with a view on the sea.

they were. Only some strategically located important land and coastal forts were improved. Concrete and thick layers of sand were then added to masonry, in order to cover and protect vulnerable parts such as the roofs and walls of ammunition magazines, personnel shelters, and combat chambers.

The Stelling of Amsterdam, 1880–1920

Generalities

The "Stelling" (fortified position) of Amsterdam was an ambitious program that was decided as a part of the Vestingwet of 1874. The construction started in 1880 and the whole scheme was constantly improved, modernized (particularly after the introduction of new high explosive projectiles in 1885), garrisoned and refurbished during World War I, and completed in the 1920s. The purpose was to create a strong defense in depth, and the Stelling was the last ditch in that scheme.

Should the neutrality of the Netherlands be violated by a foreign power (Germany was then regarded as the most probable aggressor) the Royal Dutch Army would first resist in the various lines established at or near the borders. Then if needed it would retreat to the Holland Waterline in order to defend the *Randstad*—the economic and political heart of the kingdom, the conurbation that stretches in a horseshoe shape from Dordrecht and Rotterdam around Utrecht and Amersfoort via The Hague, Leiden, Haarlem, and Amsterdam. The majority of the people of the Netherlands live in this area. Then should the odds be overwhelmingly in favor of the aggressors, the Dutch army would withdraw in the Stelling of Amsterdam for a final combat. The whole strategy was intended to repulse any invasion, or to delay the advance of the enemy, to win time for diplomatic negotiations, and perhaps for a Franco-British military intervention.

The Stelling was 135 km (83.7 miles) in length, and was established about 15 to 20 km (10 or 11 miles) from the center of Amsterdam. It enclosed a perimeter of 14,953.3 ha (57,735 square miles).

Located in the low, flat and floodable grounds of the province of North Holland, the Stelling obviously comprised large inundations, and about 45 main forts (plus a number of smaller strongholds and infantry positions) placed on, across or near the accesses (dikes, highways, waterways, railroad lines, and the like).

The precursors of the Stelling of Amsterdam were the already discussed Kraijenhoff posts built during the French Era, but the new general project was much more sophisticated. The tracé of the Stelling was drawn up in the late 1870s, and was intended to be a much improved, and much stronger version of the New Holland Waterline. The Stelling comprised three components: First the reduit, the perimeter enclosed by the Stelling including Amsterdam and its populations; second the linear girdle of forts; and third the peripherical inundations zones and a wide glacis.

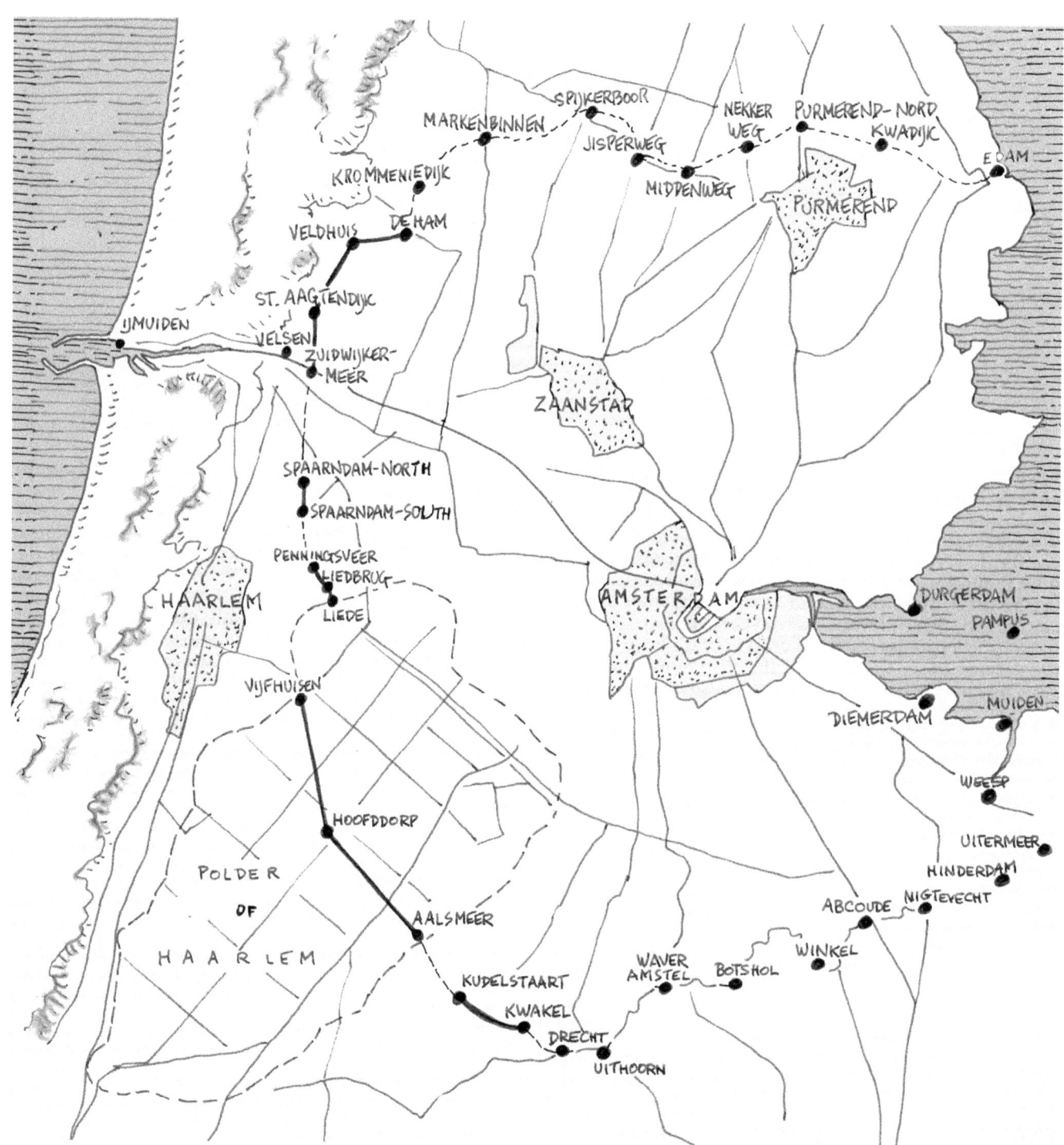

Map of the Stelling of Amsterdam.

The reduit of Amsterdam

In the perimeter enclosed by the inundations and the fortifications, many ammunitions, military materiel, fuel, supplies and food of all sorts were stored for the population and the garrison—enough, it was hoped, for a siege of maximum six months. By the end of the 19th century, the region of Amsterdam was much less populated than today, and there were many open countryside areas with meadows for rearing animals, as well as orchards, and agricultural lands where food could be produced. Reserves of coal, and grain silos were built. Drinking water was supplied by sources, tanks and cisterns collecting rainwater; additional wells with pumps and filter installations were built. At Hemveld, Ouderkerk and Muiden as well as along the Noordzeekanaal (North Sea Canal) black powder manufactures and arsenals were established. In the 1910s and 1920s airplanes appeared and two airfields were created at Schiphol and Schellingwoude for both military and civilian use.

Fortunately Amsterdam and the Stelling were never besieged. Whether these measures would have permitted the survival of all military personnel and civilian population remains an open question. It should be noted that during the siege of Paris by the Prussians, lasting from September 19, 1870, to January 28, 1871, a catastrophic situation and famine occurred. The same happened in Amsterdam and in the largest cities in the densely populated Randstad—not because of a siege but during the particularly cold *Hongerwinter* (hunger winter) of 1944–1945; between 18,000 and 22,000 deaths occurred when the German occupiers cut all communication and food supplies lines.

Inundations

The most essential features, advantages, weaknesses and drawbacks of flooding have already been amply discussed. Suffice it to say that the flooding level was designed to give a depth of about 50 to 80 cm (20 to 30 inches), too shallow for boats to cross, and too difficult and too dangerous to wade. The huge quantities of water needed to create large inundations came from rivers, and from the Zuiderzee, and were regulated by dikes, canals, dams, watergates, weirs, sluices, locks, culverts, pumps, and various basins. Further, all around the Stelling a large glacis was created in order to deny cover. Any buildings within one kilometer of the line had to be made of wood so that they could be dismantled or burnt and all obstruction removed in case of emergency.

Flooding was on the whole a successful method of defense, but it presented also an enormous and sometimes ruinous burden for the civilian peasant population. Since the 16th century there was always resistance and sabotage from victimized farmers whose lands were flooded. In April 1896 a special law (called the *Inundatiewet*) was issued stipulating that farmers would be indemnified by the state for all damages suffered in their land. Although this law brought a rather fair measure of compensation, flooding remained a very unpopular scheme amongst agriculturists.

The Forts in the Stelling

Organization

The Stelling of Amsterdam had a roughly circular/oval outline consisting of inundations and forts placed about ten to twenty kilometers (six to 12 miles) from the center of the city. The forts in that enclosing circle were able to provide mutual support and could fire on one another to repulse attacks. In addition to the principal forts, smaller works were provided to support the infantry in the intervals between forts. For example the position at Spaarndam included Fort North Spaarndam, one defensive wall connecting to the Fort Spaarndam South, forts and walls being fitted with a moat. Ahead of the line there was a continuous advanced infantry earth wall with redan and a dry ditch.

The line of Kudelstaart (south of the present-day Schiphol Airport near the municipality of Aalsmeer) was composed of the Fort Kudelstaart, an earth wall running to the Fort Kwakel, Fort on the Drecht, and Fort Uithoorn. Such linear works included combat positions and shelters for infantry during bombardment and could also contain reserve artillery.

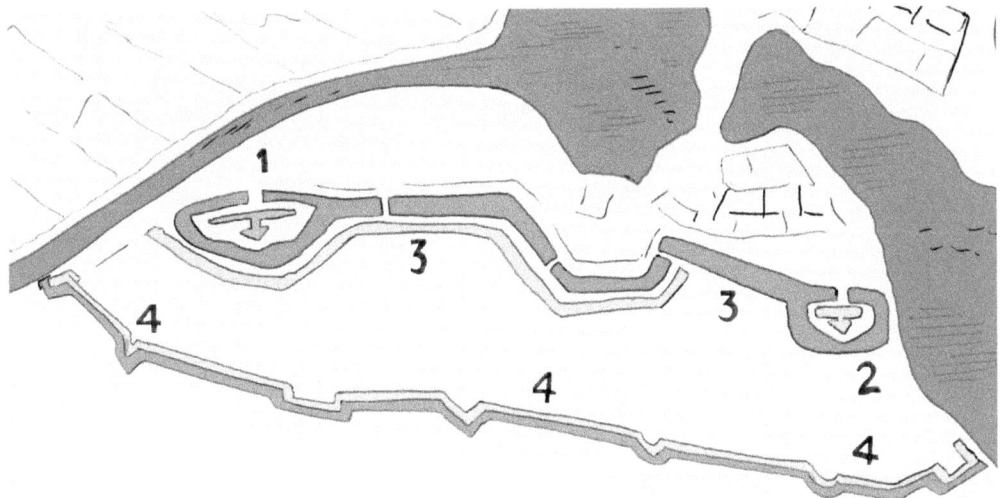

Position of Spaarndam. Located northeast of Haarlem in the Stelling of Amsterdam, the fortified position of Spaarndam included: 1: North Fort. 2: South Fort. 3: Main defensive wall. 4: Advanced defensive earth wall.

Standardized models

After many studies, trials, experimentations and designs, the construction of the forts in the Stelling of Amsterdam started in 1883. The first fort was built at Abcoude. It was an experiment made of brick masonry and sturdy mass of piled earth with shellproof barracks and store places.

The fort with its outdated open artillery emplacements rapidly proved obsolete, however, when new high explosives were introduced in 1885. It was then decided to construct it in concrete but the Dutch engineer corps had little experience in that domain, and so extensive tests had to be performed. Designers went back to their drawing board, and the construction of the Stelling made a fresh start. Forts after 1885 were built in concrete from the beginning, but after budget cuts they were smaller in scale than what had been planned in 1874. All forts were different as they had to fit to peculiar places and ground but all were designed according to several standardized patterns.

The first forts built in the period 1885–1907 were so-called model A. Made of concrete covered with dirt they included a rather low and long main building placed at the back of the fort for sheltering quarters, stores and all facilities for the garrison. The building featured two protruding gorge casemates (the so-called *traditore batteries*) intended to protect the entrances of the fort (short range flanking armed with machine guns) and to defend the neighboring works (middle-range flanking armed with quick firing 10 cm guns). The main building included a *poterne* (a rather broad underground passage that could be used as refectory, assembly, meeting and appel place) leading to an advanced underground combat building. This included an ammunition store, a small metal observation cupola, and two domed armored turrets armed with light quick-firing guns.

After 1907 fort model B came in use. It was basically the same design but each armored gun turret was connected to the main building by a protected passage thereby making its communication much safer for the troops. The so-called Type C concerned only the last two works constructed in the Stelling of Amsterdam: Fort Middenweg and Fort Jisperweg, completed between 1912 and 1914 in the north of the Stelling. They did not include the advanced combat building but instead had a single armed turret.

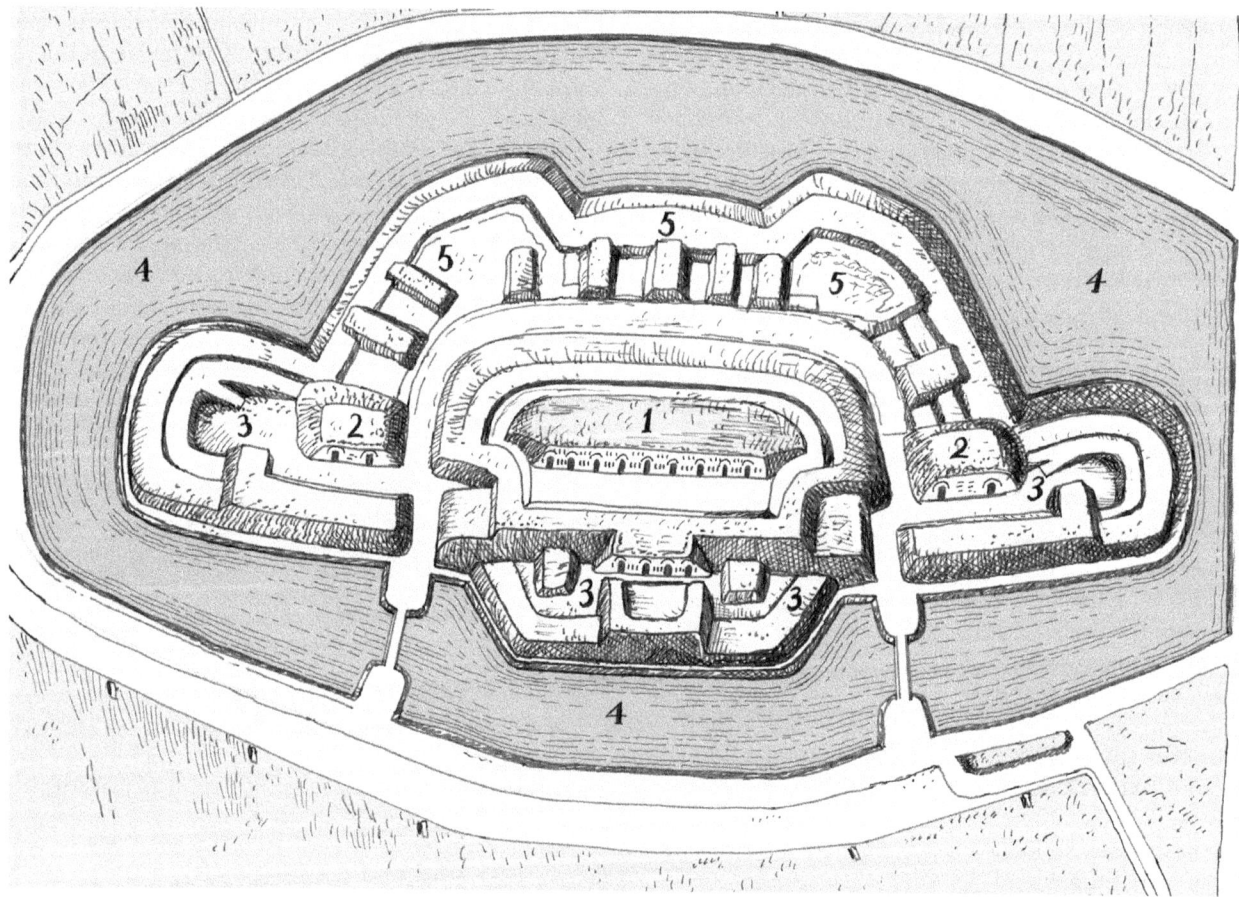

Fort Abcoude. Located on the Gein River near the small town of Abcoude (province of Utrecht), the fort was built in 1883–1885. It was a prototype, actually the first work constructed in the Stelling of Amsterdam. It covered about 8 hectares. With open artillery emplacements, open breastworks for infantry, open flanking batteries, and vaulted brick masonry buildings, the fort was obsolete before completion. It had a garrison of 353 soldiers, and was decommissioned in 1959: 1: Main building (two stories high) with Kazerne (barrack), services and stores. 2: Remise (garage) for artillery pieces. 3: Flanking battery emplacement. 4: Moat. 5: Long range battery.

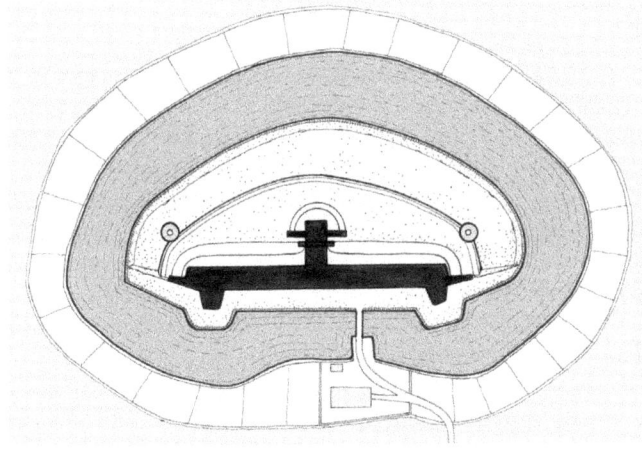
Fort Type A (Plan).

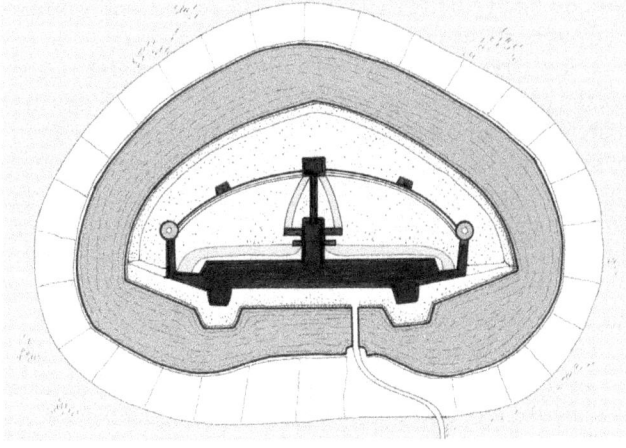
Fort Type B (Plan).

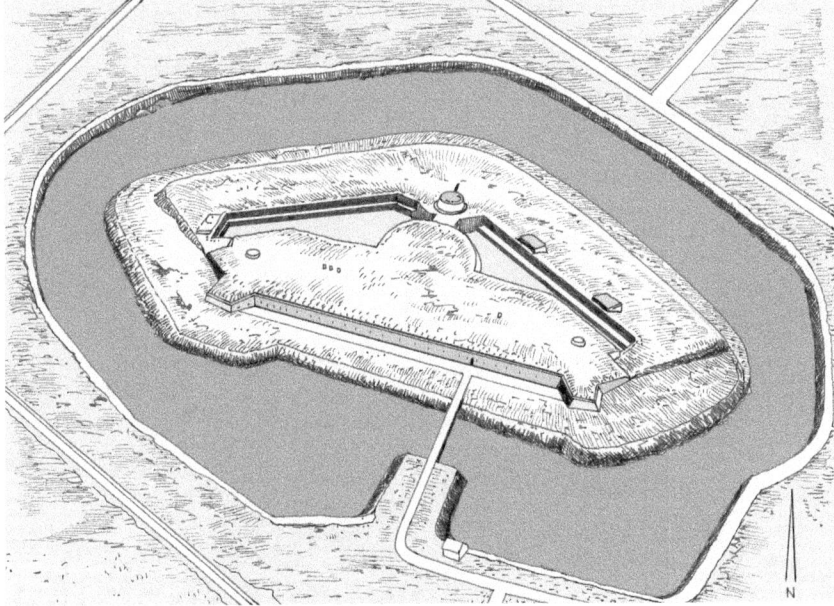

Type C. Fort Middenweg. Located in a polder near the village of Beemster in the north of the Stelling of Amsterdam, Fort Middenweg was intended to defend two roads leading to Amsterdam called Middenweg and Zuiderweg. It also protected a sluice permitting the water of the Noordhollandsch Kanaal (North Holland Canal) to flood the polder of Beemster. Because of budget restrictions the fort (designated model C) included only one gun placed in a central armored turret (connected to the main building by a poterne), and two small armored turrets armed with machine guns. Today the fort is managed by the association Natuurmonumenten.

Miscellaneous types

Next to these more or less standardized models there were also a number of forts with variant features, and later additions in order to fit peculiar natural conditions, and adapt to specific situations. For example the main building of Fort Hoofddorp included two stories because of the small perimeter available. Fort Spijkerboor's main building, too, was two stories high because of the elevated dike, and was armed with two 10.5 cm (4 inch) gun placed in a rotating armored turret. Fort Velsen also featured extra weapons, notably three 15 cm (6 inch) cannons. Further there were coastal seaforts like the Durgerdam Battery, Fort IJmuiden, and Fort Pampus that were unique designs, powerfully armed, and established on artificial islets.

Common features of the forts

All forts in the Stelling of Amsterdam were different of course, as they had to fit to peculiar places and ground but all included the following features: Many of the forts were built in sandy or boggy polders with soft ground. So very often strong foundations were needed. The walls at the front of the main building were 1.5 meters thick, and all other walls were 1 meter. All forts were surrounded by a wide defensive moat filled with water, and all were made of reinforced concrete covered with thick layers of dirt.

In wartime, the garrison totaled about 300–350 soldiers. The main building included shellproof sleeping accommodations (large dormitories for 24 to 36 men), as well as storerooms, kitchens, washing rooms, infirmaries, telegraph and later telephone exchanges. Rooms were fitted with ventilation systems with armored windows, pipes, vents and chimneys. Rooms were lighted by petroleum lamps, and later fitted with electricity. All forts had latrines, and water in cisterns, wells and tanks fitted with pumps for drinking and washing.

The most vulnerable locations in the fort were the ammunitions magazines, which were often dispersed and more deeply buried. Each fort included one *genieloods* (engineering shed and workshop). During peacetime the forts were generally not occupied, but there was always a *fortwachter* (a guard, often a retired serviceman) who lived permanently with his family in the vicinity or inside the fort, and who carried out small maintenance tasks, and security duty.

Artillery was removed from open emplacements on the ramparts and placed into concrete flanking casemates and into rotating armored turrets. While the number of artillery pieces declined, the new weapons (notably the rifled repeater rifles and carbines, the machine guns and the quick firing guns) were as effective as the former batteries.

A few forts were equipped with long-range weapons but most forts in the Stelling of Amsterdam were armed only for short-range fire—within one kilometer. Armament included Gardner M90 machine guns, and several kinds of breechloading rifled guns with caliber ranging from 6 cm (2.5 inches) to 15 cm (6 inches). The coastal forts were always armed with heavy weapons because in wartime they would have to deal with strongly armed—and armored—enemy warships.

Aftermath

The Stelling of Amsterdam was occupied and put in a state of readiness during World War I and in 1939, but it has never seen combat service. The use of tanks and aircraft rendered it obsolete after World War I. During the period 1940–1945 some of the forts of the Stelling of Amsterdam were occupied by the Germans and used as training grounds, anti-aircraft batteries, ammunition depots, or prisons and concentration camps for political opponents. After 1945, although obsolete, the Stelling was maintained and kept in service until it was decommissioned in 1963.

Many of the forts are now under the control of town councils, or culture and nature associations. Some of them have become private properties, some may be visited, some are museums, others are nature and animal preserves. In 1996, the complete Stelling was designated a UNESCO World Heritage Site.

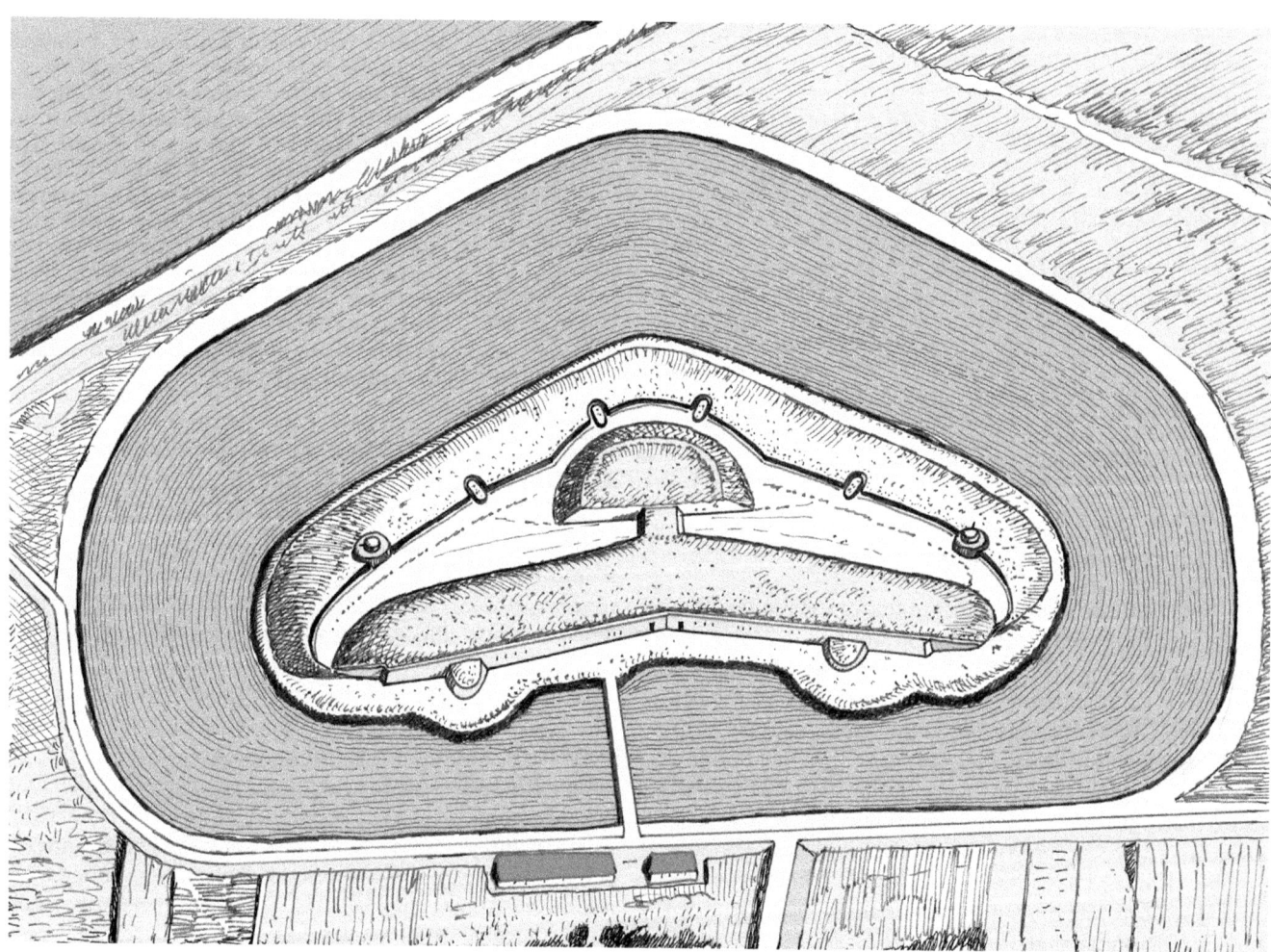

Fort Nigtevecht. Located on the Amsterdam-Rhine Canal (aka Merwedekanaal), near the village of Stichtse Vecht in the province of Utrecht. Built between 1893 and 1903, Fort Nigtevecht was of Type A. It was tasked with defending the accesses formed by the Merwedekanaal en de Oude Vecht River, and protecting various installations used for flooding the flat land around it. It had a garrison of 280, and was armed with two retractable turrets armed with 6 cm guns, and rear caponiers with machine guns. Today the fort is owned by the Vereniging Natuurmonumenten and is a reserve for birds.

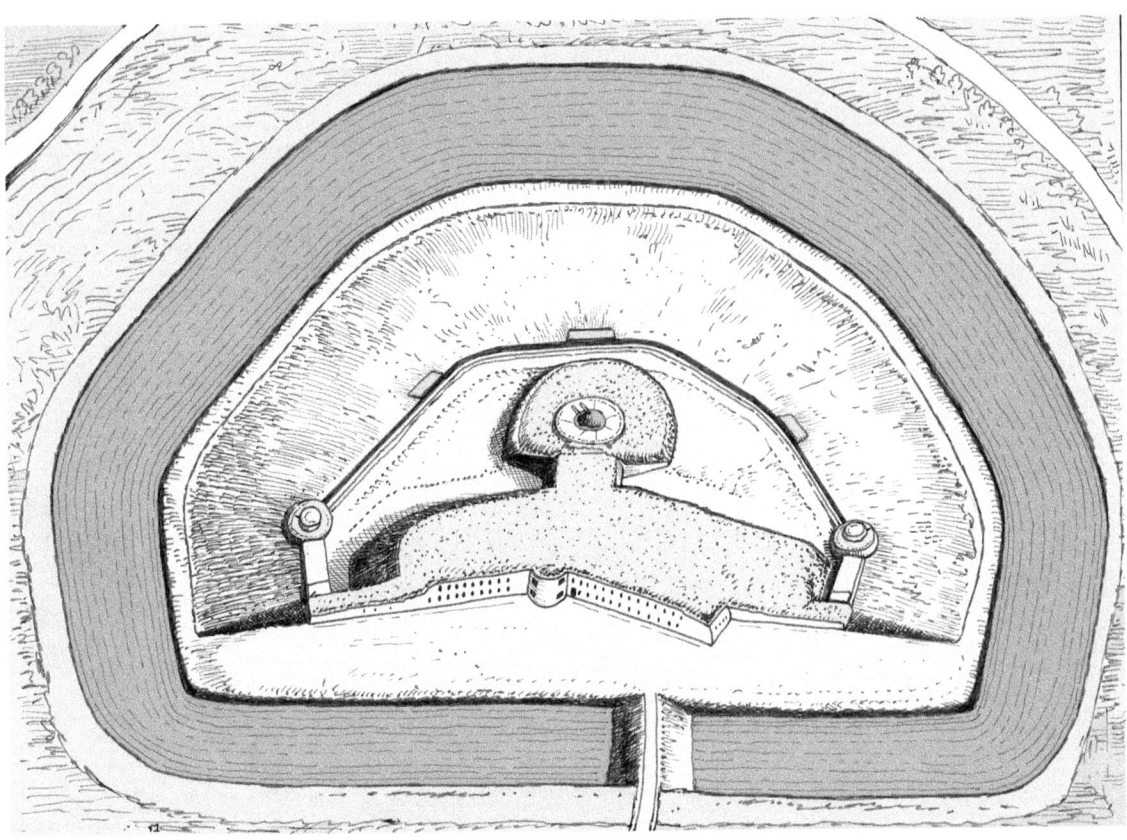

Fort Spijkerboor. One of the largest, most modern, and most expensive forts in the Stelling of Amsterdam, Fort Spijkerboor was built between 1911 and 1913. It was intended to defend the canal called the Beemsterringvaart. It is 205 m in length, 125 m in width, and surrounded by a 28 m wide moat. Spijkeboor was one of the few forts fitted with a modern armored rotating turret armed with two 10 cm guns. Further armament included two eclipsable turrets each with one 6 cm gun covering the neighboring forts of Marken-Binnen and Jisperweg, respectively at a distance of 4 and 2 km. In addition there were about 10 machine-guns emplacements for close-range defense. The fort had a garrison of about 200 soldiers both infantry and artillerymen accommodated into a two-story high shellproof barrack. After 1918 the fort became a prison, until 1951 when it was decommissioned.

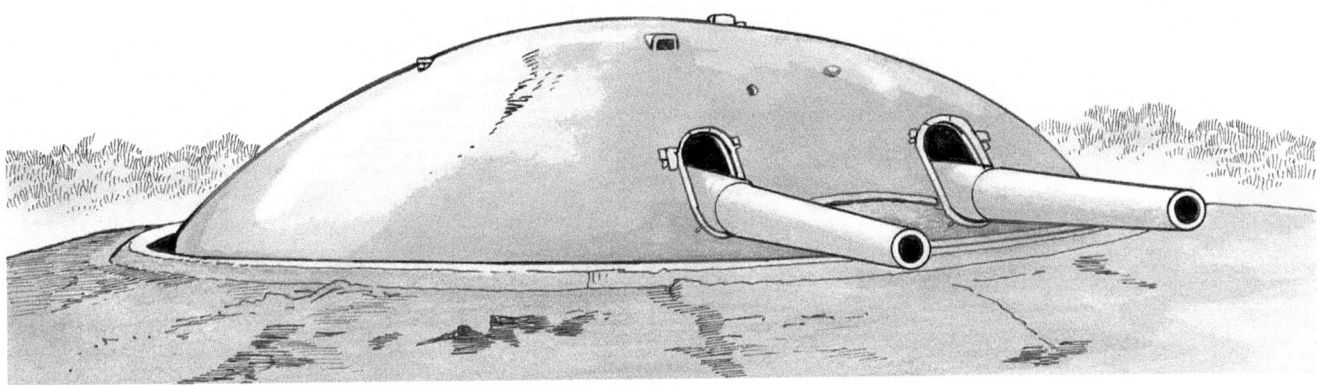

Armored turret Spijkerboor. Fort Spijkerboor, completed in 1913, featured a steel (12–15 cm thick) armored cupola with two heavy 10.5 cm guns fitted with a hydraulic recoil system, periscope and modern aiming devices, and a lift for bringing up shells from the ammunition store. The guns, served by a team of 15 artillerymen, could fire projectiles weighing 18 kg to a range of 10 km, and had a rate of fire of 20 rounds per minute. Armored cupola and guns were purchased from the German Krupp-Gruson Company. Originally developed for armored steamships, rotating gun turrets with spherical cupolas rapidly became effective elements for land fortifications. Gruson's chilled iron armored turrets, casemates and cupolas proved to be a success and were purchased by Italy, Austria, Belgium and the Netherlands for both coastal and land fortifications. In France the military engineer Henri Mougin, in association with the Saint Chamond Company, designed, built and sold similar armored turrets and casemates. When all Dutch fortifications were plundered by the Germans during World War II, Fort Spijkboor was miraculously spared, and today still displays its armored turret.

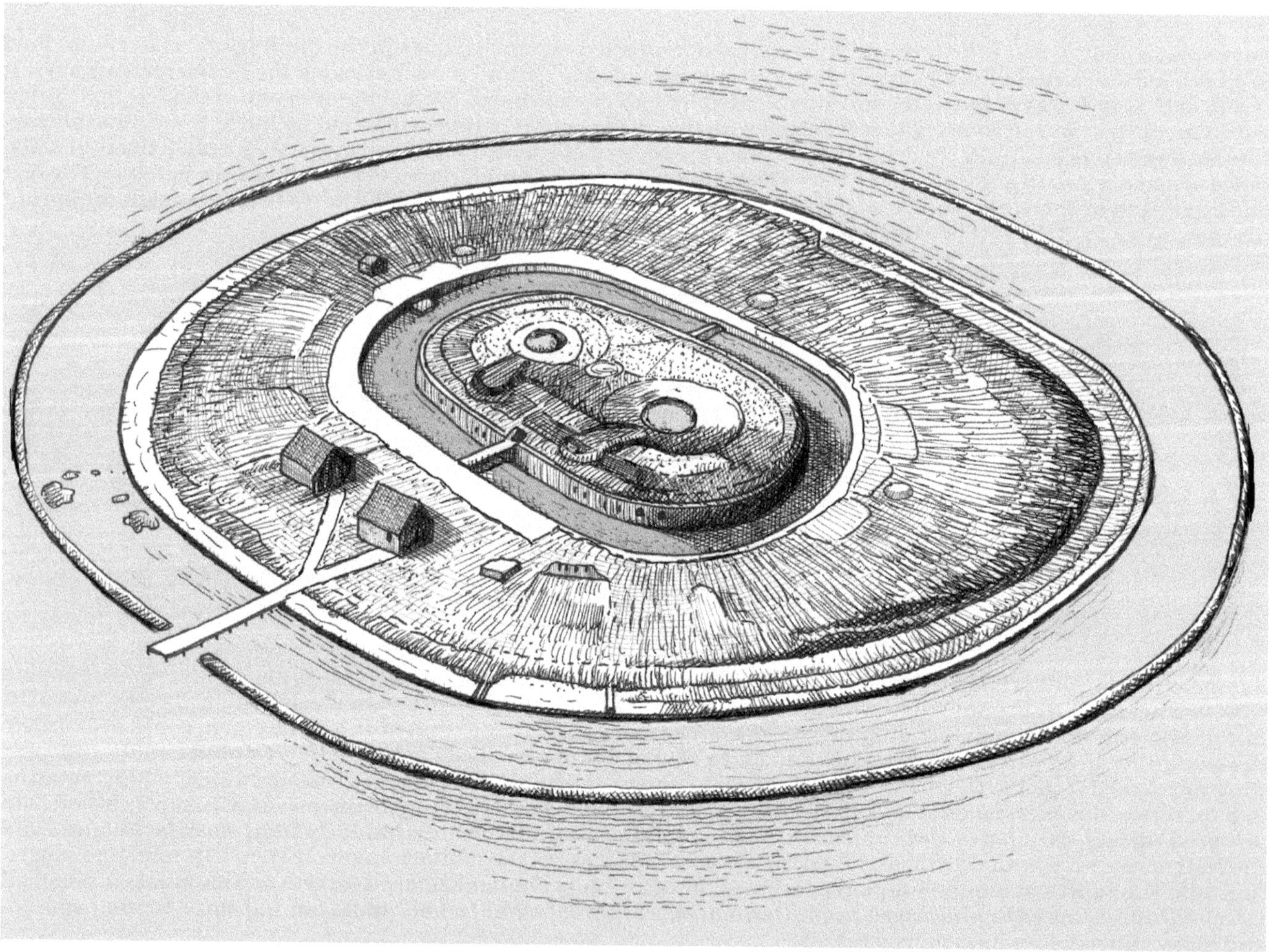

Dutch Fortifications

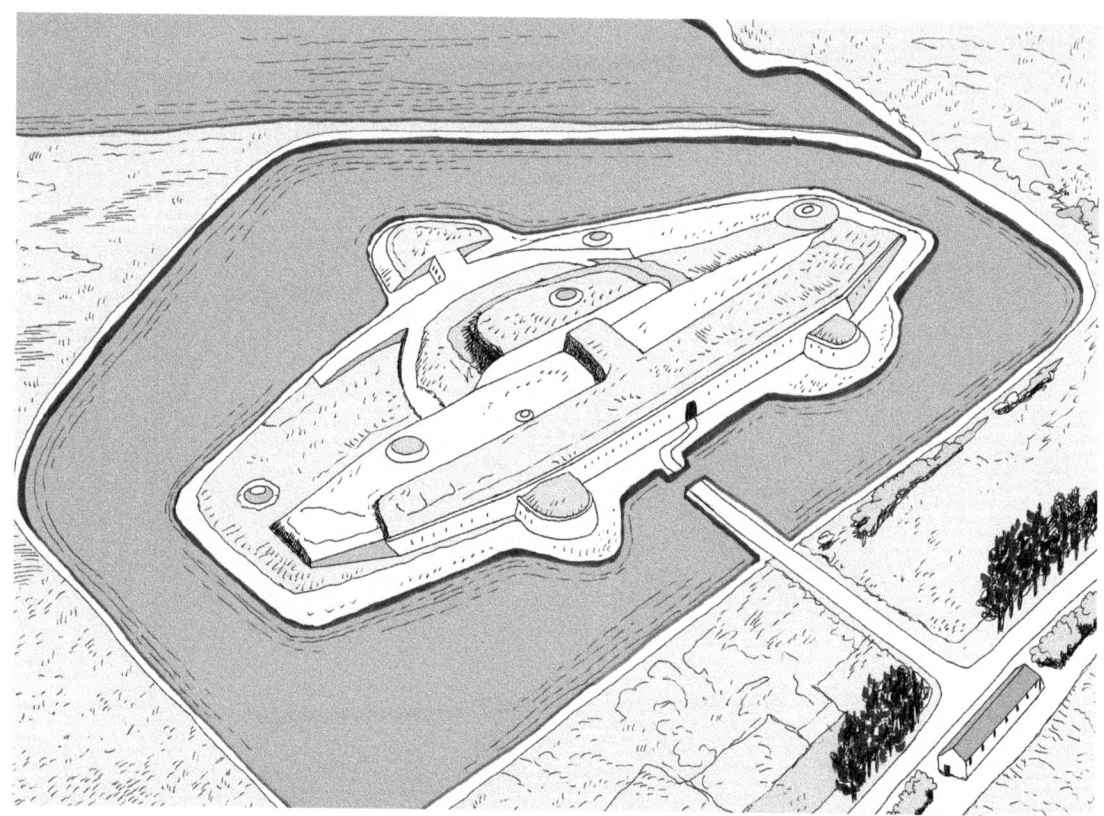

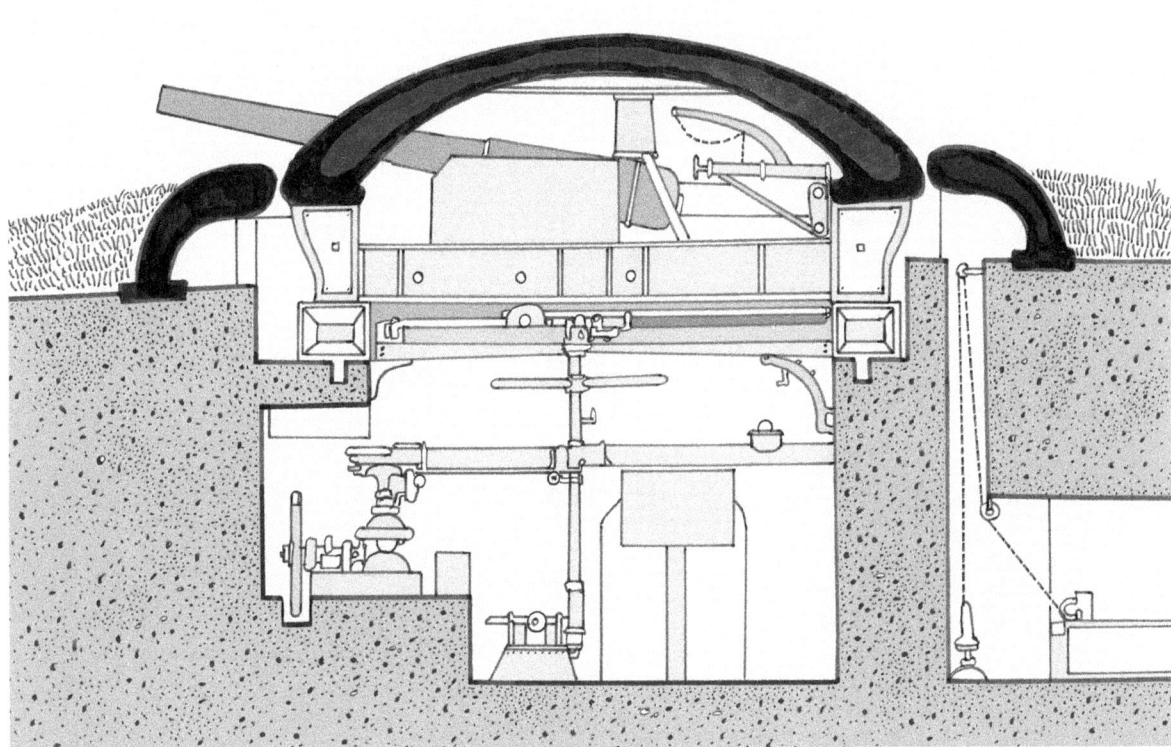

Gruson rotating turret at Fort Pampus (cross-section). The German Gruson Company from Magdeburg-Buckau (state of Saxony-Anhalt) was a branch of the large Krupp industrial firm.

Opposite, top: Fort St. Aagtendijk. The A style Fort St. Aagtendijk is located near Beverwijk in the province of North Holland. Built between 1895 and 1899, it had a garrison of ca. 300 soldiers, and was intended to protect the Saint Aagten Dike and the installations allowing the flooding of the polder of Zuidwijkkermeer. Fort aan den Sint Aagtendijk (just like most of the other forts of the Stelling of Amsterdam) was occupied by the Germans in the period 1940–1945, and was used as an ammunition depot. The fort was decommissioned in 1956, and is now used by various associations.

Opposite, bottom: Fort Pampus. Located in the IJsselmeer near Amsterdam, Fort Pampus was a unique sea fortress in the Stelling of Amsterdam. Together with the artillery battery on the lighthouse island near Durgerdam and the battery at the Diemer seawall, Pampus protected the entrance to the harbor of Amsterdam. The fort is constructed on an artificial island (about 204 m in length and 125 m in width) situated on what was the Pampus shallows or sandbank in the then open Zuiderzee. Work commenced in 1887. Creating the island and fort required the sinking of 3,800 piles and the importation of 45,000 cubic meters of sand. It took the Dutch eight years and lots of effort and funds to construct both the island and the fort. The fort is built of bricks and concrete. It has a symmetrical oval shape and the main building has three floors with a total of 80 rooms that include troops' quarters, various storing places, kitchen, laundry, two coal-fired steam engines of 20 hp, two dynamos, telegraph, first aid station, and magazines. An eight-meter dry moat surrounds the building. Tunnels on the north and south connect the ground floor of the building to the counterscarp. The counterscarp is made of concrete and contained a gaol, the forge, and several supply rooms. On top of the counterscarp there is a parapet to provide close-in defense. A wide glacis surrounds the whole fort. Commissioned in 1895, Fort Pampus was armed with four Krupp 240 mm (9.4 in) L35 (35 calibers long) guns emplaced in two hydraulically operated cupolas of two guns each. Electric lifts brought shells and cartridges up from the ammunition magazines. These guns could fire shells weighing 280 kg (617 lb) to a range of eight kilometers (5 miles). Each gun had a crew of one NCO and six gunners, who could get off one shot every six minutes. Pampus was one of only four forts in the Stelling of Amsterdam armed with large caliber guns. (The other three forts were the forts near IJmuiden, Velsen, and Spijkerboor.) Pampus also had two positions for 57 mm (2.2 in) quick-firing guns for close-in defense. There were emplacements on the counterscarp for four M90 Gardner machine guns on garrison mounts for the defense of the moat. Pampus had facilities for a permanent garrison of 200 men, but the only time it achieved that strength was during the First World War. In 1926, airplanes had become a serious threat and Fort Pampus received three anti-aircraft guns. The fort never saw action and the completion in 1932 of the Afsluitdijk severed the IJsselmeer from the open sea. At this point Pampus lost its strategic role. In July 1933 armament was removed and the fort was abandoned. During World War II, the German occupation forces used the fort for target practice, before installing some 88 mm (3.5 in) anti-aircraft guns. After the war the fort was neglected until the 1990s when it was partially restored and opened to the public.

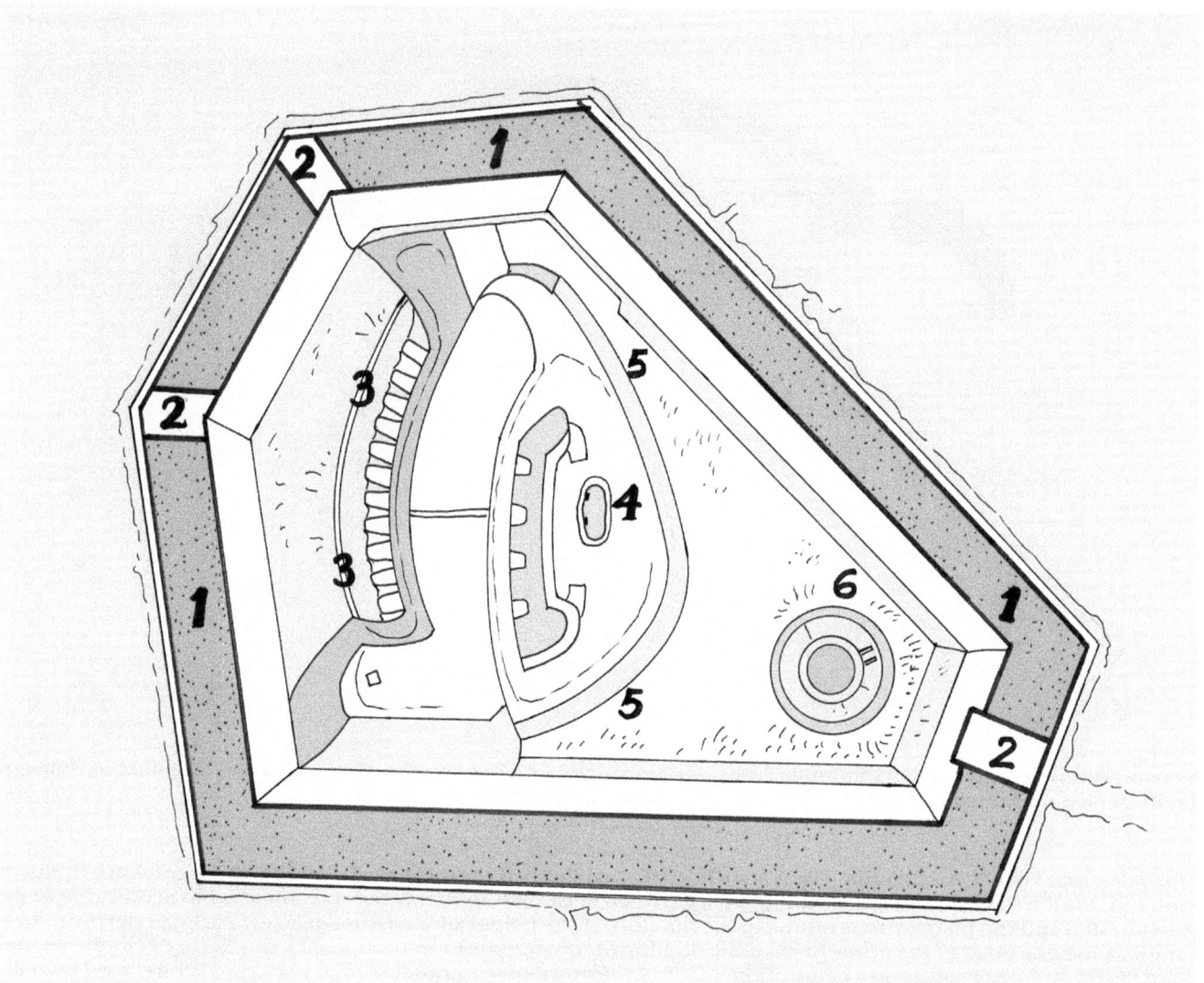

Fort IJmuiden. Fort IJmuiden aka Forteiland (Fortress Island) is a coastal fort belonging to the Stelling of Amsterdam. It is located at IJmuiden at the mouth of the Noordzeekanaal (North Sea Canal), completed in 1876 and connecting Amsterdam to the North Sea. The fort was built between 1881 and 1889 originally on the bank of the canal, but with the widening of the canal in 1929, its location became an artificial rectangular island of four hectares. With the widening of the canal in 1967, the island was further reduced to the shape we see today. The fort proper is roughly triangular, and featured a deep dry ditch, underground quarters and all facilities for the garrison of ca. 287 soldiers. Its main armament placed on top of the building consisted of an armored battery with five 24 cm guns facing the open sea, and a rotating armored turret with two 15 cm guns. There were several machine-gun emplacements placed in caponiers in the dry ditch, and a close-range infantry gallery. During World War II the fort was incorporated as Festung (fortress) into the German Atlantic Wall. All metal parts were looted by the occupiers, and several concrete bunkers were added to the original Dutch defenses. The fort with its remarkable site can be visited and is now listed as a UNESCO World Monument: 1: Dry ditch. 2: Caponier. 3: Armored gun gallery. 4: Observation cupola. 5: Infantry rifle gallery. 6: Armored turret with 15 cm gun.

Chapter 7. The 19th Century

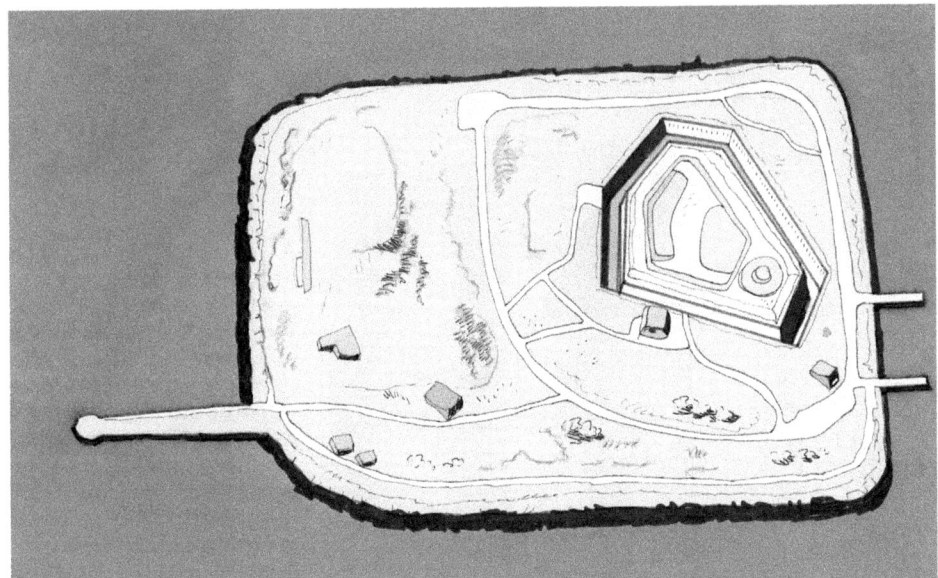

Forteiland IJmuiden—a plan of the island with the fort today.

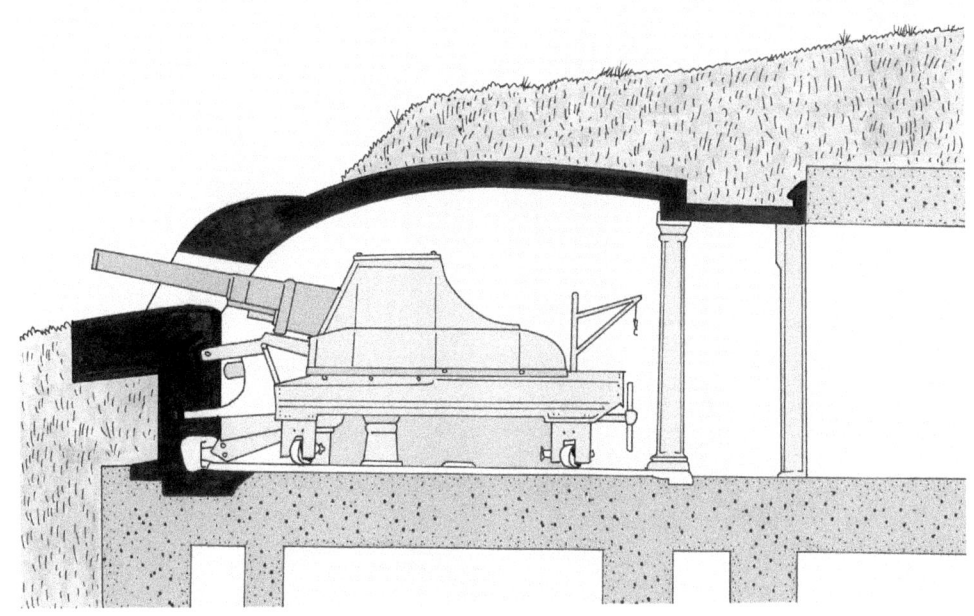

Fort IJmuiden. Cross-section of the armored casemate. The armored gallery housed five 24 cm L30 guns.

Observation cupola at Fort Waver-Amstel. The armored cupola protruded ca. 40 cm above the concrete. The observer could watch the vicinity of the fort through a small opening and could report what he saw by means of a speaking tube, later telephone.

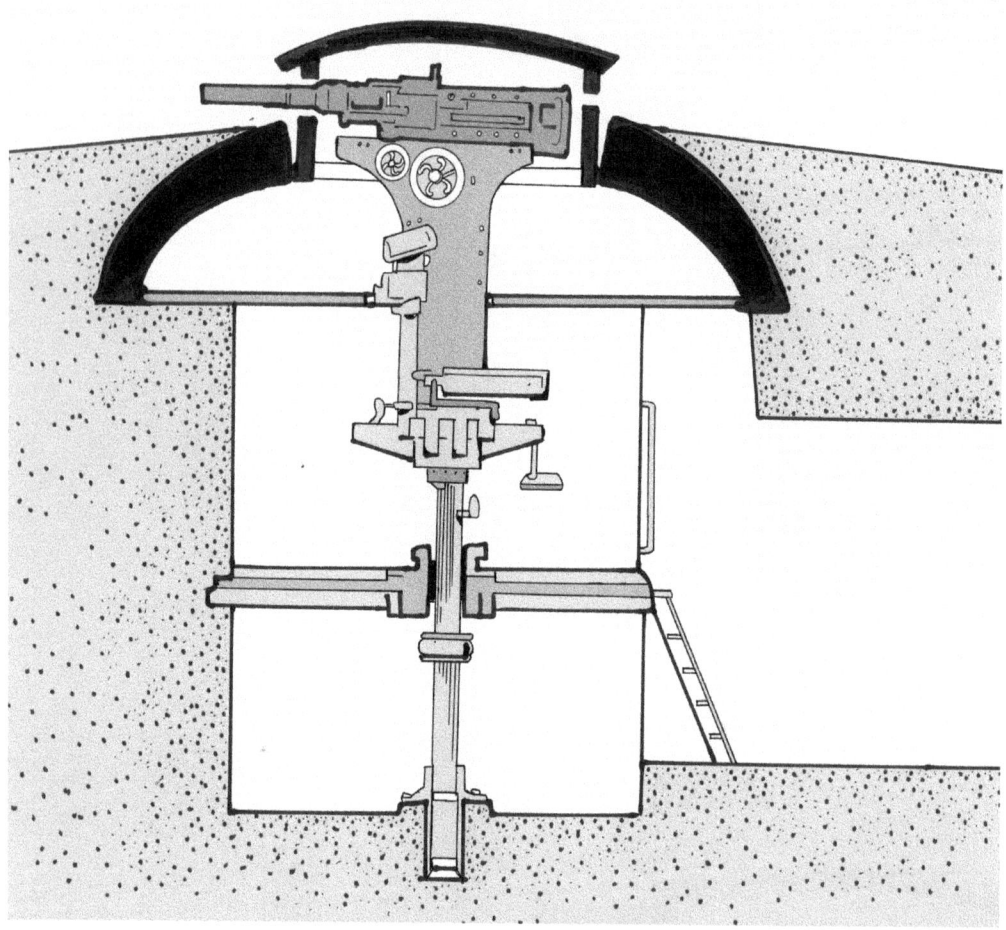

Armored cupola Krupp-Gruson for 6 cm gun (cross-section). The small domed firing chamber on top of the turret was retractable or "disappearing." It was raised by counterweights for firing and lowered for reloading or when not in operation. As a result crews and guns were perfectly protected, and the only visible feature from a distance was a grassy mount with the metal dome flush with the concrete surface.

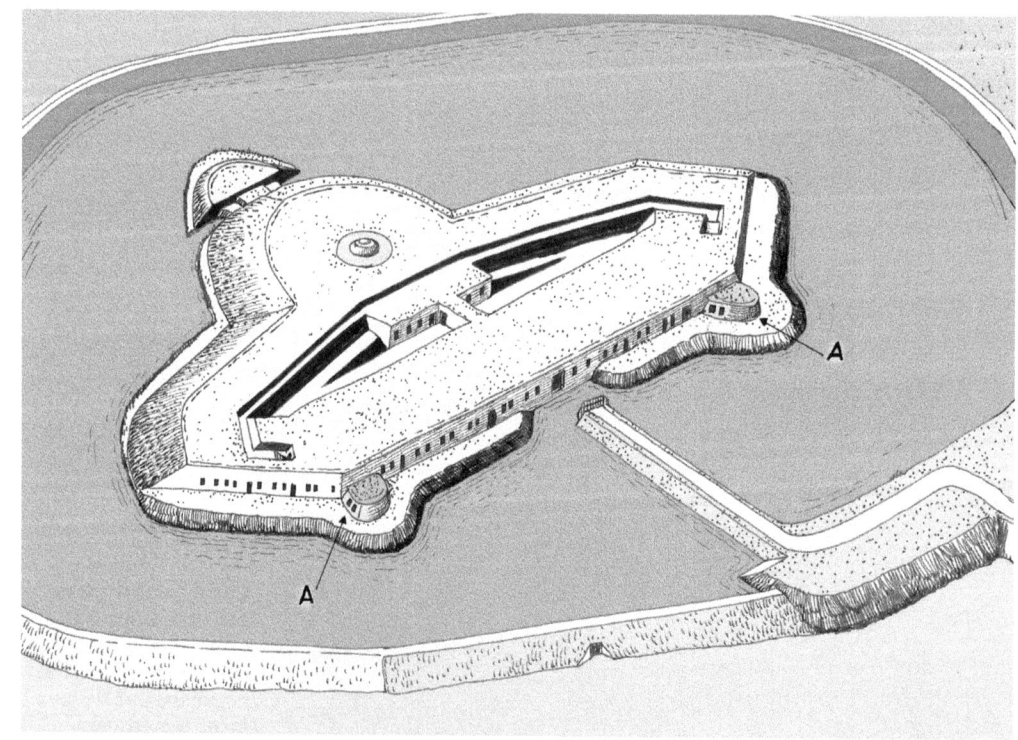

Traditore casemate. The so-called "traditore" casemate (or battery) derivates from the Italian term *traditore* meaning traitor. It was a flanking armed casemate (A on the illustration) "treacherously" placed at the back of a fort and intended for short range firing at enemy attacking the entrance. It was also designed for long range flanking, and covering the intervals between forts in the Stelling of Amsterdam, in order to create a continuous belt of fire around the forts.

Traditore casemate at Fort Hoofddorp.

Fortifications in the Colonies

Under the Republic (1588–1795), the Dutch East India Company and West Indies Company had gained many footholds around the world. In the 19th century many colonial posts and possessions had been lost (mainly to Britain). In the second half of the 19th century, the Dutch focused on the East Indies, which later became Indonesia. The Dutch started the conquest and control of the hinterland of the huge archipelago including large islands like Java, Sumatra, Borneo, Bali, Celebes, and many other islands like the Moluccas.

In the Caribbean, the Dutch possessed several islands (including Bonaire, Aruba and Curaçao) and a colonial territory in South America—Suriname (capital Paramaribo), near Brazil. The Dutch government received considerable revenue from the colonies; in some years as much as a quarter of the nation's income came from them. Exotic produces like spices, indigo, sugar and coffee were cheaply produced in the colonies and sold with enormous profits in Europe by the *Nederlandsche Handel-Maatschappij* (NHM Netherlands Trading Company).

Obviously the highly profitable colonial empire once conquered had to be defended. Basically the fortifications in the colonies had two main functions: Firstly defend the Dutch colonies against any attack from European competitors. The archipelago of Indonesia still possesses today a great number of fortifications built by the Portuguese and the Dutch from the 16th century to the first half of the 20th. These legacies include colonial forts, coastal batteries, and fortified trading posts and cities.

The second purpose was subduing, suppressing and impressing any local resistance, unrest or insurgence. In order to defend and maintain order in the colonies the Dutch created in 1832 a special colonial army: the *Koninklijk Nederlandsch Indisch Leger* (KNIL; the Royal Dutch East Indies Army). The KNIL was a professional army especially administrated by the Ministry of Colonies, and not dependent on the National Royal Armed Forces. By 1850 the KNIL counted about 27,000 men of which 10,000 or so were European, the rest being locally recruited auxiliaries.

There were indeed several important and protracted revolts like the so-called Java War (1825–1830), and the Aceh War (1873–1903).

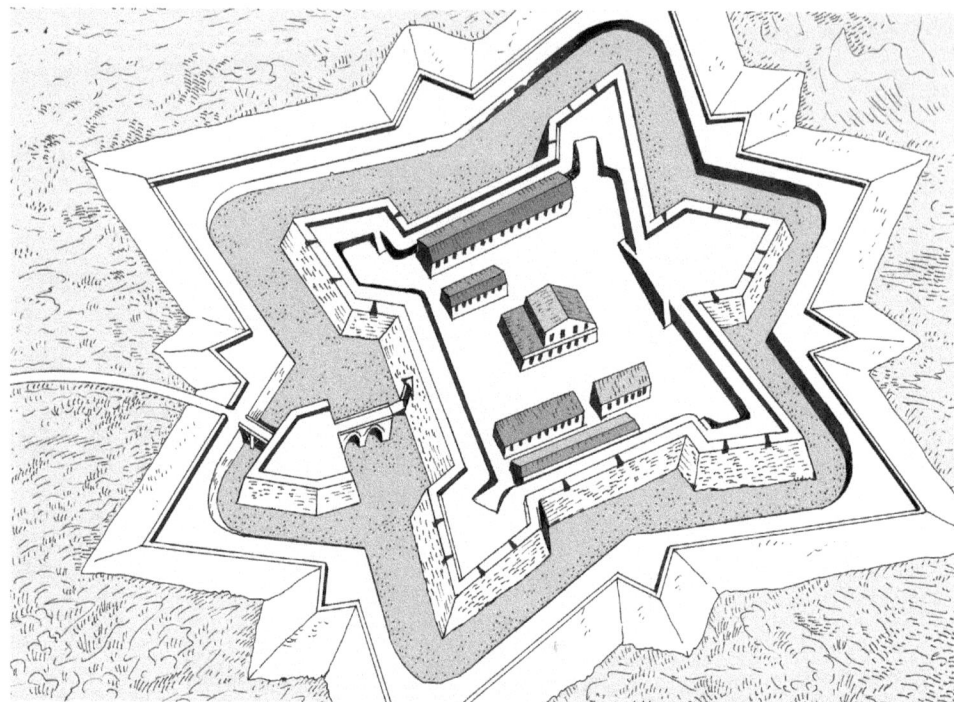

Fort Rotterdam Makassar (Sulawesi). Fort Rotterdam is a 17th-century fort in Makassar located on the island of Sulawesi (then called Celebes) in the Greater Sunda Islands east of Borneo in Indonesia. It is a Dutch fort built on top of an existing fort of the Gowa Kingdom. The first fort on the site was constructed by a local sultan around 1634, to counter Dutch encroachments. The site was ceded to the Dutch under the Treaty of Bongaya, and they completely rebuilt it between 1673 and 1679. It had five bastions and a ravelin defending the entrance, and was surrounded by a seven-meter high rampart and a two-meter deep moat. The fort was the Dutch regional military and governmental headquarters until the 1930s. It was extensively restored in the 1970s and is now a cultural and educational center, a venue for music and dance events, and a tourist destination.

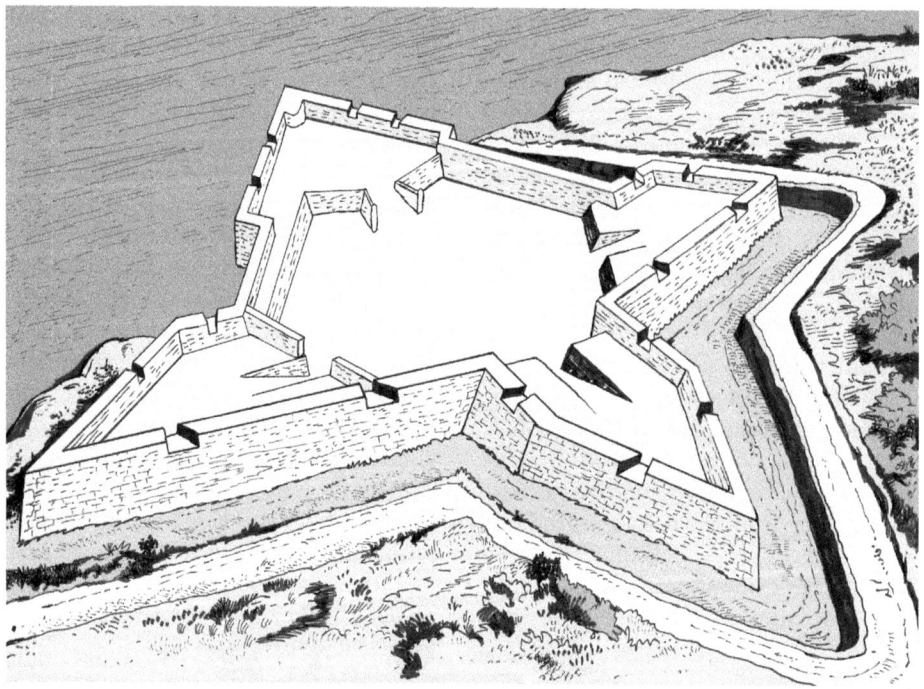

Fort Kalamata. Fort Kalamata is located at Ternate, the most important city of the Maluku Island (aka Moluccas) in the Seram Sea between Sulawesi and New Guinea in Indonesia. Originally built by the Portuguese, it was initially known as Kayu Merah or Fort Red Wood. In 1824 Ternate became the capital of a residency (administrative region) covering the Halmahera Islands, the entire west coast of New Guinea, and the central east coast of Sulawesi.

Chapter 7. The 19th Century

Torenfort Palau Cipir. Torenforts were also built in the colonies, notably on a small island (known as Kuiperseiland) located in the Bay of Batavia (today Jakarta in Indonesia). The depicted torenfort was built in 1823–1825 and replaced a former fortlet constructed in 1608.

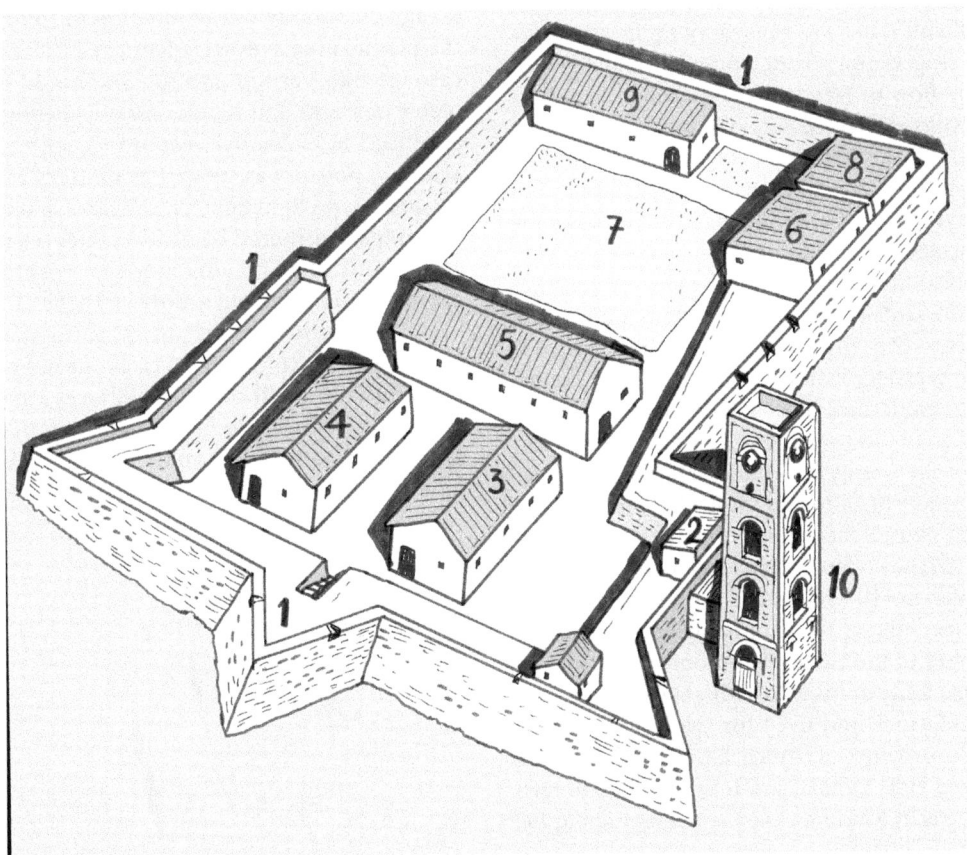

Fort Zoutman Aruba. Located at Oranjestad, the capital of the island of Aruba in the Caribbean, Fort Zoutman was built in 1798. Intended to protect the harbor of Oranjestad, the fort was named after the Dutch Rear Admiral Johan Arnold Zoutman (1724–1793). The plan shows the enclosing masonry walls and flanking redans (1), the kitchen (2), the guardhouse (3), the quarters for the troops (4), the storehouse (5), the prison (6), the central parade ground (7), the latrines (8), and the guest-house (9). The square tower Willem III (10) that served as a clock tower and lighthouse was added to the entrance on the west side of the fort in 1868. The fort and tower were restored in 1983 and became the Historical Museum of Aruba.

Chapter 8

The Period 1914–1940

The First World War, 1914–1918

Dutch neutrality

At the outbreak of World War I in August 1914, the Netherlands declared neutrality. Although not directly involved in the conflict, and not subjected to desolation and large-scale carnages, the people of the Netherlands suffered between 1914 and 1918. The disruption of international relations engendered many problems, notably a large reduction and collapse of import and export. The difficulty to maintain contact with global trading clients and the colonies in the Caribbean and Indonesia caused enormous harm to the Dutch economy based on commerce.

Another difficult issue was caused by no fewer than a million Belgian refugees (for a Dutch population of 6.5 million). To stop the flow of refugees the German occupiers installed a 2,000 volts electrified barrier known as *De Draad* (The Wire) all along the border between both countries. Later in the war many refugees returned to their country but about 100,000 stayed in the Netherlands until the end of the war. In 1917 a shortage of food adding to the tensions.

Because a violation of the neutrality by the Germans was always possible, the Dutch were in a precarious situation. It was an armed neutrality. The army (ca. 200,000 men) was mobilized for four years, the existing fortifications were refurbished and hastily modernized, occupied and manned, and measures were taken to carry out inundations at short notice.

By the end of the war in November 1918, the main victorious Allies (France, Britain, and the United States) considered the Dutch neutral stance unsympathetic, the more so when the Netherlands offered a safe exile to the abdicated German Kaiser Wilhelm II, who evaded the punishment that the Allies had wished to impose on him. The German Kaiser (who was a member of the German Teutonic Order) benefited from the hospitality of his fellow knight Godard Count van Aldenburg Bentinck and lived for two years at castle Amerongen (in the province of Utrecht). Later the abdicated emperor purchased Huis Doorn, and retired in that luxurious castle near Utrecht where he lived a pleasant and quiet life until his death in June 1941.

Fortifications in World War I

In the domain of fortifications, the First World War brought to the fore the obsolescence of the large late–19th century forts. Indeed the Germans had developed super heavy 42 cm (16.6 inch) guns that smashed the Belgian forts of Liège, Namur and Antwerp in 1914. By then the huge 19th century forts' value had dramatically decreased, and even the modern concrete forts of the Stelling of Amsterdam were rather outdated. Work on the Dutch forts was therefore stopped. The last fort near the village of Winkel remained thus what it was: an earthen incomplete defensive work without shellproof barracks.

Ultimately what stopped the German advance in northern France in 1914 were not permanent fortifications but an improvised system of trenches and field fortifications. For four years, there was a kind of stalemate, and troops fought a long and bloody war of attrition in trenches. So now and then large offensives were launched that only added to the carnage like Verdun and the Somme in 1916. In order to pierce defenses and repulse assaults belligerents developed and used new deadly weapons: machine guns, powerful artillery, poison gas, as well as military aviation (bombers for attack, and light fighters, and aircraft for observation). In 1916 tracked armored combat vehicles (aka tanks) appeared at the front. They were intended to cross the no man's land, crush barbed wires and get over trenches.

The Dutch were well aware of the changes brought by trench warfare in the Western Front in France. As early as 1916 and 1918, small concrete pillboxes armed with machine guns and shelters for the troops were designed and constructed around the forts of the Stelling of Amsterdam, at the navy base of Den Helder, and in the Holland Waterline.

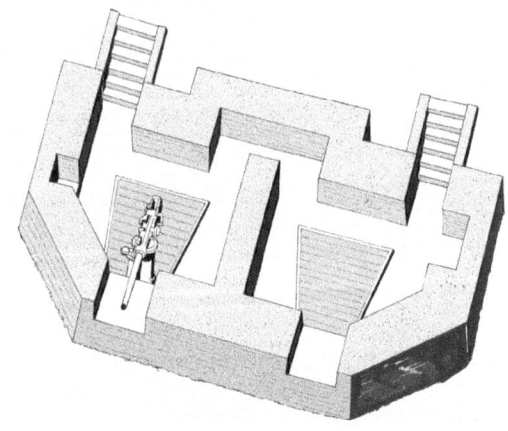

Double artillery pillbox 1916 (Den Helder).

Chapter 8. The Period 1914–1940

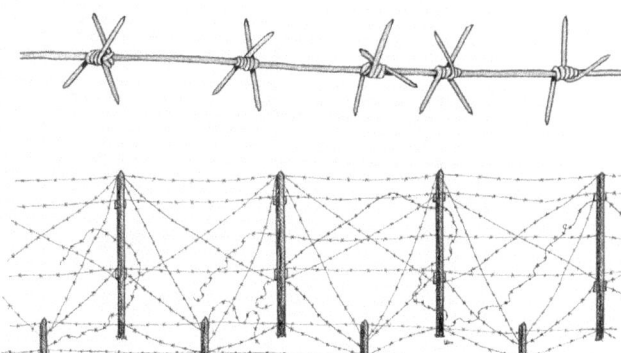

British Mark IV tank. Although the first tanks made a rather poor showing in 1916, the concept of mobile armor later caused a revolution in warfare. The British Mark IV was first used in June 1917 at the Battle of Messines Ridge south of Ypres in Belgium. The Mark IV had a length pf 26 ft 5 in (8.05 m), a width of 13 ft 6 in (4.12 m), a weight of 32 tons (28,4 tonnes), and was operated by a crew of 8. It was protected by 0.25–0.47 inches (6.1–12 mm) armored plates and was armed with either two 6-pounder (57-mm) 6 cwt QF guns with 332 rounds, or five .303 Lewis machine guns. It had an operational range of 35 mi (56 km), and a maximum speed of 4 mph (6.4 km/h). A total of 1,220 Mk IV were built.

Barbed wire. Invented in the 1870s in the United States of America, barbed wire was originally a farm fencing equipment intended to restrain cattle. Barbed wire consists of artificial metal thorns, made of sharpened wire barbs twisted at intervals around a central strand. It was first adapted to military purposes by the end of the 19th century, and its full significance was appreciated during the First World War. Rather cheap, easy to install, effectively dangerous, and bringing spaces under control, barbed wire can be used to make a fence enclosing a base, a camp, an airfield, but also around a POW camp, a prison or a concentration camp, which has made it symbolic of oppression and denial of freedom. On the battlefield it can be placed in line to form a dangerous obstacle intended to hinder the progress of infantry or channel attack to defiles in defended areas.

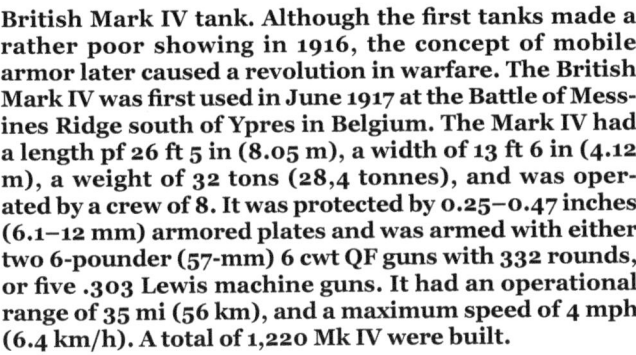

Concrete shelter for infantry model 1918 (Schuilplaats Type 1918/I) located near Spaarndam in the Stelling of Amsterdam.

Kleine Waarnemingspost. This concrete sentry box or "small observation post" was built near Fort Kijkduin in the Stelling of Den Helder in 1916. It could shelter a single observer, and was fitted with an armored steel door at the back.

Concrete shelter for infantry 1916.

Dutch Fortifications in the Interbellum, 1918–1939

The interwar years

In the interwar period of 1918–1939, the Netherlands was a member of the League of Nations, and pacifist ideas dominated the policy of most political parties. In that period the kingdom underwent substantial improvement in its infrastructure. Great water projects saw a more efficient use of land. The Afsluitdijk (causeway spanning the north of the Zuiderzee) was completed in 1932, connecting Frisia directly to North Holland. At the same time ambitious and extensive land reclamation programs were launched. The wet and swampy Wieringermeer region in the province of North Holland was dried and given for agriculture use in 1934. Work on the Northeast polder started in 1936 and was completed in 1942, gaining 1,650 square km (about 640 square miles) from the IJssel Lake. However, as elsewhere the Great Depression following the October 1929 Wall Street crash in the USA started a profound and lasting economic crisis in the 1930s. This caused unemployment, hardship, austerity policies, and social and political turmoil that soon fed the growth of radical populist parties, on both the far right (Fascism) and the far left (Communism).

Interwar fortifications

Militarily the lessons learned from World War I were numerous. The full potential of aviation and armored vehicles had not been fully exploited during World War I but in the interwar years 1918–1939 military airplanes and tanks saw important development. They gradually became standard and essential attack weapons in modern armies, allowing for rapid offensives with great mobile firepower. The post–1918 period heralded a change of patterns in warfare, and as a reaction a new phase in the design and use of permanent fortifications.

The era of the large forts inherited from the 19th century was over. To face machine guns and artillery fire—newly mobile, extremely lethal and destructive—a new concept appeared. The new fortifications emerging from World War I experience were small concrete units (called bunkers) including shelters for troops, ammunitions and supplies storage places, observation and command posts, artillery casemates, and pillboxes for automatic weapons.

The bunkers were partly underground, they could feature ventilation and filters to counter the use of toxic combat gasses, and their vicinity was systematically defended with anti-personnel barbed wires and anti-tank obstacles. Spreading over a large area or along a line of defense, a complex of small concrete bunkers was less vulnerable than a large fort, and also much cheaper to build. Scattered bunkers made a defense in depth possible, and they added a great flexibility: the elimination of a single bunker did not mean that the rest of the line was put out of action. It also became easier to respond to new developments, as it was rather simple, quick and cheap to add new reinforcing bunkers to any existing position.

In the interwar period, and particularly in the 1930s when war threat re-appeared, permanent lines of such concrete defensives works were erected all over Europe.

Dutch fortifications

World War I belligerents had respected the Netherlands' neutrality, and although the postwar period heralded a period of peace, the Dutch national defenses were modernized in the 1920s and more particularly in the late 1930s when the threat of a new European war appeared. As ever, the Dutch relied upon their traditional system combining flooded zones and lines of fortifications defending the accesses to the Fortress Holland. The tiny Dutch army was intended to withdraw to the National Réduit, the Fortress Holland, and several lines of defense were set up in order to delay any enemy advance. After the seizure of power by the Nazis in January 1933, Germany was never mentioned as a foe but it was clear that an invasion from the East was the most likely.

In the late 1930s work on those lines was gradually accelerated.

When France opted for a strong line of elaborate, expensive, subterranean forts at their borders (called the Maginot Line), Germany—and to a lesser extent the Netherlands—built a variety of small, independent, and standardized bunkers and pillboxes. Indeed the interwar Dutch fortifications were much less sophisticated than the formidable French Maginot Line. While the latter demanded highly trained specialist fortress-troops for a complete static underground defense, the former aimed to active surface strategy. The various Dutch lines were intended to provide protection to the regular Dutch infantry, to repulse or delay any attack from the east, and to resist until negotiations would take place.

To make such a temporary strategy practicable, the lines were constructed in depth, and wherever possible all bunkers were positioned in order that each of them was always covered by fire coming from a neighboring one. Gaps in the system were filled with field fortification including prepared infantry trenches, shelters, dugouts, machine gun nests and mortar pits, barbed wire fences, and obstacles of all kind. The Dutch bunkers were simple units generally designed to fulfill only one (occasionally two) function. All casemates and military installations were camouflaged with nets holding fresh cut vegetation and painted in patterns that matched the nearby natural environment.

Weakness of the Dutch Army

However, the kingdom of the Netherlands was militarily speaking in a weak position. The country—accustomed to and relying upon a policy of neutrality since the short and ill-fated war with Belgium in the 1830s—had invested much more in static defensive fortifications than in modern mobile weapon systems like armored vehicles. After 1918, the defense budgets were trimmed to such low levels that funds for modern weapons and updated gears were hardly made available. The Dutch had only a modest army and a limited weapon industry. In fact the kingdom was largely dependent on purchasing military equipments, heavy guns, machine guns, light weapons, and ammunition from foreign countries (Germany, France, Britain, and the USA).

On the whole, the Dutch army was neglected. It had a few modern homemade Fokker and Koolhoven airplanes, but it had no heavy armored vehicles except for a few homemade DAF 6-wheeled *Pantserwagen* (six-wheeled armored trucks), and a few American Marmon-Herrington four-wheeled armored cars. They also had a few Vickers-Carden Loyd tankettes (small lightly armored tracked vehicles) purchased from Britain. In the domain of artillery the Dutch army had a limited number of heavy and field guns—mostly obsolete as they dated from the late 19th century.

The raise and threat of Nazi Germany after 1933 woke up the successive Dutch governments. In the late 1930s, new funds were allotted to the defense, but these last minute efforts could not compensate for years of neglect. In 1940 the Dutch army had an insufficient numbers of soldiers, inadequate weapons, materiel, and equipment. Although the Dutch military command did secretly draw up plans with Belgium, France and Great Britain, commitments to strict neutrality prevented the implementation of clear military planning and thorough co-ordination.

DAF M-39 armored truck. Designed and constructed by the Dutch Van Doorne's Aanhangwagenfabriek NV (aka DAF trailer company) from Eindhoven, the 6-wheeled armored truck M-39 was operated by a crew of five. It was powered by a Ford Mercury V8 engine, had a speed of 75 km/h (47 mph), and an operational range of 200 km (124.27 mi). Armament included one 37 mm Model 1939 Bofors gun (1.45 in) placed in the turret and three Lewis 7.65 mm machine guns (0.30 in), one placed co-axially with the gun in the turret, the second placed in the front hull, and the third at the rear of the hull. Its armor was 10 mm (0.39 in) thick. After 1949 the Germans captured Dutch armored vehicles, and used them as training vehicles under the designation Panzerspähwagen DAF 201 (h). Other DAF military trucks included the six-wheeled Trado type whose chassis was used for several variants; e.g., troop transport, command vehicle, artillery tractor, engineering truck, or searchlight carrier.

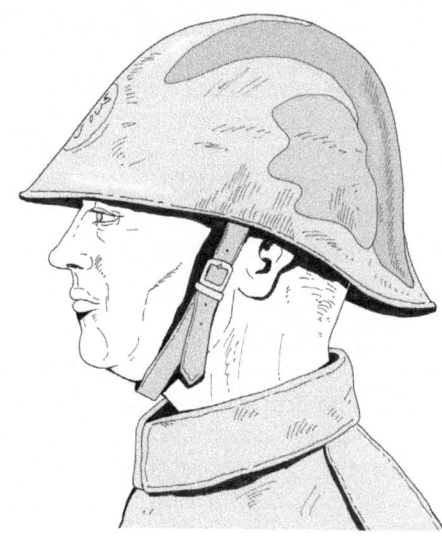

Dutch helmet 1940. The Dutch steel helmet had a pronounced neck protection, and was decorated with an oval shield at the front bearing the heraldic arms of the Netherlands (a standing lion wearing a crown and armed with a sword and a bunch of small arrows).

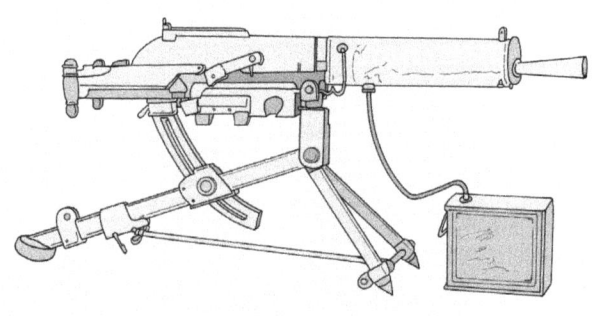

Dutch infantryman 1940. The uniform included a greyish light blue tunic, matching baggy trousers, and puttees. The collar badges and piping displayed the regiment identification and the arm of service. Equipment included a backpack, a leather belt with breadbag, bayonet, canteen and ammunition pouches for the dienst geweer (rifle). Dutch soldiers were mainly armed with foreign purchased weapons, including for example the German imported Mauser M.95 carbine, the Austrian Mannlicher M1895 rifle, the British automatic Lewis M20 light machine gun, the water-cooled Austrian-made Schwarzlose M.08 heavy machine gun, the British Vickers M.018 machine gun, and the German Spandau M.25 heavy machine gun. Anti-aircraft light quick firing guns included the Austrian-made Böhler 4.7 cm, and the Swiss made Oerlikon 2 cm TL No1.

Schwarzlose 08 Machine gun. The heavy Schwarzlose MG machine-gun was originally designed by the Dutchman Andreas W. Schwartzlose (1867–1936). The weapon was further developed by the Österreichische Waffenfabriks Gesellschaft in Steyr (Austria) in 1908. It was used as a standard issue firearm in the Austro-Hungarian Army throughout World War I. It was utilized by the Dutch, Greek, Italian, and Hungarian armies during the Interbellum and World War II. Derived from the Maxim design, it was cloth-belt fed, and water-cooled. It used a toggle-delayed blowback action, was 1.20 m in length, and weighed 41.2 kg (tripod included). In Dutch service it fired a 6.5 × 53 mm R bullet, and had a rate of fire of about 450 round per minute. The Schwarzlose could also be used as a fortress, anti-aircraft, or naval weapon in which case it was placed on a variety of heavy and specialized fixed mountings.

Fortifications of the Afsluitdijk (IJsselmeer Dam)

During the interbellum a number of concrete weapon positions had been built, usually at strategic points. In the early 1930s, the large bay named the *Zuiderzee* (South Sea) was separated from the North Sea by a long dam, thereby becoming the *IJsselmeer* (Lake of IJssel River). At the same time, parts of the low-lying submerged and swampy lands were reclaimed for agricultural use, which increased the surface of dry land. The 32 km (20 mile) Afsluitdijk represented a formidable achievement in civilian engineering. It created a welcome economic link and a communication highway between the northern provinces and North Holland—and it posed a great problem to the military and the Ministry of Defense.

The new dam (that could be used by vehicles by September 1933) created a dangerous access toward the naval base of Den Helder, and formed a treacherous weakness as a direct connection between the northern provinces (Groningen, Drenthe and Frisia) bordering Germany and North Holland—the economic and political heart of the kingdom. In order to thwart a possible aggression coming from the east, the Army demanded the creation of three defensive blocking positions at Wons, Kornwerderzand and Den Oever. These positions comprised locks, sluices and elaborate concrete bunkers (built between 1932 and 1936). Designed by the engineering corps and the Ministry of War, the high cost was financed entirely by the Ministry of Public Works.

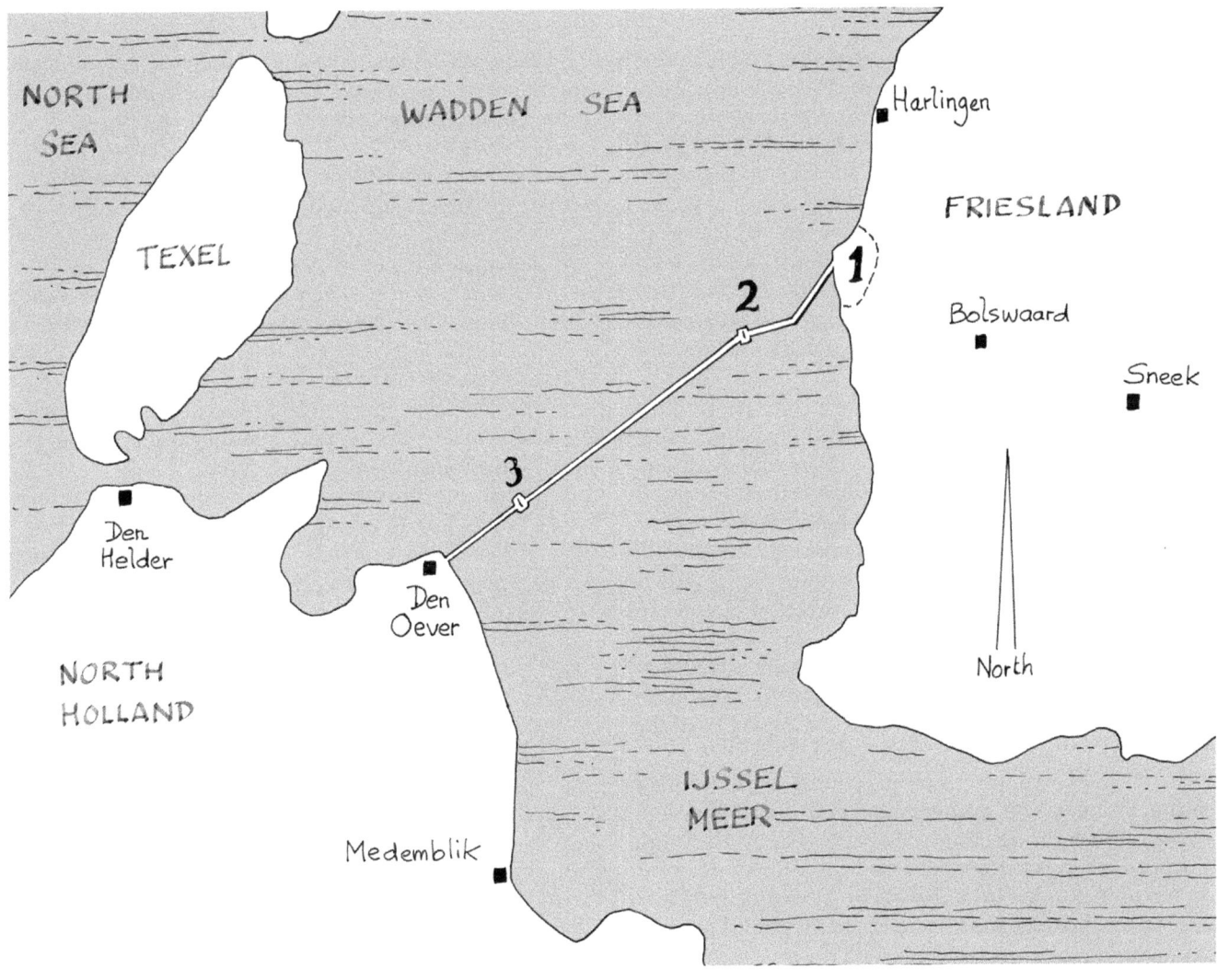

Map of the IJsselmeer Dam: 1: Wons line. 2: Fortified position of Kornwerderzand. 3: Fortified position of Den Oever.

Line of Wons

On the Frisian mainland near the village of Wons at the head of the closing dike a small line of defensive works was installed running along Zurich, Gooium, Hajum, Wons and Makkum. Occupied by about 650 soldiers, it included roadblocks, infantry field works (trenches and wire obstacles), about ten mobile 3.7 cm cannons, and a few flooded meadows. The purpose was to regroup withdrawing troops, and offer a first resistance in the case of an assault by a foe from the east. For the sake of neutrality Nazi Germany was never mentioned, but everybody knew what the euphemism "eastern foe" meant. If facing too heavy odds, the position of Wons was to be evacuated with troops retreating to the main center of resistance: the fortified position of Kornwerderzand.

Kornwerderzand

The fortified complex of Kornwerderzand was installed upon an artificial island intentionally created near the large sluices, which played a major role in the flooding control of the waterlines and fortified areas in the south of the IJssel Lake. The island also included locks allowing the passage of ships. Kornwerderzand was a heavily fortified belt of strong, permanent, modern and technically advanced reinforced concrete works. The complex featured 16 large concrete bunkers called *Kazematen* (casemates) or *Werken* (works). They were exceptional for their advanced design, and high quality of construction—actually quite comparable to some of the French Maginot Line *ouvrages*.

They were extremely strong. Their concrete external walls were 2 to 3 meters (9.8 feet) thick. They were especially designed to resist 210 mm (8 inch) rounds (direct fire), and 280 mm (11 inch) rounds (indirect fire). The internal walls were 1.5 meters thick, and most casemates were covered with a layer of 1 meter thick dirt for extra protection. All armed bunkers held the dam under fire as well as their neighbors. All bunkers were constructed in order to respond to an attack coming from the east. So all works had their weapons facing that direction, and all entrances were placed in the west, except Werken XII and XIII that were intended to oppose any presumed enemy coming from the west.

All casemates were multifunctional, and included entrances closed with thick steel armored doors. They had a low profile, and were partly sunk in the ground. They were independent units surrounded with barbed wires, and connected to each other by a network of zigzagging trenches. They were concealed, and camouflaged with painted patterns that matched the local vegetation and the greyish building material employed for the construction of the dam. Phony gun embrasures were painted to suggest weapon openings that actually did not exist. All works featured a cistern collecting rainwater, a periscope for observation around the bunker, and a telephone connected to the main central command post.

The garrison totaling 230 men operating heavy Schwarzlose 7.9 mm (0.311 inch) M08/15 water-cooled machine guns; 5 cm (1.97 inch) anti-tank guns; 2 cm Swiss-made Oerlikon anti-aircraft cannons; and 5 cm shore-based naval guns. The greatest shortcoming in the defenses of the dam was financial restraints in the armament budget: it was the weakness of effective anti-aircraft guns, and the lack of heavy long-range guns. Actually the top quality concrete bunkers housed only small caliber weapons with relatively weak firepower and short range. In fact they could not reach the mainland of Frisia. To remediate this deficiency, the Royal Dutch Marine (Navy) had agreed to dispatch gunships armed with medium-range artillery in case of emergency—the defense of the IJsselmeer Dam being within the command area of the nearby naval base of Den Helder. Besides, the basic design also included open machine gun emplacements (protected only by concrete breastworks), which made them extremely vulnerable to artillery and air attacks.

The Kornwerderzand complex included two lines, one placed ahead east of the sluices, and a second one behind them. The first line included the following works:

Kazemat I: armed with two machine guns, one inside the bunker and another in open emplacement outside;

Kazemat II: armed with one anti-tank 5 cm "long" 50 cannon emplaced on a fortress-mount inside the bunker, and one machine gun in open emplacement outside;

Kazemat III: not armed but fitted with a powerplant (a Diesel engine providing electricity for the whole complex); and an infirmary (for a team of three medical orderlies);

Kazemat IV: command station comprising a telephone central, an observatory (in an armored cupola placed on the roof), a command/firing coordination room, and one firing chamber with one machine gun;

Kazemat V: armed with four machine guns, one inside a firing chamber in the bunker, and three in open emplacements outside behind a thick concrete breastwork;

Kazemat VI: armed with two 5 cm anti-tank "long" 50 cannons emplaced inside the bunker, and one heavy machine gun in open emplacement outside; this bunker also comprised a lower basement floor providing additional troop shelters, and store rooms deep underground;

Kazemat VII: garage for one mobile searchlight, and one firing chamber with one heavy machine gun:

Kazemat VIII: armed with two machine guns, one inside the bunker and another in open emplacement outside;

Kazemat IX: armed with two machine guns, one inside the bunker and another in open emplacement outside;

Kazemat XIV: kitchen for the garrison, and garage for two 2 cm LT mobile anti-aircraft quick firing guns, and shelter for personnel serving the anti-aircraft battery.

The second line included the following bunkers.

Kasemat X: armed with one machine gun in firing chamber;

Kazemat XI: armed with two machine guns, one inside the bunker and another in open emplacement outside;

Kazemat XII: armed with two machine guns facing rearward, one inside the bunker and another in open emplacement outside;

Kazemat XIII: equipped with an additional communication room, and armed with two machine guns facing rearward, one inside the bunker and another in an external open emplacement;

Kazemat XV: garage for two 2 cm LT mobile anti-aircraft quick firing guns, and shelter for personnel serving the anti-aircraft battery;

Kazemat XVI: armed with one machine gun placed inside the bunker in a firing chamber.

In addition scattered amidst the main works, there were smaller partly subterranean bunkers used o.a. for fuel and water storage.

Today the defensive complex still exists, and is managed by a group of volunteers. It has become a museum and several *Kazematten* are open to the public.

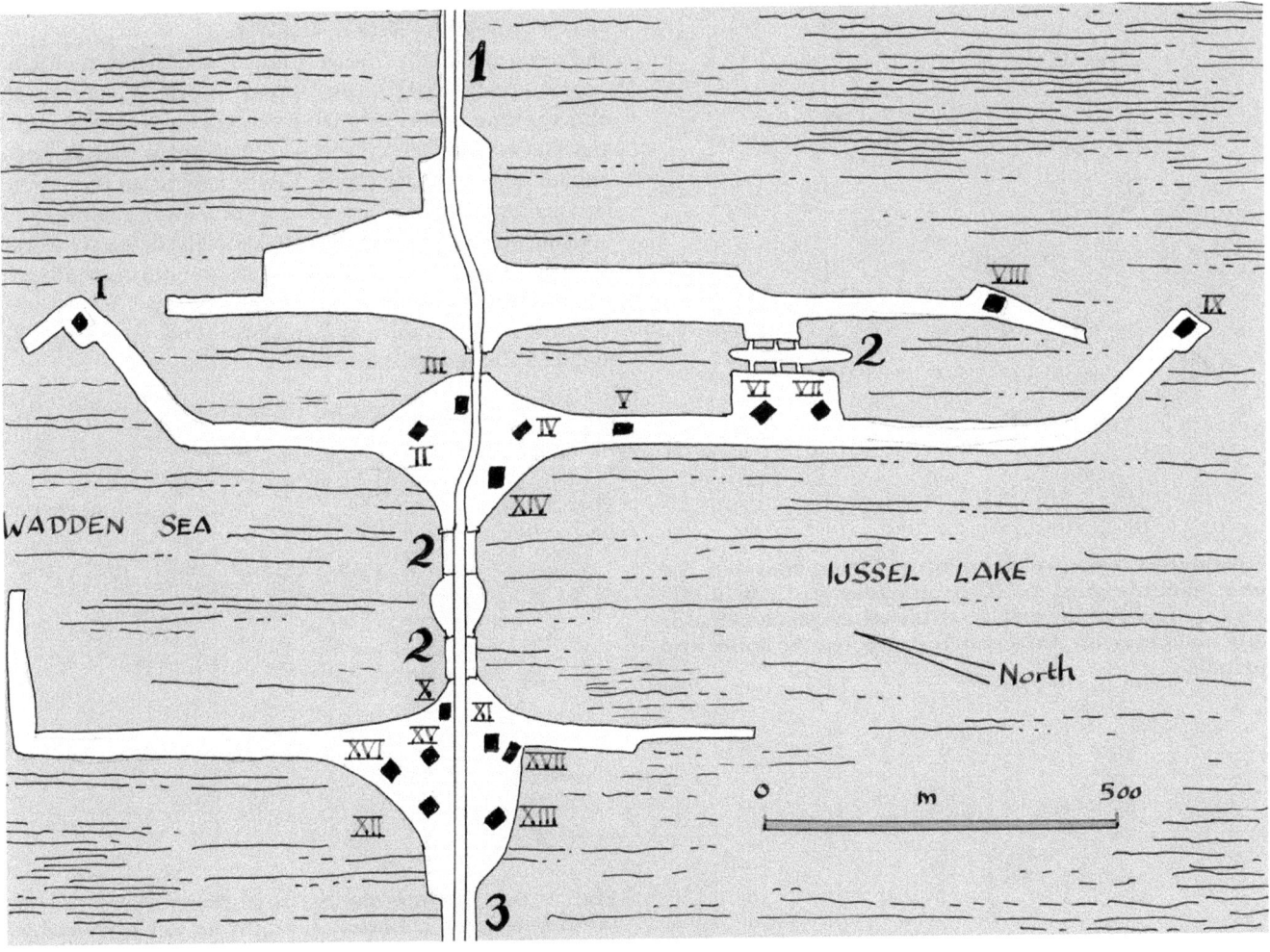

Map of Kornwerderzand position: 1: Highway to Wons and Frisia. 2: Sluices. 3: Highway to Den Oever and North Holland. The bunkers are indicated by Roman numerals.

Casemate IV front view (Kornwerderzand).

Den Oever

At the other end of the dam near Den Oever in the province of North Holland, there was another defensive complex established upon a triangular artificial island. It was intended to defend a system of sluices and locks called *Stevinsluizen*—after the great 17th century engineer Simon Stevin. The Den Oever position was tasked to deny any enemy advance—should the position of Kornwerderzand be overrun. The position's garrison totaled 7 officers and 211 soldiers. It included 13 bunkers, which on the whole were quite similar to those at Kornwerderzand with only slight differences.

Just like Kornwerderzand the position of Den Oever comprised two groups of kazematten: One ahead of the sluices on a triangular artificial island, and the other lined up behind them.

The dam was defended with anti-tank obstacles, and stood under fire coming from casemates armed with four machine guns (Werk I, Werk II, Werk VIII and Werk XII); two machine guns (Werk IX and Werk X); two machine guns and one searchlight (Werk V and Werk VI); armed with one cannon (Werk III); two cannons (Werk IV, featuring an underground cellar); featuring garages for anti-tank guns, and shelters for their crews (Werk VII and Werk XI); powerplant and infirmary (Werk XIII). In addition there were six smaller partly underground works (designated A to F) housing powerplant and fuel. Here too, the defensive works were ventilated, heated, sunk in the ground, camouflaged and connected by a network of deep trenches.

Today the complex of Den Oever is not open to public and is totally absorbed into natural vegetation.

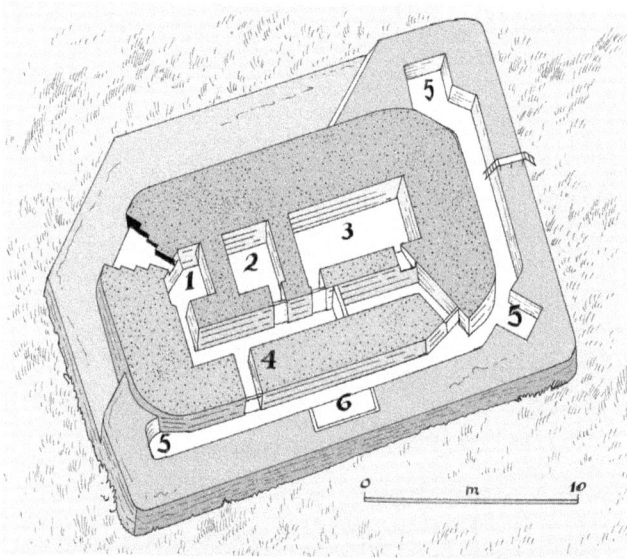

Casemate V Kornwerderzand: 1: Firing chamber for one machine gun. 2: NCOs quarters. 3: Crew quarters. 4: Emergency exit. 5: External emplacement for one machine gun, thus poorly protected. 6: Toilet and urinals.

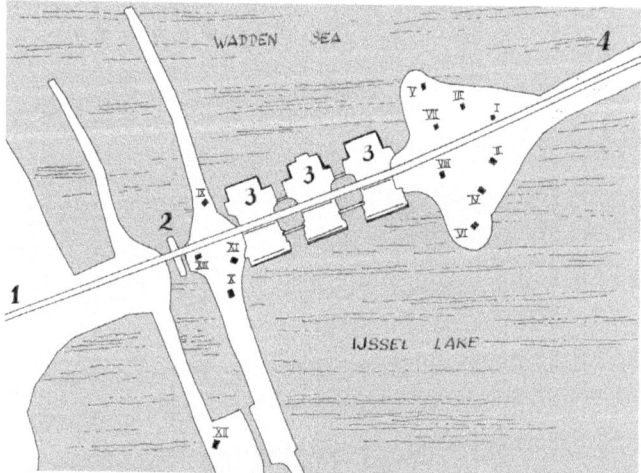

Plan of Den Oever complex: 1: Highway to North Holland. 2: Sluice. 3: Locks. 4: Highway to Kornwerderzand, and Frisia. The bunkers are indicated by Roman numerals.

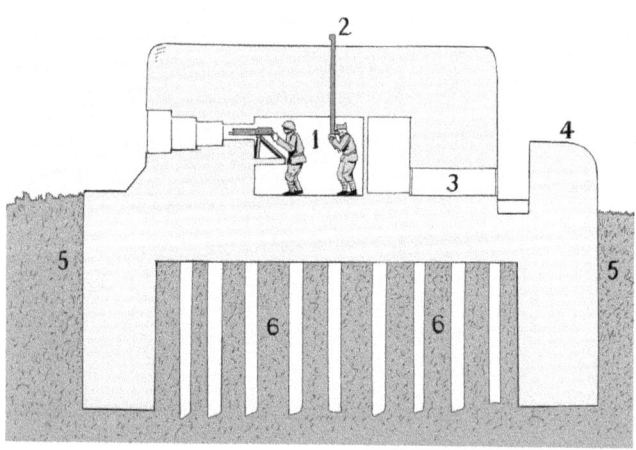

Left: **Casemate V Kornwerderzand (Cross-section): 1: Firing chamber for one machine gun. 2: Periscope. 3: Emergency exit. 4: External breastwork. 5: Underground vertical concrete slab for lateral protection against enemy projectiles. 6: Foundation beams.**

Bunker IX Den Oever. Werk IX was armed with one machine gun placed in a firing chamber, and another poorly protected installed outside behind the concrete breastwork. Further it included a troop chamber, and an observation armored cupola placed on top of the roof.

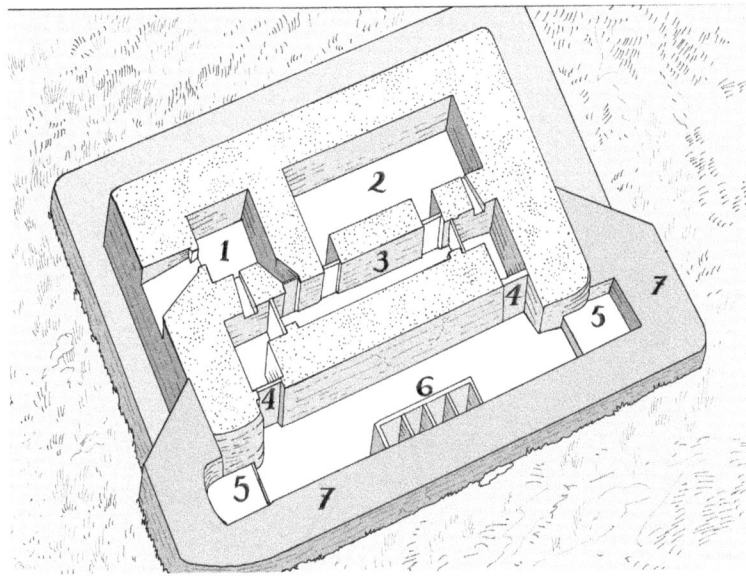

Werk XI (MG casemate) Den Oever: 1: Firing chamber for one machine gun. 2: Crew quarters. 3: Central corridor. 4: Entrance defended by a loophole. 5: External emplacement for one machine gun. 6: External toilet and urinals. 7: Concrete breastwork.

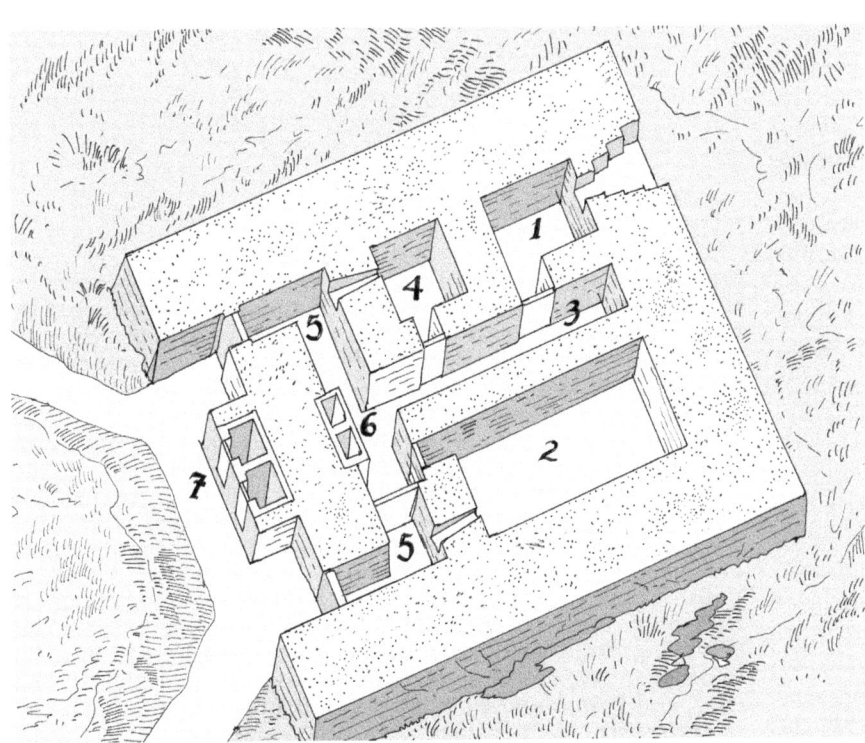

Werk XII (MG casemate) Den Oever: 1: Firing chamber for one machine gun. 2: Troop chamber. 3: Central corridor. 4: Ammunition store. 5: Entrance. 6: Internal toilet and urinals. 7: External toilet and urinals.

Lines of Defense, 1937–1940

After the seizure of power by the Nazis in 1933, and particularly with the increased tension due to the German aggressive diplomatic conquests (Austria and Czech Sudetenland in 1938–1939), national funds for military purposes grew. The Dutch hastily established advanced defensive lines. These lines included advanced inundations (50 or 80 cm deep) and comprised concrete machine gun pillboxes, artillery casemates, infantry shelters and trenches, barbed wire fences and various anti-tank obstacles intended to thwart a surprise attack. These lines generally consisted of three defensive systems: the *frontlijn* (front line) would take, delay and slow the first brunt of the attack; the *stoplijn* (stop line) was intended to repulse the enemy, and be the place whence a counterattack would be launched; the *ruglijn* (back line) formed the last ditch resistance with reserve forces.

Behind the *ruglijn* were local and regional command centers, logistics and other reserve forces, as well as field artillery taking the advancing enemy under fire. The lines of defense were placed along natural obstacles like rivers, swamps, lakes, and canals. Bridges crossing those waterways were defended and prepared for direct destruction by explosive should the Germans violate the Dutch kingdom's neutrality.

Fortress Holland

As discussed, Vesting Holland ("Fortress Holland") was an ensemble of fortifications and military flooding built and established in the 19th and 20th centuries. Fortress Holland was a *National Réduit* intended to defend the *Randstad* (the most populated and most industrialized part of the Netherlands including the important ports of Rotterdam, and Amsterdam, and the large cities of The Hague and Utrecht). There—protected behind flooding and fortifications—the Dutch army was to regroup, fight and resist until negotiations would take place or when French and British forces would intervene and rescue the besieged.

In February 1940 it was decided, however, that Fortress Holland would not be defended. Despite the floodings and the fortifications, the area was too bare and too flat, and to tell the truth the government did not wish to expose the densely populated *Randstad* to enemy fire, notably long-range artillery and bombardment aviation. The defenses of Fortress Holland consisted of floodings in the New Dutch Waterline (stretching from Naarden near the Zuiderzee in the North, passing along Utrecht, and ending near Moerdijk in the South), and forts around the Stelling of Amsterdam built in the end of the 19th century and at the beginning of the 29th century. In the 1930s, due to the tremendous development of air power and armored vehicles these fortifications were no longer regarded as practicable defense lines. As early as 1922 the system had lost its capability and status as a last-ditch defense. Nevertheless the concrete forts around Amsterdam and in the New Waterline were left unaltered, and remained in use as troop quarters, shelters, command centers and supply bases.

Instead it was decided to defend Fortress Holland from a new advanced front composed of several additional eastern, southern, and northern facing lines resting on rivers and canals (allowing flooding) wherever possible. The defenses were thus shifted a bit easterly along the IJssel and Meuse rivers.

For the sake of neutrality, token defenses were established in Zeeland at Bath and in South Beverland, officially in order to repulse a hypothetical British or French landing.

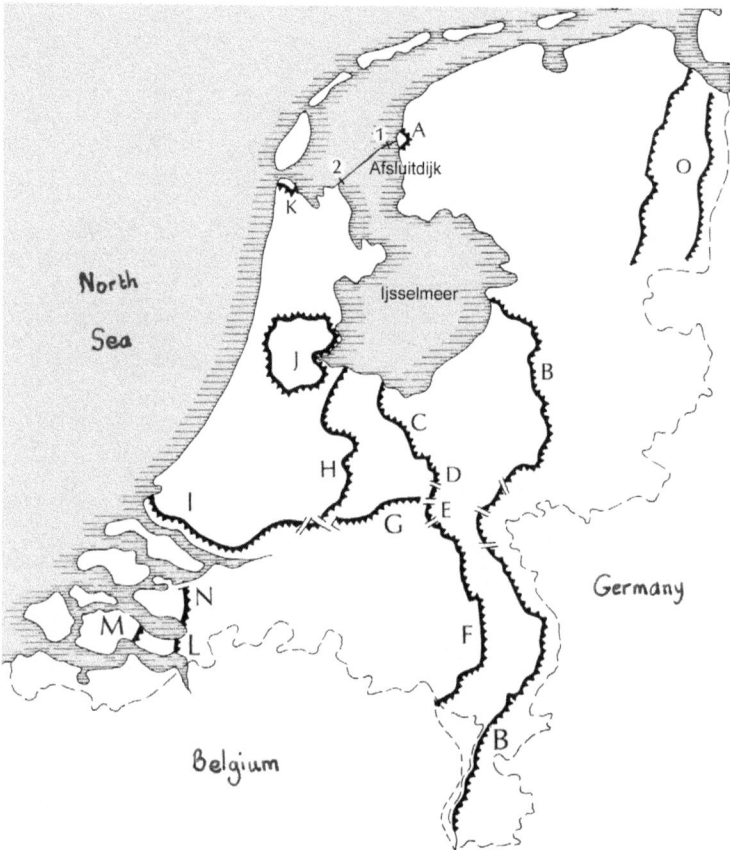

Map of Dutch defenses in 1940. The heart of the Netherlands called Fortress Holland (National Redoubt) was protected by flooding and fortifications in the following advanced lines of defense. A: Afsluitdijk with defenses at Wons, Kornwerderzand (1) and Den Oever (2) closing the north-east access to the province of North Holland and the capital Amsterdam. B: IJssel and Maas advanced lines. C: The Grebbe line (main defense line). D: The Speesline with Betuwe forts. E: Maas-Waal forts. F: The Peel-Raam fortified line. G: The Waal-Linge line. H: The New Dutch Waterline. I: Southern front of the Stelling of Amsterdam. J: The Stelling (forts ring) of Amsterdam. K: Fortifications of naval base Den Helder (aka Stelling of Den Helder). L: Line of Bath. M and N: lines of South Beverland; note that L, M, and N were token defenses intended to protect Zeeland and Vlissingen (Flushing) from attack by a "foreign power" (Britain or France) in order to give credit to the official policy of neutrality. O: The O, Q, and F lines along the border of provinces Groningen and Drenthe with Nazi Germany.

The IJssel line and the Maas line

The IJssel line and the Maas line were defenses extending in a southern direction along the banks of rivers IJssel and Maas (Meuse), the rivers being used as a very wide anti-tank ditch and anti-personnel obstacle. Work on those positions had already started in 1936 with the establishing of armed bunkers near all the bridges crossing the rivers. With the increase of international tension in 1938 many more flooding zones, armed casemates, troop shelters and obstacles were added with a total of 855 bunkers constructed along the front. Running from Kampen in the North down to Maasbracht in the South, both lines were the most advanced lines in the East. They were intended to absorb the first attack (from Germany), and to delay as long as possible the advance of the invaders.

The Grebbe line

The Grebbelinie (Grebbe line), placed behind the IJssel line, was the main line of resistance facing an aggression from the east. Established in 1745 it was never used until the 1930s. In 1815, the construction of the New Dutch Waterline started, and the Grebbe Line was no longer needed. It was soon left uncompleted, neglected and decommissioned in 1926. However, with international tensions growing in the late 1930s the Dutch army prepared for an enemy invasion. In 1940 the decision was made to reactivate the Grebbe line and make it the main advanced defensive line. The Grebbe line stretched from the southern part of the IJsselmeer at Spakenburg to Rhenen on the great rivers Rhine and Meuse.

It was defended by numerous vast inundations, ancient 19th century forts, to which modern obstructions, roadblocks, barriers, obstacles, infantry trenches, command posts, artillery positions, and 278 concrete bunkers were hastily built. It was garrisoned by 44 battalions totaling some 26,000 soldiers. There were plans for more concrete combat works but these were not yet started when the Germans launched their attack in May 1940. The Grebbeline had two vulnerable points. Notably in the vicinity of the city of Amersfoort, and near the village of Rhenen where the Grebbeberg, as a 150 foot high hill, made inundations out of question. To remedy those weaknesses both sectors had been additionally fortified. Instead of inundations advanced fortified positions were placed as main defenses.

Peel-Raam line

In 1938 work started south of the Meuse River. The defense line was installed behind the Maas Line at a distance varying from 9 to 21 km (up to 13 miles). It had a length of 75 km (47 miles). It started at Grave, where a barracks complex was built. From there, the line ran along Mill, Peel, and the Zuid-Willemsvaart (canal) until the Belgium border nearby Weert. In the North the defense line was linked with the Grebbe line.

The defense line made use of natural swamps, rivers and canals in the area. In the northern part, an artificial obstacle in the form of a canal was created called the Defensiekanaal (Defense canal). Like all the other lines, the Peel-Raam Line included flooded meadows, machine gun pillboxes, cannon casemates, infantry trenches, troop shelters, barbed wire obstructions and anti-tank obstacles.

The Dutch wanted to link the Peel-Raam defense line with the Belgian defense works installed along the Albert Canal. But the Belgian military authorities preferred a separate linear defense position to be called the Dutch Oranjelinie between Tilburg-Waalwijk and the Bergsche Maas River. Because of this lack of coordination the defense line was vulnerable as the enemy could pass around it by crossing into Belgium.

The Waal-Linge line

Between the southern part of Fortress Holland and the Betuwe defense, there was a gap. That was closed by the establishment of the southern Waal-Linge line, running along the meandering Linge River from Asperen, Leerdam, Oosterwijk, and Arkel down to Gorinchem on the Waal River.

Fortifications of Den Helder and Texel

Before World War II Den Helder—the most important Dutch military harbor, was still defended by land- and coastal fortifications dating from the Napoleonic era with three large obsolete bastioned forts and a number of coastal batteries and forts established in the 19th century. Between 1916 and 1930 numerous concrete infantry trenches, shellproof personnel shelters, machine gun pillboxes, artillery casemates, observation posts, anti-aircraft batteries, and coastal batteries bunkers were added to these outdated permanent fortifications. Located north of Den Helder in the Wadden Sea, the island of Texel, too, received concrete fortifications, notably several coastal batteries.

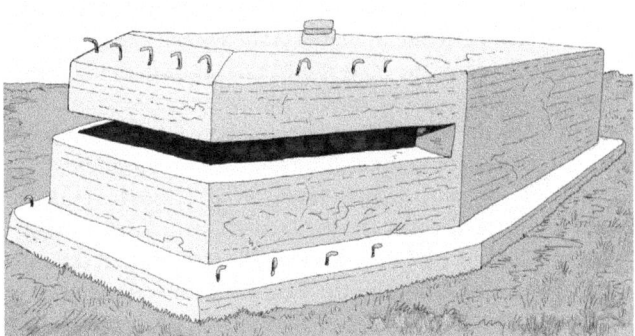

Fire directing post at coastal battery Den Hoorn on Texel Island. The battery Den Hoorn was located in the dunes overlooking the beach west of the village of Den Burg on Texel Island (province of North Holland). Built in 1938–1939 it included the depicted concrete fire directing post (equipped with a range finder), three 15cm L35 coastal guns, three ammunitions stores, and a shelter for the gunners.

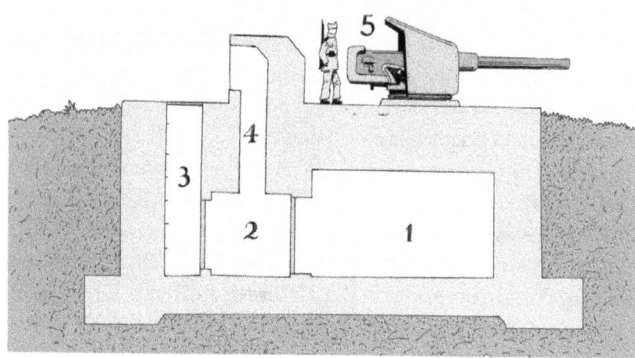

Gun bunker (cross-section) at battery Den Hoorn on Texel Island: 1: Troop chamber. 2: Ammunition store. 3: Entrance pit fitted with climbing rungs. 4: Ammo lift. 5: Naval cannon in armored gunhouse.

O-, Q- and F- lines

In the north-east, right at the border with Germany, the Dutch had built three prepared defense lines, the O-line, the Q-line, and the F-line. Hastily established in 1939 and 1940, they were intended to protect the three northern provinces Groningen, Drenthe and Frisia. However, those three "lines" were actually rather thin and weak. Composed of obstacles, roadblocks, infantry field fortified positions and a few machine gun concrete casemates, and *rivierkazematten* (see below), they were designed only as small delaying positions around bridges crossing rivers and canals. If meeting heavy odds, troops were ordered to withdraw to the South or to the stronghold of Kornwederzand after destroying all bridges behind them.

Interwar Standardized Pillboxes

The defense lines previously described were composed of controlled inundations, obstacles and standardized casemates and pillboxes. A small number of pillboxes had been designed and constructed in the First World War, and wherever possible these were integrated into the interwar defense plans.

In 1928 the Dutch Ministry of War issued a manual entitled *Bouw van zware gewapend beton-schuilplaatsen—Voorschrift Inrichting Stellingen no. 77* (Construction of heavy reinforced concrete shelters—Instruction Arrangement of Position # 77). Destined to the attention of the military, it was published by the Royal Military Academia of Breda and soon became known as VIS77. The book gave a complete overview of the guidelines for modern warfare based on World War I experiences. Part VII particularly was designed for military engineers as it focused upon ferro-concrete casemates.

The Dutch VIS casemates, shelters and pillboxes presented some similarities with the British concrete FW works, designed by the Directorate of Fortifications and Works, which were built in a hurry in 1940 when Nazi Germany threatened to invade England. In both cases the purpose was to provide a number of basic, very cheap, and simple defensive works that could be quickly constructed by unskilled soldiers and local civilian building companies.

Although the Dutch engineering corps never reached the high and strict level of standardization comparable to the German *Regelbau* (standard types of bunkers), the VIS instruction book presented ten standardized works including shelters for men, pillboxes for machine guns, casemates for anti-tank guns, observatories, and command posts. Each model could be employed in locations where geographical and tactical conditions were similar. Where slightly different situations were met, the VIS instruction offered alternative variants, which differed from each other by internal arrangements, firepower, and various concrete thicknesses. The creation of standard designs and alternatives made construction projects easier and calculable in planning.

All VIS designs were made of reinforced concrete of various quality. The reinforced concrete used in construction was generally conventional, making use of steel rebars with floor, walls and roof all mutually bonded. However, several instances were known where improvised scrap metal was used, such as parts of fences, rods or park railings. In some cases bricks essentially were formed a mold into which concrete was poured, the bricks being left in place on purpose as another revetment, an additional skin of protection. The Germans used the same technique in the last phase of construction of the Atlantic Wall.

The instruction book VIS77 indicated three regular coded concrete thicknesses. Note that *W* stands for *weerstand* (Dutch for "resistance").

(a) W12–15 (1 meter thick) was intended (in theory) to resist shots fired by 12 to 15 cm artillery;
(b) W15–21 (1.5 meters thick); and
(c) W21–28 (1.8 meter thick) were to resist heavier artillery of caliber 15 to 28 cm.

These thicknesses were used for frontal walls and roofs exposed to enemy fire. Other walls were to be constructed with thicknesses varying from 0.8 to 1.2 and 1.5 meters. The degree of protection offered by the different VIS types of pillboxes varied as the designers had not taken into account aerial bombs, and modern high velocity armor-piercing projectiles (such as those fired by the modern German 88 mm FlaK gun, and the 37 mm PaK gun). This negligence was a serious flaw during the battles of May 1940.

Compared to the French Maginot Line's underground forts and the German Westwall bunkers (carefully planned, designed and built with advanced technology, and the backing of comfortable budgets in the prewar years), Dutch pillboxes were rather primitive, inadequate, and vulnerable. With the exception of the river casemates, most lacked sanitary facilities, air filtering devices against poison gasses, equipment storehouses, replacement weapons, spare water (e.g., for the cooling of machine gun barrels), communication means such as telephone, as well as any fumes exhaust system—obliging the crews to wear a respirator when firing their weapons. Soldiers often considered the pillboxes as "death traps" whose occupants had few chances of survival. They were indeed only semi-permanent works hastily built with limited means and poor budget.

The small concrete troop shelters, artillery bunkers, and armed pillboxes were built in or in the vicinity of existing obsolete fortifications as last minute reinforcement. All works were carefully camouflaged. Most of them featured small iron hooks allowing covering with camouflage nets. Some pillboxes were half sunk so that the embrasures were as low as ground level for machine gun grazing fire. Others were raised up to give a better view and a wider arc of fire. Many were dug or inserted into a hedgerow, a sloping or a talus or into a dike for providing the lowest possible profile. They often had layers of dirt piled up on the roof and/or amassed on their sides as extra protection and camouflage.

Natural vegetation like ivy and grass, as well as painted patterns, camouflage netting, and other artifices were used to help break up the outline, blur shadows, and blend in with the surroundings. In built-up areas pillboxes were made inconspicuous by the addition of realistic deceiving artifices (e.g., a tiled roof and wood revetment, painted bricks and windows in trompe l'oeil) to look as if they were harmless civilian constructions or adjacent buildings carefully matching their vicinity.

Internally, Dutch pillboxes were cramped and spartan with low ceilings, small rooms, and no or few accommodation for the crews. They were not designed for enjoying a pleasant and comfortable sojourn but merely and simply to providing protective combat emplacements during a (hopefully short) period of intensive crisis (bombardment and fighting). Only the already discussed modern Kornwerderzand casemates built to defend the IJssel Dam were strong, rather adequately equipped, and well designed.

Most shelters, pillboxes and casemates built in the defense lines had no or few heating and ventilation devices, no or few protection systems against poison gas. Soldiers were equipped with respirators and some casemates featured hand- or battery-operated air filters. Not all works were provided with electric light, so many had only niches for oil, or acyteleen or carbid lamps. Some internal concrete or wooden shelves and T-shaped tables were provided to support weapons and some were whitewashed inside.

Many casemates were deprived of modern communication systems like a telephone. So when it came to actual fighting during the German offensive in May 1940 each small team of soldiers occupying a pillbox was rather isolated, missing coherent order, and deprived of a clear view of the general situation.

The Dutch engineering corps built about 1,500 small bunkers during the late 1930s in the various defensive lines, but in May 1940, the program of fortification was far from complete.

The following VIS77 models were the most frequently used.

Type IIB

The type IIB, also designated VIS77 (W15–21) was one of the first standardized models designed by the VIS Committee. It was a tiny mono-functional concrete pillbox intended to house one machine gun and its operators. It included an entrance at the back, a small corridor defended by a rifle loop, and a combat room (3 × 3 meters, and only 1.5 meters high so it was not possible for the average Dutch soldier to stand up in it). That cramped room was fitted with an embrasure for firing the machine gun, and a hand-operated ventilation system to evacuate toxic fumes. The walls (1.5 meters thick) were made of a revetment of bricks and a filling of concrete. The roof was 1.2 meters thick and covered with a layer of earth for camouflage and additional protection.

Type IIB pillbox; type VIS 1937 Haarlemermeer: 1: Entrance with armored door. 2: Ammunition store. 3: Firing chamber for one MG. 4: Embrasure.

Model P Pyramid

The so-called pyramid was a mono-functional concrete shelter officially designated *Groepsschuilplaats* (shelter for a squad), and colloquially nicknamed *pyramid* because of its sloping roof intended to deflect shells and bombs. The unarmed, and passive personnel shelter was ca. 8.2 meters long (9 yards), 6.5 meters wide and 4.85 meters high. It included a single entrance at the rear fitted with a thick armored door, a staircase leading down to a small corridor defended by a rifle loophole, and to the confined *afwachtingsruimte* (stand-by room 3.5 × 3 meters) for a squad of 8 soldiers. A variant existed with a larger room for a group of 12 riflemen.

The room was equipped with a ventilation system, telephone, and a periscope for observation—although this was often omitted due to lack of funds. They were intended for first line defensive positions and therefore constructed in W15–21 thickness. The concrete thickness of the roof was 2.15 meters, and that of the walls was 1.8 meters, allowing the shelter to withstand impact up to 21 cm (8 inch) caliber shells. The monobloc Pyramid shelter was rather heavy and when established on soft or swampy ground it usually rested on thick foundation beams. More than 700 were built right before World War II, and many (ca. 400) are still extant in the Dutch countryside. Exceptionally, two pyramids were set together for sheltering 16 men as can be seen today in a meadow at Maarssen (Province of Utrecht).

Worthy of mention is the *doorgezaagde Groepsschuilplaats 599* (sawn-through shelter) that lies at the Diefdijk near the town of Culemborg (province of Gelderland). This 1940 pyramid bunker was cut in half in 2010 as a tourist attraction and piece of open air art.

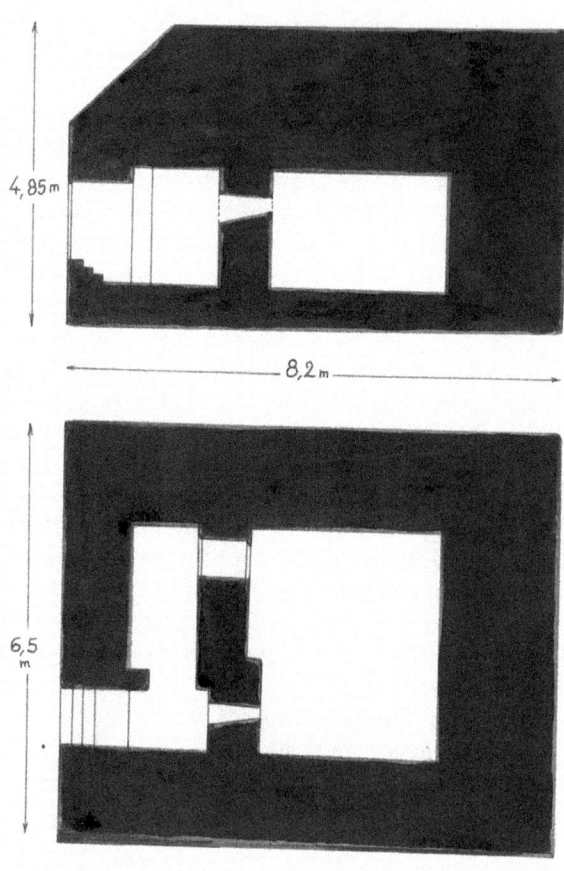

Pyramid—profile (*top*) and plan (*bottom*).

Pyramid.

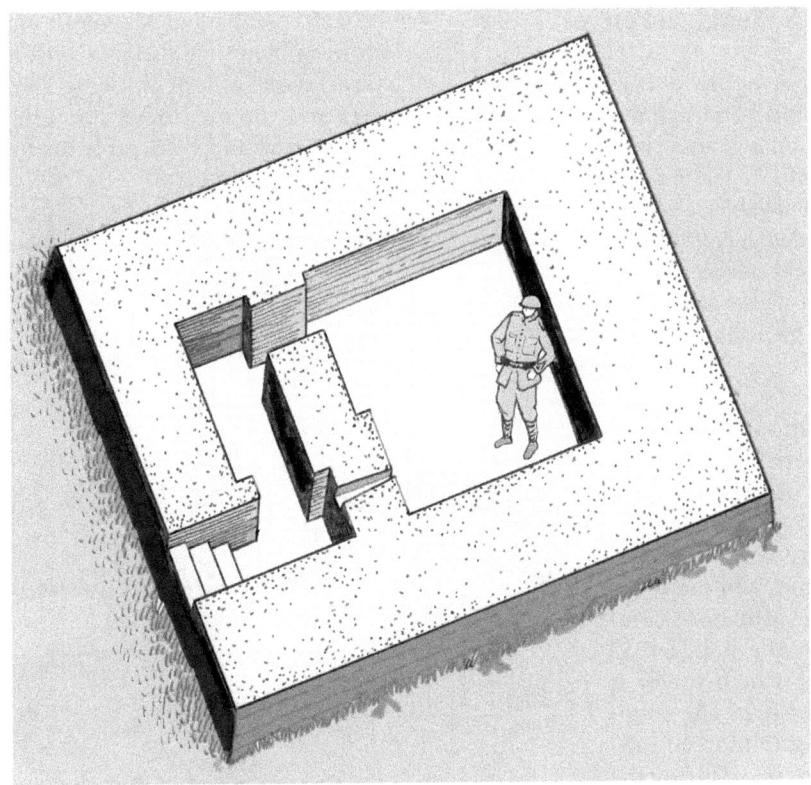

Pyramid (plan).

Double Pyramid shelter (with concrete foundation poles) near Maarssen (province of Utrecht).

Model S Stekelvarken

The *Stekelvarken* (porcupine) also known as *Spinnekop* (spider head) was a small and simple mono-functional trapezium-shaped rifle and machine gun concrete pillbox. Designed in the late 1930s, it featured no facilities, not even a fume exhaust system for the weapons. It included only one small combat firing chamber, and an access at the rear, either a vertical pit with a hatch or a small armored door in the back wall. The reinforced concrete walls were only 80 cm (32 inches) thick.

The S model came in several variants according to the number (1, 3, 4, 5, 7, or 9) of embrasures and rifle holes respectively coded S1, S3, S4, S5, S7 and S9. About 800 S-types were constructed in the Dutch defensive lines, and many are still extant to these days. The most commonly encountered was the standard S3 with three embrasures, which together had a 190 degree arc of fire. The loopholes were made of steel reinforced profiles anchored in the concrete. During the ill-fated combats of May 1940, the model S3 (and all the more so the other variants with more than three embrasures) made a poor showing since the number of openings and the mediocre concrete thickness made them rather weak and vulnerable to enemy fire.

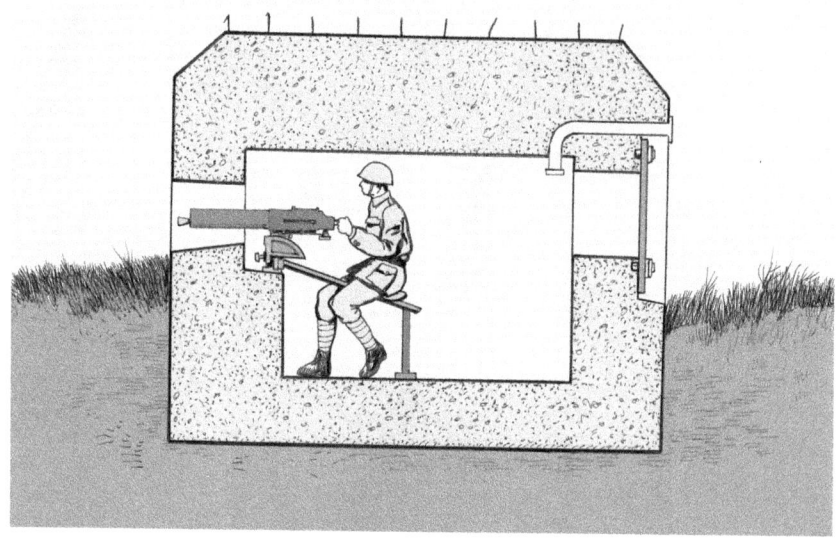

Cross-section of Stekelvarken.

Stekelvarken with trench.

Model B

Introduced in June 1939, the pillbox model B was a small flanking machine gun casemate, which means that the direction of fire from the pillbox was not frontal, but directed into the side of the advancing enemy force. It had a length of 4.70 meters (5 yards), and a width of 3.85 meters. It front wall was 1.25 meters thick and the roof often covered with one meter of additional dirt. It came in several variants, and featured only one cramped combat room for one machine gun and two or three operators. It had a small lateral flanking wall (placed right or left) protecting the embrasure from enemy fire, and an entrance at the rear with an armored door. Only a few hundred were constructed, often behind front defenses or at strategic positions such as crossings or road hubs. Enlarged versions of this steel-concrete work were planned to house anti-tank Bohler 4.7 cm guns, but their construction had not started when the Germans invaded the Netherlands in May 1940.

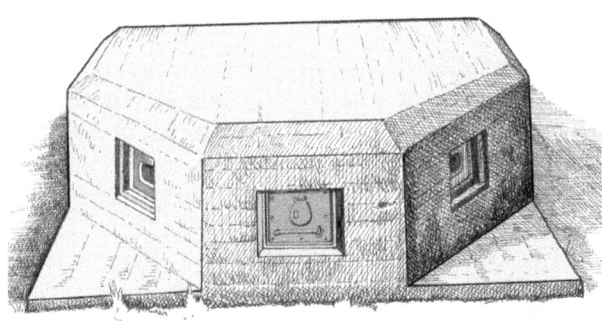

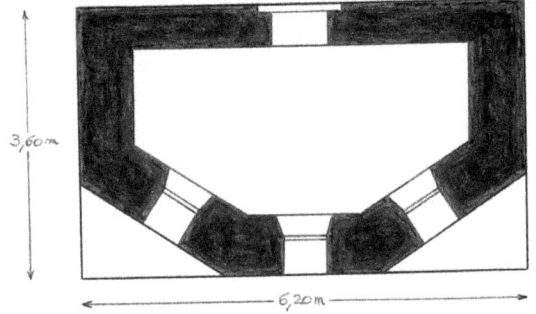

Stekelvarken S3. *Top:* **Front view.** *Bottom:* **Plan.**

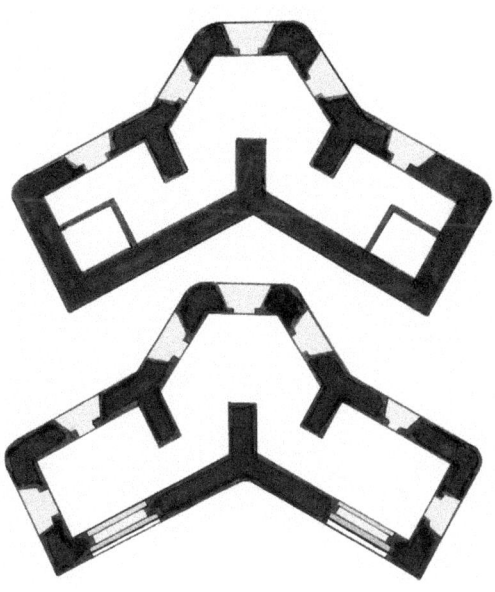

Stekelvarken variants: *Top:* **Hatch.** *Middle:* **S5 with five embrasures.** *Bottom:* **S7 with seven embrasures.**

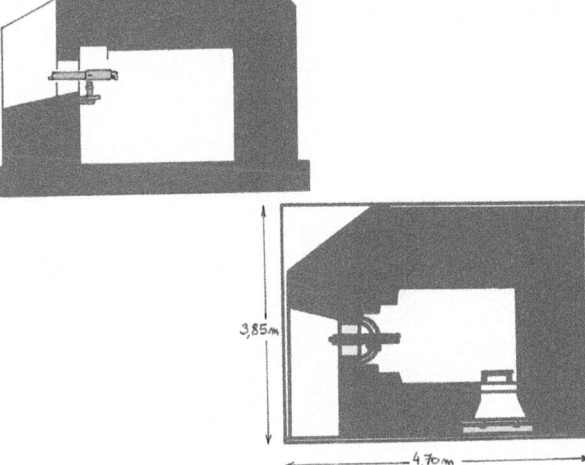

Pillbox model B: *Top:* **Front view.** *Bottom left:* **Cross-section.** *Bottom right:* **Plan.**

Model G

Model G (*Gietstalen kazemat* = Steel armored pillbox), designed in March 1939, was a small bunker comprising a square concrete base (6.5 × 6.5 m) housing a fixed armored cupola for one frontal-firing heavy machine gun operated by two or three servants. The domed cast steel cupola was 10 cm thick, and naturally fitted with an embrasure with an angle of fire of 35 degrees for a heavy Schwarzlose or Lewis machine gun.

There existed several variants adjusted to the specifics of the envisaged position. So there were alternative embrasures allowing an arc of fire from 45 to 90 degrees. Also the access to the armored cupola could differ, either by means of a small armored door placed on the side of the bunker, or by means of a hatch and a narrow pit with rungs placed vertically behind the cupola. All Model G pillboxes featured an air filtering system against poison combat gas and automatic exhaust fume gear. Although cramped and uncomfortable, it was squat and strong, gave a good protection to the weapon and the three-men crew, and easy to camouflage being small and low. A total of 702 were built.

River casemates

The so-called *rivierkazematen* (river casemates) were specifically designed for placement behind strategic points such as dikes, accesses and bridges, sluices, pumping stations and other hydrological installations that were used to carry out the inundations. Between 1936 and 1939, about 850 units were built along the most important junctions and passages of the IJssel, Waal, Nerderrijn and the Meuse rivers from the IJssel Lake to South Limburg. Generally built in the W12–15 strength, the river casemates were permanently manned by professional soldiers of the Special Police Troops, and were more elaborate than the mono-functional S, G, and B pillboxes.

The dikes and the bridges defended were not identical so river casemates were tailor-made to each situation. They were thus not standardized works but however presented many similar features. They often comprised two or more stories. Depending on local situations, there existed also tower-like river casemates with three and even four levels allowing to fire above a dike. In the upper floor there was one or more combat rooms (often armed with Schwarzlose machine guns or 5 cm anti-tank guns), and an observation post and sometimes a monitoring armored metal cupola placed on top. Some were equipped with a searchlight. In the lower floor there was the entrance, a resting/stand-by crew chamber, an ammunition storeroom for HE and AT projectiles, a storeroom for the bridge demolition charges, and sometimes a small powerplant.

Pillbox model G Front view.

Model G Cross-section.

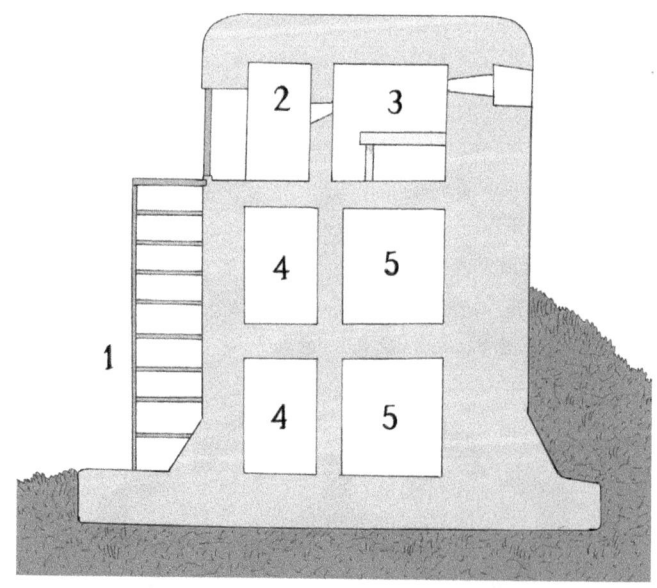

River casemate cross-section. The casemate was about 5.80 m in width, 7.10 m in length and 8.70 m high: 1: Ladder or stair. 2: Entrance. 3: MG combat chamber. 4: Ammunition stores. 5: Troop chamber.

Chapter 8. The Period 1914–1940

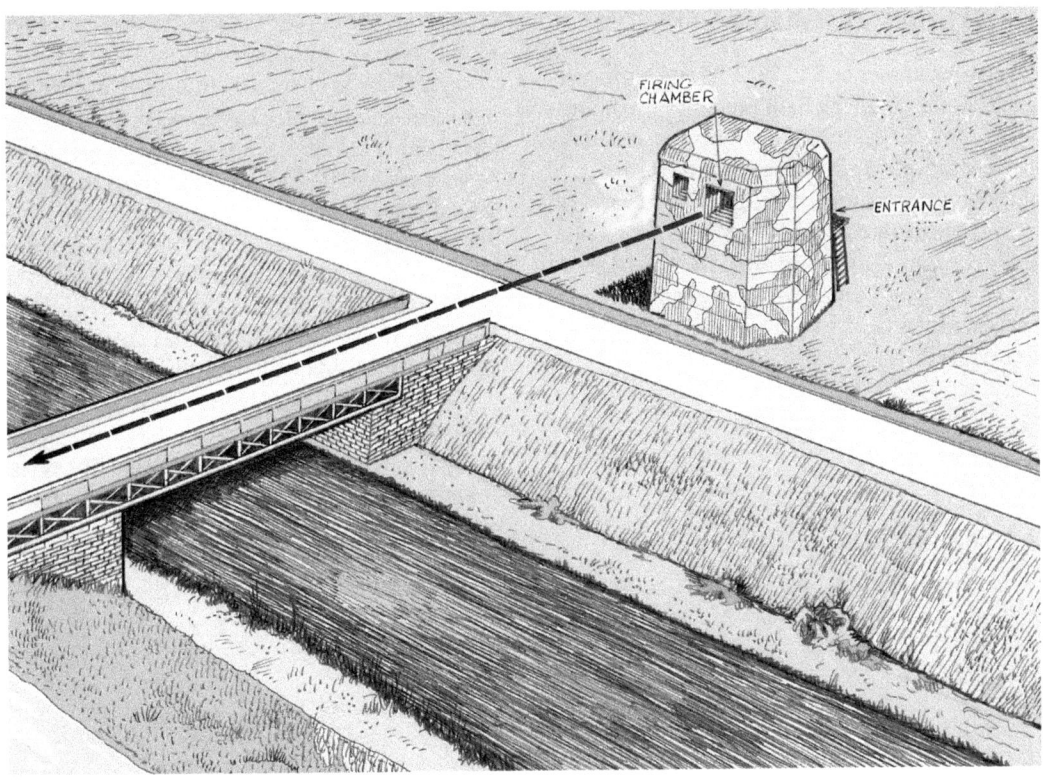

River casemate. Schematic implantation showing a river casemate defending a bridge along a dike.

River casemate front view (Grave). Note the fake painted embrasures.

Artillery casemates

There were also a number of VIS W12–15 strength casemates designed to house artillery field pieces (6 cm), anti-tank weapons (notably Bohler 4.7 cm), and coastal navy cannons. They were simple, rather cheap, and box-shaped concrete works comprising only a firing chamber with embrasure, an ammunition store, and sometimes a small stand-by room for the crew. They were often part of a battery and connected to a fire-leading station. Only a few were ever constructed, and examples can be seen at Den Helder, Haarlemmermeer and Weesp.

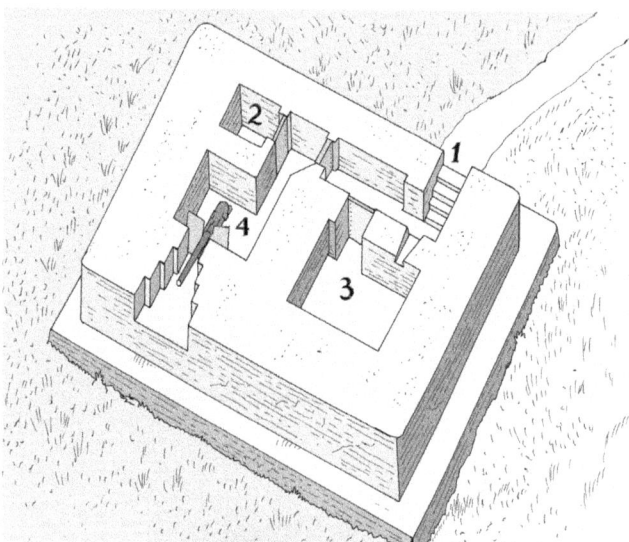

Artillery casemate VIS 1937: 1: Entrance. 2: Ammunition store. 3: Troop chamber. 4: Firing chamber for one gun.

Obstacles

In the 1930s a national program of new road and highway construction was undertaken, in order to connect the western and northern provinces with the rest of the country. Roads offered the enemy fast routes to their objectives and consequently they were blocked at strategic points. The need to prevent tracked tanks and wheeled vehicles from breaking through was of key importance. Consequently, the defense lines consisting of bunkers and obstacles generally ran along pre-existing barriers such as rivers and canals; railway embankments and cuttings; thick forests; and all other natural hindrances. And naturally, wherever possible, well-drained land was allowed to flood making the ground too soft to support even tracked vehicles. Several obstacles, hindrances, obstructions and roadblocks were designed in order to oppose the progression of enemy infantry, wheeled vehicles and armored tracked vehicles on smooth highways, and well-metaled roads.

Obstacles against infantry included inundations, ditches filled with water, various sorts of barbed wires, and mobile Frizzy horses (aka knife stand or *cheval de Frise*: a mobile fence or infantry obstacle).

Anti-tank obstacles included a number of concrete pyramid-shaped blocks, as well as sections of rail, and steel beams called *asparagus*. They were placed in concrete bases and installed into holes prepared on roads. The anti-tank ditch was an excavation intended to interdict the progression of tanks. The ditch had sloping sides, and was ca. 6 meters deep and 8 meters wide. The anti-tank ditch was either a simple V-profiled excavation dug into the ground or occasionally a permanent and costly installation made of concrete. It could be filled with water.

Prepared demolitions were also planned. Bridges, important installations and any other key points were made ready for demolition at short notice by preparing sabotage chambers and niches filled with explosives. Prepared demolitions had the advantage of being undetectable from the air—the enemy could not take any precautions against them, or plot a route of attack around them.

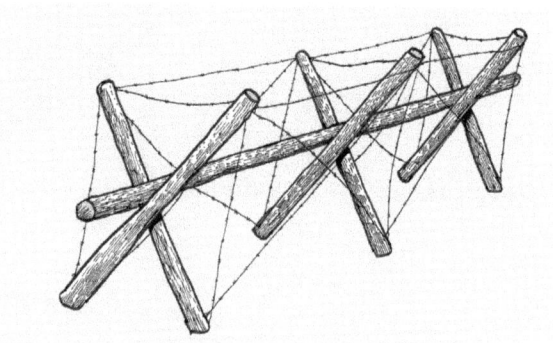

Frizzy horse. This was a mobile obstacle, often used as roadblock, composed of a baulk of timber or a metal beam about 4 m (13 ft) long with pointed stakes protruding from the side. The stakes (often wrapped with barbed wire) acted as legs while the remainder formed an effective obstacle against infantry and cavalry. In the 1930s the Dutch produced a modern and heavy type all made of metal.

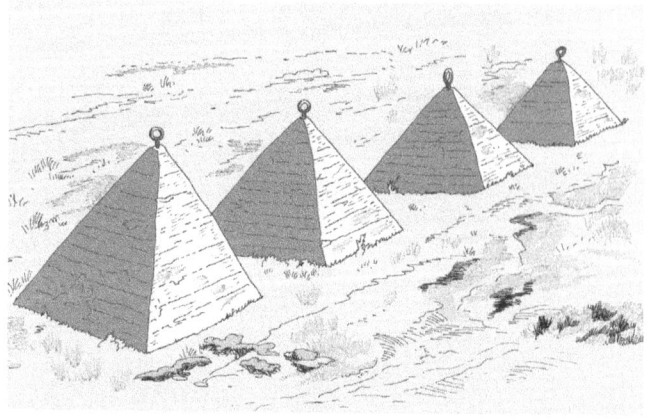

Anti-tank Pyramid blocks. Pyramid-shaped concrete blocks were common anti-tank obstacles. The metal ring placed on top of the pyramid was intended to take a hook from a crane for moving the obstacle.

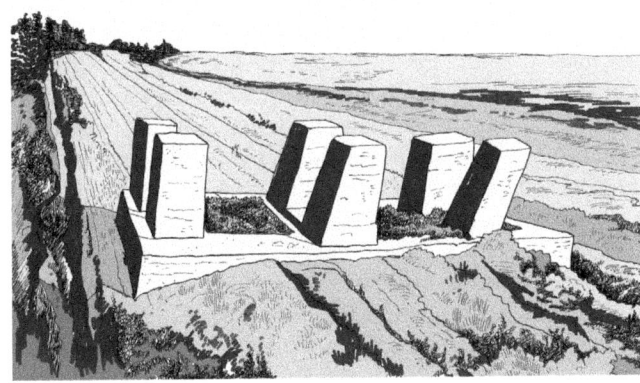

Anti-tank obstacles made of concrete blocks on the dike of Asschatterkade near Leusden in the Grebbeline.

Asparagus on railroad track. This could also hinder the progression of armored combat vehicles using the railroad track.

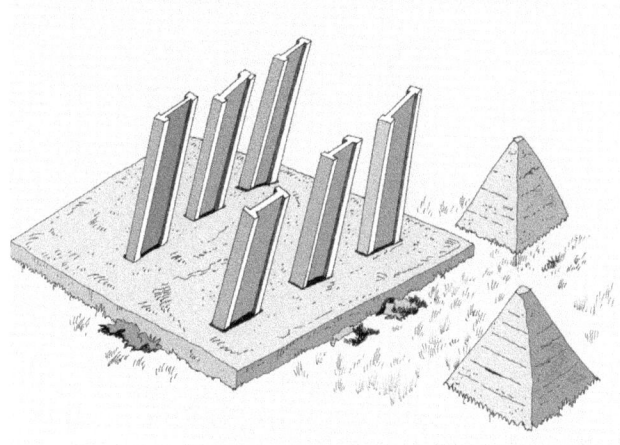

Anit-tank "Asparagus." An asparagus was an anti-tank obstacle composed of inclined strong steel beams (or pieces of rail) planted at 45° angles facing the enemy in prepared holes in thick concrete bases, sometimes with explosive caps.

Anti-armored train obstacles

At the start of the Second World War the German army was equipped with seven armored trains numbered 1 to 7. As a result of their successful diplomatic moves in 1938 and their victorious campaigns in 1939, the Germans were able to add Austrian, Czech and Polish armored trains to their own stock. In 1940 four more armored trains, numbered 21 to 24 were available. In the opening phase of World War II, German armored trains were used as offensive weapons tasked with seizing strategic positions. The surprise attack on Poland in the early morning of September 1, 1939, which started World War II, included attempts to seize key stations and facilities in the Pomeranian corridor employing armored trains. The Germans used armored trains also during the May–June 1940 offensive in the West. For example Panzerzug No. 6 rushed through the Dutch border at Nieuweschans in the northern province of Groningen but it got stuck when the Dutch blew up a rail bridge at Winschoten.

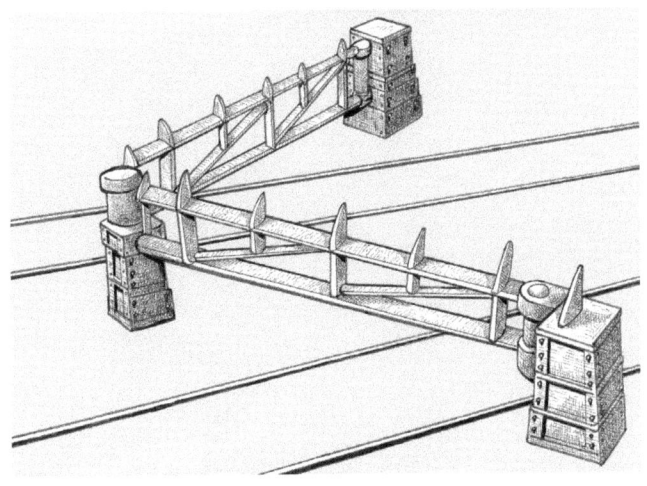

Pantserhek (Anti-train fence).

In all, ten attacks were carried out by armored trains in the campaigns of 1939–40 and only two were successful. In order to impede enemy armored trains, the Dutch had designed several obstacles including a strong openable steel fence known as *pantserhek*; and asparagus obstructions that could be placed in prepared excavations across the railway track when needed.

Assessment

If one would wish to summarize the value of the Dutch fortifications and the reason for their failure in May 1940, the following could be said.

The Dutch fortifications were founded on obsolete notions. The Dutch attempts to place fortified positions in the way of the German assault displayed an unfortunate error in judgment, both of the German power and of their own situation. The hastily established installations of the border defense position were in part not garrisoned adequately, and in part not at all, so that they could hardly be defended against the "lightning" assault of the German forces. Not one of the major defense positions could be held until French and British help would arrive. The expected help of the French and the British, on which the Netherlands had counted in planning the strength of their installations, failed. Even under the assumption that the fortifications would only have to hold until the arrival of outside aid, about 4 to 5 days, the main defense positions were too weak.

Counterattacks were limited, only consisting of weak delaying attempts, and the offensive potential of the German Wehrmacht had been greatly underestimated. The Dutch fortifications were not eased on a fundamental, and clear plan. In part for political reasons, they were much too extensive for the little country and the weak army.

The Meuse-IJssel position played no role whatever, any more than did the Peel-Raam position, which was broken through at Mill on May 10-11. The Grebbe Line held up the attack at Grebbeberg on May 12 only for a matter of hours. In the new positions there was hardly any possibility of an orderly combat leadership, since at no time did they have appropriate communications; every defensive position was on its own. The situation of the Fortress of Holland was already critical on May 10 when the German parachute troops took possession of the undamaged Moerdijk, Dordrecht, and Rotterdam bridges.

CHAPTER 9

The Second World War, 1939–1945

Defeat and Occupation

Fall Gelb

Although some senior military personnel were alert to the Nazi German threat, the Dutch were quite unprepared for a war. In the late 1930s, many works and installations in the defense lines were not yet completed, and perhaps the Dutch authorities had underestimated the military odds that Hitler was ready to unleash on them. As was the case in other Western European countries, the Dutch were long reluctant to recognize the real nature of Hitler's racist, criminal and bellicose regime. In September 1939 when Germany invaded Poland, the Dutch army was mobilized. For six months the armies on the western front remained motionless. It was the "Phony War." The German plan coded Fall Gelb (yellow case) was designed in the winter 1939–40 and relied upon an attack on the Netherlands, with a thrust through the Belgian Ardennes mountains. Its details were several times changed, and its actual execution was postponed until spring 1940 because of bad weather.

On the early morning of May 10, 1940, Nazi Germany launched the surprise attack on the neutral kingdom of the Netherlands. The rather well prepared offensive started abruptly and violently with air attacks followed by paratroopers landing on Dutch airfields and bridges weakening the Dutch defenses from the inside. The German Fallschirmjäger—both parachuted and glider-borne units—seized strategic points at The Hague, Rotterdam and Moerdijk but with high casualties. Of the 2,000 men committed to this operation, 40 percent of the officers and 28 percent of the men were killed. There were also many losses of the transport airplane trimotor Junkers Ju 52s. Of the 430 Ju 52s engaged in the offensive about two-thirds were destroyed or badly damaged. However the paratroopers' achievements justified the methods developed by Generaloberst Kurt Student (of the 7th Air Division), and greatly stimulated Hermann Göring's support and Hitler's interest in paratroopers.

At the same time the German 18th Army progressed in the direction of Deventer and Nijmegen. In the North, the German 1st Cavalry Division supported by field artillery, and Junkers Ju 87 Stukas (ground support dive-bombers) made a rapid advance through the weakly defended F-, Q- and O-lines in the provinces of Groningen. However—a fact worthy of mention—the small force defending the IJsselmeer Dam at Kornwerderzand headed by Captain Christiaan Boers managed to repulse the invaders. The garrison (totaling 224 soldiers reinforced with the motor gunboat *Hr. Ms. Johan Maurits van Nassau*) held the position, and surrendered only with the capitulation of the Dutch army on May 15, 1940. This remarkable achievement was unfortunately only local. Indeed on the other front in the south, things went badly wrong.

From the start the small Dutch army was doomed, the German forces had extensively studied the Dutch defenses and discovered several weak spots. Also the Germans simply had much more soldiers, much more armored vehicles and much more attack airplanes than the Dutch. After the German Luftwaffe (air force) had destroyed many aircraft by surprise on the ground, the German Heer (ground force) enjoyed total air supremacy. Their combined forces made use of a fast, brutal and devastating mobile tactic known as *Blitzkrieg* (lightning war). They attacked the weakest parts of the Grebbe Line with most dramatic fighting taking place at Scherpenzeel as well as on the vulnerable Grebbeberg hill at Rhenen. The German troops broke through the line on May 13, and the Dutch army retreated behind the New Dutch Waterline. The Germans bypassed the defense lines by the south but were stopped by a stubborn Dutch resistance at Ypenburg and Valkenburg.

By then Hitler had become impatient. The German advance was delayed, casualties increased, the tiny Dutch army held out longer than expected, so he decided to strike with ruthless savagery in order to break the Dutch resistance. On May 14, 1940, although negotiations were taking place for a cease-fire, the Luftwaffe launched a terror-bombing raid that destroyed most of the center of the old city of Rotterdam killing about 900 civilians and making over 80,000 people homeless. This highly controversial and criminal air attack sealed the fate of the defenders' resistance. The Dutch Army considered its strategic situation to have become hopeless and feared further destruction of such Dutch cities as Utrecht or Amsterdam. They surrendered on May 15, 1940. However, Queen Wilhelmina (1880–1962, reign from 1890 to 1948), the Royal family and the government managed to escape to Great Britain. They joined the British camp, and the legal government in exile organized from London the resistance against the Nazi occupiers.

In the meantime the Germans had moved into Belgium,

and started their successful invasion of France. They trapped the British Expeditionary Force and parts of the French army in the region of Dunkirk and moved on their final offensive into Paris. On June 21, 1940, it was all over. France was defeated, and Hitler had conquered all of Western Europe. Badly mauled Britain was left all alone to face a triumphant Nazi Germany. The new British Prime Minister Winston Churchill (1874–1965) refused to negotiate and decided to continue the hostilities.

Occupation

The defeat was followed by the occupation of the Netherlands. The anti–Semitic and racist Nazi regime considered the Germanic Dutch people as *Volksdeutsche* ("racial cousin volk" or "fellow Aryans"). So in the long term the Dutch, and the Scandinavian people would find a permanent place in the Greater German Reich. The early time of the occupation was "mild" with many prisoners of war released, and German occupiers behaving correctly toward the population. The Nazis installed a *Reichskommissar* who ruled through the existing administrative machinery. The terms of occupation and the pressure, however, became increasing rigorous as the war dragged on. The condition and standard of living gradually declined, and German plunder of the economy greatly impoverished the country. The Dutch suffered more and more under the burden and restrictions of war, and increasing German terror.

From 1941 on, the members of the Dutch Jewish community experienced the harshness of the occupation. The Jews were particularly harassed, persecuted, despoiled, and deprived of basic liberties and civil rights. Finally they were isolated, regrouped and parked in overcrowded ghettos and "transit camps," soon deported by train and murdered by gas by the hundreds of thousands in extermination camps in Poland. It should be noted that the population of Amsterdam vigorously protested against the arrest of the Jews, and went on strike in February 1941.

In April and May 1943 another massive general strike broke out when the Germans decided to re-arrest ex-soldiers of the Dutch army. Both strikes were quite unique in Nazi occupied Europe. They were quickly and ruthlessly repressed by the Germans. Two senior Nazi leaders were particularly hated by the Dutch for their ruling of the land: the Austrian Arthur Seyss-Inquart (1892–1946) who was *Reichskommissar* (Commissioner of the Reich), and SS Brigadeführer Johann Albin Rauter (1895–1949) who was *General-Kommissar für das Sicherheitswesen und der SD* (head of Security service).

Meanwhile in Asia, the Dutch East Indies (present day Indonesia) was attacked in February 1942, and occupied by the Japanese. The Nippon occupation was a disaster for the white Dutch colonists who were arrested and detained in atrocious conditions in concentration camps. As for the local native population, their nationalist and independence aspirations were ignored, repressed, and brushed aside. All the Japanese wanted was to exploit the archipelago's strategic materials including rubber, tin and more particularly the substantial oilfields.

The German Atlantic Wall, 1940–1945

Purpose

After the victorious campaign of 1940, the Germans used all Dutch prewar fortifications. However, the occupiers found them in many cases too weak, too vulnerable, and not correctly located. Besides they were not adapted to the exigencies and equipments of the Wehrmacht. So the Germans plundered and recycled everything that could be useful (weapons, and armored parts, notably the metal cupolas). In 1941 they started to construct a new system of coastal defense with their own designs.

The *Atlantikwal* (Atlantic Wall, AW in short) was an extensive system of coastal fortifications built by the German Third Reich between 1941 and 1945 along the whole western littoral of continental Europe. Extending from the French-Spanish border at Hendaye in the South to the Polar Circle in Norway in the North, it was an enormous and colossal undertaking on a continental scale intended to defend against an anticipated Allied invasion coming from Great Britain. Indeed the main purpose of those coastal fortifications was to guard the western Atlantic façade against any Allied intervention while the bulk of Germany's forces were engaged against the Soviet Union in the East.

Construction

The Atlantic Wall was not constructed in one time but successively in different phases related to the fluctuating military situation on the western and eastern fronts. From the start the Atlantic Wall was an amalgam of various projects carried out separately by the three branches of the *Wehrmacht* (German armed forces) including *Heer* (army), *Kriegsmarine* (navy), *Luftwaffe* (air force) to which must be added the *Organisation Todt* (OT)—a conglomerate of building contractors created and headed by the Nazi engineer Fritz Todt (1891–1942).

In December 1941 the USA entered the war at Britain's side. It then became clear that sooner or later the Allies would attempt an attack or even an invasion on the Atlantic façade. Adolf Hitler called for the official creation of the Atlantic Wall by issuing the secret *Führerweisung No. 40* (directive) on March 23, 1942. The British attack on the French port of Saint Nazaire (March 1942), and the failed landing at Dieppe (August 1942) convinced the Germans that a new front could be opened somewhere on the coasts of the Atlantic Ocean or of the North Sea.

In a first phase the Führer ordered naval and Kriegsmarine submarine bases to be heavily defended. Fortifications remained concentrated around important ports until late in 1943 when defenses were increased in all other

Atlantic Wall map. The "Wall" stretched from Norway along Denmark, northern Germany, the Netherlands, Belgium, and France. There was an extension, known as the South Wall, on the French south shores.

areas. This new project, launched in autumn 1943 was called *Schartenbauprogramm* (construction program for casemates). It specified that artillery positions had to be placed under concrete, and that the beaches defenses had to be reinforced.

In January 1944, Hitler, finding the situation too weak on all fronts, decided to reshape the Atlantic Wall once again. Several important seaports were designated *Festungen* (fortresses). In the Netherlands these were Vlissingen (Flushing) on Waalcheren Island, Den Helder, Hoek van Holland and IJmuiden. Another last minute building project was launched in May 1944 called *Gesamte Sommerausbauprogramm* (general summer construction program) with priority given to a newly designed generation of bunkers called *Kleinschartenstände* (series 700) which were small, cheap, but well-armed casemates. By the spring of 1944, the Atlantic Wall was not finished, as only 10,500 of the 15,000 planned bunkers were completed.

Marshal Erwin Rommel

The Atlantic Wall primarily consisted of a juxtaposition of defensive "hedgehogs," infantry positions, machine gun and mortar nests, anti-tank gun emplacements, medium and long-range coastal artillery batteries, observation posts and radar stations. In addition the Germans held several armored divisions as mobile strategic reserve.

Early in 1944, General Field Marshal Erwin J.E. Rommel (1891–1944) was assigned to inspect and improve the Wall's defenses. Under his direction, the number of bunkers and general strength of the Atlantic Wall were increased by additional reinforced concrete pillboxes built along the beaches to house machine guns, anti-tank guns, and anti-aircraft cannons, as well as light and heavy artillery intended to destroy any fleet offshore supporting a landing.

Minefields and a multitude of obstacles (anti-personnel, anti-tank, anti-landing craft) were established on the beaches themselves, and underwater obstacles and mines were placed in waters off shore. More gun emplacements and minefields extended inland, along roads leading away from the beaches. In likely landing spots for gliders and parachutists, the Germans placed mined poles known as asparagus. Low-lying zones, marshes, swampy grounds and estuaries were flooded. The much celebrated and decorated Field Marshall Edwin Rommel firmly believed that Germany would inevitably lose the war unless the Allied landing could be repulsed on the beaches.

The Atlantic Wall was never completed, and in reality, the much-vaunted Atlantic Wall, against which the invading Allies were intended to dash themselves into ruins, was only a figment of Hitler's imagination. However, owing to Marshal Erwin Rommel's frantic efforts to create a strong defense system, the Atlantic Wall was relatively sturdy along the Dutch, Belgian and northern French coasts facing the English Channel and the North Sea. Ultimately it was only in the Pas-de-Calais (the northern part of France facing the Strait of Dover), and in the occupied British Channel Islands that the "Wall" existed in anything like its intended form. Those heavily fortified sections were naturally avoided and instead the Allies chose to land in the less strongly defended Normandy. The Atlantic Wall did not stop the invasion on June 6, 1944, which marked the start of the liberation of Europe from the Nazi yoke.

As for Rommel who seemingly was implicated in the failed 20 July 1944 plot to assassinate Hitler, he was forced to commit suicide. Due to his status as a national hero, Hitler eliminated him quietly, leaving his reputation intact and his family unharmed.

Structure and composition of the German Atlantic Wall

The Atlantic Wall was completely heterogenic. It displayed many differences depending on many factors: strategic value of the site (distance between England and the supposed landing points as well as presence of important harbors); geographical specifics (tide, currents, reefs, shoals, cliffs or shores suitable for a landing, hinterlands favorable to movement warfare); symbolical and psychological importance of some occupied places such as the British Channel Islands for example.

The Atlantic Wall was divided into *Heeresgruppen* (HGr, army groups), which was composed of *Armeeoberkommando* (AOK, army command). Every AOK was divided into *Armeekorps* (AK, army corps). Every AK occupied a section of shore called *Küstenverteidigungabschnit* (KVA, defensive coastal sectors). KVAs were divided into *Unterabschnitten* (UA, sub-sectors); some KVAs with long coastlines were further subdivided into *Küsten Verteidigung Gruppe* (KVGs coastal defense group).

Actually the "Atlantic Wall" was not an accurate name. It covered not only the western Atlantic Ocean coasts in France but also the English Channel and the North Sea along Belgium, the Netherlands, Denmark and Norway. It also had an extension in southern France called *Südwall* (South Wall) intended to repulse a landing on the French Riviera and Languedoc Mediterranean shores. The term "wall" suggested a continuous barrier, an uninterrupted fence or a defensive line, but the German coastal defenses were constituted of separate fortified positions, varying in density of implantation, size, importance, firepower and strength. From September 1942 onward the terminology of these fortified positions was standardized, and the theory of defense was further determined with the following strongpoints.

Widerstandsnest

A *Widerstandsnest* (in short W or WN or Wn, resistance nest) was the smallest unit, a small Army fortified post, called *defense post* by the Americans. WNs were generally placed on heights, or on top of dunes overlooking beaches. In the first period of development of the Atlantic Wall, these small posts were usually occupied by a well-armed unit of platoon size equipped with light infantry weapons; these were protected by field fortifications, excavations reinforced by sand-bags, earth entrenchments, temporary troop shelters and rudimentary facilities. These positions were often built in a makeshift fashion by the troops themselves. Elements were camouflaged and defended by barbed wires and mines.

Progressively, when the effectiveness of the Atlantic Wall was developed, WNs were given a more permanent character, and firepower was increased. Howitzers, field and anti-tank guns were placed in bunkers designed for the Siegfried Line (series type 100 and 500), later in new modern designs (standardized bunker series 600). Observatories were placed, barbed wire networks were widened out, anti-tank ditches were dug, mine-fields were installed, beach obstacles were erected, concrete shelters were built for men, material and ammunitions.

A Wn had in theory enough supplies to hold out for one week. Not all Wns were permanently manned by a full garrison though. For example Wn 212H located at De Punt on Goeree-Overflakkee (the southernmost delta island of the province of South Holland, Netherlands), was intended to be occupied by 60 men, but actually there were only 18 soldiers permanently occupying the 15 bunkers. In case of alarm reinforcement would come to Wn 212H from the local headquarters located in the village of Ouddorp.

Schematic view of a WN. WNs and StPs were enclosed defensive positions combining concrete shelters and bunkers armed with artillery and infantry weapons. The beach in front of the position was protected by obstacles, barbed wires and mines (1). Flanking casemates (2) were placed sidewise to enfilade the beach. Infantry guns or anti-aircraft guns were placed in open emplacements (3). Firing was conducted from an observatory (4). The garrison could stand by and find refuge in shelters (5). Ammunitions were protected in an underground store (6). The rear and the sides of the position were defended by trenches (7), reinforced by MG and mortar nests, Tobruks (8), barbed wires and minefields (9).

Stützpunkt (StP)

In March 1942, several seaports were declared *Stützpunkten* (support or strong points, in short StP). Those harbors had to have defenses against aggression from sky, sea or land. Therefore important works of fortification were undertaken. Later on they were declared *Verteidigungsbereiche* (VB defensive areas). Curiously, the term *Stützpunkt* did not disappear with this new creation. The StP became—in theory at least—the reunion of two or more Widerstandsnester. In practice, however, Wns and StPs became very flexible units widely used in the Atlantic Wall. Actually a small StP could be as strong as a large Wn. The surface—enclosed by barbed wires and mines—could vary a lot. An StP could be occupied by a unit of battalion strength, or less, or more. Armament was also variable: for example a Stützpunkt could receive army units, a navy coastal battery or Lufwaffe Flak guns. Even a radar station was possible. It all depended on the strategic importance of the site.

StPs and WNs were generally placed along shores, on points and capes, by a river mouth or near a harbor. In the hinterland they defended important crossroads, hubs, local industrial sites, power plants or airfields for example.

Where flat and low grounds were encountered, the Germans carried out flooding. That was the case in certain parts of France, notably in Normandy, and particularly in the provinces of Zeeland and Holland in the Netherlands.

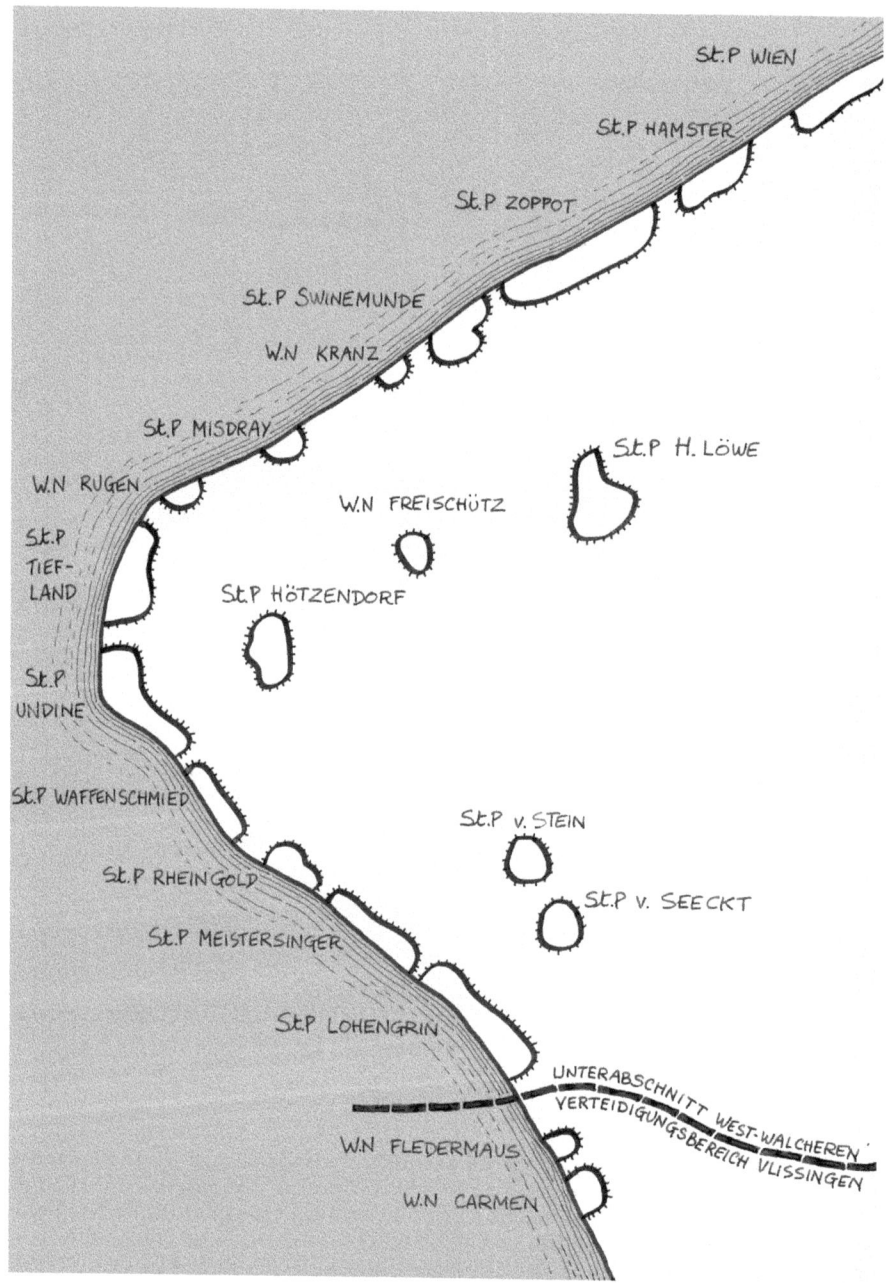

Map of StPs and Wns in the sector West Walcheren Island in Zeeland in March 1943.

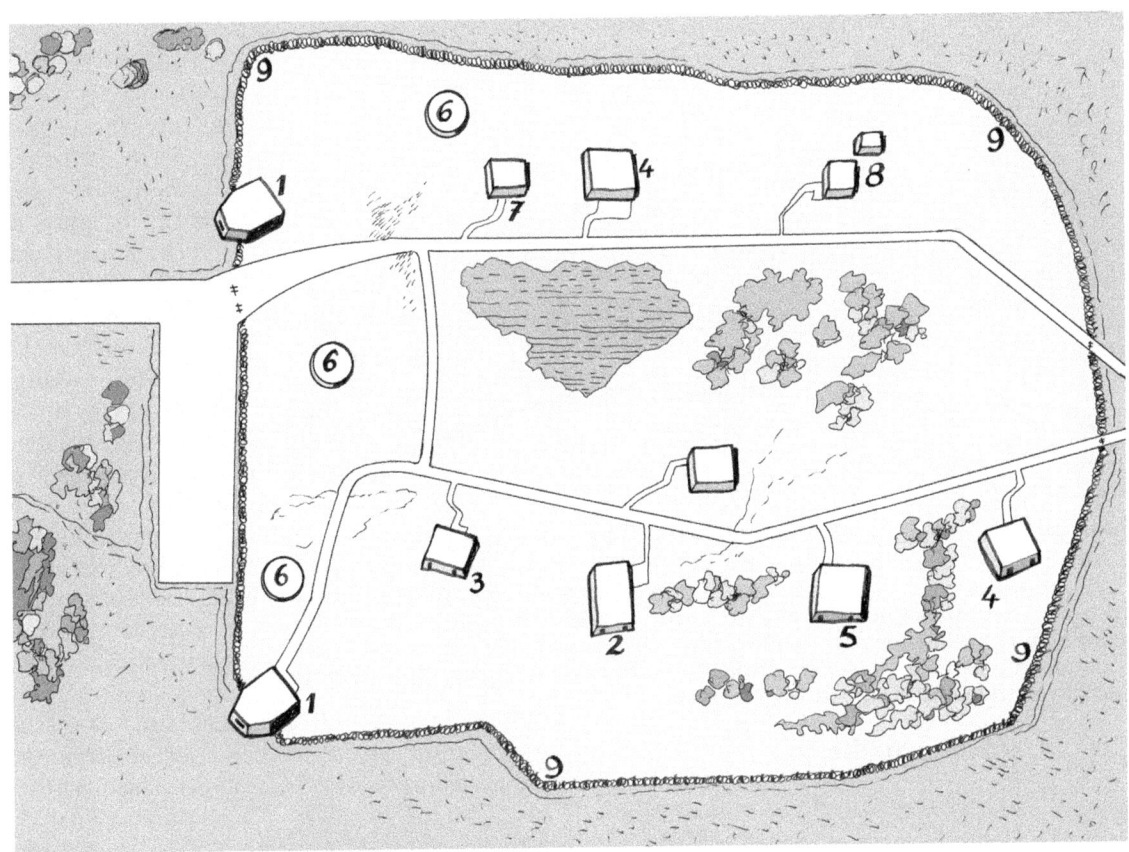

StP Groede. The German strongpoint was situated north of the village of Groede near Sluis, about 5 km north of Oostbug in the province of Zeeland. Established in early 1943, StP Groede was armed with four captured Czech 10.5 cm K 53 (t) field guns. The StP was manned by the 3th battery of the 652nd battalion (III/652) of the 712nd infantry division until August 1944. After this time the battery was occupied by units of the Grenadier Regiment No 1037 of the 64th infantry division. The StP/battery included the following: 1: Artillery casemate type 669. 2: Troop shelter type 502. 3: Troop shelter type 501. 4: Ammunition bunker type 134. 5: Infirmary type 134 S. 6: Open artillery emplacement. 7: Infirmary type Vf 57a. 8: Ammunition bunker type Vf 52a. 9: Barbed wire fence. The battery was placed in open ground and much attention was paid to camouflage. Actually it was disguised as a small village with bunkers concealed as houses, barns, shops, pubs, etc. The garrison of StP Groede (2 senior officers, 15 officers and NCOs, and 64 soldiers) surrendered without a fight to the 7th Canadian Infantry Brigade on October 25, 1944. Today the former German StP serves a peaceful purpose: it is arranged as a natural forested park for deer and other wild animals.

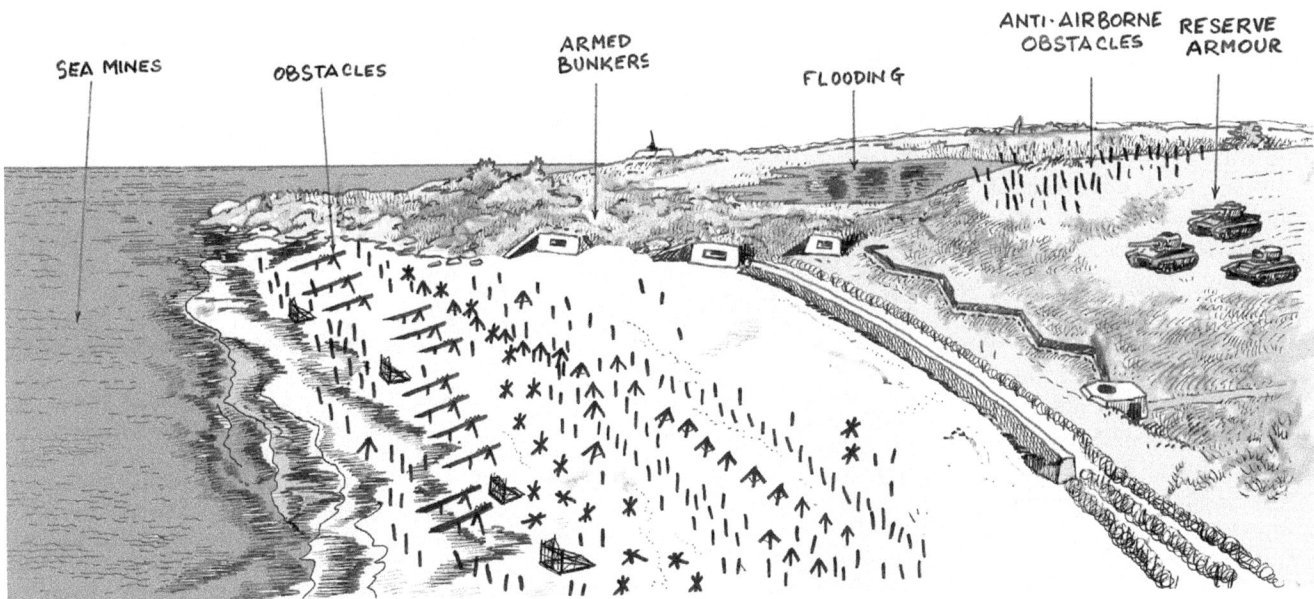

Coastal defense according to Marshal Erwin Rommel, combining mines, obstacles, fortifications and mobile intervention reserve troops.

Bunker type 623. The flanking MG-casemate (MG Schartenstände) type 623 was issued in April 1942 in order to replace the older Westwall bunker model Regelbau 105. Type 623 measured 11.50 m in length (flanking shield not included), 9.60 m in width and 5.10 m in height. It included a crew chamber for one NCO and five men, and a firing chamber for one machine gun with a porthole protected by a 6 cm thick armored plate. The Germans built 17 bunkers type 623 most of them in France and the Netherlands (e.g., at Vlissingen).

Coastal artillery

The coastal batteries formed the core of the German Atlantic Wall. The main purpose of coastal artillery was to break up an Allied invasion fleet before it reached the inshore approaches, by firing upon the ships supporting either an amphibious invasion or a raid. However the quality of the weapons was actually a weak point. Whereas the bunkers of the Atlantic Wall were standardized *Regelbau*, the artillery was not. It included 61 models of all calibers and ca. 4,000 captured foreign artillery pieces, which created enormous problems for finding spare parts and of course appropriate ammunitions.

In the spring of 1944, about 700 heavy army and navy coastal batteries were deployed in the *Atlantikwall*. Beneath millions of tons of ferro-concrete, German long-range guns seemed impressive, but they were prime targets for Allied air and naval forces before and during D-Day. On the whole their role was poor and they did not succeed in repulsing the Allied invaders in June 1944.

Coastal batteries were generally arranged within an StP (stronghold) or part of a StPGr (group of strongholds).

The Germans distinguished two kinds of long-range coastal artillery: those manned by the navy (called, in German, *Marineküstenartillerie*, MKA in short) and the army ones (designated *Heeresküstenartillerie* aka HKA). The general structure of all coastal batteries (both MKA and HKA) was rather similar. It included a firing control post giving instructions and orders to four, six or even ten guns placed in various types of open emplacements or roofed bunkers. Various shelters, ammunitions stores and service facilities were installed behind the firing line.

Both Army and Navy heavy coastal artillery also included long-range ordnance fixed on special railway carriages. This heavy railway artillery, called *Eisenbahnartillerie*, could be either mobile or established as permanent batteries. In this last case, the guns were placed in huge rotative concrete pits allowing a 360° traverse.

View of a navy battery (MKB). A navy coastal battery was composed of several works grouped together in a functional complex. The center of the unit was the fire directing post, called Feuerleitstand (1); its crew calculated and coordinated four (sometimes six) guns sheltered in bunkers (2) placed in line facing the sea. Behind those active elements, other bunkers were scattered: ammunitions stores (3), gunners shelters (4) as well as various service and logistical buildings, latrines, canteen, kitchen, stores and garages (5). The battery was arranged as a Stützpunkt with barbed wire fences (6), mines, beach obstacles (7), anti-tank ditch (8), trenches, and army flanking pill-boxes (9). The number and quality of the bunkers, firepower of the guns, the density of the batteries depended of the strategical importance of the site.

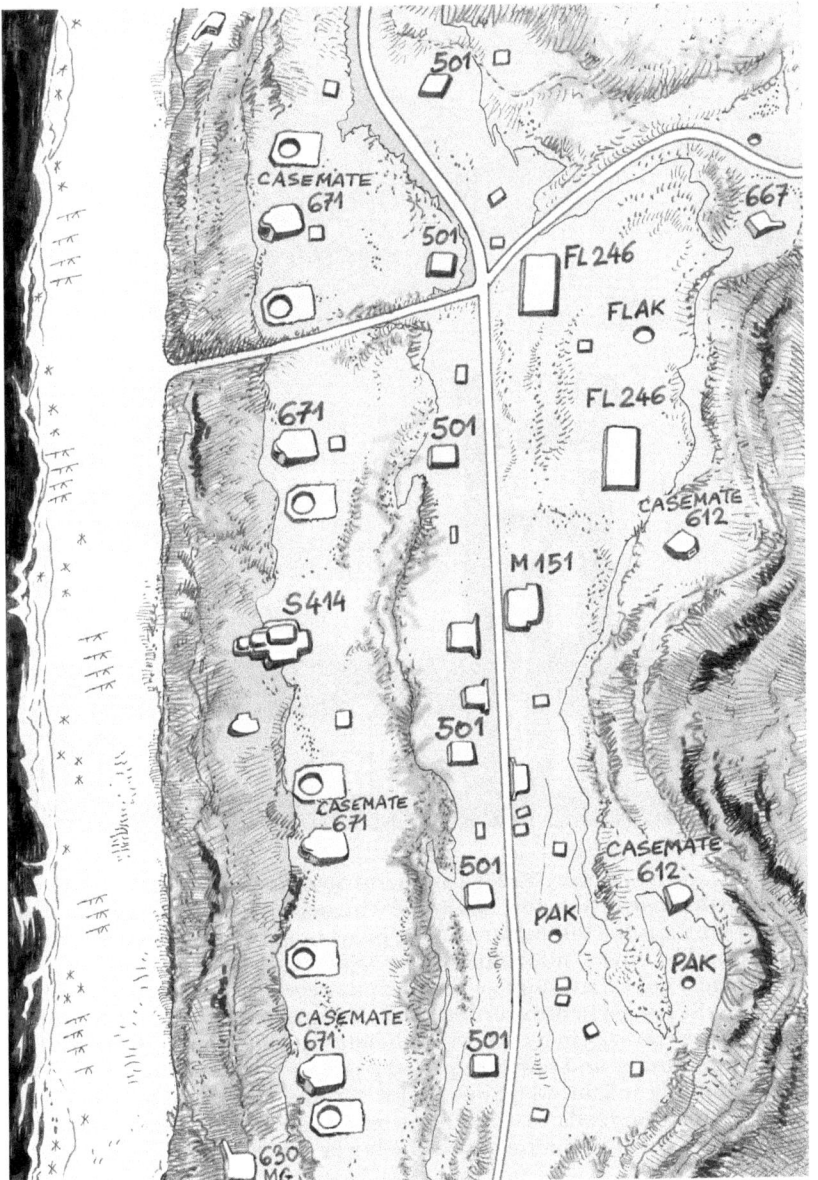

Top: Leitstand type M178. Issued in April 1943, the Leitstand M178 was designed for heavy and medium Navy coastal batteries. Length of the bunker was 22.80 m, width was 14.80 m, height was 9.80 m and mass of concrete needed was 2,300 cubic meter. Four Leitstände M178 were built notably at Pointe Saint-Gildas near Saint-Nazaire (France) and at MK battery Heerenduin south of IJmuiden (Netherlands). This Seezielbatterie (Navy coastal battery) was installed in WN 81 and included four heavy guns placed inside concrete casemate model M272. Firing leading stations were equipped with modern optical instruments (binoculars, telescopes, range-finders) and included rooms for preparing and controlling coastal artillery fire. The observation rooms were characterized by slits, which were horizontally very wide for a broad panoramic view and vertically very narrow for good protection. In addition, Leitstände always included calculation-rooms and communication centrals as well as various chambers for troops, officers, ventilation, heating and so on. The Kriegsmarine had a long experience of firing at sea target, and had numerous types of Leitstände. Medium range battery Leitstände were characterized by one observation room while long range Leitstände were fitted with two. The German navy also designed non-standardized Leitstände and impressive high Sonderkonstruktion (SK special construction) concrete towers.

Bottom: MKB coastal battery Scheveningen-North. This Navy battery was part of StPGr. Scheveningen located near the Hague in the province of South Holland. Interestingly it was composed of mixed Navy and Army bunkers. The battery included a Navy fire-directing station type S414 and open gun emplacements, which were later replaced with Army roofed casemates type 671. Ammunition was stored in two large Navy Flak bunkers (Fl246) and gunners were sheltered in Army bunkers (type M151 and 501). The close range defense of the battery included various armed Tobruk pits and Army anti-tank and MG bunkers (type 630, 612 and 667).

FlaK artillery

Against airplanes, the Germans had a specialist artillery branch called *FlaK* (short for *Flugzeugabwehrkanone*) controlled by the Luftwaffe (air force). An important radar system and numerous FlaK artillery batteries were deployed in the Netherlands to detect and combat Allied bombers on their way to Germany. A FlaK battery could be mobile or permanently established. In this case, it generally included one or more fire-directing posts, a variable number of medium and heavy guns, searchlight units supplied by power plants, various shelters for men, material and ammunitions. The FlaK fire-directing post was connected to local radar stations and regional command centers. Guns were always placed in pits. Obviously the vertical nature of FlaK firing was not compatible with roofed casemates.

Making matter more complicated, the Heer had its own divisional AA artillery. The Kriegsmarine, too, had its own FlaK called *Marineflakabteilung* (shortened as MaFla). The MaFla was intended to protect the air space above harbors, navy installations and submarine bases. Navy FlaK batteries (called *Marineflakbatterieen*, short as MFLB) were placed on ships or permanently installed as ground units. A MFLB was quite similar to a standard Luftwaffe FlaK battery. It was generally composed of a fire control station, various concrete open gun-platforms, ammunitions store, a garage and platform for a mobile radar, one or more shelters for searchlights, and a power plant. The battery was arranged as a Stützpunkt (StP) and fenced with barbed wires. Guns, searchlights, measuring instruments, range finders and radar were the same for all branches of the Wehrmacht, but each arm designed its own specific bunkers.

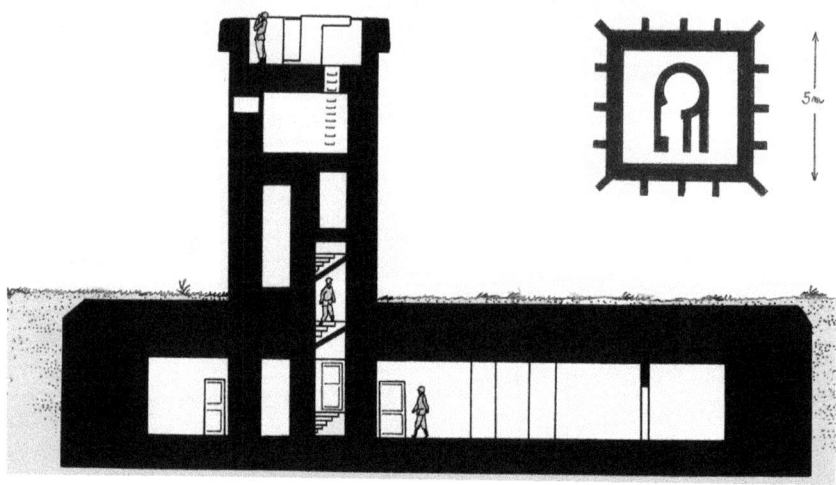

FL250 Flak command post (Flakgruppenkommandostand). The Navy MaFla command posts (called Flakgruppenkommandostände or Flakgruko in short) coordinated the defense in a sector. Equipped with sophisticated calculation and communication instruments, they connected radar-stations to Flak-batteries. The heart of the Flakgruko was a room where data (strength, speed, height, direction followed by enemy aircrafts) were projected on vertical glass maps in order to determine which targets were going to be attacked. When supposition became certainty, alarm was given to the menaced sectors where Flak batteries were hurriedly issued fighting orders. The Flakgruko FL250 was introduced in March 1942. The work was crowned by an observation tower, 5 m by 5 m with a height varying from 10 to 20 m. The bunker length was 25 m, width was 23.80 m and both bunker and tower required a concrete volume of 2,500 cubic meter. Two Flakgrukos FL250 were built in France (notably at Saint-Marc near Saint-Nazaire), three in Denmark, and four in the Netherlands (notably at Den Helder and Hoek van Holland); another was situated in Walcheren Island in Zeeland, and comprised two towers and an observatory. The illustration shows the tower (*top*), and the cross-section (*bottom*) of the bunker.

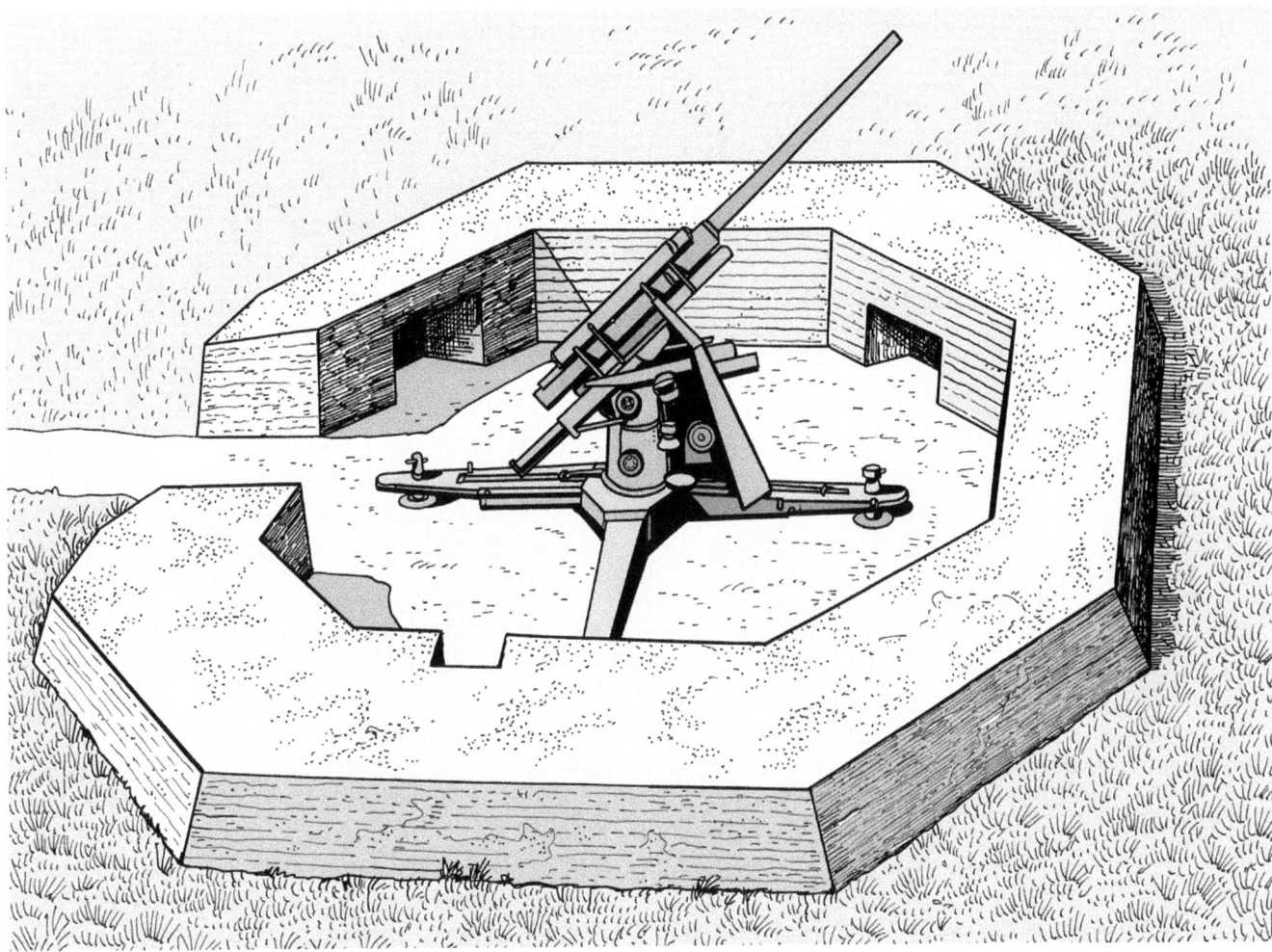

FlaK emplacement. The armed AA emplacements were called Flak-Kampfstände. FlaK guns were always placed in a Bettung or a Flakstellung (open pit) for one obvious reason: anti-aircraft fire is impossible from a roofed casemate. The Flak emplacements were therefore vulnerable but they enabled what FlaK guns needed: a full 360 degree arc of fire and a rapid rotation. Rather few navy FlaK armed emplacements were designed. Most guns were placed in non-standardized Vf (field fortification) pits. It should be noted that the German Flak artillery could be engaged as much against aircrafts as against naval and ground targets.

Stützpunktgruppe (StPGr)

An *Stützpunktgruppe* (in short StPGr) was a group of StP (strongholds). In certain strategically important sectors, all existing units were regrouped to form a powerful and continuous position with a length of several kilometers along the shore. The StPGr generally constituted a strong multi-arm position including army StPs and WNs, but also navy, army and air force coastal or anti-aircraft batteries, radar stations, command posts, military hospitals, numerous stores and everything needed to withstand a siege. An StPGr generally maintained enough supplies for resisting about two weeks.

According to Hitler's orders they were to be tenaciously defended in the event of enemy penetration, act as a reserve of defense and—should the enemy break through—serve as hinges and corner stones and positions in the main line of battle from which counterattacks could be launched—at least that was Hitler's theory. The status of *Stützpunktgruppe* was usually given to secondarily important harbors.

Verteidigungsbereich (VB)

After the repulsed attack at Dieppe in August 1942, the German high command was convinced that the most important European harbors would become priority objectives when the Allied invasion happened. Den Helder, IJmuiden, Hoek van Holland and Vlissingen in the Low Countries, Ostend in Belgium, Boulogne, Le Havre, Cherbourg, Brest, Lorient, Saint-Nazaire, Royan, La Rochelle and Bordeaux in France were then declared *Verteidigungsbereich* (VB), which meant defensive area. These places, soon to have the status of *Festung* (fortress), received full priority regarding garrison, weapons and fortifications. A typical VB included a large perimeter, comprising a variety of strongholds (Wn, StP and StPGr) linked together with various barriers, minefields and obstacles.

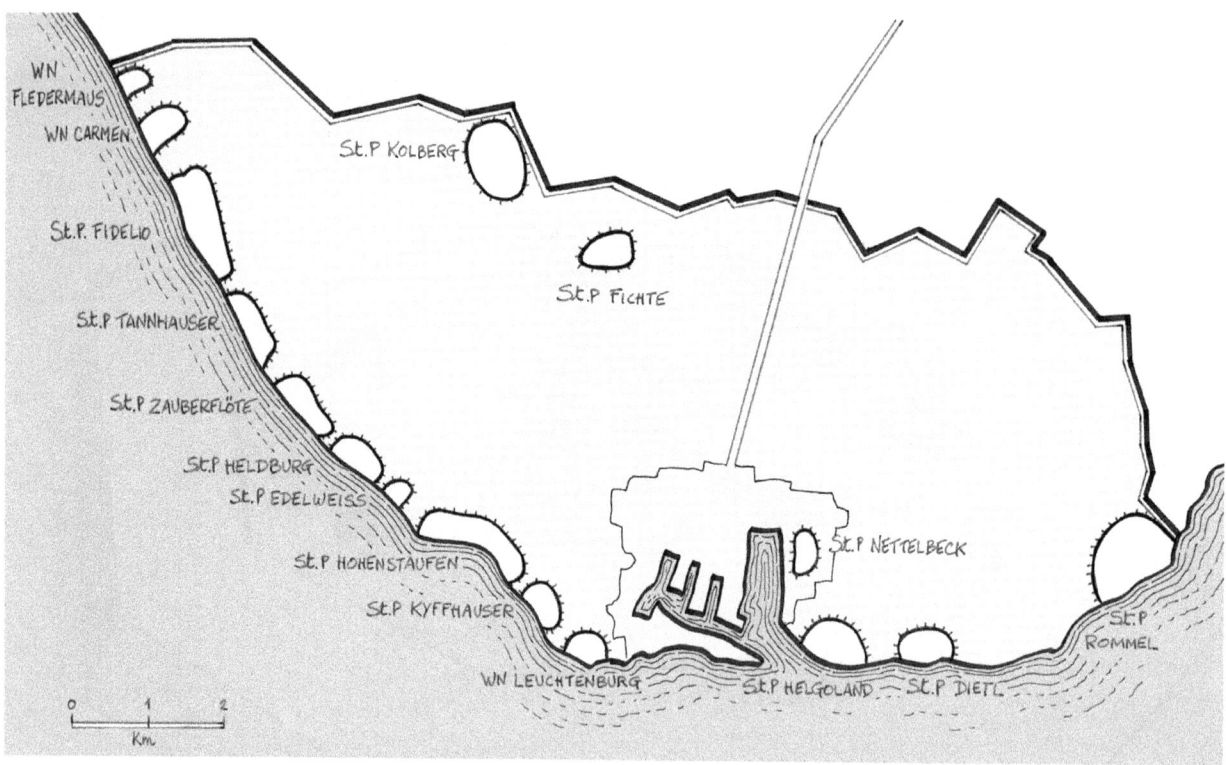

Verteidigungsbereich Vlissingen in March 1943. Vlissingen is situated in the south of the Walcheren island in Zeeland. The Verteidigungsbereich was intended, together with StPGr Breskens, to defend the access to the Belgian harbor of Antwerp.

Festung (F)

As the Allied threat increased, Hitler decided to restructure and complete the Atlantic Wall. In January 1944, he ordered the creation of the so-called *Festungen* (fortresses). This measure affected Saint-Malo, the British Channel Islands, Le Verdon and all *Verteidigungsbereiche* (VB) listed above. In the Netherlands that concerned VB Den Helder (Dutch military port and naval arsenal); Festung IJmuiden (protecting the access to Amsterdam); Festung Hoek van Holland (defending the canal to the port of Rotterdam); and Festung isle of Walcheren and VB Vlissingen defending the access to the Belgian port of Anwerp.

The *Festungen* were given total priority in manpower, means and weapons. They were to be impregnable citadels. Hitler had theorized that no invasion could succeed unless a major seaport was captured to lend logistical support. A *Festung* was supposed to have at least a garrison of two divisions, and ammunitions, food and supply stocks to support all of its positions for three months. Some Festungen (e.g., Hoek van Holland) were fitted with a Kernwerk. A *Kernwerk* was a réduit (redoubt), a kind of citadel within the fortress having the same function as the keep in a medieval castle. It formed the last entrenched nucleus, the ultimate place of resistance where combat still could be continued, even when the rest of the Festung was taken.

Although the fortress program was not completed in June 1944, some *Festungen* however formed powerful hedgehogs. They were the most formidable strong points in the Atlantic Wall. After the victorious D-Day, the Battle of Normandy and the breakout in France in August 1944, Hitler's Festungen—referred to as *Atlantic pockets* by the Allies—were isolated, besieged or taken one by one. Some surrendered like Ostend in Belgium, where no resistance was given, and the harbor installations were only slightly damaged. Others were taken after hopeless combats and pointless demolitions.

Zeebrugge in northern Belgium and Breskens in south Netherlands—both key positions defending the mouth of the Scheldt leading to Antwerp—were defended by 14,000 men commanded by General Eberding who held out until November 1, 1944. The Scheldt Festung, including South Beveland and Walcheren Island defending the access to the port of Antwerp, was garrisoned by about 8,000 men commanded by General Dasser who faced a fierce and desperate resistance during the so-called Battle of the Scheldt from October to November 1944.

Bunker design

The bunkers of the Atlantic Wall were ordered by the OKW (Supreme Command Staff), and designed by the German engineering corps. Construction was carried out by the Nazi related Organisation Todt (OT). As can be easily imagined the task was so huge, and the deadlines so short that the OT could not do it on its own. Therefore they subcontracted many worksites to local civilian building companies.

The construction of the Atlantic Wall was a fruitful business, and many subcontracted entrepreneurs made huge, rapid and scandalous fortunes. The work force included voluntary workers (lured by good wages), but also many forced laborers and over-exploited political detainees, and prisoners of war who were ruthlessly treated. Many of them died from exhaustion, exposure, malnutrition and mistreatment.

Regelbau

The bunkers of the Atlantic Wall were *Regelbau* (standardized construction), allowing precise planning of required materials, manpower, and time of construction. Regelbau types existed in sub-models of bunkers with variants, options, and prefabrication of most elements (e.g., armored doors and cupolas, furniture, ventilation, communication, and observation devices).

Concrete thickness

German fortifications were classified in several categories according to standardized thicknesses. *Feldmäßiger Ausbau* (shortened as F) was the name given to semi-permanent field fortifications made of earth reinforced with sandbags, wooden planks and beams (trenches, machine gun nests or simple artillery emplacements). It protected only against small projectiles and shell splinters. *Verstärkt Feldmäßiger Ausbau* (shortened as Vf) was reinforced field fortification. This category included masonry and concrete defense works divided into several sub-categories (D, C, B-1, and B-2) enabling to resist up to 15.5 cm projectiles.

The more elaborate *Ständig Ausbau* (permanent construction, *St* in short) included all concrete works able to resist heavy bombardment. This was divided into four subcategories.

(1) *Baustärke E* was 5 meters thick. Very expensive, this was reserved for exceptional bunkers, e.g., U-Boote shelters, and important headquarters.

(2) *Baustärke A (StA)* was 3.50 meters thick, making the expensive A-works capable of resisting up to 1,800 kg (4000 pound) aerial bombs and 52 cm (20 inch) artillery rounds.

(3) *Baustärkte A-1* was the non-frequent intermediary thickness of 2.50 meters of concrete.

(4) *Baustärke B (StB)* was the standardized 2 meters concrete thickness, designed to resist up to 500 kg (1,100 pound) aerial bombs and 22 cm (8.5 inch) artillery fire. Most of the Atlantic Wall bunkers were constructed in *StB*.

Some bunkers could not be built according to a standard model. In this case a unique work was constructed adapted to the very situation in a special category called

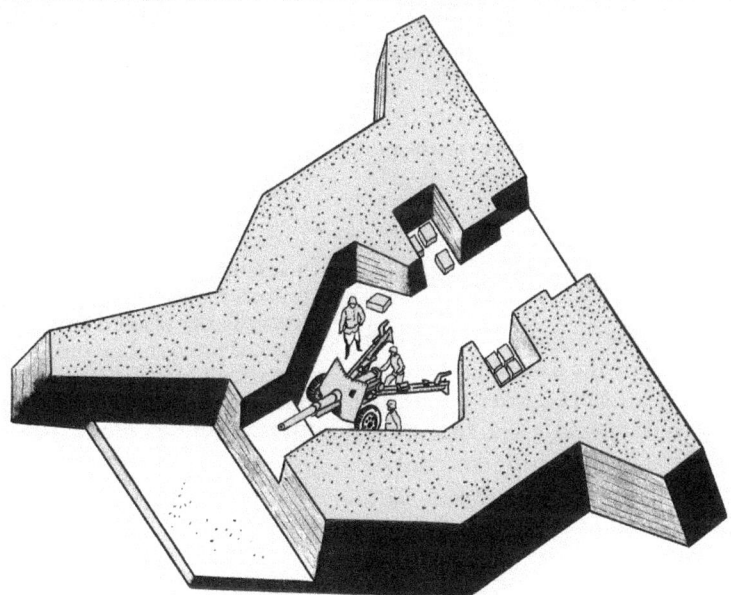

Artillery PaK casemate 680. *Top:* Front view: *Bottom:* Plan. The Schartenstand für 7.5 cm PaK 40 ohne Nebenraüme type 680 was a simple casemate for housing an anti-tank field gun. Issued in March 1943, it was a comparatively cheap artillery emplacement including only a firing-chamber and two small lateral ammunition stores each containing 300 shells. Length was 9 m, width was 8.80 m and construction required 440 cubic meters of concrete, 22 tons of reinforcement rebar and 3.3 tons of metal plates for ceilings. There were 212 type 680 bunkers built. Specimens are to be seen at Soulac and Lacanau-Plage (France). Exceptionally a casemate type 680 could be fitted with a flanking wall (placed where needed to the right or left) to protect the embrasure, as still can be seen today on the beach of Saintes-Maries-de-la-Mer in the Camargue (southern France), and at Vlissingen in the Netherlands.

Sonderkonstruktion (exceptional construction, *SK* in short). Just like the term indicates, SK bunkers were out of the ordinary and reserved to very strategically important parts of the Atlantic Wall such as *Verteidigungsbereiche* (fortified zones), *Festungen* (Fortresses) or *Kernwerke* (redoubts) and other locations, which were regarded as unsuitable for normal Regelbau.

The Germans had also various standardized *Panzer* (armor) parts for their bunkers, which included doors, gas proof doors, protective plates and shutters, observation cloches and armed turrets. Each item was standardized and given a code for identifying year of issue and type of part.

All pre-war strategically placed fortifications, which had not been sabotaged, destroyed or damaged during the 1940 campaign were re-utilized by the Germans.

In the Netherlands instances show German Vf troop-shelters added to Dutch 1937 Stekelvarken combat casemates. The bastioned Fort Frederick-Hendrik—situated on the left bank of the western Scheldt in Breskens in the Netherlands—was built by the French in 1811 and named Fort Imperial. After Napoléon I's defeat in 1815, the fort was called fort Frederik-Hendrik. After 1940, it was used by the Germans as support to a navy coastal battery (MKB Breskens) defending the approaches to the port of Antwerp.

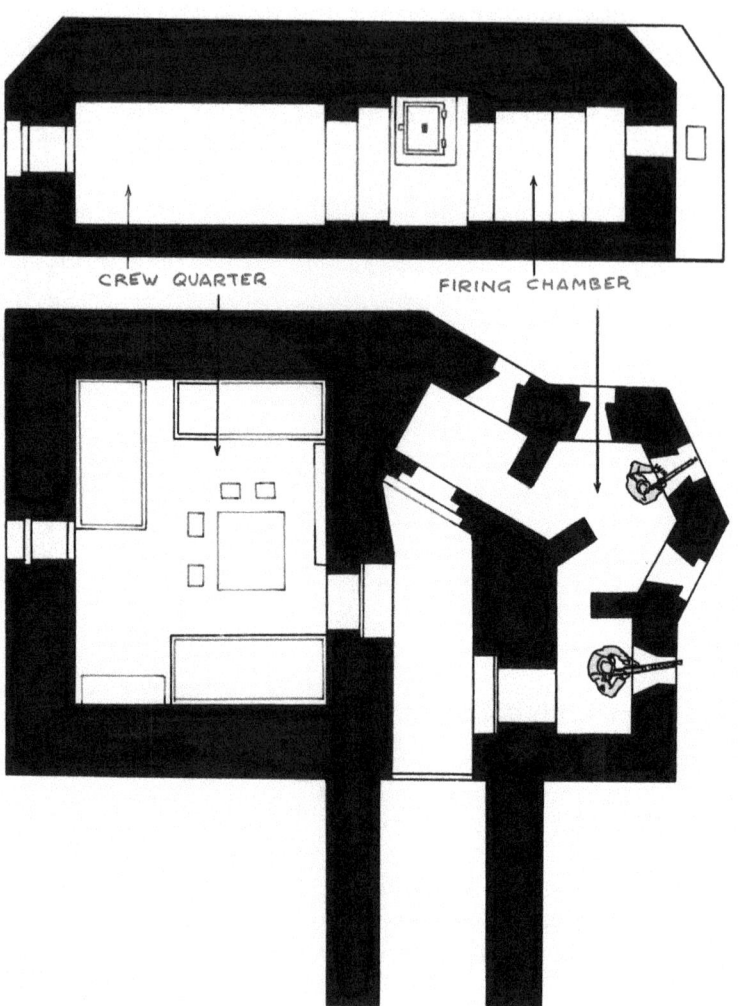

Dutch 1937 S-5 infantry casemate with added German Vf troop shelter. Cross-section (*top*), and ground plan (*bottom*). All pre-war strategically placed Dutch fortifications, which had not been sabotaged, destroyed or damaged during the May 1940 campaign were re-utilized by the Germans. For example they added an 80 cm thick Vf troop-shelter to an existing Dutch Stekelvarken five loop-holed infantry pillbox model 1937.

Bunker internal arrangement

There were a wide range of Regelbau bunkers, but many shared the following standardized features. The access to a bunker was usually a path or a trench leading to an entrance always placed at the rear of the work. The entrance was usually defended by a jutting out *Nähkampfraum* (a caponier) or a small Tobruk-like open pit called *Offener Beobachter* (open observatory).

After a few steps down and a small and narrow bent corridor, there was the *Gasschleuse* (gas lock)—a small chamber closed with heavy armored and airtight doors isolating the inside of the bunker from its environment.

Then there was a *Bereitschaftsraum* (stand-by room or troop chamber) for a various number of soldiers. This room was quite sober and rather cramped but it was lit, heated and ventilated, it was fitted with air filters, standardized furnitures, folding bunks and a telephone. In bunkers featuring only one entrance there was a *Notausgang* (emergency exit), in the form of a small crawling gallery pierced through the wall, leading to a pit fitted with climbing rungs emerging outside the bunker.

In bunkers designed for active combat service there was a *Kampfraum* (firing chamber pierced with an embrasure) whose dimensions and arrangement matched the weapon used: machine gun, mortar, anti-tank gun, field howitzer, fortress cannon, or large heavy ordnance. The combat chamber could also be a steel armored turret housing a machine gun or a mortar. Next to the combat chamber there was often a *Munitionsraum* (ammunition store). For anti-aircraft fire the typical combat emplacement was a *Bettung* (an open, roofless platform).

Large bunkers (e.g., field hospitals, command stations, fire control posts, and observatories) could feature larger quarters, a ventilation room, a computing room, store places, observatories, a powerplant, and other specialized rooms. Of course all bunkers and related semi-permanent field fortifications around them were camouflaged, and sunk into the ground or in the dunes bordering the beaches.

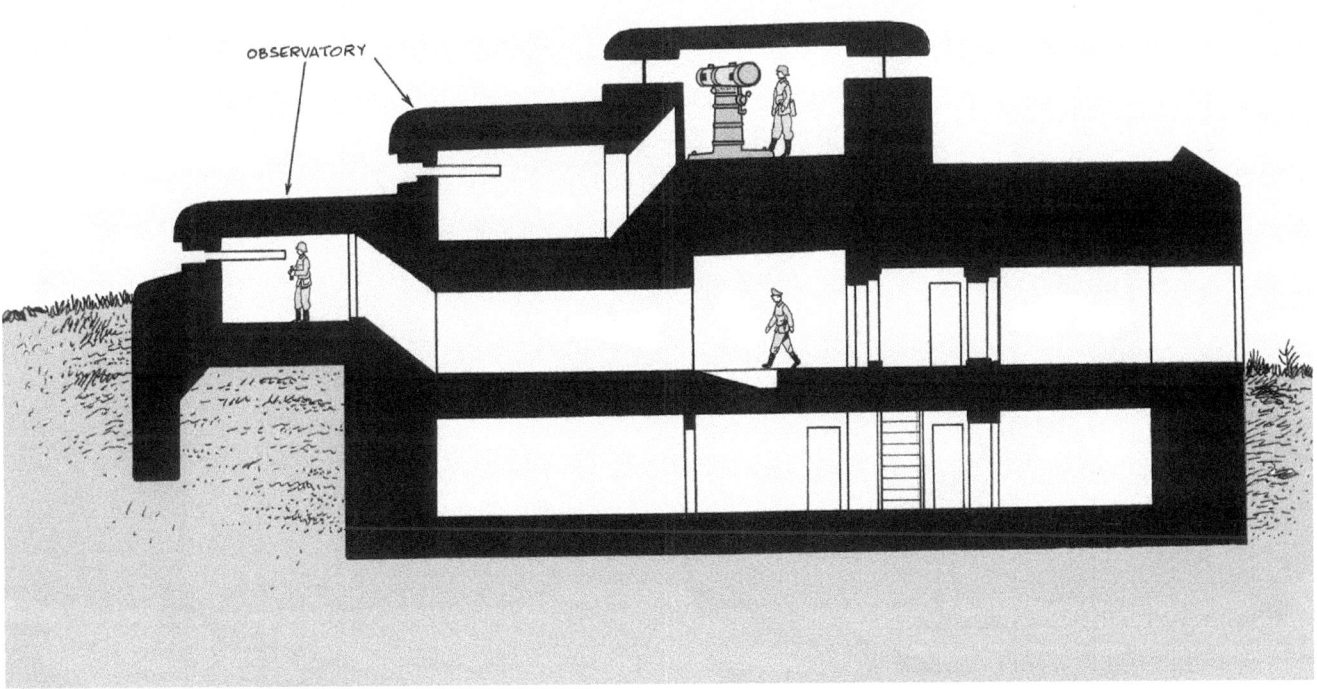

Leitstand S414. *Top:* View. *Bottom:* Cross-section. The navy fire leading station S414, issued in March 1943, was a large three-story work designed for heavy navy coastal batteries. Length was 25.60 m, width was 15.10 m and height was 12 m. The mass of concrete needed was 1,800 cubic meters with 90 tons of steel reinforcement rebars. It featured quarters for the crew, calculation and computing rooms, a telephone exchange, observatories, and a covered top platform for a range finder. Four Leitstände S414 were built in the Atlantic Wall, one notably at battery Scheveningen-Nord in the Netherlands.

Front view of bunker Type 625B at Battery Nordmole in the Festung Hoek van Holland. Batterie Nordmole was a Navy anti-aircraft battery located on the North Pier. The battery was surrounded by a protective concrete anti-tank wall. Type 625 B was a variant of Type 625 (Casemate for a 7.5 cm PaK field anti-tank gun) intended to be incorporated into a concrete anti-tank wall.

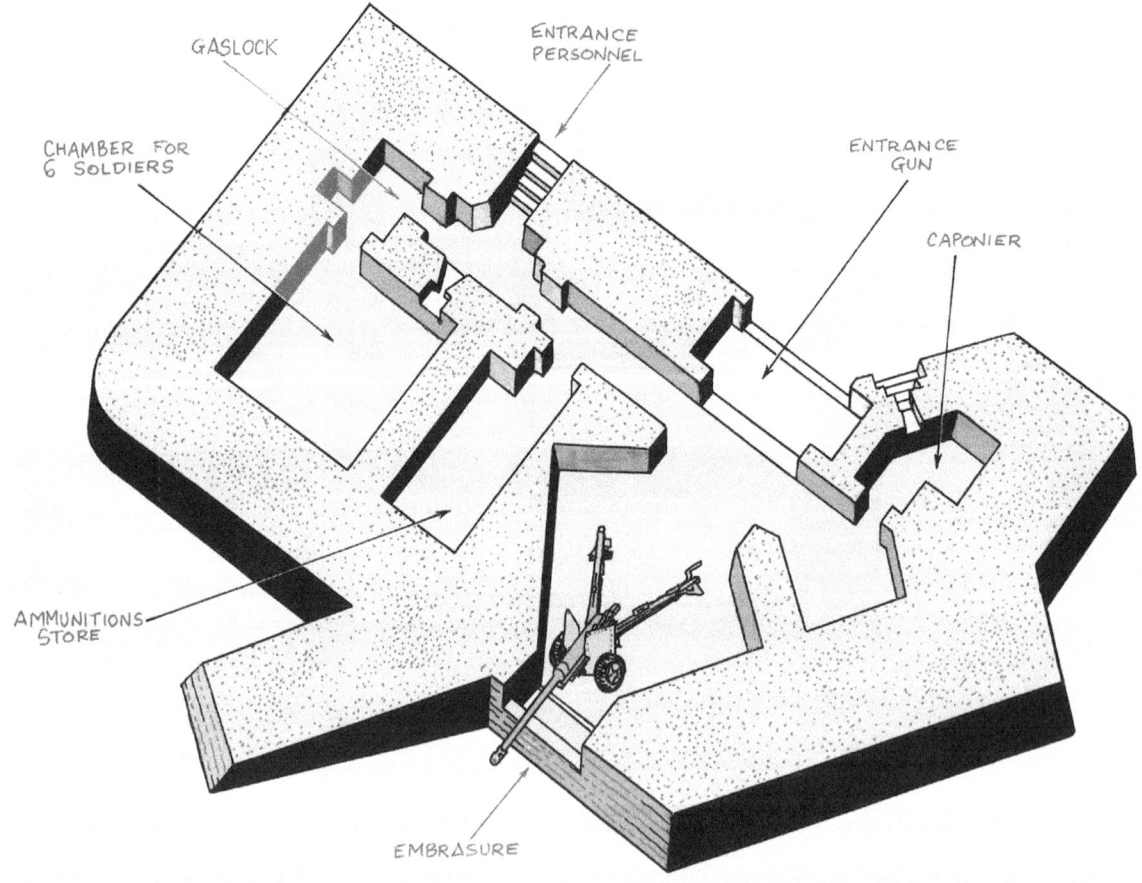

Type 625B plan. Bunker type 625B was a casemate for a field AT-gun (Schartenstand für 7.5 cm PaK 40 gun). It was a variant of type 625 intended to be embodied into an anti-tank wall. Length was 18.40 m and width was 9.20 m. Nine types 625B were built in the Netherlands.

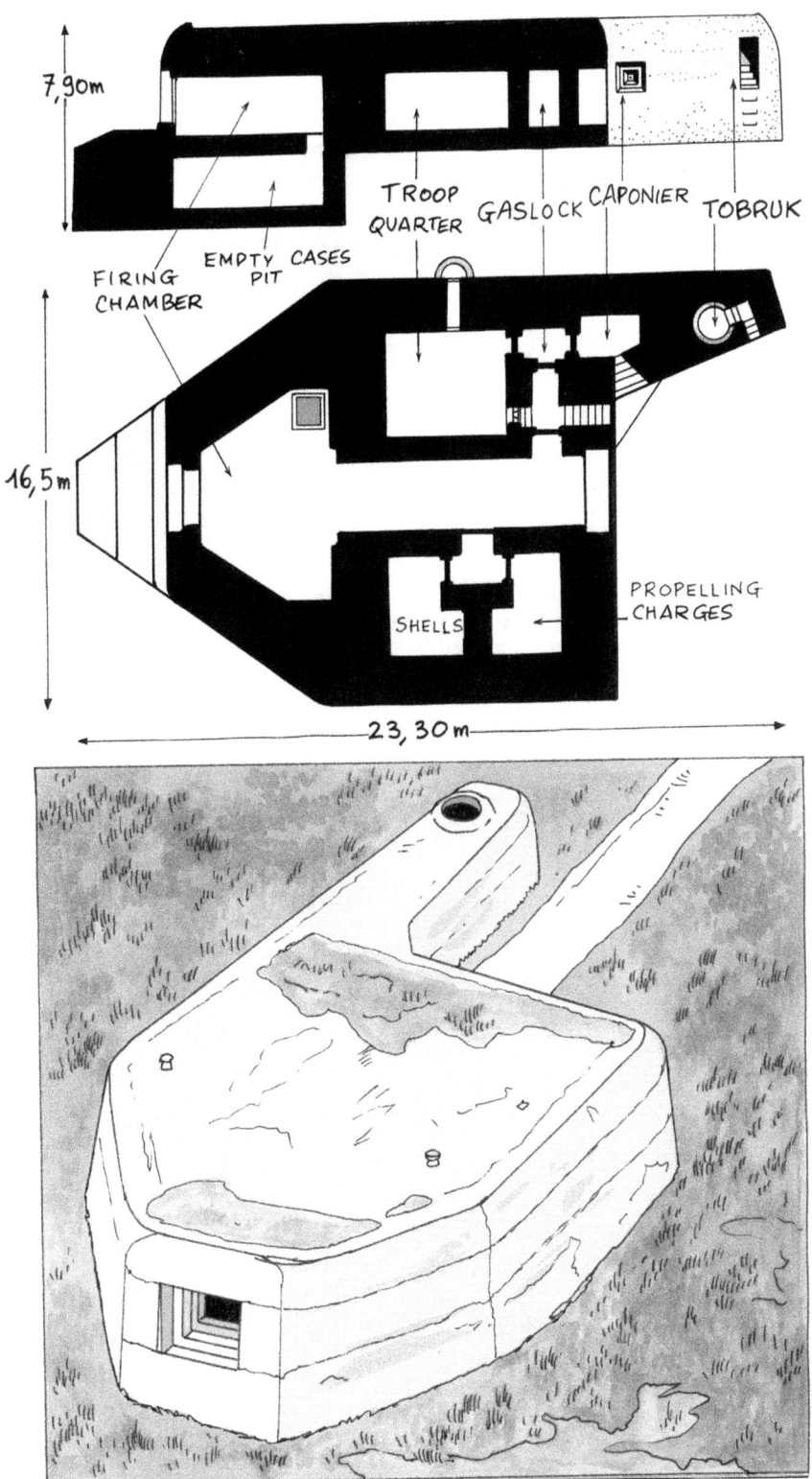

Bunker type 611. Regelbau 611 was an embrasured emplacement for a field gun (Schartenstände für Feldgeschütze). Issued in April 1942, it was designed by the insomniac Adolf Hitler himself to house mobile field artillery from 7.5 cm to 15.5 cm caliber. The bunker 611 was re-designed by the engineer corps and included a Tobruk, a caponier, two armored backdoors, a 2.50 m wide corridor at the rear, a chamber for nine gunners, two ammunition stores, a case-pit under the fire-chamber and an embrasure at the front allowing for a 60° arc of fire. The construction of the bunker needed 1,330 cubic meter of concrete and 63 tons of reinforcement rebar. Eighty-eight bunkers 611 were built, notably at Merville in Normandy (France). Three bunkers type 611 were built in the Netherlands, notably at the former Battery von Kleist in VB Vlissingen.

M170 Casemated gun-emplacement (Geschützschartenstand). The Geschützschartenstand M170, issued in January 1943, was designed to house a fortress long-range gun firing through a 90 degrees embrasure. The bunker included one entrance defended by a close range defense caponier, two ammunition stores, and one firing chamber. Length was 18.80 m, width was 12.80 m and height was 6.80 m. The mass of concrete needed was 1,335 cubic meters. Thirty-two M170s were built. At fort IJmuiden, two M170s were coupled with army troop-shelters type 656 that could accommodate 15 soldiers.

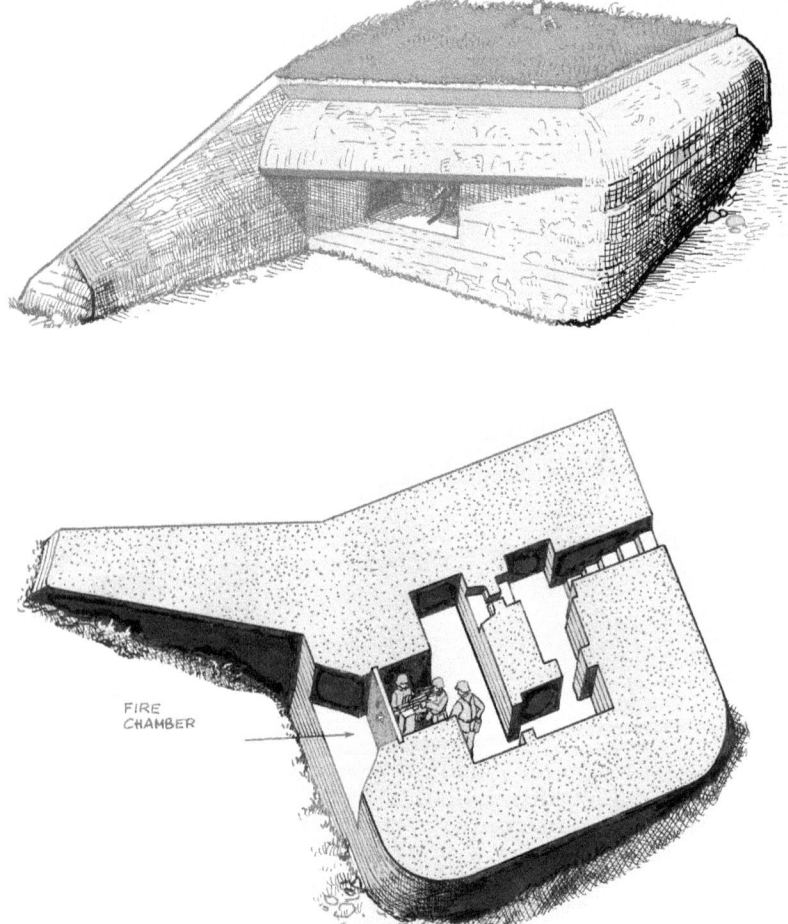

Bunker type 681. The small flanking MG Schartenstand 681 (MG pillbox without crew chamber), issued in April 1943, had a length of 8 m. Its construction necessitated 280 cubic meters of concrete, 13 tons of reinforcement rebars, and 2.5 tons of metal plates for ceilings. Seven were constructed in the Netherlands.

Bunker type 667. The embrasured small casemate for 5 cm fortress AT-gun (Kleinschartenstand für 5 cm KwK) was issued in January 1943. It was a simple artillery emplacement including only a firing chamber and a small ammunition store containing 144 shells. Roof and walls were built in 1.50 m B-1 new thickness and thus its construction only necessitated 165 cubic meters of concrete, 7.5 tons of reinforcement rebar and 1.3 tons of metal plates for ceilings. Length was 6.80 m, width was 6.40 m and height was 4.60 m. The small, cheap, rapidly built, easily camouflaged and well armed Kleinschartenstand für 5 cm KwK type 667, was very frequent: 651 of them were constructed in the Atlantic Wall. Many specimens are still to be seen today in the Netherlands; e.g., at Kornwerderzand, and at Wassenaar Slag near The Hague.

Tobruk and Koch bunker

The *Ringstand* (aka *Tobruk*) was a small concrete Vf foxhole, composed of an open pit (ca. 1.60 meters high) in which a soldier could stand upright. It was both an observatory post and a combat emplacement for one machine gun. Very often a small room was added to serve as ammunitions-storage and shelter. It was accessible by means of a small and narrow curved corridor/staircase.

The Tobruk existed in numerous variations, armed with a mortar, or a light Flak gun, or a 5 cm (2 inch) KwK anti-tank gun or a rotating discarded tank turret. The standardized and small Tobruk was cheap, rapidly built, easy to camouflage, adaptable and multifunctional. Widely used in the Atlantic Wall (between 10,000 and 15,000 were constructed), thousands of them still exist today everywhere along the European shores.

Designed by the German army in the last months of 1944, the so-called *Kochbunker* was a simplified variant of the Tobruk. The Kochbunker (aka K-Rohrenstand) was a small, circular, prefabricated, concrete combat/observation pit usually similar to the Tobruk but without shelter. Suitable for only one man, the pit could be open, but it could also be covered with a protective concrete dome of ca. 140 cm (55 inches) in diameter. The pit with cover was ca. 2.40 meters high. The circular concrete wall was ca. 20 cm (8 inches) thick. There was of course an embrasure for a weapon, and an access hatch about 60 × 75 cm at the rear.

Kochbunkers were often part of a trench system and thus partly underground, they could be coupled, and each one could house a sentry or an armed soldier. Kochbunkers were hastily constructed in Silesia (Poland), in the Netherlands—e.g., in the region of Doetinchem, Assen and at Arnhem, and probably elsewhere in the Atlantic Wall.

244 Dutch Fortifications

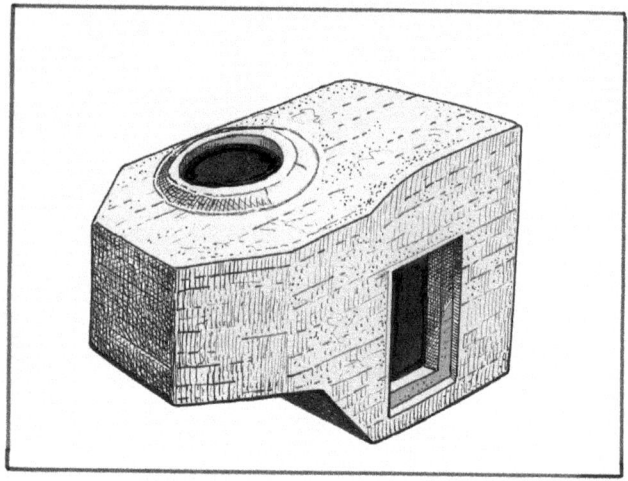

Tobruk. *Top:* **Opening with MG.** *Bottom:* **View.**

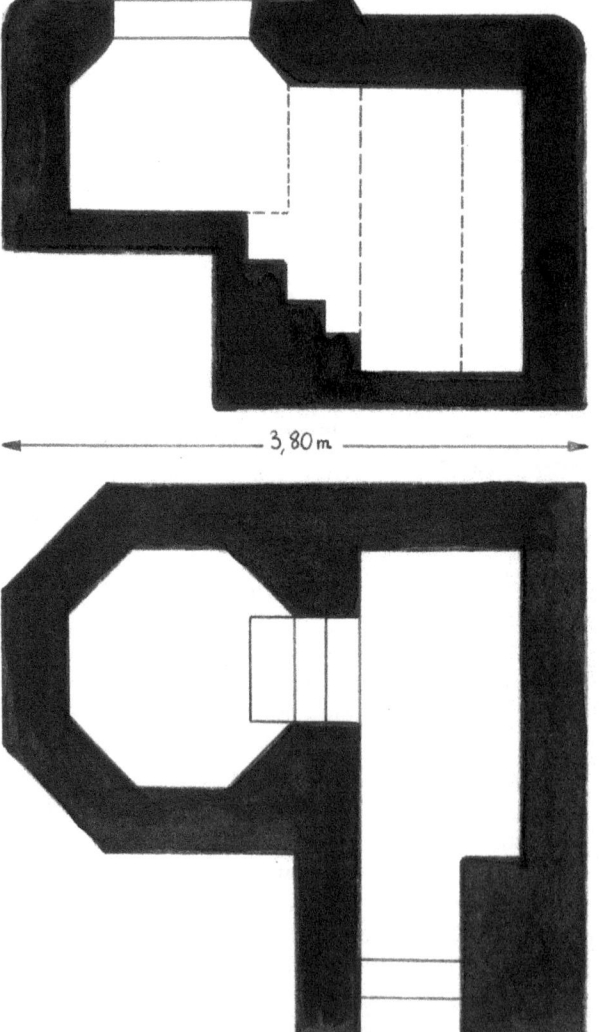

Tobruk (cross-section, and plan).

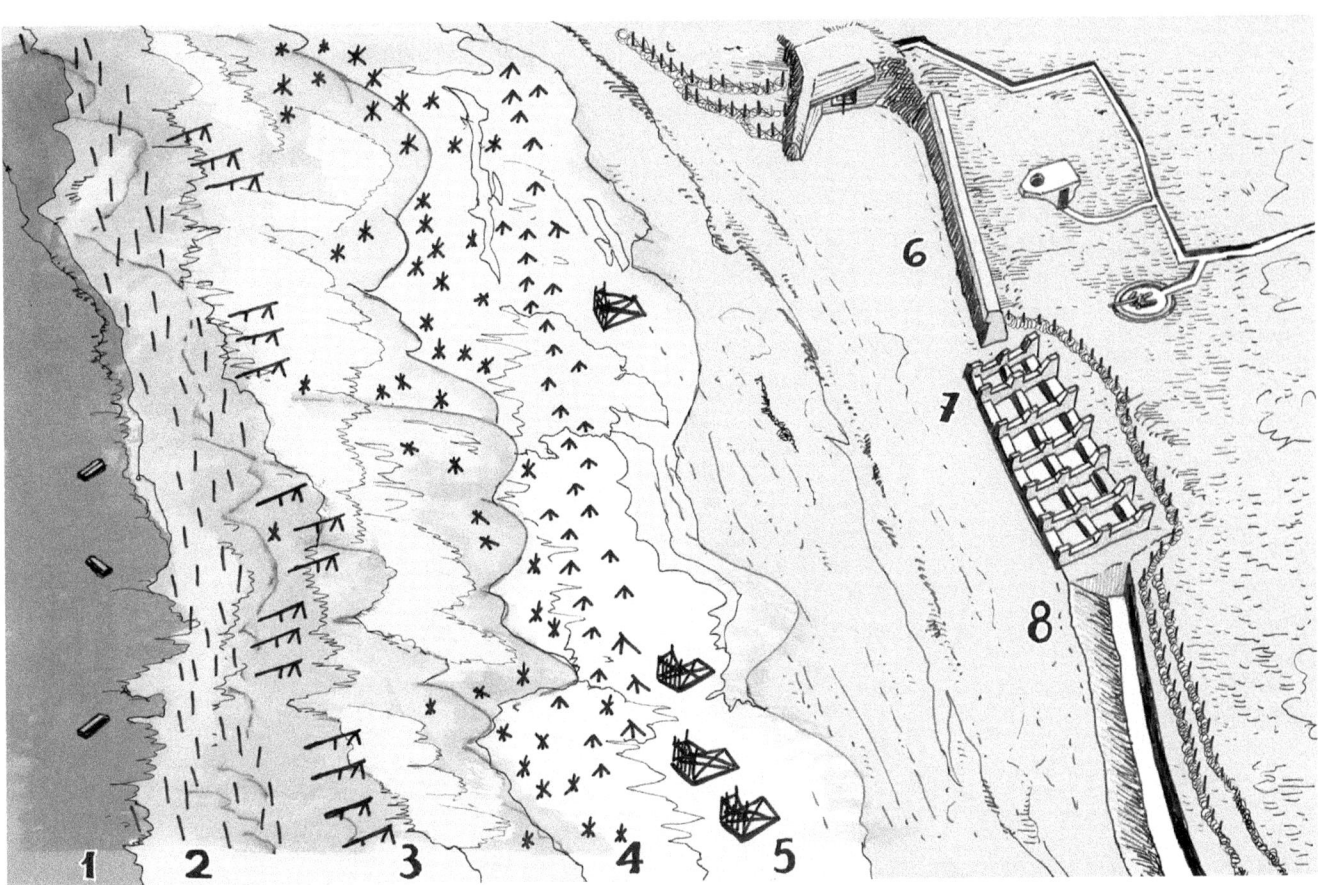

Obstacles. The Atlantic Wall included many coastal obstacles intended to block and hinder the Allies' advance. The Germans established three main kinds of obstacles: anti-landing craft; anti-tank; and anti-personnel: 1: Sea mines and mined rafts. 2: Mined beams and poles. 3: Hemmbalken (strand beams). 4: Tetraeders (formed of three wooden, metal or concrete pieces assembled as a pyramid) and Czech edge-hogs (metal angled bars). 5: Rollbock or Belgian Gate (device composed of a 3 m wide and 2.5 m high strong metal gate firmly fixed on a 3.30 m long carriage moved by three rollers). Against tanks there were specialized and extremely strong hindrances. 6: Concrete anti-tank wall. 7: Dragon's teeth (rows of concrete blocks). 8: Anti-tank ditch. Anti-personnel obstacles included various forms of ditches and barbed wires, as well as hidden AP mines.

The Atlantic Wall in the Netherlands

The Netherlands, occupied from May 1940 onward, could actually have formed a practicable point for an Allied landing. The possibility of an invasion of Europe via the Netherlands was not a fantasy. The Netherlands is relatively close to England. Although treacherous shoals and heavy breakers would have been a hazard to a landing force, the Dutch coasts are formed of large, flat and continuous beaches bordered with gentle sand dunes. Although the country is crossed by numerous rivers, canals, waterways and ditches, the hinterland is flat and suitable for rapid progression (at least the parts that could not be flooded).

Following the Rhine, the Allies could have reached Germany's industrial heart, the Ruhrgebiet. Three modern and well equipped harbors could have been used by the Allies for building up supplies, troops and weapons: Amsterdam and Rotterdam, as well as Antwerp in Belgium. So the occupiers took the possibility of an Allied landing in the Netherlands seriously.

The German forces stationed in the Netherlands were commanded by Luftwaffe General Friedrich Christiansen. His headquarter was situated at Hilversum near Utrecht. Friedrich Christiansen (1879–1972) was originally a seaman. During World War I he was a seaplane pilot, and in the interwar period a merchant ship's captain. In 1929, he entered service as a pilot in the Claudius Dornier Company and flew the huge, iconic, and famous 12-engined Dornier Do X flying boat on its maiden Atlantic flight to New York in 1930. From 1933 to 1937 he was employed at the *Reichsluftfahrtministerium* (RLM, Reich Aviation Ministry headed by Hermann Göring). In 1936 Christiansen was promoted to the rank of General Major, and in April 1937 appointed Korpsführer (leader) of the NSFK (Nazi Flyers Corps).

From May 1940 until April 1945, Christiansen was *Wehrmachtbefehlshaber in den Niederlanden* (WBN, Supreme Commander of the Wehrmacht in the Netherlands). Christiansen was responsible for the food embargo in winter 1944, causing famine in western Holland resulting in the death of about 22,000 Dutch civilians. Right after the war Christiansen was arrested, and sentenced to 12 years imprisonment for war crimes. He was nevertheless released in 1951, and died in 1972.

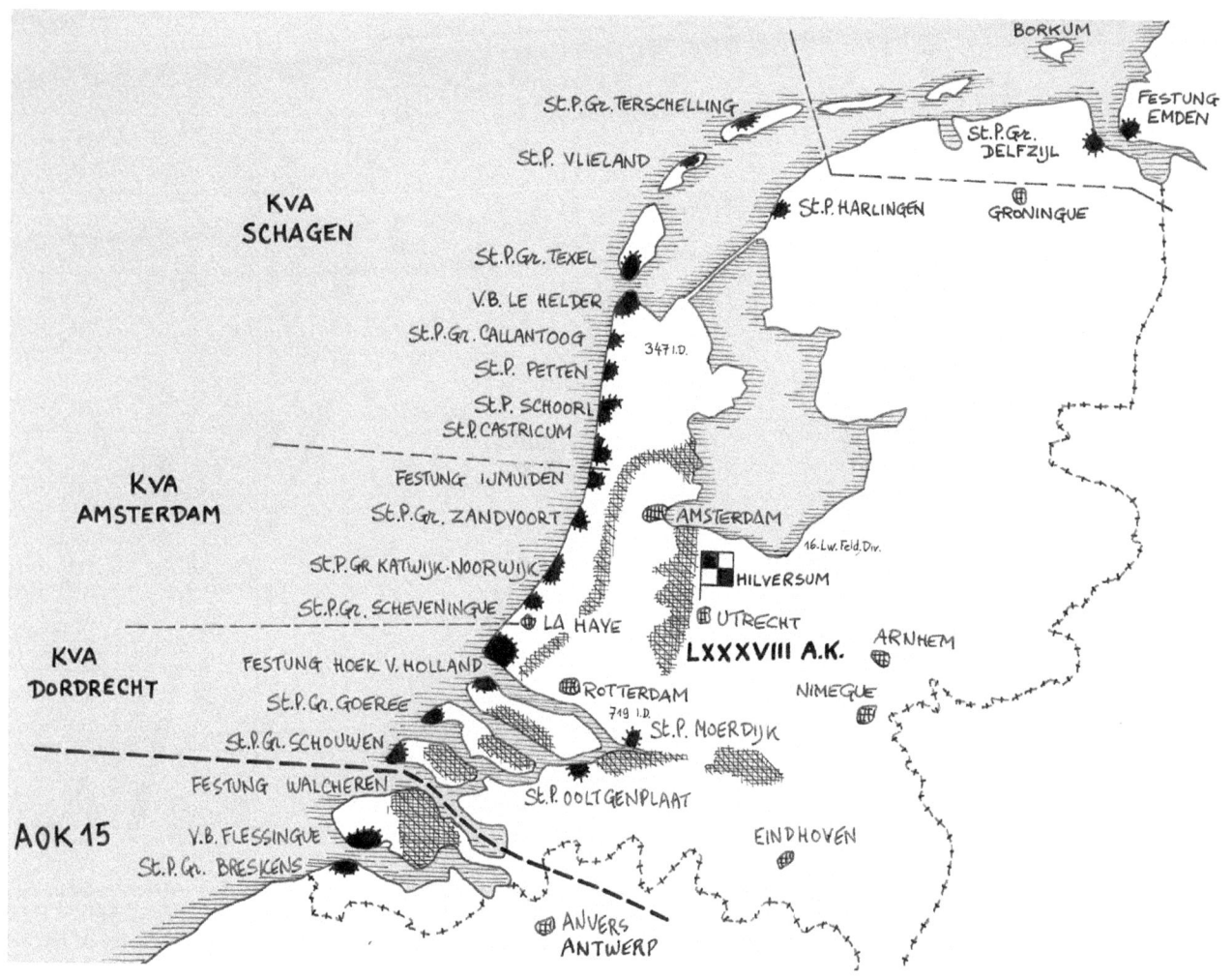

Map of the Atlantic Wall in the Netherlands.

Coastal fortifications

As already pointed out, before World War II the Dutch had fortified several coastal points with forts and batteries, notably the navy base Den Helder, as well as IJmuiden, Hoek van Holland and Vlissingen (Flushing). Of course, these strategically important strongholds were occupied by the German forces, and improved with new fortifications.

At first the German defenses along the Dutch coasts were simple, but from the summer 1942 they were constantly enlarged, reinforced and modernized. An important reason for strong defense of the Dutch ports was their role in basing S-Boote (torpedo boats), which were very active in naval operations in the North Sea.

The Dutch navy base Den Helder, as well as IJmuiden (protecting the access to the port of Amsterdam), Hoek van Holland (defending Rotterdam) became *Verteidigungsbereiche* (VB defensive zones) and later had the status of highly defended *Festungen* (F fortresses). VB Vlissingen (Flushing) and VB Island of Walcheren (both defending the access to the Belgian port of Anwerp) were incorporated into the territory of the 15th Army in Belgian and Northern France.

Between those fortified positions, the coast was divided into three KVA (defensive coasts sectors), each comprising several *Stutzpunktgruppen* (groups of strongholds): KVA Schagen in the north, KVA Amsterdam in the middle, and KVA Dordrecht in the south.

The northern coasts along the provinces of Frisia and Groningen were well protected by nature, namely the shallow Wadden Sea, which was unsuitable for a landing being full of shoals, sandbanks and strong tidal currents. On the Wadden Islands of Texel, Terschelling, Vlieland, Ameland and Schiermonnikoog, a few Wns and StPs with radar stations and artillery batteries were established.

At Kornwerderzand on the Afsluitdijk (IJssel Lake closing dam) the Germans reinforced the Dutch prewar defensive positions by adding a kitchen, Vf troop shelters, three bunkers (type 612 and 667), and anti-aircraft gun emplacements near Kazematen IV and XIV.

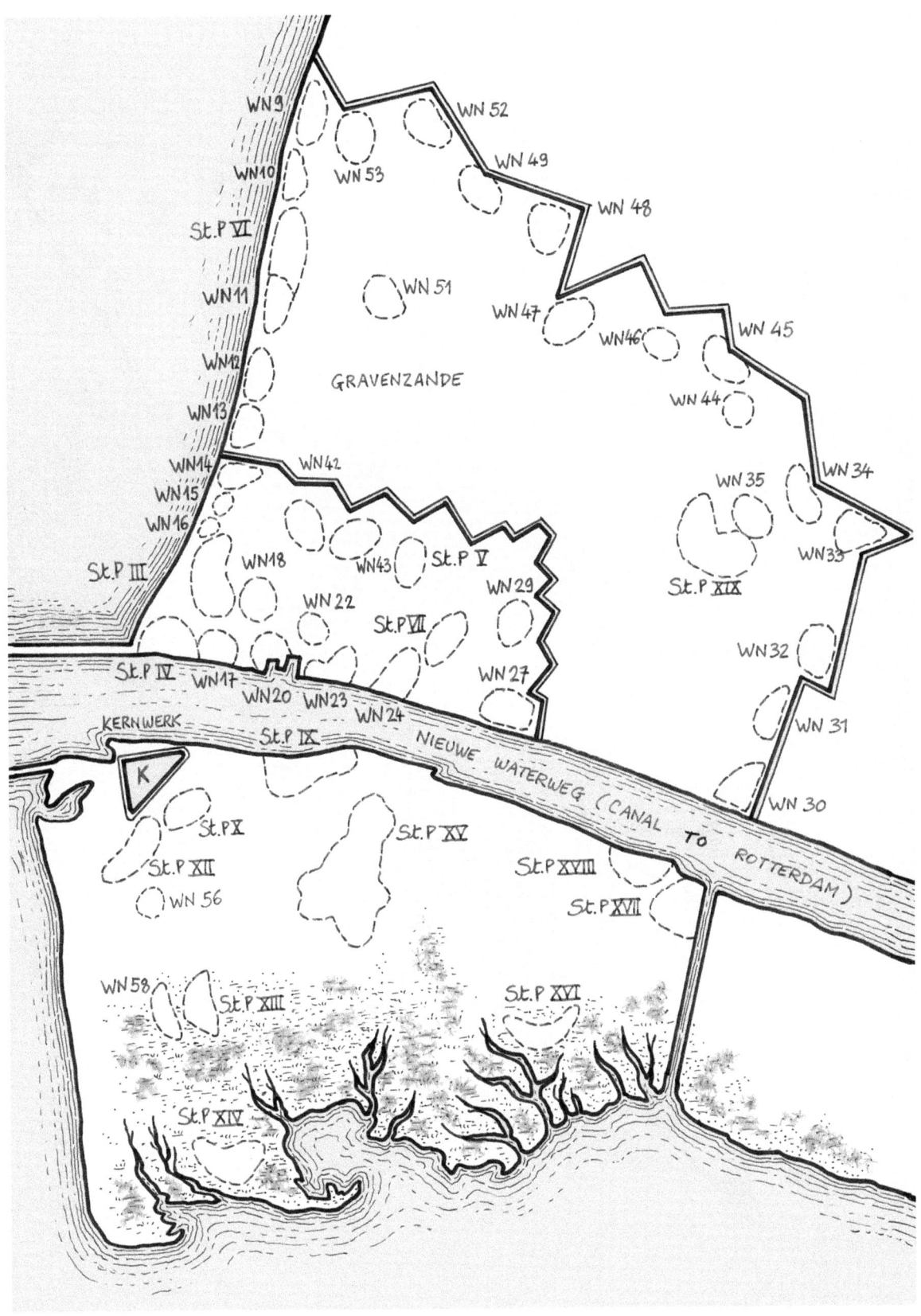

Festung Hoek van Holland (in March 1945). The powerful Festung Hoek van Holland was intended to protect the canal (Nieuwe Waterweg) connecting the important harbor of Rotterdam to the North Sea. It included a continuous anti-tank ditch. The strongest points in the Festung were MKB Vineta in Stützpunkt III, MKB Brandenburg in Stützpunkt X, MKB Rozenburg (armed with three 280 mm guns from the turrets of the retired battle cruiser *Gneisenau*) in Stützpunkt XIII, and the strongly fortified Kernwerk (K).

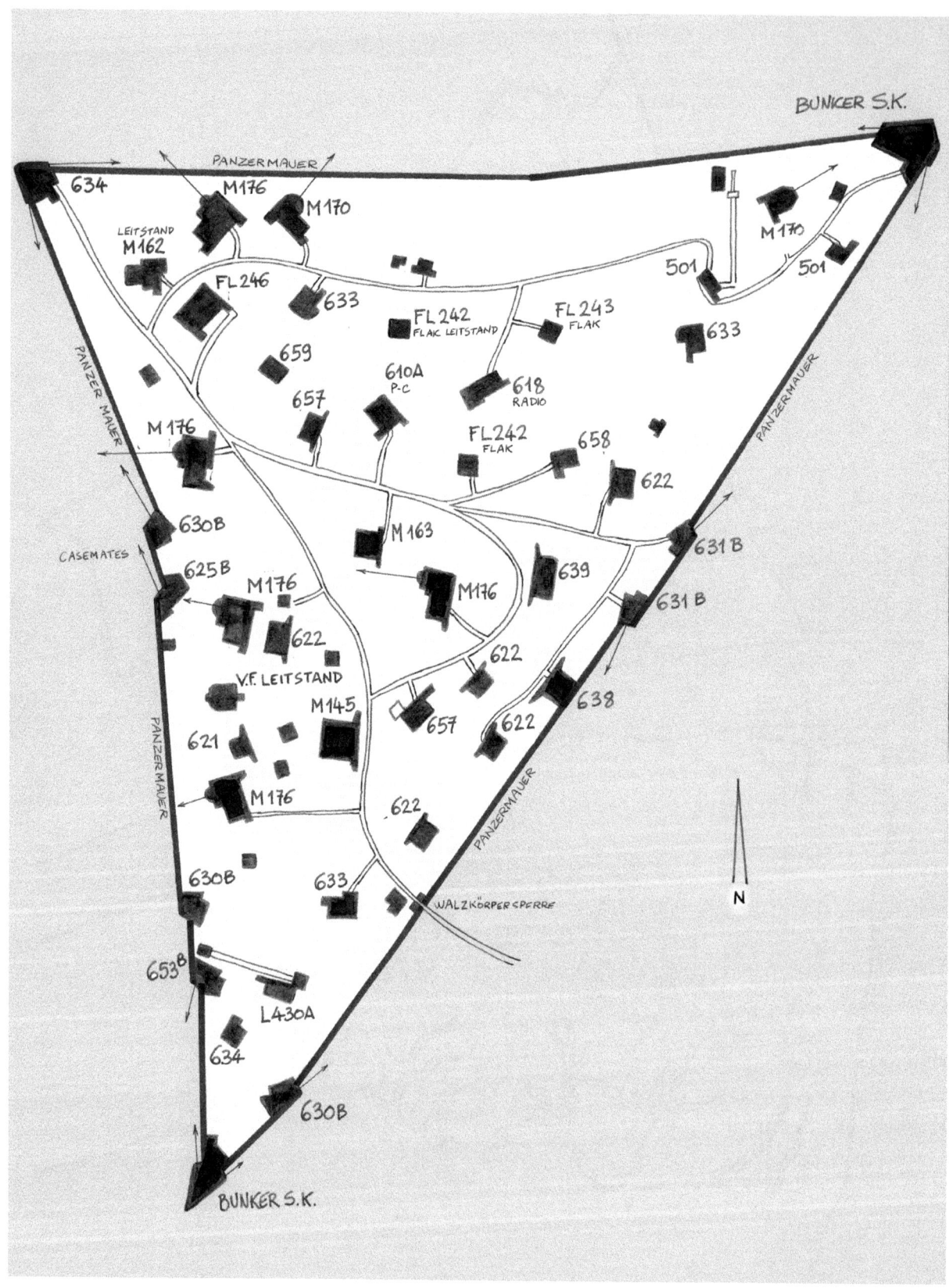

Kernwerk Hoek van Holland. Work at the Kernwerk (core work) of the defense at Festung Hoek van Holland started in November 1942. By the end of the war the heavily fortified complex was a large triangle 300 × 250 × 200 m surrounded by an anti-tank wall and ditch. It totaled 52 permanent bunkers featuring a naval coastal battery with fire control station, a FlaK battery, as well as mortar, AT-gun, machine-gun casemates, searchlight shelters, supply stores, kitchen, power plant, field hospital and infirmary, troop shelters, radio and command bunkers.

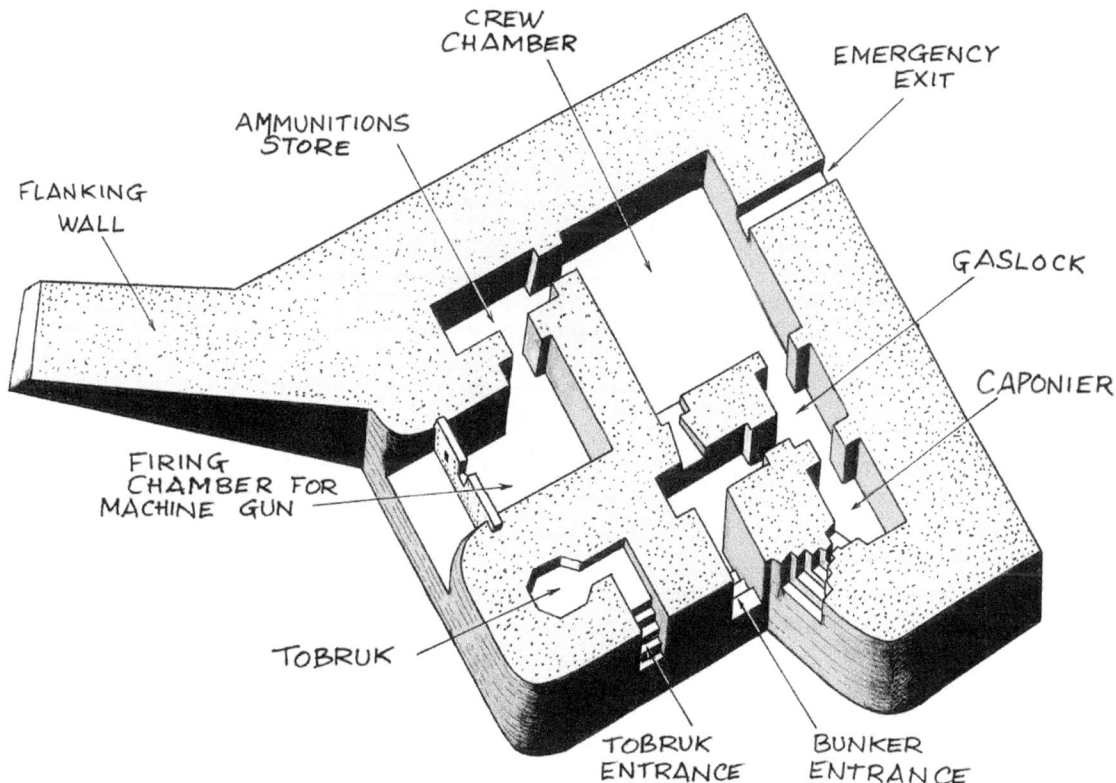

MG bunker Type 630. Front view (*top*) and plan (*bottom*). The MG-casemate (MG Schartenstände) type 630 was issued in April 1942. It was 11.50 m long (flanking shield not included), 9.60 m wide and 5.10 m high. It was manned by one NCO and five men, serving one MG, protected under a carapace of 610 cubic meters of concrete, 29 tons of reinforcement rebar and 3.8 tons of metal plates for ceilings. The Germans built 297 casemates type 630, most of them in France and the Netherlands (e.g., at Vlissingen).

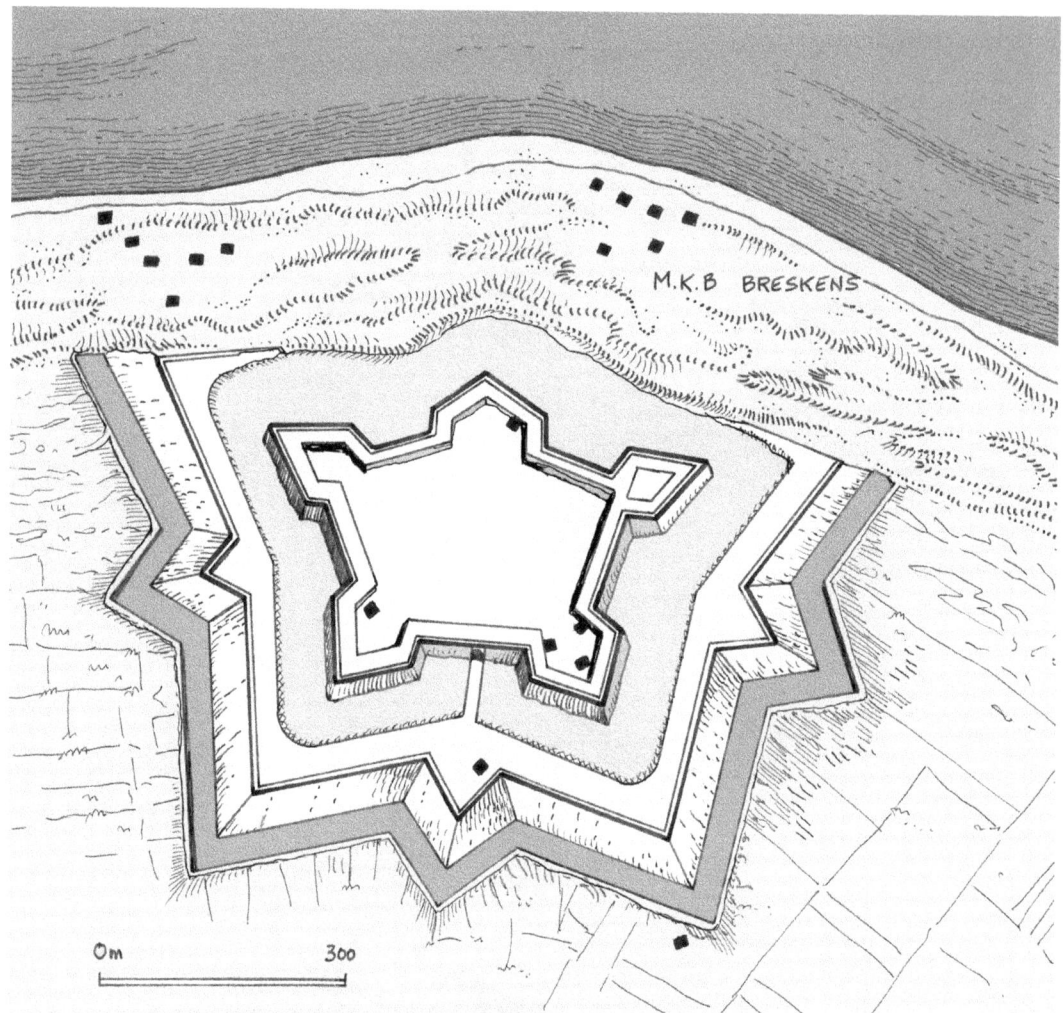

Fort Frederick-Hendrik. The bastioned fort—situated on the left bank of the western Scheldt in Breskens in the province of Zeeland—was built by the French in 1811 and named Fort Imperial. After Napoléon's defeat in 1815, the fort was called fort Frederik-Hendrik. During World War II, the Germans used it as a support position to a navy coastal battery (MKB Breskens) defending the approaches to the port of Antwerp, Belgium.

Military zones

Beaches and coasts were declared military off-limit zones. Of course, all harbor installations, communication, bridges and other facilities were mined because nothing useful was to be left to fall intact into Allies' hands. Private houses, villas, and seaside hotels were requisitioned, transformed into German barracks or simply destroyed if they stood in the way of batteries' arc of fire and observatories vision. Many dwellings, seaside boulevards, and other civilian buildings were ruthlessly demolished. About 15,000 residences were shattered in order to create a field of fire for the defenders.

In Den Haag (The Hague) for example, the installation of the coastal defensive *Stützpunktgruppe* caused the destruction of 92 hectares (227 acres) of forest, 19 ha (47 acres) public gardens, seven ha sport-ground, 20 ha streets, seven bridges, 1,900 meters (2077 yards) draining sewage, three churches, two hospitals, seven schools, 117 various buildings (stores, work-shops, offices) and 3,180 dwelling houses.

Heer

The German Heer (ground army) provided soldiers and bunkers in practically all infantry defenses (Wns, Stps, and StPGrs) established around and within the coastal batteries in the VBs, and in the Festungen. The number of troops that Hitler could allocate to the defenses of the Atlantic Wall depended primarily on his ability to stabilize the Italian and, foremost, the Russian front. By early June 1944, the divisional total in Western Europe—for both Heeresgruppen G and B—stood at 60. According to the historian Hans Umbreit this force totaled 1.4 million soldiers, including 865,000 soldiers of the Army, 326,000 men of the Luftwaffe, 102,000 men of the Kriegsmarine, and about 100,000 soldiers of the Waffen SS and SS police services. They were widely disseminated.

In the Netherlands five divisions (including four infantry and the rump of 19th Panzer Division) were positioned and scattered in the various KVAs. Troops in the Atlantic Wall were not only dispersed, the majority was ill prepared and of poor quality. They included rather aged reservists born between 1903 and 1915, very young draftees without combat experience, and ill-motivated foreigners. The German High Command had to make a large comb-out to fill the ranks, and some Atlantic Wall divisions were curiously composed. The 70th Infantry Division, for example, was made up of infantry regiments 1018, 1019, 1020 and artillery regiment 170. Most recruits had stomach complaints, therefore the unit was nicknamed *Weißbrotdivision* (white bread division) illustrating the special diet soldiers were on. The 70th ID was totally destroyed in November 1944 in Festung Walcheren in the Netherlands. It will come as no surprise that morale and pugnacity among those units were rather low.

The Germans also employed many foreigners in their static occupation army. On the Wadden Island of Texel (province of North Holland) the 822nd Georgian Infantry Battalion (about 600 men strong) manned the bunkers of the Atlantic Wall. To escape the terrible conditions of their captivity, many Soviet Red Army prisoners-of-war saw themselves more or less forced to serve in the German army. Others opted to do so voluntarily, in the hope of ousting communism in their home country. When it looked like the Germans were going to lose the war, the Georgians on Texel island feared for their lives: the Soviets considered all soldiers, voluntary or otherwise, as traitors to their country.

The only possibility the Georgians thought of to rehabilitate themselves for their betraying service in the German army was a violent mutiny. In early April 1945 the Georgians revolted and killed about 450 Germans. However, they failed to get control over the North and South batteries, and the Germans were quick to send reinforcements to the island to crush the uprising. It took five weeks of heavy fighting before the Germans were able to subdue the Georgian mutineers. Although Germany had already surrendered unconditionally on May 5, 1945, combat on Texel Island continued until May 20. Consequently, the Texel Georgian Uprising is referred to as "Europe's last battlefield." The battle claimed the lives of 565 Georgians, 120 Texel civilian islanders, and about 800 Germans.

Marine

There was a strong presence of the Kriegsmarine (Navy) in the Netherlands. The navy and the Organisation Todt built coastal MKBs all along the shores and within the VBs/Festungen. The naval base of Den Helder in the northern end of the province of North Holland was defended by obsolete Napoleonic forts, Dutch 19th century forts, coastal batteries, and pre-1940 Dutch pillboxes. The Germans added four coastal batteries, three heavy FlaK batteries, a number of radar and radio stations, and many infantry armed bunkers.

The German Navy also constructed two large Motor Schnellboot (torpedo boat) concrete shelters in the Haringhaven—the estuary of the North Sea Canal at IJmuiden. The first bunker—completed by November 1942, could shelter ten torpedo boats, and included diverse workshops, ammunitions and fuel stores. It was bombarded and heavily damaged in 1944, and never completed. Work on the second bunker, intended to shelter 14 torpedo boats, started in 1943, but was so regularly attacked by the Allied air force that it was never operational. In the vicinity there was another large concrete bunker in which torpedoes were stored, and various workshops and storage places.

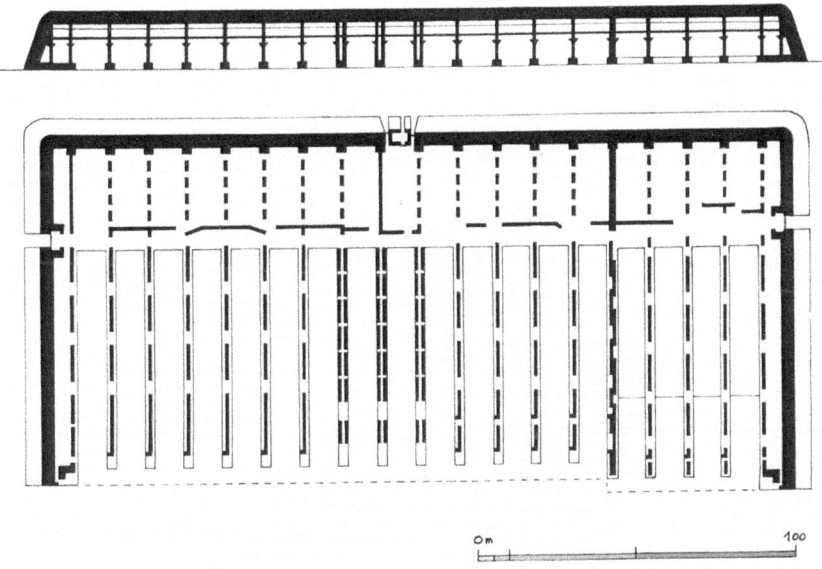

Raumbootbunker at IJmuiden (profile and plan). Located on the southwest side of the Haringhaven at IJmuiden, the concrete bunker for mine sweeping boats had a length of 117.65 m, a width of 247 m, a height of 17.68 m, and walls and roof 5 m thick. The building included 14 pens, and several repairing docks.

Luftwaffe

Since 1940, the Luftwaffe (air force) occupied the main military and civilian Dutch airfields (e.g., Schiphol, Leeuwarden, Bergen, Valkenburg, Hilversum, Waalhaven, Gilze-Rijen, Ypenburg, Eindhoven, Venlo, Twenthe and several others). These airfields were used during the Battle of Britain in 1940, and later served as defensive bases particularly for fighter aircrafts intended to intercept Allied bombers (American during the day, and British at night) on their way to attack Germany. The Luftwaffe constructed headquarters, flight control posts, weather stations, radio and guiding stations, warning and radar installations, searchlight emplacements, and anti-aircraft batteries.

The widespread deployment of Navy and Luftwaffe AA artillery and radar installations was in part due to the need to defend Dutch harbors from Allied attacks, but the Netherlands was also in the path of Allied heavy bombers heading towards Germany. So the batteries were used both in harbor defense and against the bomber streams. Used by both the Marine and Luftwaffe, five Mammut radar stations were installed in the Netherlands, notably at Den Helder, Wijk aan Zee, IJmuiden, Oostkapelle, and Den Haag. The long-range early warning radar Mammut FuMo51 was designed by the Gema Company. It had a range of 400 km (about 250 miles), and comprised a huge antenna with a width of 27 meter and a height of 10.5 meter placed on top of a large concrete bunker.

The German air force also installed command centers. For example, Diogenes aerial command center was located at Schaarsbergen near Arnhem (province of Gelderland). It was a giant three-story concrete *Grossraumgefechtsstand* (large command center) intended to coordinate the operation of night fighters in northwest Europe. This huge rectangular bunker was 40 meters in width, 60 meters in length, 16 meters in height, and had walls 3 to 4 meters thick. It was connected to numerous radar stations, and its core consisted of a large control room with a huge vertical glass map (12 × 9 meters) representing northwest Europe on which colored light beams were projected by female auxiliaries to materialize the positions and flights of friend and foe airplanes.

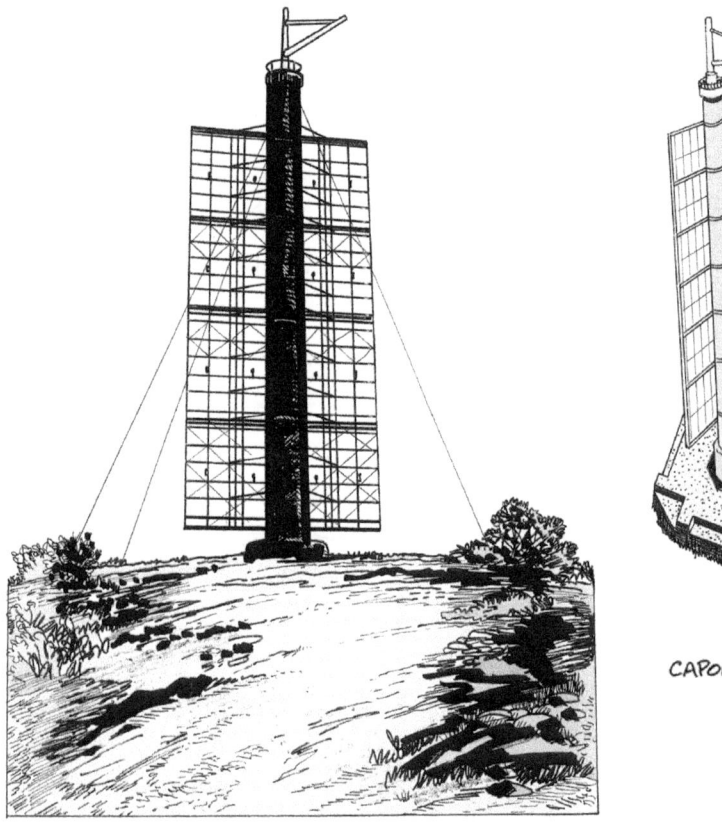
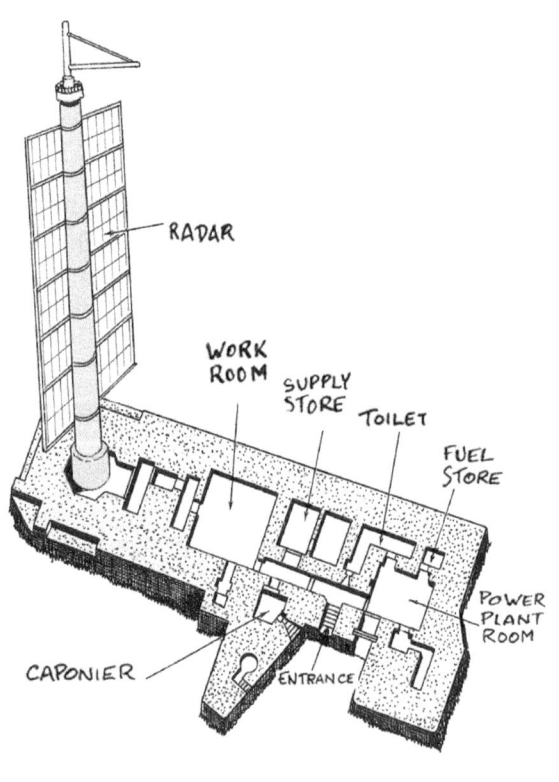

Wassermann radar. The Funkmeßgerät FuMG 402 Wassermann radar (*left*) was designed and produced by the Siemens company. It was an early warning device with a range of about 200 km. It was composed of a strong 40 m high mast supporting a flat vertical rectangular metal aerial. This framework was 30 to 37 m high and 13.50 to 21 m broad. The mast and thus the aerial fixed to it could fully rotate giving a complete 360 degree angle of action. The bunker type L480 bunker (*right*) was issued in April 1942. It was designed by Luftwaffe engineers especially for operating the Wassermann radar. Its length was 27.40 m, its width was 12.20 m, its height was 6.80 and its construction demanded 1,460 cubic meters of concrete. Seven bunkers type L480 were built in the Atlantic Wall, notably in radar-station Auerhahn at Antifer (France). A specimen (deprived of its former mast and antenna) may also be visited today on the Wadden island of Schiermonnikoog in the Netherlands.

Flakstand L422A. The anti-aircraft bunker L422A (FlaK-Geschützstand) was designed, issued, and made available for construction, or came at hand, or was ready for use from February 1943 onward. It included an open platform (*left*) designed to house either a heavy 8.8 cm or a 10.5 cm FlaK guns. The underground bunker (*right*) had a length of 11.20 m, a width of 11.10 m and a height 6.50 m. Its construction called for 850 cubic meters of concrete.

Inland fortifications

In the Netherlands the focus was on coastal fortification but there were also several inland defenses. The Germans established a number of strongholds intended for controlling the strategic railway and highway bridges—for example, an StPGr (group of strongholds) at Moerdijk over the Waal and the Rhine River. Important bases, supply depot, headquarters and regional command posts, airfields, radar and radio control stations, waterway passages, and all vital military installations were guarded and protected, too, with more or less strong fortifications.

The so-called Panther Stellung, established in the winter 1944–45 was a linear defensive position in central Netherlands spreading between the Lower Rhine and the IJssel Lake. It actually comprised parts of the 1938–39 Dutch Grebbe Line, which was reactivated and modernized. Its purpose was to resist and repulse Allied forces that would come from the South. After the June 6, 1944, Normandy landing it was clear that an attack on the Dutch North Sea beaches would not happen. It was then certain that the Allies would come by land from the South, from France and from Belgium. The Panther line was a northern lengthening of the Weststellung (a part of the prewar Westwall aka Siegfried Line defending West Germany).

The Panther Stellung started in Germany near Kleef, running parallel to the Rhine River at Arnhem, Wageningen, and Veenendaal. Then it ran in northern direction (using a part of the tracé of the Dutch prewar Grebbe line) along Amersfoort, Woudenberg, and Nijkerk and ended south of the IJssel Lake. The Panther line was hastily designed by the Organisation Todt, and constructed by about 12,000 Dutch rounded-up forced laborers and deported Russian prisoners-of-war. It comprised existing canals, rivers, and inundations as well as various anti-personnel and anti-tank obstacles and mines, infantry positions, troops shelters and about 20 concrete armed bunkers—notably type 703 armed with a powerful 88 mm anti-tank gun.

In late 1944 the Germans established the so-called Frieslandriegel aka Assenerstellung (position of Assen). It ran along the canal of Drenthe and comprised anti-tank ditches, infantry positions like trenches and Kochbunkers (small prefab concrete combat circular pits for only one man).

It should be noted that the leadership of the German

occupiers had concrete shelters constructed for their own safety particularly against air attacks. So the Reich Commissioner Arthur Seyss-Inquart had an enormous bunker built for himself at Clingendael near Wassenaar (in the province of South Holland). Designed by the architect August Kubitza, the Reich Commissioner's bunker was both a command center and a private shelter completed in late 1943. It measured 61 meters (67 feet) in length, 30 meters in width, and 20 in height. A part of the bunker consisted of a reinforced shelter (actually a double kitchen Regelbau type 645) with a roof 4 meters thick and walls 2 meters thick.

The whole bunker was camouflaged as a large farmhouse with a thatch roof and brick patterns painted on the external concrete walls. It included several living and sleeping rooms, work places, a large basement, a powerplant and a ventilation room. All doors and windows were armored and gas-tight, and there were observatories and two Flakstände (emplacement for 2 cm anti-aircraft guns) installed on top of the roof and camouflaged as chimneys. In the vicinity there were several Tobruk pits, four artillery bunkers, type 625, housing PaK 40 guns, and a Polizei Kazerne (police barracks). The whole complex was a part of Stutzpunktgruppe Den Haag.

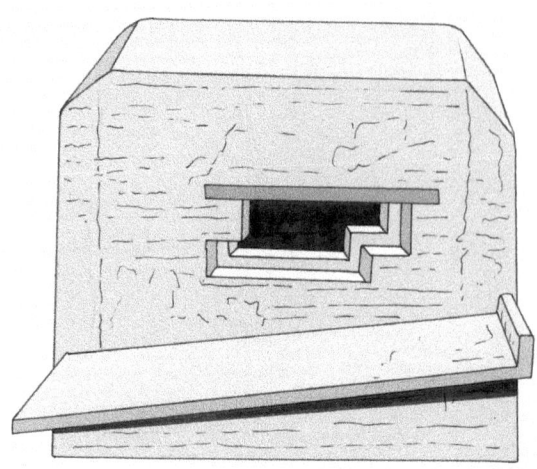

Bunker type 703 front view. This front view of type 703 shows the irregular stepped embrasure, and the inclined Mündungsgasgrube concrete plate.

Mobile V2 launching units

During the last year of World War II, the Germans eagerly defended the Netherlands. Indeed since Northern France and Belgium had escaped their control in the summer of 1944, the Netherlands had become the last place they occupied facing England. There they could install (both fixed and mobile) launching sites for V1 flying bombs (notably at Harfsen, Wesepe, Rijssen, Borne and Nijreesbos) and for the supersonic V2 rockets (notably in the region of The Hague). By then Hitler ordered a retaliative offensive against England.

In the autumn of 1944, a new strategy was organized with revolutionary weapons. The V2s were fired from the southern Netherlands (e.g., Walcheren island in Zeeland, Staveren, The Hague, Harlem and Wassenaar) by small and mobile units called *Schieß-Züge* (launching trains) intended to fire at London. Other mobile units were deployed at Zwolle, Zutphen, Almelo, and Enschede in the eastern Netherlands as well as at Bonn in western Germany to fire V2s to Antwerp. Indeed, one of the principal advantages of the V2 rocket was that it could be fired from practically any clear space of a few square yards, even from a street corner in a town.

An *Schieß-Zug* (firing unit) was composed of a special trailer called *Meillerwagen* (serving both to transport and to vertically fire the V2), several trucks carrying fuel and equipment, lorries transporting the servants, and an armored command-halftrack containing launching controls. Rapidly put in position, and quickly withdrawn, such a mobile launching unit was difficult to locate and destroy. It only took one hour to prepare, load and fire the V2, and less than half an hour for the *Schieß-Zug* to move away.

In March 1945 the Allies had located one such mobile launching site in a small forest, named Haagse Bos, near The Hague in the Netherlands. An air raid was promptly launched by the RAF; unfortunately the V2 Schieß-Zug was already gone, and moreover the airplanes did not aim accurately; instead of the wood they destroyed a part of the city of The Hague resulting in about 500 civilian casualties.

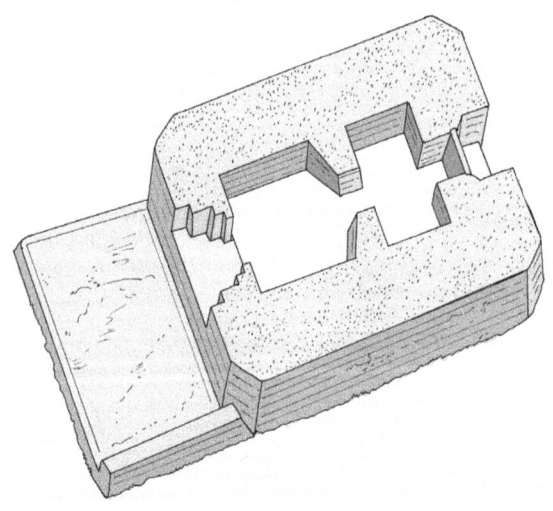

Bunker type 703. The Kleinschartenstand für 8.8 cm PaK 43/41 (small casemate standard type 703) was especially designed to house one 8.8 cm 43 anti-tank gun with a range of 4 km. The casemate was issued in April 1944. It was a simplified artillery emplacement including only a firing chamber, a small recess for ammunition (with a capacity of 180 rounds), and a narrow entrance at the back with a thick armored door. Right under the embrasure there was a so-called Mündungsgasgrube, a large sloping concrete plate intended to deflect the enormous fumes and gasses produced by firing the gun. The walls and the roof were 2 m thick. The construction of the casemate required 380 cubic meters of concrete, 17 tons of reinforcement rebar and 3 tons of metal plates resting on I-profiled beams for the ceilings. Length was 9.30 m, width was 7.50 m, and height was 5.30 m. Several were constructed in the Pantherstellung in the Netherlands.

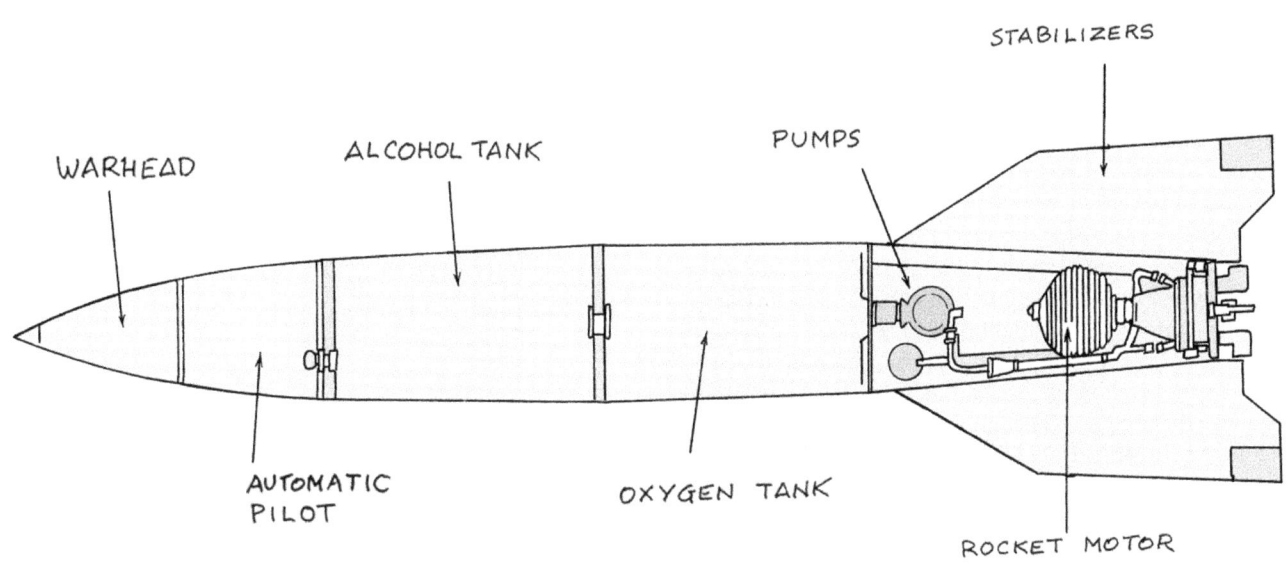
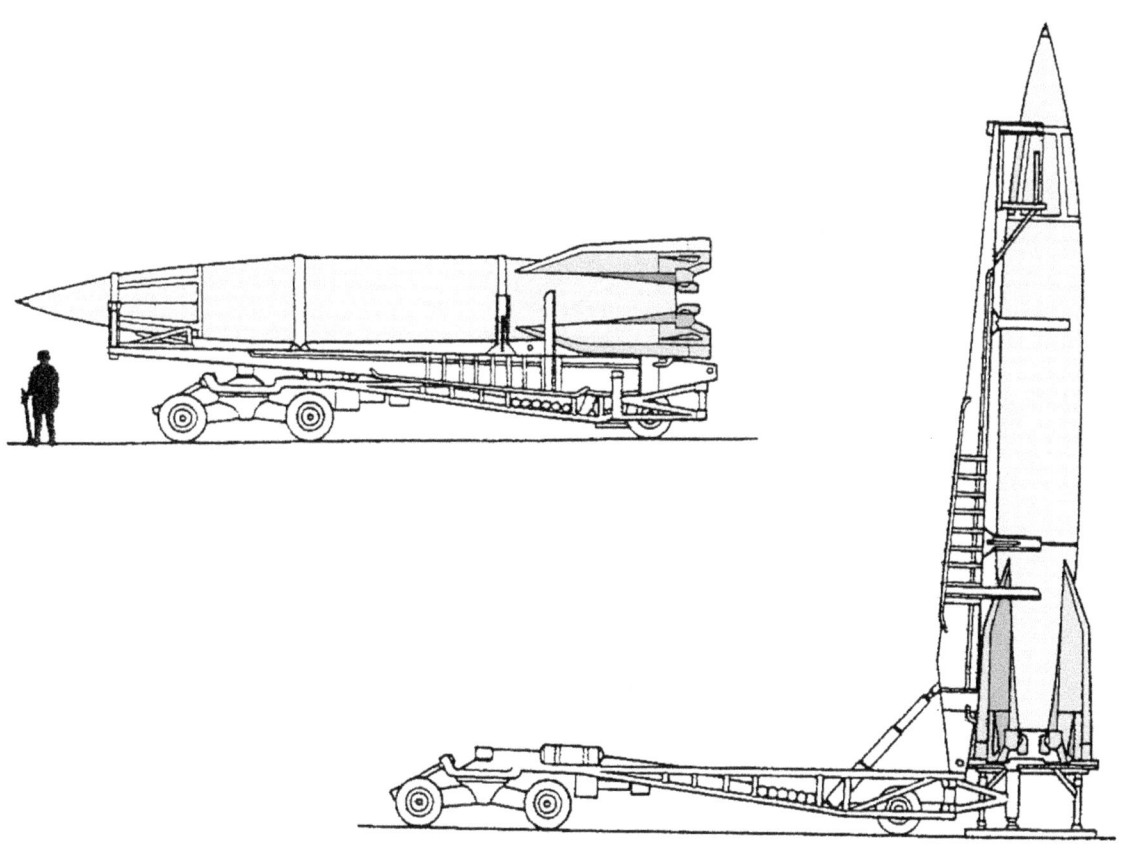

V2 rocket. *Top:* Cross-section. *Middle:* V2 rocket on Meillerwagen trailer towed by a halftrack or truck for transport. *Bottom:* V2 in firing position on erected trailer.

Liberation

Combat in 1944 and 1945

The Atlantic Wall in the Netherlands saw fierce combat in late 1944. Following the successful Battle of Normandy (June–August 1944) the Allied made a series of rapid advances deep into France and Belgium. However, in September 1944, most of their supplies still had to be brought in over the beachheads of Normandy as many important ports in France were either heavily damaged and useless, or still held by the Germans (the so-called "Atlantic Pockets").

To solve crucial problems of logistics, the Allied focused on Antwerp in Belgium, the first large and relatively intact liberated port available. But the Germans held the mouth of the river Scheldt—the harbor opening to the sea in the Dutch province of Zeeland. It was then clear that forcing a way through the German defenses was essential. From October to November 1944 Canadian, Polish and Norwegian troops launched a daring series of amphibious operations in Zeeland. It was a grim, difficult, and costly operation fought in the very bitter cold winter.

The German defenses were strong, and had to be taken one by one. The Germans had destroyed sections of the dikes, and large parts of the region (notably the Island of Walcheren) were totally flooded and devastated. Nevertheless the Allies won the battle, the mouth of the Scheldt River was liberated, and the port of Antwerp became available for supplying the Allied Army in northern Europe.

However the Netherlands was not yet liberated. In September 1944 the large-scale airborne Operation Market Garden (Battle of Arnhem) was launched with the aim of securing the crossings over the wide Dutch rivers, Maas (Meuse), Waal, and Lower-Rhine, in order to open a way for a decisive advance into northern Germany before the winter. Because of the failure of the operation at Arnhem, the northern part of the Netherlands was not liberated until April and May 1945 after a very cold winter during which the urban civilian population suffered severely from famine, exposure and misery.

In March 1945 the Panther line had lost a great part of its usefulness and could not repulse the Allied's offensive (Operation Plunder) that allowed the crossing of the Rhine River and the decisive advance into desintegrating Nazi Germany.

After pointless destruction and hopeless fighting in the city and province of Groningen, the Georgian uprising on Texel, and skirmishes in the Wadden Islands, Germany surrendered on May 8, 1945. The Second World War was over, at least in Europe.

In Asia, the capitulation of Japan was declared on August 15 and formally endorsed on September 2, 1945, ultimately bringing the hostilities of World War II to an end.

The liberation of the Netherlands cost the Allies (mainly British, Canadian, Polish and Norwegian) about 13,000 casualties. As for the Germans casualties totaled between 15,000 and 20,000.

Assessment

On the military level, the Atlantic Wall was neither an impregnable concrete and steel barrier spitting fire as the propaganda proclaimed, nor a ridiculous and pointless waste of energy and means as General Feldmarschall von Rundstedt thought. Indeed, the Atlantic Wall was armed with disparate armaments and defended by rather old soldiers deprived of martial enthusiasm. Indeed the Atlantic Wall was not perfect and was uncompleted in many areas, but it was not a useless fence at all. At least it played a deterrent role. The strong German defenses in the Netherlands, in Belgium and in the northern French Pas-de-Calais obliged the Allies to choose Normandy for Operation Overlord, the landing and invasion of Europe that started on June 6, 1944.

It is reasonable to figure out that should the landing have taken place on the North Sea beaches in Belgium and in the Netherlands, the Allies could then have been able to march rapidly toward the Reich industrial heart, the Ruhr. In this eventuality the liberation of the Netherlands, and the final victory could possibly have been achieved in a shorter period of time, thereby saving a lot of lives.

CHAPTER 10

Dutch Fortifications in the Cold War

Cold War

After World War II a new catastrophic war seemed imminent between Communist East and Capitalist West. During World War II, the United States and the Soviet Union fought together as allies against Nazi Germany and Japan. However, the relationship between the two nations was a tense one. Americans had always been wary of Soviet godless, anti-bourgeoisie, anti-capitalist communism and concerned about Russian leader Joseph Stalin's tyrannical, and ruthless rule of his own country.

Postwar Soviet expansionism in Eastern Europe fueled many U.S. fears of a Russian plot to take control of the whole world. For their part, the leadership of the Soviet Union resented the Americans' decades-long refusal to treat the USSR as a legitimate part of the international community. They also felt bitter about America's delayed entry into World War II, which—they said—had caused the deaths of millions of Russians. The USSR also came to take offense at what they perceived as American officials' bellicose rhetoric, arms buildup, greedy and selfish imperialism, and aggressive interventionist approach to international relations.

After 1945, these grievances ripened into an overwhelming sense of mutual distrust and enmity, resulting in international tensions known as the "Cold War." Both countries possessed atomic bombs, and as a result, the stakes of the Cold War were perilously high. This hysterical period was marked by a number of serious crises: the Berlin blockade (1948–49); the Korean War (1950–1953); the Hungarian Revolution (1956); the Berlin Wall (established in 1961); the Cuban missile crisis (1962); the Russian invasion of Czechoslovakia (1968); and the Vietnam War (1955–1975). The neurotic Cold War came to an end in 1991 when the Soviet Union collapsed.

Dutch Postwar Reconstruction

In the first post–World War II years, the succeeding Dutch governments endeavored to clear both the trauma and the rubble of the war. The devastating World War II changed a lot of things in Dutch society. Apart from the tremendous personal losses and material damage, large parts of the economy were wrecked. What had once been taken for granted, the colonial Dutch East Indies for example, entered into revolt. After an unsuccessful attempt to maintain their dominance by force in Indonesia, the Dutch were forced to grant independence to the former colony. Dutch East Indies became the independent United States of Indonesia in 1949.

Aided by the European Recovery Program (ERP—aka Marshall Plan) the Netherlands recovered with surprising rapidity. The ERP was an American economic and assistance scheme (from 1948 to 1952) intended to help rebuild Western Europe, and thereby hinder the spreading of Communism. After difficult years of reconstruction, the Netherlands sustained in the second half of the 20th century a continuous and fast economical growth.

As a small country that had suffered a lot from the upheaval between the great powers, the Dutch eagerly joined the international organizations (notably the United Nations) that strove to work for a situation in which war would not happen again. Indeed the Netherlands was deeply grateful to the Allies and the United States and particularly Canada for their wartime liberating efforts, and also for the postwar Marshall Plan that contributed to the rebuilding of Europe.

Militarily, the main European defense against the Communist East Block was guaranteed by the North Atlantic Treaty Organisation (NATO) and the U.S. Army. Created in 1950, NATO was an intergovernmental political alliance and a system of military collective defense agreed between the USA and several western European countries. In a background of hysteric Cold War the Netherlands became a member of the North Atlantic Treaty Organization.

Dutch Fortifications During the Cold War

New weapons

The post–World War II era saw the advent of fast and powerfully armed jet airplanes as well as the tremendous development of large jet bombers and guided intercontinental ballistic missiles. These modern system weapons (most of which could deliver nuclear bombs) considerably reduced the role of permanent fortification and coastal artillery in defending a country against air and sea attacks. Nuclear fire rendered both fixed military emplacements and concentrations of large armies vulnerable to enemy strikes. As a result most Dutch fortifications lost their military value after 1945. A few new secret special command centers were, however,

Sherman tanks dating from World War II were discarded and replaced with modern types, but a number of turrets with their guns were recycled and placed on purpose on top of concrete bunkers for additional firepower to Cold War fortifications. This unit is still to be seen at Kornwerderzand.

Ridderzaal The Hague. Construction of the Ridderzaal (Hall of the Knights) started in 1248 under Count Willem II of Holland, and was completed under his son, Count Floris V of Holland, around 1280. The counts of Holland used the building for festive occasions. Later it was used as a meeting hall for the States General delegates of the Republic of the Seven United Provinces. By the middle of the 19th century, the building was in ruins. In 1904 it was restored in neo–Gothic style, and ever since has been used for the State Opening of Parliament, for royal weddings, large-scale official receptions and dinners, and for international conferences.

Chapter 10. Dutch Fortifications in the Cold War

Muiderslot. The popular castle of Muiden from the 13th century is open to the public all year.

Fort Rammekens. The 16th century coastal fort Rammekens can be visited by guided tours.

built. Strictly reserved for important VIPs such as top military officers, governmental and provincial leaders, and a few administrative personnel, these bunkers were made of concrete and buried deep in the ground where they could withstand nuclear fire. Those few special centers were equipped with all facilities, and stores allowing for the short survival of these "happy few." At Bilthoven near Utrecht, for example, the Dutch army had a large two-story concrete building built with walls 2 to 3.5 meters thick. In case of nuclear attack it could house 40 persons for two weeks. In 1960 the Nederlands Spoorwegen (NS National Railway Company) had a nuclear bunker constructed at Maarssen north of Utrecht. This co-ordination shelter could house 30 to 40 people for a few months.

In 1975 an emergency underground civil residence bunker was started under the former Ministry of Home Affairs at the Schedeldoekshaven in The Hague. Intended to shelter the prime minister and members of the government, it was completed in 1979. It had a surface of 2,000 square meters, and could accommodate 200 persons for a short period of two weeks.

Drawbridge. This kind of ophaalbrug (drawbridge) moving by means of counterweights is still widely used today.

As for the purely military fortified lines built before World War II, they were declared obsolete and many were dismantled in the 1950s. For example the pre-1940 Grebbe Line was permanently decommissioned by the Dutch government in 1951. However during the Cold War, a number of fortified positions were kept in activity for the defense of the West against a Soviet aggression. Indeed inundations and fortified lines could still be useful against an enemy using conventional non-nuclear weapons.

The IJssel Line

The IJssel Line (IJssellinie in Dutch) was a line of defense established in the 1950s and 1960s. As the name indicates, the fortified line was placed along the IJssel River valley. Starting at Kampen in the North, it ran along Zwolle, Olst, Deventer, Zutphen, Dieren, and down to Arnhem in the south. It was about 127 km (79 miles) in length and had an extension along the Waal River near Nijmegen and the Lower Rhine River between Arnhem and Oosterbeek. As a part of NATO defense it obviously faced to the east in order to resist or at least delay a conventional non-nuclear

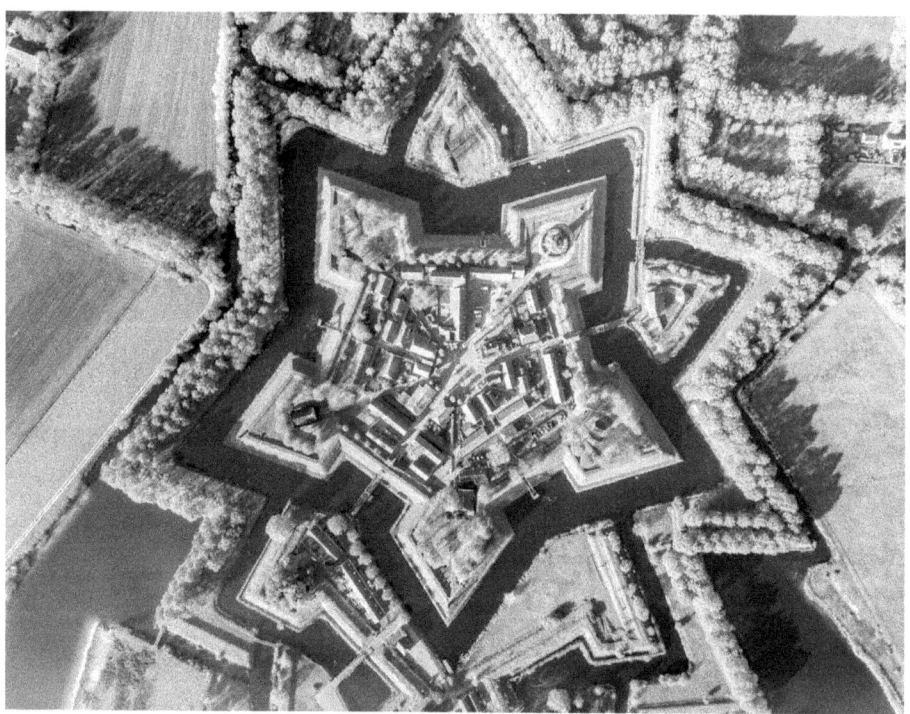

Bourtange (province of Groningen). Built in the 1590s, Fort Bourtange was abandoned in the 19th century and restored in the 1960s. It is now a profitable tourist attraction.

Chapter 10. Dutch Fortifications in the Cold War

Citadel of Den Bosch. The citadel was constructed between 1637 and 1645 by order of stadhouder Frederik Hendrik van Oranje in order to protect the city of 's-Hertogenbosch, but also to keep a watchful eye on the newly conquered Catholic population. It was later used as barracks, a military store and depot, and a prison. Today the citadel houses the Information Center of Brabant History. Originally the citadel was a regular pentagon, but one bastion was demolished in 1880 when the canal (called Zuid Willemsvaart) was widened.

Moat at Naarden.

aggression by Soviet and Warsaw Pact forces, thereby providing time for NATO Allies to reach the Netherlands.

Just like the pre–World War II defense lines, the IJssel Line consisted of observation posts, infantry troop shelters, hospital bunkers, command posts, pillboxes armed with machine guns, anti-tank weapons and guns, anti-aircraft batteries, as well as many passive anti-tank and anti-personnel obstacles and mines. There were also discarded Ram and Sherman tank armored turrets encased in concrete bunkers providing a 360 degree traverse field of fire.

The line included large and sophisticated inundation systems allowing for the flooding of the whole IJssel valley. The key elements of the IJssel Line were dikes, dammed-up reservoirs, sluices and three movable floating dams (90 meters long at Olst and Arnhem and 240 meters at Nijmegen) constructed in the Lower Rhine and Waal rivers. Flooding waters were to be released from the IJssel River by means of 15 water intake works and 750 other water regulating installations. Behind the line of defense, five motorized infantry divisions of the Dutch field army were to be positioned.

The IJssel Line was intended as a northern extension of NATO's main defenses along the River Rhine, preventing an almost immediate occupation of the Randstad (comprising the three provinces of Holland, Utrecht, and Zeeland). Fortunately the IJssel Line was never engaged in any combat action, but preparations were made on several occasions in anticipation of a possible war, notably in 1956 during the Hungarian Uprising against the Soviet regime; in 1961 during the construction of the Berlin Wall; and in 1962 during the Cuban missile crisis. In all three cases about 70 bunkers in the Olst/Welsum sector were fully manned, and supplied to defend the flooding installations.

Fort Napoleon. Located at Oostende in Belgium, Fort Napoleon was constructed during the French Era. It has been restored, and transformed into a restaurant, a museum and a tourist attraction.

The IJssel Line was established under strict military top secret. Only a few civilian authorities knew the details of what was intended with it. The Parliament and the more than 400,000 valley inhabitants who would have had to evacuate if the IJssel Valley had been flooded knew nothing about the scheme. Only in 1990 was the secrecy concerning the IJssel Line lifted. The Line was discarded and partly dismantled after 1968, when Western Germany became a member of NATO. Then NATO grand strategy adopted a forward defense along the Federal German border.

Kornwerderzand

In the 1950s, the fortified complex of bunkers at Kornwerderzand (built in the 1930s) was occupied by the Dutch army. The prewar bunkers were refurbished, improved, notably their armament, living accommodations, communication and

Fort Napoleon (Oostende) Ditch with caponier.

ventilation systems. A Sherman turret was also placed on top of a small concrete pillbox for additional firepower. The defenses of the IJssel Dam were included in the Dutch defenses based on floodings along the IJssel River. The bunkers of the complex of Kornwerderzand were demilitarized in the late 1960s. Since 1985 a part of the complex has been restored, and open to the public. It is now a very interesting historic museum and tourist attraction known as Kornwerderzand Kazemattenmuseum.

Korps Luchtwachtdienst (KDL), 1950–1968

In 1950 in the background of the Cold War and under the threat of a possible Soviet attack from the air, the Koninklijke Luchtmacht (Dutch Royal Air Force) created a special observation and warning formation known as the *Korps Luchtwachtdienst* (KLD Air Watch Service Corps). This unit was composed of volunteers who manned observation posts all through the country. Between 1951 and 1955 some 276 posts were installed. There were two sorts: a simple observatory positioned on top of a high existing building in towns; or a specially built observation tower (ca. 10 meters high) made of prefabricated perforated concrete plates (a few were made of bricks) placed in the countryside.

In both cases the observatory included a small top platform where teams of two observers (most of them were part-time civilian volunteers who lived in the neighborhood) stood guard equipped with binoculars and also using their ears. Their task was to report the presence, direction and approximate speed of any low flying non-identified aircraft. Should an intruder be spotted the guards would then telephone the nearest regional air force command post or the Sector Operations Centre established at Driebergen in the province of Utrecht. An alarm signal would be given to anti-aircraft batteries, to fast interceptor airplanes and to the *Bescherming Bevolking* (BB, Civil Defense).

Fortunately no Soviet air attack ever occurred, and the whole makeshift scheme was only used for a few training

Muiden Muizenfort. The casemated Muizenfort (Mouse fort) at Muiden was a part of the 19th century Stelling of Amsterdam.

Muizenfort. Made of masonry covered with dirt, Muizenfort had artillery crew quarters, a room for the guard, a kitchen, a room for officers, and storage for food, water and ammunitions.

actions. Besides, in the 1960s, detecting rapid jet airplanes by using human eyes and ears was a thing of the past. By then air attack warnings were carried out on an international scale with NATO radars and electronic devices placed in West Germany farther in the East. In 1964 the rather anachronistic and mostly useless KLD was considerably reduced and finally disbanded in 1968.

Remnants of Dutch Fortifications Today

The passage of more than 2000 years have inflicted enormous changes on Dutch fortifications. Many medieval urban fortifications had been dismantled when bastioned defenses were erected, and these in their turn were removed during the Industrial Revolution in the second half of the 19th century when more space was needed. Large-scale destructions took place during World War II. The post–World War II era also saw further reconstruction, enlargements, and modernization. Sadly, many urban canals and waterways have been filled in the 1950s and 1960s to make room for avenues, broad streets and parking lots for automobiles.

Fortunately several Dutch city centers have preserved appealing, charming and delightful sights including cathedrals, churches, narrow streets, old brick houses, squares, canals, bridges and many other historical, architectural and cultural places of interest such as old municipal buildings like *stadhuizen* (town halls), as well as remains of tower, sections of wall or original gatehouses.

Remise. In 19th century Dutch forts, a remise was a bombproof vaulted garage covered with dirt intended to protect artillery.

Roman era

None of the wooden military architecture that prevailed before the Roman conquest has come down to us. Most vestiges of the Roman period have vanished. It was common practice in the Middle Ages to re-use building materials (stones, bricks, timber and the like) taken from ancient abandoned buildings. Until today some artifacts and foundations are occasionally discovered, and they are eagerly surveyed by archaeologists.

At Nijmegen (Province of Gelderland) a fragment of the old city wall can be seen near the casino and the foundations of the Roman amphitheater are traced in the paving of the present-day Rembrandtstraat.

There is a distinctive route (about 200 km—125 miles—in length) for cycling and hiking called the *Romeinse Limespad* (Roman Limes Path) running along the antique Roman frontier from Katwijk aan Zee up to Nimegue with reconstructed camps at Zwammerdam, De Meern, and Valkhof in Nimegue. At Alphen aan den Rijn (Province of South Holland) there is a park called *Archeon*—an open-air educational museum about the Netherlands in prehistory, in Roman time and in Middle Ages. See http://www.archeon.nl

Middle Ages

Several Dutch towns have kept interesting and impressive vestiges of medieval walls, towers and city gates notably at Zutphen, Zwolle, Utrecht, Hoorn, Amersfoort and Maastricht, to name a

Fort Edam. Located in the province of North Holland, Fort Edam was constructed between 1885 and 1908 as a part of the Stelling of Amsterdam.

Fort Pampus. Located on an artificial small island near Amsterdam, Fort Pampus is now a museum and a major tourist attraction.

Type S-3 Stekelvarken at Leusden (Province of Utrecht). Some 88 S-3 Stekelvarken MG casemates were built in the period 1939–1940 in the defensive Grebbeline that stretched about 40 km between the hills of Utrecht and the Veluwe district.

few. In Nijmegen there are a few noteworthy sights on the Valkhof hill: a Carolingian chapel (8th–9th century CE) and a small remainder of the imperial Carolingian castle that was demolished in 1798. In Amsterdam medieval remnants include several towers and gatehouses, e.g., Munttoren, Schreierstoren and Montelbaanstoren. Amsterdam fortifications were demolished between 1839 and 1848, but all moats have been turned into canals, giving the city its particular concentrical configuration and charming appeal.

About 1,200 medieval castles still exist today, either in more or less original state, or rebuilt, or in ruins. Interesting examples can be seen and visited, such as the famous castles of Muiden, Lovestein, Groesberg, Middelburg, Doornburg, Ammersoyen, Hernen, Keppel, and many others.

The celebrated neo-gothic castle De Haar, located in the village of Haarzuilens (Province of Utrecht), was built in 1892 on the ruins of a former castle. It is actually a romantic fantasy building designed by the architect Pierre Cuypers for the owner, Baron Etienne van Zuylen van Nyevelt van de Haar. The reconstitution of De Haar castle is historically rather reliable but in fact it displays Cuypers's enthusiasm for medieval architecture, and is an expression of late 19th century romantic attitudes regarding the Middle Ages.

Renaissance

From the Burgondian and Hapsburg era only a few fortifications have come down to us. The best preserved and most spectacular is without doubt Fort Rammekens near Vlissingen. In the Binnenstad (Old City) at Utrecht several

Type S-3 Stekelvarken MG pillbox at Cuijk. Stretching in the province of North Brabant along the left bank of River Meuse, the Maaslinie (Line of River Meuse) was intended to defend the southern access to Amsterdam.

Sawn-through bunker. Located on the Diefdijk on the Lek River near Culemborg (province of Gelderland), the Dutch Pyramid type group shelter designated #599 was built in 1940. In 2010 it was sliced down in the middle, and made into a sculptural open-air art object.

early Italian bastions (e.g., Lucasbolwerk and Servaasbolwerk) still can be seen.

Bastioned fortifications

Although many bastioned fortifications were demolished after 1874, there are a number of towns still displaying 17th and 18th century bastions in various Dutch styles, notably Heusden, Hulst, Willemstad, Dokkum, Bourtange, Den Helder, and Maastricht. The fortress of Naarden is perfectly preserved and now houses a Dutch fortifications museum.

In many cases former fortifications have been turned into *plantsoenen* (public gardens or parks)—e.g., the Noorderplantsoen in Groningen or the Weteringsplantsoen and Marnixplantsoen in Amsterdam. Many concentric defensive *grachten* (ditches) have been preserved or turned into canals, adding to pleasant sightseeing in a lot of Dutch towns.

Many independent *schansen* (sconces) have disappeared but a few are still to be seen, notably on Texel Island, at Ouddorp, Een, and Groenlo, for example. All but one bastioned *dwangburchten* (citadels) have been dismantled—only that of Den Bosch still exists.

19th century forts

Many New Holland Waterline torenforts and other strongholds are still extant, notably at Muiden, Vuren, Asperen, Nieuwersluis, Uitermeer, Honswijk, Everdingen and Weesp. Most of the forts around Utrecht are well preserved and some, such as the impressive polygonal fort at Rijnauwen, are open to the public.

Most of the concrete forts of the Stelling of Amsterdam are also well conserved too. Some are still used as training grounds by the military forces, fire brigades, police forces or Scouts associations. Others are now under the control of town councils or the nature department. Some are museums opened to the public. The various Dutch waterlines and Stellingen are unique. Since 1996 the Defense Line of Amsterdam has the status of a UNESCO World Heritage site.

Dutch 20th century bunkers

Many Dutch bunkers of the interwar period 1918–1939 are still in existence, and can be seen in the Dutch countryside. For example the Peel-Raam Line is for most part intact. The stretch between Griendtsveen and De Peel Air Base and the spot nearby Mill features several visible remains. The fortifications and casemates in the municipalities of Deurne, Venray and Mill and Sint Hubert are now protected as National Monuments. The bunker complex of Kornwerderzand is perfectly preserved and now houses a museum.

German Atlantic Wall

The German Atlantic Wall has had a confused fate. After World War II, all mines, ammunitions, obstacles and barbed wires were removed. All metal parts (doors, turrets and other armor) were scrapped and recycled for further use. When all was safe and sound, the beaches were given back to the public for recreation, but nobody knew what to do with the thousands of concrete bunkers along the Dutch (and European) coasts, which were considered as hated symbols of the Nazi occupation.

After World War II, removing bunkers, relieving the housing shortage, and rebuilding the economy were priorities. Cleaning-up was first undertaken in ports and along boulevards (for tourism), railway lines, and road and waterways. As the years passed, natural erosion and modern construction further destroyed or damaged many structures. Many bunkers have succumbed to road improvements and for many of those that remain, neglect and the attentions of nature have achieved a degree of camouflage greater than that during the Second World War.

Casemate XIV at Kornwerderzand (province of Frisia) is now a part of the Afsluitdijk (dam and causeway) defenses museum.

For a few decades now, there has been a renewal of interest in German World War II bunkers, and associations have risen to preserve them. It is important that this period of history, which is ever less available from primary sources, be properly conveyed to the next generations.

In the Netherlands there are several Atlantic Wall Museums, notably at Huisduinen and Noordwijk (province of North Holland), Hoek van Holland (province of South Holland) and at Zouteland (Walcheren province of Zeeland). A large and interesting open-air museum can be visited at Raversijde near Oostende on the Belgian coast.

Cold War remnants

Remnants of the Cold War are rather few. Some underground nuclear bunkers are still in military use and cannot be visited. There are still about 18 KLD *luchtwachttorens* (air observation towers). These few remnants include Post 5K3 Strijensas (South Holland), Post 3W3 Eede (Zeeland), Post 8O2 Posterholt (Limburg), and Post 7O1 Den Hoorn (Warfhuizen, Gronigen) for example.

Armored door Casemate XIV at Kornwerderzand.

Casemate IV was the command post of the Kornwerderzand defensive complex.

Observation cupola placed on top of command bunker IV at Kornwerderzand.

Chapter 10. Dutch Fortifications in the Cold War

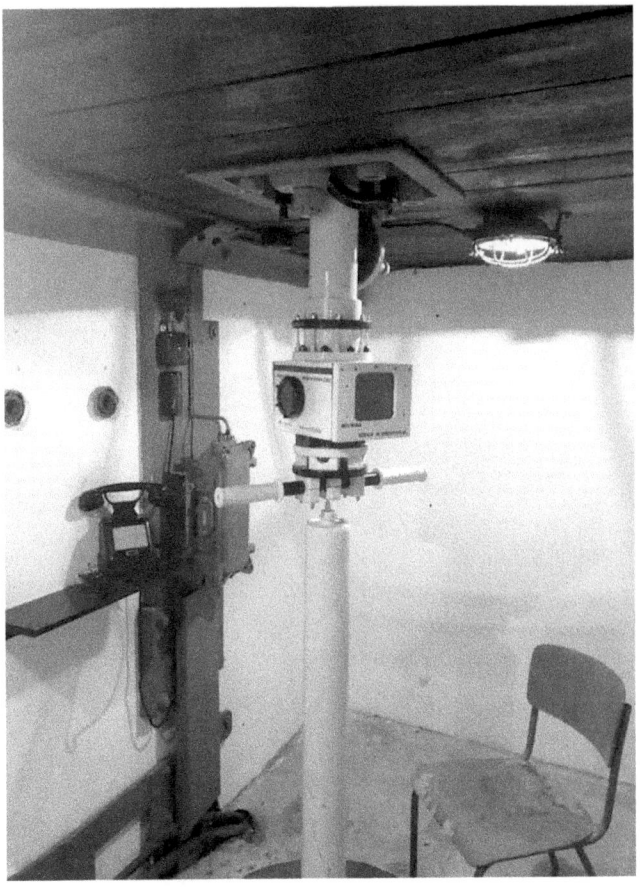

Periscope in bunker at Kornwerderzand.

Top, left: Embrasure for machine gun at Kornwerderzand.

Left: Crew quarters with bunks at Kornwerderzand.

Asparagus.

Frizzy horse. Dutch: Friese ruiter; French Cheval de Frise; American: knife stand.

German Atlantic Wall type 667 bunker. Issued in late 1943 type 667 was a simplified artillery emplacement for one anti-tank 5 cm PaK gun.

Atlantic wall artillery bunker type 671 at MKB Hundius along the Spinoladijk at Oostende, Belgium.

Chapter 10. Dutch Fortifications in the Cold War

German MKB Navy coastal battery Heerenduin (Wn 81) near IJmuiden (province of North Holland). The battery included one Leitstand (firing control station) type M178, and four concrete gun casemates type M272.

German Atlantic Wall artillery bunker type 671 (front view).

Caponier. Also known as Nähkampfraum (close combat room)—a caponier was a small concrete building part jutting out of the rear façade of a bunker allowing defenders to deliver fire on enemies approaching doors, ventilation pipes and entrance.

Tobruk. The so-called Tobruk (aka Ringstand) was a small open observation and combat emplacement made of concrete.

Tobruk could also be incorporated into a bunker, and in this case designated an Offener Beobachter (open observation pit).

Range finder. A range finder was an optical instrument for calculating the distance from a gun to a target. It was a fragile, vulnerable and expensive device often placed in specialized concrete bunkers.

Tetraeder. The tetraeder (aka porcupine) was an anti-tank obstacle/road-block used in the 1930s and during World War II. It was formed of three wooden, metal or concrete beams assembled together as a pyramid.

Conclusion

Remnants of fortifications in the Netherlands are numerous and present the great advantage of being located in a small country. Often at the same place there is a lot to see. Medieval vestiges, 17th century bastioned fortifications, 19th century forts and batteries, as well as 20th century concrete bunkers are sometimes to be found in the vicinity of each other.

Many forts and military installation have been renovated, made accessible and opened to the public. Many former strongholds now serve a variety of purposes as museums, exhibitions places and hosts of all forms of cultural events in art, theater, films and music. Many can be used as conference centers, business events, care organizations, wedding venues, offices, restaurants and hotels. The surrounding sites have been arranged for tourism with the possibility of parking your car, visiting the forts, as well as hiking, cycling or canoeing in the peaceful variegated landscape of woodland, hedgerows, meadows, canals and rivers.

The former fortifications help create jobs, and boost

Naval mine. This explosive device was placed at sea to damage or destroy surface ships or submarines. Unlike depth charges, World War II contact sea mines were deposited and left to wait until any vessel bumped into them.

the economy while forming at the same time an important basis for conservation, and a living asset for understanding the rich and interesting history of the Netherlands. Worthy of mention and definitely meriting a visit is the *Nederlands Vesting Museum* (Dutch Fortress Museum) located at Naarden near Amsterdam in the province of North Holland.

Associations

The Dutch association *Stichting Menno van Coehoorn*—named after the great military engineer Menno van Coehoorn (1641–1704) was founded in 1932. Located at Utrecht it is a national association composed of volunteers whose aims are the safeguarding of historical fortifications including medieval city gates and urban enceintes, fortresses, forts, casemates, pillboxes, and any other military buildings related to fortifications. The Association regularly organizes congresses, trips and visits to still existing fortifications not only in the Netherlands but also in former colonial territories and in foreign countries. The Stichting has an Internet website, and a documentation center at Utrecht. It publishes a trimestral magazine called *Saillant*, as well as books, maps, booklets, dictionaries, glossaries, catalogues and brochures, all related to fortifications.

The *Simon Stevin Vlaams Vestingbouwkundig Centrum* (VVC) is an association that does the same for Flemish fortifications in Belgium. Named after the influential mathematician and engineer with a broad range of interests Simon Stevin (1548–1620), it was founded in 1964. It is presently located at Antwerp, Belgium, and publishes a trimestral bulletin in Flemish called *Vesting*.

Bibliography

Ampt, Kees (ed.). *Verre Forten, Vreemde Kusten. Nederlandse Verdedigingswerken Overzee.* Leiden: Sidestone Press, 2017.

Asselbergs, Fons (and various authors). *Op Weerstand Gebouwd. Verdedigingslinies als Militair Erfgoed.* Zwolle: Waanders b.v. Uitgevers, 2004.

Augusta, Pavel. *Encyclopédie de l'Art Militaire.* Paris: Ars Mundi Editions, 1991.

Beekmans, J.R., C. Schilt (and various other authors). *Drijvende Stuwen voor de Landverdediging.* Utrecht: Stichting Menno van Coehoorn, 1997.

Beemsterboer, W. *Wereld in wording, van 1715 tot heden.* Den Haag: Van Goor Zonen, 1968.

Benevolo, Leonardo. *De Europese Stad.* Amsterdam: Agon BV, 1993.

Blijdenstein, Roland. *De Hollandse Waterlinie.* Amsterdam: Buijten & Schipperheijn, 1990.

Blockmans, Willem. *Oorlog door de Eeuwen heen.* Hilversum: HD Uitgeverij, 1977.

Boxer, C.R. *De Hollanders in Brazilië, 1624-1654.* Alphen aan de Rijn, 1977.

Bragard, Philippe, Termote, Johan, and Williams, John. *A la Découverte des Villes Fortifiées (Kent, Côte d'Opale et Flandre Occidentale).* Dunkerque: Syndicat Mixte de la Côte d'Opale, 1999.

Brand, Hans, and Jan. *De Hollandse Waterlinie.* Utrecht: Veen Publishers, 1988.

Braun, Georg, and Franz Hogenberg. *Civitas Orbis Terrarum (Steden van de Wereld, Europa-Amerika).* Amsterdam: re-print by Atrium, 1990.

Braun, Georg, and Franz Hogenberg. *De Hollandse Steden 1574.* Groningen: BV Foresta (re-print), no date.

Braunstahl, H.J. *Schoolatlas voor de Vaderlandse Geschiedenis.* Kampen: J.H. Kok NV, 1964.

Braure, Maurice. *Histoire des Pays-Bas.* Paris: Presses Universitaires de France, 1974.

BunkerSite.

Corvisier, André. *Dictionnaire d'Art et Histoire Militaires.* Paris: Presses Universitaires de France, 1988.

Cotterel, Geoffrey. *Amsterdam, The Life of a City.* Farneborough: Saxon House, D.C. Heath, 1973.

Crefeld, Martin van. *The Art of War.* London: Smithsonian Books-Cassell, 2002.

De Roy van Zuydewijn, Noortje. *Neerlands Veste.* The Hague: SDU Uitgeverij, 1988.

De Voogd, Christophe. *Histoire des Pays-Bas.* Paris: Hatier, 1992.

Diderot, Denis. *Encyclopédie. Arts Militaires.* Paris: 1751-1772.

Dijk, P.J.J. van (ed.). *Vestingbouw Overzee.* Utrecht: Stichting Menno van Coehoorn and Walburg Pers, 1996.

Faucherre, Nicolas, Pieter Martens and Paucot Hugues. *La Genèse du Système Bastionné en Europe 1500-1550.* Navarrenx: Cercle Historique de l'Arribère, 2014.

Fisher H.A.L. *A History of Europe.* London: Edward Arnold, 1938.

Gaag, Arie van der. *Fort bij Rijnauwen.* Bunnik: Henk Reinders Uitgeverij, 1990.

Génicot, L., and P. Houssiau. *Le Moyen Age.* Paris: Casterman, 1959.

Goossens, Allert M.A. (website) www.War Over Holland.nl

Haslinghuis, E.J., and H. Janse. *Bouwkundige Termen.* Leiden: Primavera Pers, 2005.

Haye, T. *Het Dagelijks Leven van onze Voorouders in het Midden van de 17e Eeuw.* Utrecht: Pictura Boeken, 1962.

Heer, Friedrich. *The Medieval World: Europe from 1100 to 1350.* London: Weidenfeld & Nicolson, 1961.

Het Hart, Marjolein. *The Dutch Wars of Independence. Warfare and Commerce in the Netherlands 1570-1680.* Abington-on-Thames: Rootledge Taylor & Francis Group, 2014.

Historische verdedigingswerken in Noord-Holland 1915-1940. Author unknown. Haarlem: Provinciaal Bestuur N-H, 1990.

Hofdijk, W.J., and J. Van Lennep. *Merkwaardige Kastelen in Nederland.* Groningen: BV Foresta, 1983.

Hoof, J.P.C. van. *Langs Wal en Bastion.* Utrecht, 1992.

Jansen, H.P. *Middeleeuwse Geschiedenis der Nederlanden.* Antwerp: Prisma Boeken, 1965.

Kalkwiek, K.A. *Atlas van de Nederlandse Kastelen.* Alphen a.d. Rijn: A.W. Sijthoff, 1980.

Kamps, P.J.M. (ed.). *Terminologie Verdedigingswerken, Inrichting, Aanval en Verdediging.* Utrecht & Zutphen: Stichting Menno van Coehoorn-De Walburg Pers, 1999.

Kaufmann, J.E., and W.H. Kaufmann. *The Atlantic Wall History and Guide.* Barnsley: Pen & Sword Books, 2011.

Keegan, John. *A History of Warfare.* London: Hutchinson, 1993.

Kielich, W., and J.Z. Zwaan. *Aanzien 40-45 Vijf jaar bezetting in Nederland and België.* Amsterdam: Amsterdam Boek B.V., 1975.

Koch, H.W. *Het Krijgsbedrijf in de Middeleeuwen.* Amsterdam: Elsevier B.V., 1980.

Koen, Douwe, and Roland Blijdenstein. *De Hollandse Waterlinie.* Amsterdam: Buijen & Schipperheijn Uitgeverij, 1993.

Koenigsgerger, H.G. *Medieval Europe 400-1500.* Harlow: Longman Group, 1987.

Leegwater, Dick. *Fort bij Rijnauwen.* Utrecht: Walburg Press, 1995.

Lepage, Jean-Denis G.G. *Castles and Fortified Cities of Medieval Europe.* Jefferson, N.C.: McFarland, 2002.

Lepage, Jean-Denis G.G. *Vestingbouw stap voor stap. Het bastion hoekpunt in oude stadomwalligen.* Zutphen: Stichting Menno van Coehoorn, 1992.

Lepage, Jean-Denis G.G. *Vestingen en Schansen in Groningen.* Utrecht: Stichting Matrijs, 1994.

Lewis, Archibald R. *Emerging Medieval Europe, AD 400-1000.* New York: Alfred A. Knopf, 1967.

Libal, Dobroslav. *Châteaux Forts & Fortifications en Europe du Ve au XIXe Siècle.* Paris: Ars Mundi, 1993.

Loeb, Herbert. *Castles of the Netherlands.* Laren: Uitgeverij Andries Blitz, no date.

Lohnstein, M., and A. Hook. *Royal Netherlands East Indies Army 1936-1942.* Osprey Publishing (no date).

McEvedy, Colin. *The Penguin Atlas of Medieval History.* Baltimore: Penguin Books, 1971.

Merian, Matthaeus. *Die Schönsten Europäischen Städte.* Hamburg: Hofmann und Campe Verlag, 1963.

Mohr, A.H. *Vestingbouwkundige Termen.* Zutphen: Stichting Menno van Coehoorn, 1983.

Montgomery, Bernard, Law. *A Concise History of Warfare.* Ware: Wordsworth Editions, 2000.

Montsma A. *De Vaderlandse Geschiedenis (Leerboekje 3).* Rotterdam: Uitgeverij Nijgh & Van Ditmar, 1967.

Nimwegen, Olaf van. *The Dutch Army and the Military Revolutions, 1588-1688 (Warfare in History).* Woodbridge: Boydell Press, 2010.

Nossov, Konstantin. *Ancient and Medieval Siege Weapons.* Staplehurst: Spellmont Limited, 2005.

Painter, Sidney. *A History of the Middle Ages, 284–1500*. London: Macmillan, 1972.

Parker, Geoffrey. *Spain and the Netherlands, 1559–1659 (Ten Studies)*. Glasgow: Collins Sons, 1979.

Peyen, Paul. *Middeleeuwse Kastelen*. Bussum (no date).

Philippart F., Peeters D., and A. van Geetruyen. *De Atlantik Wall van Willemstad tot de Somme*. Tielt: Lannoo Uitgeverij, 2004.

Pols, Ruud. *Seefront IJmuiden*. Published by the author, Leo de Vries, and Bunker Museum IJmuiden (no date).

Poppema, Simon, and Jean-Denis G.G. Lepage. *Historische Verdedigingswerken*. Amsterdam: Stichting Open Monumentendag, 1995.

Roe, Derek. *Prehistory*. London: Paladin-Granad, 1970.

Rolf, Rudi. *Atlantikwall, Batteries and Bunkers*. Middelburg: PRAK Publishing, 2014.

Rolf, Rudi. *Bunkers in Nederland*. Den Helder, 1982.

Rolf, Rudi. *Napoleons Forten aan de Schelde. Franse Vestingbouw in Zeeland*. Middelburg: PRAK Publishing, 2019.

Rolf, Rudi. *Torens, Wallen en Koepels*. Middelburg: PRAK Publishing, 2007.

Rolf, Rudi, and Peter Saal. *Fortress Europe*. Shrewsbury: Airlife Publishing, 1986.

Rorive, Jean-Pierre. *La Guerre de Siège sous Louis XIV*. Bruxelles: Editions Racine, 1998.

Saal, Peter, Jaap de Zee, and Rob Schimmel. *De Stelling van Amsterdam. Vestingwerken rond de Hoofdstad, 1880–1920*. Beetsterzwaag: AMA-Boeken, 1990.

Sakkers, Hans. *Atlantikwal in Zeeland en Vlaanderen*. Middelburg: Published by the author, 1990.

Sakkers, Hans. *Enigma en de Strijd om de Westerschelde. Het falen van de Geallierde opmars in September 1944*. Soesterberg: Aspekt Uitgeverij, 2011.

Sakkers, Hans. *Festung Hoek van Holland*. Middelburg: Published by the author, 1992.

Sakkers, Hans. *Generalfeldmarschall Rommel*. Koudekerk: Zeelucht Uitgeverij, 1993.

Sakkers, Hans. *Luchtwachttorens in Nederland. Industrieel Erfgoed uit de Koude Oorlog*. Middelburg: Stichting Natuur en Recreatie-Informatie, 1989.

Schukking, W.H.S. *De Oude Vestingwerken van Nederland*. Amsterdam: Allert de Lange Uitgeverij, 1941.

Schulten, C.M. and J.W. *Het Leger in de 17e Eeuw*. Bussum Fibula, 1969.

Stichting Menno van Coehoorn. *Atlas van Historische Vestingwerken in Nederland*. Original version not published for public; date unknown.

Stier, Hans-Erich. *Grosser Atlas zur Weltgeschichte*. Braunschweig: Georg Westermann Verlag, 1956.

Strayer, Joseph R. *Feudalism*. New York: Van Nostrand Rheinhold, 1965.

Treu, Herman, M. Tydeman, and Jaap Sneep. *Vesting Vier Eeuwen Vestingbow in Nederland*. Zutphen: Stichting Menno van Coehoorn/Walburg Press, 1982.

Van Dijk P.J. (ed.). *Kustverdediging*. Zutphen: Stichting Menno van Coehoorn, 1992.

Van Straalen, T., and H. Janse. *Middeleeuuwse Stadswallen en Poorten in de Lage Landen*. Zaltbommel: Europese Bibliotheek, 1974.

Vermaseren, B.A. *Atlas Algemene & Vaderlandse Geschiedenis*. Groningen: Wolters-Noordhoff, 1970.

Viollet-le-Duc, Eugène-Emmanuel. *Histoire d'une Forteresse*. Paris: Berger-Levrault, 1978 (reprint of the 1874 edition).

Index

Abatis 124
Abcoude fort 150, 190
Aceh War 199
Act of Abjuration 79
Advanced work 84, 119, 131, 142, 178
Afsluitdijk 1, 112, 195, 204, 207, 246, 247
Afwachtingsruimte 216
Air Watch Service Corps 263
Aix-la-Chapelle 17, 71, 141, 142
Alkmaar 23, 84, 98, 105, 148
Allure 21, 51
Alva, Fernando, Duke of 80, 98, 99, 122
American Declaration of Independence 79
Amersfoort 34, 54, 72, 105, 140, 175, 187, 213, 253
Ammersoyen Castle 37, 41, 265
Amsterdam 1, 2, 3, 18, 23, 31, 38, 50, 73, 85, 98, 99, 111, 115, 117, 119, 120, 140, 146, 150, 178, 189, 191, 202, 212, 246, 263, 265, 266
Anthonisz, Adriaan 80, 81, 94, 98, 99, 100, 105, 106, 110, 122
Anti-aircraft artillery *see* FlaK
Anti-landing obstacle 228, 245, 253
Anti-personnel weapon 77, 184, 204, 213, 245
Anti-tank obstacle 204, 206, 222, 228, 245, 247
Anti-train fence 223
Antwerp 70–75, 80, 95, 107, 146, 152, 155, 156, 161, 202, 236, 238, 245, 250, 254, 256, 274
Appanage 59
Approach 77, 124, 125, 142
Archeon Museum 264
Arçon lunette 146, 148
Arkel 83
Armored truck 205, 254, 255
Arrowloop 47, 48, 51, 62
Asparagus 222, 223, 226, 269
Asperen fort 164, 170, 213, 256
Atlantic Pockets 236, 256; *see also* Festung
Atlantic Wall 185, 226–256
Austrasia, realm of 16

B-casemate 219
Bailey 22, 24, 26, 28, 30, 31, 37, 38, 43
Barbed wire 203, 222, 231, 245, 267
Barbican 28, 37, 40, 61, 65
Barrier 52, 107, 140, 161, 184
Bartizan 36, 37
Bastion 66–76, 83, 90, 120
Batardeau 53, 85, 93
Batavia 114, 117
Batavian Republic 145

Battlement 34, 47, 51, 54
Belfry 58
Belgian gate 245
Bereitschaftsraum 238
Bergen-op-Zoom 130, 136, 142
Berm 6, 51, 81, 82, 84
Bescherming Bevolking 263
Bettung 235, 238, 253
Beverwijk 148, 149, 151, 195
Bonaparte, Charles-Louis 145
Bonaparte, Louis-Napoléon 145
Bonaparte, Napoléon 145
Bonifatius evangelist 16, 17
Borsele 24, 30, 156
Boulevard 63
Bourgeoisie 46, 120, 160, 257
Bourtange fort 105, 134, 151–154, 260
Brattice 36–38, 147
Breach 66, 76, 77, 125
Breastwork 6, 34, 84, 129, 154, 210
Breda 70, 143, 144, 178
Brederode 43
Breskens 152, 158, 236, 238, 250
Brick 4, 8, 31, 214
Brockhurst fort 181
Brune, Guillaume 148
Bulwark 61, 63, 64–66, 84, 141
Burg 19, 20

Cambrai 70, 71
Camouflage 208, 215, 238, 254
Cape of Good Hope 113, 116
Caponier 61, 65, 86, 159, 162, 180, 181, 183, 184, 187, 238, 242, 262, 272
Caribbean 80
Carolus Magnus *see* Charlemagne
Casemate 61, 63, 131, 146, 162, 199, 208, 210, 219, 220, 229, 232, 237, 240, 243, 249
Castellum 7, 12, 14, 22
Castrum 7, 12
Cavalier 68, 83, 86, 94, 96, 124
Cavalier de tranchée 125
Charlemagne 15, 17, 18
Charles II the Bald 17
Charles V 60, 71, 78
Charter of Freedom 28, 44, 78
Chasseloup-Laubat, François 146
Christiansen, Friedrich 245
Churchill, Winston 226
Cicignon, Johan, Caspar de 110
Cippe 6
Citadel 22, 71, 73, 74, 119, 132, 261
City right 28, 44, 57
Civitas 8, 10, 73
Cloche 197, 238
Clovis 16
Coehoorn, Menno van 122, 126–144, 274

Coehoorn mortar 136
Coevorden 126, 128, 130, 155
Cofferdam 53, 93
Colbert, Jean-Baptiste 119
Colonia 8, 9
Colonia Claudia 3
Colonialism 56, 160
Concrete 184, 185
Congress of Vienna 160, 161
Constantinople 16
Continental Blockade 145
Copenhagen 110
Cormontaigne, Louis de 146
Coucheron, Anton 110
Counterguard 86, 97, 123, 131, 133, 146, 152
Counterscarp 69, 84, 131, 168; gallery 173, 185, 187
Courtyard 14, 22, 26, 28, 37, 38, 41, 72, 180
Courtyard-castle 38
Covered way 69, 81, 84, 86, 88, 96, 129, 131, 135, 162
Crenel 6, 34, 40, 51
Cross-and-orb 61
Crownwork 84, 86, 97, 119, 131, 153, 158
Curtain 35, 40, 46, 51, 67, 68, 74, 81, 135

Dark Age 10, 14, 16, 19, 57
De Bilt fort 165, 166
Defence in depth 37, 146, 162, 204
De Gomme, Bernard 108, 109, 110
De la Haye, Abraham 110
De Leethe 154
Delfzijl 154
Demi-lune 69, 81, 84; *see also* Ravelin
Den Bosch 102, 121, 261, 266
Den Haag 10, 250, 252, 254; *see also* The Hague
Den Ham tower 45
Den Helder 146, 151, 153, 212, 234, 251, 266
Den Oever 209, 210, 211, 212
De Roovere fort 107
Dien Bien Phu 125
Diogenes command center 252
Directorate of Fortifications and Works 214
Ditch 3, 22, 69, 82, 97, 247, 262
Doesburg 134, 135, 136
Dominium 31
Donjon 22, 28, 31, 35
Doornenburg castle 28
Doorwerth castle 25
Dorestad 19
Dragon's teeth 245
Drawbridge 19, 56, 61, 88, 89, 180, 260
Dufalga fort 149, 152

Du Moulin, Carel, Diderik 143
Du Puy de l'Espinasse, Charles 133
Dussen castle 33
Dutch Antillen 117
Dutch Brazil 60, 116, 199
Dutch neutrality 161, 177, 187, 202, 204, 205, 206, 212
Dwangburcht 73, 92; *see also* Citadel

Edam fort 264
Emergency exist 210, 238
Enceinte 24, 46, 63, 136
Enkhuizen 100, 105
Enumatil schans 92
Entrenched camp 135, 149, 159
European Recovery Program, ERP (aka Marshall Plan) 257
Ewssum castle 67

Fausse-braie 61, 65, 81, 82, 85, 97, 108, 121, 131, 135
Federati 16
Festung 196, 235, 236, 240, 247, 248, 250, 276
Feudalism 18, 22, 57, 160
Fief 18, 22, 70, 160
FlaK 233, 234, 235, 243, 248, 251, 253, 254
Flank angle 69, 85, 118, 131, 135, 181
Flanking 36, 65, 69, 96, 131, 190, 219, 229, 232, 237, 242, 249
Flooding 1, 2, 15, 36, 46, 84, 85, 106, 107, 108, 120, 121, 189, 206, 212, 230, 262, 263
Fortification Perpendiculaire 162
Fortress Holland 204, 212
Fortwachter 182, 191
Forum 8
Frederick-Hendrik of Nassau 79, 102, 164, 238, 250
Free arts 75
Freitag, Adam 97, 98
French Revolution 79, 116, 143, 144, 145, 151
Frizzy horse 222, 270
Furnes 141

G-model casemate 220
Gardner machine gun 192, 195
Gaslock 238
Gatehouse 19, 52, 53, 65, 85, 86, 137, 265
Gaul 3, 16
Gay de Vernon, Simon, François 148
Geelkercken, Isaac van 110
General States 79
Generaliteitslanden 79
Genieloods 191
Georgian uprising 251, 256

277

Germania Inferior 3, 14, 15
Germersheim 163, 164
Gillet, Claude 148
Gorge 67, 68, 72, 84, 86, 90, 135, 148, 152, 165, 168, 174, 180, 183, 190
Gracht 81, 266
Grazing fire 61, 63, 82, 215
Grebbe Line 140, 176, 177, 212, 223, 224, 225, 253, 260, 265
Groningen 31, 32, 44, 46, 52, 66, 79, 86, 90, 101, 107, 120, 128, 133, 135, 136, 176, 207, 212, 214, 256, 260, 266
Gruson Company 186, 193, 195, 198
Guild 44

Hagendorp fort 168
The Hague 8, 10, 19, 23, 110, 117, 137, 145, 187, 212, 225, 233, 243, 250, 254, 258, 260
Hale, John 66
Hall of the Knights 258
Hamei 85
Hanseatic League 46, 54, 99
Harderwijk 46, 105
Haxo, François 146, 149
Haxo casemate 146, 149
Hellevoetsluis 128, 129, 131, 134
Helmond castle 41, 44
Henricus fort 108
Het Hemeltje fort 166, 172, 182, 183
Heusden 199, 106, 123, 134
Heuvel, Charles van den 74, 99
Hitler, Adolf 225, 226, 228, 235, 236, 241, 250, 254
Hoarding 37, 51, 53
Hoek van Holland 185, 186, 235, 247, 248
Hondius 98
Honswijk fort 170, 266
Horizontality 92
Horn castle 26
Hornwork 86, 97, 119, 131, 187
Hunger winter 189
Hunneschans 21

Ideal city 95, 96, 126, 131
IJmuiden 2, 12, 196, 197, 233, 242, 251, 271
Improved Dutch Bastioned System 120–123
Indonesia 80, 114, 117, 199, 200, 201, 202, 226, 257
Industrial Revolution 160
Infantry Regiment Menno van Coehoorn 138
Ingolstadt 96, 163
Inundation see Flooding
Inundatiewet 189

Jakarta 114, 117, 201
Java Island 114, 117, 199
Jews 226
Joost, Mattheus 106
Jutphaas fort 166, 167

Kalamata fort 200
Kamerijk 70, 71; see also Cambrai
Kampen 28, 46, 53, 105, 213, 260
Kampfraum 238
Keep 22, 26, 37, 45, 74, 165, 174, 236; see also Reduit
Keldermans, Marcellis 73, 74
Kemp, Jacob 106
Kernwerk 236, 247, 248

Kijkduin fort 151, 187, 203
Knife stand 222, 270
Knokke fort 140
Kochbunker 242, 253
Koninklijk Nederlandsch Indisch Leger (KNIL) 199
Kornput, Johan van den 106
Kornwerderzand 207–215, 225, 243, 246, 262, 263, 267–269
Korps Luchtwachtdienst (KDL) 263
Kraijenhoff, Cornelis 146, 148, 150, 188; posts 147, 148, 150, 164, 188
Kubitza, August 254

Lasalle fort 149, 151, 152
Leiden 26, 30, 48, 75, 80, 84, 94, 97, 108, 115, 123, 187
Le Michaud d'Arçon, Jean-Claude, Elénore 146, 148
Le Vasseur des Rocques, Guillaume 139, 143
Liefkenshoek fort 156, 157
Lillo fort 156
Limes 3, 5, 10, 12, 14, 264
Loenersloot castle 28, 29
Loevestein castle 36, 105, 164, 171
Loophole 37, 47, 52, 61, 147, 163, 211, 216
Lorica 6
Lothair 17, 59
Lotharingia 17, 59
Louis I the Pious 17
Louis XIV 85, 119, 120, 121, 122, 125, 126, 133, 140, 155, 181
Louis the German 17
Lunenburg tower 51
Lunette 84, 123, 135, 146, 149, 151, 153, 159, 162, 164, 165, 170
Lunette d'Arçon 148
Luxembourg 1, 3, 59, 161

Maasmond fort 185, 186
Maastricht 9, 12, 52, 54, 79, 97, 98, 102, 121, 125, 126, 128, 143, 144, 145, 161, 180, 264, 266
Machicolation 37, 40, 47, 57, 63, 170
Majordomus 17
Mammut FuMo 51; radar 252
Marken 17
Marolois, Samuel 97, 98
Maurice Prince of Nassau 75, 76, 80, 81, 92, 94, 95, 102, 116, 134, 143
Mayor of the Palace 17
Meillerwagen 254, 255; see also V2
Menin 139, 142
Menno van Coehoorn Association 138, 274
Mercenaries 56, 58, 74, 79, 80, 118, 121, 139, 145
Merlon 6, 34, 40
Metric system 145
Middenweg fort 190, 191
Militia 56, 58, 76, 79
Mine 76, 77, 106, 125, 228, 229, 230, 231, 232, 274
Moineau 65
Monasticism 19
Montalembert, René de 137, 146, 162, 163, 164, 180
Montfoort castle 48, 49
Motte-and-bailey castle 22, 24, 26, 28, 30, 31, 141
Muiden 2, 12, 38, 41, 42, 85, 98, 123, 164, 171, 175, 176, 189, 191, 195, 196, 197, 228, 233, 235, 236, 242, 246, 251, 252, 259, 265, 266, 271, 276
Municipia 8, 10
Münster 17, 66, 79, 86, 107, 118, 120, 128
Mutiny 80

Naarden 99, 122, 123, 130, 148, 266, 274
Namur 128, 131, 132, 133, 139, 141, 202
Napoleon fort 159, 262
National Réduit 212
Nederlandsche Handel-Maatschappij 199
Neutrality 161, 187, 202, 204, 205, 208
New York 115, 117, 245
Nieuwersluis fort 137
Nieuwpoort fort 121, 123
Nigtevecht fort 15, 192
Nijmegen (Nimegue) 2, 3, 9, 14, 15, 17, 71, 80, 134, 139, 178, 179, 225, 260, 262, 264, 265
North Atlantic Treaty Organisation (NATO) 257, 260, 262, 263
Norwood, Richard 110
Nuclear warfare 257, 260, 267

Offener Beobachter 273; see also Tobruk
Oostende 106, 110, 159, 262, 267, 270
Oostvorne castle 26, 27
Operation Market Garden 25
Operation Overlord 256
Operation Plunder 256
Ora et Labora 19
Organisation Todt 226, 236, 251, 253
Orillon 67, 58, 83, 85, 121, 131
Orliens, David van 80
Outwork 40, 84, 86, 90, 97, 141, 152, 153

Palau Cipir fort 201
Palisade 6, 21
Pampus fort 147, 191, 195, 265
Pannerden fort 184, 185
Panther Line 253, 254, 256
Pantserhek 233
Pantserwagen 205
Panzer 205, 223, 238, 250
Parados 169, 180, 181
Parallel 125, 126, 142
Paramaribo 117
Parapet 6, 21, 82, 84, 129, 130, 169; see also Breastwork
Patriot 143–145
Pax Romana 3, 9
Peppin the Short 17
Periscope 193, 208, 210, 216, 269
Peters, Garwer 101
Peutinger map 5
Pfalz 15
Philip II King of Spain 78, 70
Philip II the Bold 59
Playing card plan 6, 9
Polder 1, 80, 106, 110, 124, 149, 165, 191, 195, 204
Porcupine 216, 274
Portcullis 52, 53, 56, 85
Portsmouth 108, 110, 181
Postern 37, 38, 52, 53, 68
Poterne 190, 191

Privateer 79, 80
Propugnacla 53
Protestantism 78, 119, 120
Prussian system 163, 181
Pyramid block 222, 245, 266, 274
Pyramid shelter 216, 217

Quadrangular castle 22

Racine, Jean 133
Radar 228, 230, 234, 235, 246, 251, 252, 253, 263
Rammekens fort 75, 98, 157, 158, 159, 259, 265
Rampart 6, 36, 61, 66, 69, 181, 182, 183
Rampjaar 107, 119, 120
Randstad 187, 189, 212, 262
Range finder 213, 233, 234, 239, 273
Ravelin 69, 81, 84, 97, 119
Recoil 63
Rectangular castle 36, 41, 43, 44
Redan 84, 86, 91, 105, 125, 148, 149, 154, 189, 201
Redoubt 74, 107, 124, 151, 174, 212, 236; see also Sconce
Réduit 22, 26, 37, 45, 74, 165, 174, 189, 236
Regelbau 214, 232, 237, 238, 241, 254
Remise 118, 175, 182, 190, 264
Republic of the Seven United Provinces 1, 78, 106, 144, 149, 153, 161, 258
Revetment 66
Ridderzaal 258
Rijnauwen fort 166, 172, 181, 183, 266
Rijswijk, Johan van 81, 106, 110
Rifled artillery 164, 168, 172, 175, 177, 184, 192
Ringstand 242, 272; see also Tobruk
Ringwalburg 19, 21
River casemate 220, 221
Rollbock 245
Romanitas 6
Rommel, Erwin 228, 231
Rosworm, John 110
Rotterdam 51, 145, 187, 200, 212, 224, 225, 236, 245, 246, 247
Roundel 61, 63, 66, 73, 105
Royaume de Hollande 145
Ruigenhoek fort 166, 172, 183, 184
Ruse, Hendrik 118, 119, 120, 123, 130

S-casemate (aka Stekelvarken) 218, 219, 238, 265, 266
Sabina Henrica fort 150, 176
Saint Aagtendijk fort 195
Salient 63, 67, 68, 69, 72, 83, 167
Sawn-through shelter 216
Saxons 10, 16, 17, 195
Scarp 51, 68, 85, 129, 148
Schiermonnikoog Island 246, 252
Schiess-Zug 254
Schiphol airport 1, 189, 252
Schuttersdoele 56
Schuttersgilde 56
Schwarzlose machine gun 206, 208
Sconce 77, 90, 91, 92, 107, 111, 115, 124, 143, 147, 153, 155, 266
Sea beggar 80
Sems, Johan 110
Sentry-box 68, 83, 203, 243

Index

Sersanders, Andries 110
Shell-keep 22, 26
Sherman tank turret 258
Siege warfare 50, 58, 63, 73, 76, 77, 81, 86, 92, 119, 124, 125, 126, 135, 142, 143, 189, 236
Simon Stevin Vlaams Vestingbouwkundig Centrum 274
Smidt, Hildebrandt 101
Sonderkonstruktion (SK) 233, 238
Sortie 77, 82, 131, 162
Spaarndam 150, 189, 203
Specklin, Daniel 96
Spijkerboor fort 191, 193, 195
Spinnekop 218
Sri Lanka 113, 114, 117
Stadhouder 59, 75, 90, 122, 127, 133, 143, 161, 261
Stadsrecht *see* City right
Staircase 31, 34, 37, 51, 216, 243
Stekelvarken 218, 219, 238, 265, 266
Stelling of Amsterdam 187–199
Stevin, Simon 61, 94, 95, 98, 274
Stichting Menno van Coehoorn 274
Storff de Belleville, Paul 11, 122, 130
Student, Kurt 225
Stützpunkt 230, 231
Stützpunktgruppe 235
Suzerain 18

Tambour 187
Tank 192, 202, 203, 204
Ten-Days' Campaign 161
Tenaille 86, 97, 105, 121, 130–137, 142, 165, 167, 168
Ter Apel schans 90
Ter Does castle 48
Terra Maris Museum 30
Tetraeder 245, 274
Texel 19, 149, 153, 213, 214, 246, 251, 256, 266
Thirty Years' War 79, 110
Tiel 81, 105
Tilbury 108, 109
Tobruk 229, 233, 238, 241, 243, 244, 254, 272, 273
Todt, Fritz 226, 236, 251, 253
Torenfort 164, 168, 170, 171, 172, 173, 175, 201, 266
Total War 80
Tour-modèle 147, 148, 150
Traditore battery 190
Traverse 129, 131, 135, 148, 163, 169, 172, 181
Trench warfare 76, 77, 202
Trojan horse 76
Turris 10

Uitlegger 124
UNESCO World Heritage 192, 196, 266
Union of Utrecht 78, 106
Utrecht 3, 5, 9, 12, 14, 16, 19, 28, 30, 32, 34, 38, 45, 48, 51, 70, 72, 73, 78, 79, 90, 98, 99, 106, 113, 120, 123, 125, 126, 137, 140, 141, 145, 147, 150, 164–167, 170, 171, 172, 174, 175, 178, 181–184, 187, 190, 192, 202, 212, 216, 217, 225, 245, 260, 262–265, 274

V1 flying bomb 254
V2 rocket 254, 255
Vaalserberg hilltop 1
Valckenburgh, Johan van 106
Van Doorne's Aanhangwagenfabriek NV (DAF) 205
Vassal 18, 22, 46, 57
Vauban, Sébastien Le Prestre de 77, 119, 121, 123–126, 128, 130, 132, 133, 135, 137, 140, 141, 142, 146, 181
Vavassour 19
Vechten 12, 166, 172, 174, 183
Verdun 181, 202
Verdun Treaty 17
Verteidigungsbereich 230, 235, 236, 238, 246
Verticality 31, 50
Vesting 69, 107, 274
Vestingwet 177, 178
Vicus 12, 15, 19
Vienna Congress 160, 161
Vikings 17–22, 36, 75, 112
Villa rustica 6
VIS77 214, 215
Vlissingen (aka Flushing) 21, 24, 70, 73, 75, 134, 145, 152, 156, 157, 158, 159, 212, 228, 232, 236, 237, 241, 246, 249, 265
Volksdeutsche 225
Voordorp fort 166, 172, 175, 183
Vredenburg citadel 73
Vuren fort 171, 266

Wacht 56
Waddenzee (Wadden Sea) 107, 213, 246, 251, 252, 256
Walburg 16, 19, 21
Wallwalk 6, 21, 25, 26, 40, 46, 47, 68, 82, 84, 129
Ward, Robert 110
Wassermann radar 252
Watchtower 5, 6, 10, 75, 263
Waterburcht 33, 35, 37, 44, 67, 81
Watergate 53, 54, 93, 124, 189
Waver-Amstel fort 197
Wedde castle 66
Wehrmacht 224, 226, 234, 245
Wehrmachtbefehlshaber in den Niederlanden 245
Weitzel, August 178
Wellesley of Wellington, Arthur 161
Wellington Barrier 161
Wesphalia 118
Widerstandsnest 228–230
Wierick fort 120, 123
Wijckel 138
Wijk bij Duurstede 9, 19
Wilhelmina, Queen 225
Willemstad 98, 105, 150, 176, 266, 276
William of Orange "The Silent" 78, 79, 97
Winschoten 101
Wladislawowo 111
Wolf's pit 124
Wons 207–209, 212
Woontoren 28, 31, 34, 51, 66

Zierikzee 51
Zoutman fort 201
Zuiderzee 1, 28, 99, 107, 153, 204
Zutphen 54, 120, 134, 178, 254, 264
Zwolle 30, 54, 99, 134, 254, 260, 264

www.ingramcontent.com/pod-product-compliance
Ingram Content Group UK Ltd.
Pitfield, Milton Keynes, MK11 3LW, UK
UKHW050703160426
5217IPUK00038B/2074